MULTIMEDIA COMMUNICATION SYSTEMS:
TECHNIQUES, STANDARDS, AND NETWORKS

K.R. Rao
Zoran S. Bojkovic
Dragorad A. Milovanovic

Prentice Hall PTR
Upper Saddle River, NJ 07458
www.phptr.com

ISBN 0-13-031398-X

Library of Congress Cataloging-in-Publication Data

A catalog record for this book can be obtained from the Library of Congress

Editorial/production supervision: *Nicholas Radhuber*
Composition: *Gail Cocker-Bogusz*
Cover design director: *Jerry Votta*
Cover design: *Nina Scuderi*
Manufacturing manager: *Alexis R. Heydt-Long*
Publisher: *Bernard Goodwin*
Editorial assistant: *Michelle Vincenti*
Marketing manager: *Dan DePasquale*

© 2002 by Prentice Hall PTR
A division of Pearson Education, Inc.
Upper Saddle River, New Jersey 07458

Prentice Hall books are widely used by corporations and government agencies for training, marketing, and resale.

For information regarding corporate and government bulk discounts please contact:
Corporate and Government Sales (800) 382-3419 or corpsales@pearsontechgroup.com

[TRADEMARK ATTRIBUTIONS] Other company and product names mentioned herein are the trademarks or registered trademarks of their respective owners. All the figures and tables obtained from ISO/IEC used in this book are subject to the following: "The terms and definitions taken from the Figures and Tables ref. ISO/IEC IS 11172, ISO/IEC IS 11172-2,ISO/IEC11172-3, ISO/IEC IS 13818-2, ISO/IEC IS 13818-6, ISO/IEC JTC1/SC29/WG11 Doc. N2196, ISO/IEC JTC1/SC29/WG11 N3536, ISO/MPEG N2502, ISO/IEC JTC1/SC29/WG1 Doc. N1595, ISO/IEC JTC SC29/WG11 Doc. 2460, ISO/IEC JTC SC29/WG11 Doc. N3751, ISO/MPEG N2501, ISO/IEC JTC-1 SC29/WG11 M5804, ISO/IEC JTC SC29/WG11, Recomm. H.262 ISO/IEC 13818-2, ISO/IEC IS 13818-1, ISO/MPEG N3746, ISO/IEC Doc. N2502, ISO/IEC JTC1/ SC29/WG11 Doc. N2424 are reproduced with the permission of the International Organization for Standardization, ISO. These standards can be obtained from any ISO member and from the Web site of the ISO Central Secretariat at the following address : http://www.iso.org. Non-exclusive copyright remains with ISO."

"The terms and definitions taken from the Figure ref. ISO/ IEC 14496-1:1999, ISO/ IEC 14496-2:1999, ISO/ IEC 14496-3:1999, ISO/ IEC 14496-4:2000, ISO/ IEC 14496-5:2000 and ISO/ IEC 14496-6:2000 are reproduced with the permission of the International Organization for Standardization, ISO. These standards can be obtained from any ISO member and from the Web site of ISO Central Secretariat at the following address: www.iso.org. Non-exclusive copyright with ISO." The authors and Prentice Hall are grateful to ISO/IEC for giving the permission.

All rights reserved. No part of this book may be reproduced, in any form or by any means, without permission in writing from the publisher.

Printed in the United States of America
10 9 8 7 6 5 4 3 2 1

ISBN 0-13-031398-X

Pearson Education LTD.
Pearson Education Australia PTY, Limited
Pearson Education Singapore, Pte. Ltd.
Pearson Education North Asia Ltd.
Pearson Education Canada, Ltd.
Pearson Educación de Mexico, S.A. de C.V.
Pearson Education—Japan
Pearson Education Malaysia, Pte. Ltd.

Contents

Preface	xiii
Acknowledgments	xvii
List of Acronyms	xix

Chapter 1
Multimedia Communications — 1
1.1	Introduction	1
1.2	Multimedia Communication Model	4
1.3	Elements of Multimedia Systems	4
1.4	User Requirements	5
1.5	Network Requirements	7
1.6	Packet Transfer Concept	8
1.7	Multimedia Requirements and ATM Networks	9
1.8	Multimedia Terminals	10
1.9	Concluding Remarks	12

Chapter 2
Audio-Visual Integration — 13
2.1	Introduction	13
2.2	Media Interaction	14
2.3	Bimodality of Human Speech	16
2.4	Lip Reading	18
2.5	Speech-Driven Talking Heads	21
2.6	Lip Synchronization	23
2.7	Lip Tracking	24
2.8	Audio-to-Visual Mapping	27
	2.8.1 Classification-Based Conversion	27
	2.8.2 HMM for Audio-to-Visual Conversion	28
	2.8.3 Audio and Visual Integration for Lip-Reading Applications	28
	2.8.4 Audio-Visual Information Preprocessing	29
	2.8.5 Pattern-Recognition Strategies	29
	2.8.6 Integration Strategy	29

2.9 Bimodal Person Verification — 30
2.10 Joint Audio-Video Coding — 31
2.11 Concluding Remarks — 32

Chapter 3
Multimedia Processing in Communications — 35

3.1 Introduction — 37
3.2 Digital Media — 37
3.3 Signal-Processing Elements — 40
3.4 Challenges of Multimedia Information Processing — 42
    3.4.1 Pre and Postprocessing — 44
    3.4.2 Speech, Audio and Acoustic Processing for Multimedia — 44
    3.4.3 Video Signal Processing — 46
    3.4.4 Content-Based Image Retrieval — 48
        Texture-Based Methods — 48
        Shape-Based Methods — 49
        Color-Based Methods — 49
3.5 Perceptual Coding of Digital Audio Signals — 51
    3.5.1 General Perceptual Audio-Coding Architecture — 51
    3.5.2 Review of Psychoacoustic Fundamentals — 53
        Absolute Threshold of Hearing — 53
        Critical Band Frequency Analysis — 54
        Simultaneous Masking and the Spread of Masking — 55
        Temporal Masking — 56
        PE — 57
3.6 Transform Audio Coders — 58
    3.6.1 Optimum Coding in the Frequency Domain — 59
    3.6.2 Perceptual Transform Coder — 60
    3.6.3 Hybrid Coder — 61
    3.6.4 Transform Coding Using DFT Interblock Redundancy — 62
    3.6.5 ASPEC — 62
    3.6.6 Differential Perceptual Audio Coder — 63
    3.6.7 DFT Noise Substitution — 64
    3.6.8 DCT with Vector Quantization — 65
    3.6.9 MDCT — 65
    3.6.10 MDCT with VQ — 66
3.7 Audio Subband Coders — 66
    3.7.1 Wavelet Decompositions — 67
    3.7.2 DWT-based Subband Coders — 69
3.8 Speech Coder Attributes — 69

3.9	CD Audio Coding for Multimedia Applications	71
3.10	Image Coding	71
3.11	Video Coding	74
3.11.1	TC and Subband Coding (SBC)	74
3.11.2	Predictive Coding	75
3.11.3	Motion-Compensated Video Coding	76
3.12	Watermarking	78
3.12.1	Watermarking Techniques	80
3.12.2	Main Features of Watermarking	81
3.12.3	Application Domains	83
3.13	Organization, Storage and Retrieval Issues	85
3.13.1	Streaming Issues for Speech and Audio	85
3.13.2	Streaming Issues for Video	87
3.14	Signal Processing for Networked Multimedia	88
3.15	NNs for Multimedia Processing	89
3.15.1	NNs for Optimal Visualization	90
3.15.2	Neural Techniques for Motion Estimation	91
3.15.3	NN Application to Face Detection and Recognition	91
3.15.4	Personal Authentication by Fusing Image and Speech	92
3.15.5	Subject-Based Retrieval for Image and Video Databases	93
3.15.6	Face-Based Video Indexing and Browsing	94
3.16	Multimedia Processors	94
3.16.1	Image-Processing Hardware and Software	95
3.16.2	Multimedia Processors' Classification	96
3.16.3	General Purpose Microprocessors	98
3.16.4	Microprocessors for Embedded Applications	101
3.17	Concluding Remarks	102

Chapter 4
Distributed Multimedia Systems

		105
4.1	Introduction	105
4.2	Main Features of a DMS	107
4.3	Resource Management of DMS	107
4.4	Networking	108
4.4.1	IP Networking	109
	IP Multicast	110
	Resource Reservation Protocol (RSVP)	111
	RTP	112
4.4.2	Integrated Management Architecture for IP-Based Networks	112
	Performance Management	114

		Fault Management	114
		Configuration Management	115
		Security Management	115
		Accounting and Billing Management	116

- 4.4.3 ATM — 116
- 4.4.4 Integration of IP and ATM — 119
- 4.4.5 Real-Time Multimedia over ATM (RMOA) — 120

4.5 Multimedia Operating Systems — 122
- *CPU Management* — 123
- *Memory Management* — 123
- *I/O Management* — 124
- *File System Management* — 124

4.6 Distributed Multimedia Servers — 124
- 4.6.1 Multimedia Packing — 125

4.7 Distributed Multimedia Applications — 126
- 4.7.1 ITV — 127
  - *VoD* — 128
- 4.7.2 Telecooperation — 131
  - *Telecooperation Infrastructure* — 132
  - *Telecooperative Applications* — 133
  - *Telemedicine* — 135
- 4.7.3 Hypermedia Applications — 135
  - *Basic Features of a Hypermedia System* — 135
  - *The Web* — 136

4.8 Concluding Remarks — 137

Chapter 5
Multimedia Communication Standards — 139

5.1 Introduction — 140
5.2 MPEG Approach to Multimedia Standardization — 142
5.3 MPEG-1 (Coding of Moving Pictures and Associated Audio) — 146
- 5.3.1 The Basic MPEG-1 Interframe Coding Scheme — 149
- 5.3.2 Conditional Replenishment — 152
- 5.3.3 Specific Storage Media Functionalities — 152
- 5.3.4 Rate Control — 153

5.4 MPEG-2 (Generic Coding of Moving Pictures and Associated Audio) — 154
- 5.4.1 MPEG-2 Video — 157
  - *MPEG-2 Video—The Basics* — 158
  - *MPEG-2 Video Syntax* — 163

		MPEG-2 Video Scalability	163
		MPEG-2 Video: Profiles and Levels	166
	5.4.2	MPEG-2 Audio	167
	5.4.3	MPEG-2 Systems	171
	5.4.4	MPEG-2 DSM-CC	175
5.5	MPEG-4—Coding of Audiovisual Objects		176
	5.5.1	Overview of MPEG-4: Motivations, Achievement, Process and Requirements	178
		Media Objects	180
		MPEG-4 Version 1	182
		MPEG-4 Version 2	183
		Extensions to MPEG-4 Beyond Version 2	188
		Profiles in MPEG-4	189
		Verification Testing: Checking MPEG's Performance	192
		MPEG-4 Standardization Process	193
		Requirements for MPEG-4	194
	5.5.2	MPEG-4 Systems	195
		MPEG-4 Systems Architecture	196
		Elementary Stream Management (ESM)	199
		Auxiliary Descriptors and Streams	201
		Structuring Content by Grouping of Streams	203
		Managing Content Complexity	203
		Distributed Content-Handling Considerations	204
		System Decoder Model (SDM) for ES Synchronization	204
		MPEG-4 Systems BIFS	205
	5.5.3	DMIF	210
		DMIF Computational Model	213
	5.5.4	MPEG-4 Video	214
		Shape-Coding Tools for MPEG-4 Natural Video	216
		Motion Estimation and Compensation	217
		Texture-Coding Tools	218
		Multifunctional Coding	220
		Sprite Coding	221
		Scalability	221
		Error Resilience	223
		Relationship Between Natural and Synthetic Video Coding	223
		Synthetic Images	225
		Integration of Face Animation with Natural Video	226
		FAPs	227
		Face Model	232
		Coding of FAPs	233
		FIT	235

	Integration of Face Animation and Text-to-Speech (TTS) Synthesis	235
	BIFS for Facial Animation	236
	2D Mesh Coding	237
	VO Tracking	238
	2D-Mesh Object Encoder/Decoder	239
5.5.5	MPEG-4 Audio	243
	MPEG-4 Natural Audio Coding	244
	General Audio Coding (Advanced Audio Coding Based)	244
	Twin VQ	248
	Speech Coding in MPEG-4 Audio	248
	Scalability in MPEG-4 Natural Audio	252
	Synthetic Audio in MPEG-4	252
	Audio BIFS	255
5.5.6	Profiles and Levels in MPEG-4	256
	Visual Object Types	256
	Visual Profiles	259
	Audio Object Types	260
	Audio Profiles	261
	Graphics	261
	Systems Profiles	261
5.6	MPEG-4 Visual Texture Coding (VTC) and JPEG 2000 Image Compression Standards	262
5.6.1	JPEG 2000 Development Process	263
5.6.2	Overview of Still-Image Coding Standards	267
	MPEG-4 VTC	267
	JPEG	267
	PNG	268
5.6.3	Significant Features of JPEG 2000	268
	Region of Interest (ROI) Coding	268
	Scalability	268
	Error Resilience	269
	IPRs	269
5.6.4	Architecture of JPEG 2000	269
5.6.5	JPEG 2000 Bit Stream	275
5.6.6	Compression Efficiency Comparisons	277
	Error Resilience	281
5.7	MPEG-7 Standardization Process of Multimedia Content Description	282
5.7.1	Objective of the MPEG-7 Standard	282
5.7.2	Status of the MPEG-7 Standard	285
5.7.3	Major Functionalities in MPEG-7	286
	MPEG-7 Systems	286

		MPEG-7 DDL	289
		MPEG-7 Audio	289
		MPEG-7 Visual	290
		MPEG-7 MMDSs	297
		MPEG-7 Reference Software (XM)	299
		MPEG-7 Conformance	299
	5.7.4	Applications Enabled by MPEG-7	300
5.8	MPEG-21 Multimedia Framework		301
	5.8.1	Audiovisual Content Representation Issues	303
	5.8.2	Description of a Multimedia Framework Architecture	304
		MPEG-21 Digital Item Declaration	304
		Content Representation	305
		Digital Item Identification and Description	305
		Content Management and Usage	305
		Intellectual Property Management and Protection	305
		Terminals and Networks	306
		Event Reporting	306
	5.8.3	Requirements for Digital Item Declaration	306
5.9	ITU-T Standardization of Audiovisual Communication Systems		308
	5.9.1	ITU-T Standardization Process	308
	5.9.2	Audiovisual Systems (H.310, H.320, H.321, H.322, H.323, and H.324)	310
		H.320 Standard	312
		Standards for Audiovisual Services Across ATM H.310 and H.321	312
		Standard H.322—Guaranteed QoS LAN Systems	315
		ITU-T H.323 Standard	317
		H.324 Standard	319
	5.9.3	Video-Coding Standards (H.261, H.263 and H.26L)	319
		H.261 Standard	319
		H.263 Standard	323
		H.263+ (H.263 Version 2) Standard	327
		H.263++ Standard Development	332
		H.26L Standard	333
	5.9.4	ITU-T Speech-Coding Standards	336
		Bit Rate	338
		Delay	338
		Complexity	339
		Quality	339
	5.9.5	Multimedia Multiplex and Synchronization Standards	340
		ITU-T Recommendation H.221	341
		ITU-T Recommendation H.223	341

			ITU-T Recommendation H.225	341
			Common Control Protocol H.245	343
	5.10	IETF and Internet Standards		344
		5.10.1	IETF Standardization Process	344
		5.10.2	Internet Network Architecture	347
		5.10.3	Internet Protocols	348
			Classical IP Stack	350
			IP Version 6	351
			Priority Field	352
			Flow Label	353
			IPv6 Addresses	353
			Hop-by-Hop Options Header	355
			Fragment Header	356
			Routing Header	356
			IPv6 Security	357
		5.10.4	Real-Time Multimedia Transmission Across the Internet	359
			Signaling	360
			Session Control	360
			Transport	361
			Network Infrastructure	363
			Multimedia Data for Network Use	364
		5.10.5	MPEG-4 Video Transport Across the Internet	365
			Use of RTP	365
			System Architecture	366
			MPEG-4 Server	368
			MPEG-4 Client	369
	5.11	Concluding Remarks		370

Chapter 6
Multimedia Communications Across Networks — 373

6.1	Packet Audio/Video in the Network Environment			373
	6.1.1	Packet Voice		374
	6.1.2	Integrated Packet Networks		377
	6.1.3	Packet Video		380
6.2	Video Transport Across Generic Networks			382
	6.2.1	Layered Video Coding		386
		Layered Compression		386
		Layered Transmission		388
	6.2.2	Error-Resilient Video Coding Techniques		388
		Error-Resilient Encoding		390
		Decoder Error Concealment		392
		Error-Resilient Entropy Code		393

	6.2.3	Scalable Rate Control	394
		Rate Control Techniques	396
		Theoretical Foundation of the SRC	397
	6.2.4	Streaming Video Across the Internet	400
		Video Compression	401
		Requirements Imposed by Streaming Applications	403
		Application Layer QoS Control	404
		Continuous Media Distribution Services	407
		Streaming Servers	409
		Media Synchronization	409
		Protocols for Streaming Video	410
6.3	Multimedia Transport Across ATM Networks		411
	6.3.1	Multiplexing in ATM Networks	412
	6.3.2	Video Delay in ATM Networks	413
	6.3.3	Errors and Losses in ATM	417
	6.3.4	MPEG Video Error Concealment	420
	6.3.5	Loss Concealment	420
	6.3.6	Video Across WATM Networks	421
	6.3.7	Heterogeneous Networking	422
6.4	Multimedia Across IP Networks		424
	6.4.1	Video Transmission Across IP Networks	424
	6.4.2	Traffic Specification for MPEG Video Transmission on the Internet	426
	6.4.3	Bandwidth Allocation Mechanism	427
	6.4.4	Fine-Grained Scalable Video Coding for Multimedia Across IP	428
6.5	Multimedia Across DSLs		432
	6.5.1	VoDSL Architecture	433
	6.5.2	Delivering Voice Services Across DSL	438
	6.5.3	Multimedia Across ADSL	439
		Serial Transmission: TDM	439
		Parallel Transmission Frequency Division Multiplexing	441
6.6	Internet Access Networks		441
	6.6.1	DSL Networks	443
	6.6.2	Cable Access Networks	445
	6.6.3	Fixed Wireless Routed for Internet Access	447
6.7	Multimedia Across Wireless		449
	6.7.1	Wireless Broadband Communication System (WBCS) for Multimedia	451
	6.7.2	Audiovisual Solutions for Wireless Communications	453

			Page
	6.7.3	Mobile Networks	458
		Speech Transmission in GSM	459
		Video Across GSM	460
		Mobile ATM	462
		Mobile IP	462
		Wireless Multimedia Delivery	467
		SIP in Mobile Environment	469
		Multicast Routing in Cellular Networks	470
		Broadband Wireless Mobile	471
	6.7.4	Broadcasting Networks	473
		Digital Video Broadcasting (DVB)	477
		Data Transmission Using MPEG-2 and DVB	478
		MPEG Program Stream	480
		MPEG Transport Stream	481
		Broadband Multimedia Satellite Systems	484
		Multimedia Home Platform	486
		Multimedia Car Platform	487
6.8	Digital Television Infrastructure for Interactive Multimedia Services		488
		Interactive Broadcast Data (IDB) Services	490
		Data Carousel Concept	492
6.9	Concluding Remarks		493

References — 497
Index — 537
About the Authors — 545

Preface

The past years have seen an explosion in the use of digital media. Industry is making significant investments to deliver digital audio, image and video information to consumers and customers. A new infrastructure of digital audio, image and video recorders and players; online services and electronic commerce is rapidly being deployed. At the same time, major corporations are converting their audio, image and video archives to an electronic form. Digital media offer several distinct advantages over analog media. The quality of digital audio, image and video signals is higher than that of their analog counterparts. Editing is easy because one can access the exact discrete locations that need to be changed. Copying is simple with no loss of fidelity. A copy of digital media is identical to the original. Digital audio, image and video are easily transmitted across networked information systems. These advantages have opened up many new possibilities.

Multimedia consists of `Multimedia data + Set of interactions`. Multimedia data is informally considered as the collection of three Ms: multisource, multitype and multiformat data. The interactions among the multimedia components consist of complex relationships without which multimedia would be a simple set of visual, audio and other data.

Multimedia and multimedia communication can be globally viewed as a hierarchical system. The multimedia software and applications provide a direct interactive environment for users. When a computer requires information from remote computers or servers, multimedia information must travel through computer networks. Because the amount of information involved in the transmission of video and audio can be substantial, the multimedia information must be compressed before it can be sent through the network in order to reduce the communication delay. Constraints, such as limited delay and jitter, are used to ensure a reasonable video and audio effect at the receiving end. Therefore, communication networks are undergoing constant improvements in order to provide for multimedia communication capabilities. LANs are used to connect local computers and other equipment, and Wide Area Networks (WANs) and the Internet connect the LANs together. Better standards are constantly being developed, in order to provide a global information superhighway across which multimedia information will travel.

Organization of the Book

The book is organized into six chapters:

Chapter 1 describes the concept of multimedia communication modeling. It presents a brief description of elements for multimedia systems. After that, we discuss user and network requirements together with the packet transfer concept. An overview of multimedia terminals is also given.

Chapter 2 explains that multimedia communication is more than simply putting together text, audio, images and video. It reviews a recent trend in multimedia research to exploit the audio-visual interaction and to build the link between audio and video processing. The emphasis is on lip reading, synchronization and tracing audio-to-visual mapping as well as the bimodal person verification.

Chapter 3 is devoted to multimedia processing in communication. We present and analyze digital media and signal processing elements. Next, we describe a general framework for image copyright protection through digital watermarking. We then review the key attributes of neural processing essential to intelligent multimedia processing. Finally, this chapter concludes with recent large-scale-integration programmable processors designed for multimedia processing such as real-time compression and decompression of audio and video as well as the next generation of computer graphics.

Chapter 4 deals with the issues concerning distributed multimedia systems. We give an overview: main features, resource management, networking and multimedia operating systems. Next, we identify the applications like interactive television, telecooperation and hypermedia, and we survey the important enabling technologies.

Chapter 5 focuses on multimedia communication standards. We discuss Moving Pictures Experts Group (MPEG)-1, MPEG-2, MPEG-4, MPEG-4 Visual Texture Coding (VTC), Joint Photographic Experts Group (JPEG)-2000, MPEG-7, MPEG-21, International Telecommunications Union-Telecommunication Sector (ITU-T) and Internet standards. We discuss the ITU-T standardization process in multimedia communications from the video and speech coding, as well as from multimedia, multiplex and synchronization points of view (H.320, H.321, H.322, H.323, H.262, H.263, H.26L, H.221, H.222, H.223 and H.225).

Chapter 6 concentrates on multimedia communication across networks. After an introduction about packet audio-video in the network environment, we discuss the concept of video transport across generic networks. Multimedia transport over ATM networks is described, too. We then move to multimedia across IP networks, including video transmission, traffic specification for MPEG video transmission on the Internet and bandwidth allocation mechanism. We present and illustrate the concepts of Internet access networks. In addition, we discuss special issues relating to multimedia across wireless networks such as wireless broadband communication for multimedia audiovisual solutions, mobile and broadcasting networks and digital TV infrastructure for interactive multimedia services.

Appendix/Web Site

Appendix A contains useful information available on the Internet: standardization organizations, associations, alliances, fora and consortia; documents, software and hardware reference, and a products and services list. No software is provided. The appendix can be downloaded at the following Web site: www.phptr.com/rao.

References

The references are grouped according to the various chapters. Special efforts have been taken to make this list as up to date and exhaustive as possible.

A number of forces are driving communications, such as the following:

- The evolution of communications and data networks in today's modern Plain Old Telephone Service (POTS) network and packet (including the Internet) networks, with major forces driving these networks into an integrated structure
- The increasing availability of almost unlimited bandwidth demand in the office, the home and eventually on the road, based on high-speed data modems, cable modems, hybrid fiber-mix systems, and, recently, a number of fixed wireless access systems
- The availability of ubiquitous access to the network through Local Area Networks (LANs), wireline and wireless networks providing the promise of anywhere, anytime access
- The ever-increasing amount of memory and computation that can be brought to bear on virtually any communications or computing system
- The terminals, including sophisticated screen phones; digital telephones; multimedia personal computers (PCs) that handle a wide range of text, image, audio and video signals; network computers and other low-cost Internet-access terminals and Personal Digital Assistants (PDAs) of all types that can access and interact with the network using wired and wireless connections
- The digitalization of virtually all devices, including cameras, video capture devices, video playback devices, handwriting terminals, sound capture devices and so forth

Multimedia Communication Systems provides a comprehensive coverage of various surveys of the current issues relating to multimedia communications. This book addresses the fundamentals of the major topics of the multimedia communication systems: audio-visual integration, multimedia processing in communications, distributed multimedia systems, multimedia communication standards and multimedia communications across networks.

We have focused our attention on these topics with the hope that the level of discussion provided will enable an engineer or a scientist to design multimedia communication systems or to conduct research on advanced and newly emerging topics. The objective of this book is not only to familiarize the reader with multimedia communication systems, but also to provide the underlying theory, concepts and principles related to these disciplines, including the power and the practical utility of the topics.

A major challenge during the preparation of this book was the rapid pace of development, both in software and hardware related to multimedia communication systems. We have tried to keep pace by including many of the latest developments. In this way, it is hoped that the book is timely and appeals to a wide audience in the engineering, scientific and technical communities. In addition, we have included more than 270 figures and more than 800 references. Although this book is primarily for graduate students, it can be also very useful for academia, researchers, scientists and engineers dealing with multimedia communication systems.

Acknowledgments

It is a pleasure to acknowledge the help received from colleagues associated with universities, research labs, and industry. This help was in the form of technical papers and reports, valuable discussions, information, brochures, the review of various sections of the manuscript, computer programs and more.
We give special thanks to the following people:

>Kalman Fazekas—Technical University of Budapest, Department of Microwave Telecommunications, Hungary
>Ion Marghescu—Politechnica University of Bucharest, The Faculty of Electronics and Communications, Romania
>Athanassios Skodras—University of Patras, Electronics Laboratory, Greece

We appreciate the patience and perseverance of our families during the preparation of this book.

List of Acronyms

1D .... One-dimensional
2D .... Two-dimensional
3D .... Three-dimensional
8D .... Eight-dimensional
3G .... Third Generation
3GPP .... Third Generation Partnership Project
4G .... Fourth Generation
AAC .... Advanced Audio Coding
AAL .... ATM Adaptation Layer
AAP .... ATM Access Point
ABR .... Available Bit Rate
ACE .... Advanced Coding Efficiency
ACELP .... Algebraic Code-Excited Linear Prediction
ACR .... Absolute Category Rating
AD .... Adaptive Differential
AD .... Area Directors
ADPCM .... Adaptive DPCM
ADSL .... Asymmetric Digital Subscriber Line
AH .... Authentication Header
AL .... Adaptation Layer
ALF .... Application Level Framing
ALU .... Arithmetic and Logic Unit
AMR .... Adaptive Multirate
AP .... Access Point
API .... Application Programming Interface
ARPANET .. Advanced Research Agency Network
ARQ .... Automatic Repeat Request
ARTS .... Advanced Real-Time Simple (Profile)
ASIC .... Application-Specific Integrated Circuit
ASPEC .... Adaptive Spectral Entropy Coding
ASR .... Automatic Speech Recognition
ATC .... Adaptive Transform Coding
ATDM .... Asynchronous Time Division Multiplex
ATM .... Asynchronous Transfer Mode
ATSC .... Advanced Television System Committee
AU .... Access Unit
B .... Blue
BAM .... Bandwidth Allocation Mechanisms
BAP .... Body Animation Parameter
BB .... Band-by-Band
BDP .... Body Definition Parameter
BER .... Bit Error Rate
BGP .... Border Gateway Protocol
BIFS .... Binary Format for Scene Description
BISDN .... Broadband ISDN
BoD .... Bandwidth on Demand
BQ .... Bilevel Quantization
BRAS .... Broadband Remote Access Servers
BSAC .... Bit-Sliced Arithmetic Coding
BTS .... Base Transceiver Stations
B-VOP .... Bidirectional Interpolated VOP
CAI BIOS ... Common Air Interface Basic Input Output System
CAP .... Carrierless Amplitude Phase
CATV .... Cable Television
CBR .... Constant Bit Rate
CBT .... Core-Based Tree

Acronym	Meaning
CD	Committee Draft
CD	Compact Disc
CDMA	Code Division Multiple Access
CDPD	Cellular Digital Packet Data
CDV	Cell Delay Variation
CELP	Code-Excited Linear Prediction
CGI	Common Gateway Interface
CIF	Common Intermediate Format
CISC	Complex Instruction Set Computer
CLEC	Competitive Local Exchange Carriers
CLP	Cell Loss Priority
CN	Corresponding Node
CN	Canonical Name
CNG	Comfort Noise Generation
COFDM	Coded Orthogonal Frequency Division Multiplex
CPE	Customer Premises Equipment
CPS	Constrained Parameter Set
CPU	Central Processing Unit
CRC	Cyclic Redundancy Check
CREW	Compression with Reversible Embedded Wavelets
CS	Coding Scheme
CS	Convergence Sublayer
CSCW	Computer-Supported Cooperative Work
CSP	Content Service Provider
CS-VQ	Constrained-Storage Vector Quantization
CTI	Complete Timing Information
CTI	Computer Tomography Information
CVC	Consonant-Vowel-Consonant
D	Data or Descriptor
D/A	Digital-to-Analog
DAB	Digital Audio Broadcast
DAI	DMIF Application Interface
DAT	Digital Audio Type
DAVIC	Digital Audio Visual Council
DBA	Dynamic Bandwidth Allocation
DBNN	Decision-Based Neural Network
DBS	Direct Broadcast Satellite
DCA	Dynamic Channel Allocation
DCR	Degradation Category Rating
DCT	Discrete Cosine Transform
DDL	Description Definition Language
DDM	Dense Division Multiplexing
DFT	Discrete Fourier Transform
DIS	Draft International Standard
DLC	Digital Link Control
DLC	Digital Loop Carriers
DMA	Division Multiple Access
DMIF	Delivery Multimedia Integration Framework
DMS	Distributed Multimedia System
DMT	Discrete Multitone
DNI	DMIF Network Interface
DNS	Domain Name System
DPAC	Differential Perceptual Audio Coder
DPCM	Differential Pulse Code Modulation
DRC	Dynamic Resolution Conversion
DS	Description Scheme
DSL	Digital Subscriber Line
DSLAM	DSL Access Multiplexer
DSM	Digital Storage Media
DSM-CC	DSM-Command and Control
DSPs	Digital Signal Processors
DSVD	Digital Simultaneous Voice and Data
DTH	Direct-to-Home
DTS	Decoding Time Stamp
DTV	Digital TV
DVB	Digital Video Broadcasting
DVB-C	DVB Cable
DVB-RCS	DBV Return Channel System by Satellite
DVB-S	DVB via Satellite
DVB-T	DVB Terrestrial

List of Acronyms

DVD .... Digital Versatile Disk
DVD .... Digital Video Disk
DWT .... Discrete Wavelet Transform
EBCOT .... Embedded Block Coding with Optimized Truncation
EBU .... European Broadcasting Union
EDF .... Earliest Deadline First
EEG .... Electroencephalogram
EFR .... Enhanced Full Rate
EKG .... Electrocardiogram
EM .... Expectation Maximization
EP .... Error Protection
EPG .... Electronic Program Guide
ER .... Error-Resilient
EREC .... Error-Resilient Entropy Code
ES .... Elementary Stream
ESA .... European Space Agency
ESM .... Elementary Stream Management
ESP .... Encapsulating Security Payload
ETSI .... European Telecommunication Standards Institute
EZW .... Embedded Zero-Tree Wavelet
FAB .... Face and Body
FACS .... Facial Action Coding System
FAP .... Facial Animation Parameter
FAPU .... Facial Animation Parameter Unit
FAT .... Facial Animation Table
FCA .... Fixed Channel Allocation
FCPAS .... Fault, Configuration, Accounting Performance and Security Management Areas
FCC .... Federal Communication Commission
FDDI .... Fiber Distributed Data Interface
FDDS .... Fiber Distributed Data Service
FDIS .... Final Draft International Standard
FDMA .... Frequency Division Multiple Access
FDP .... Facial Definition Parameter
FEC .... Forward Error Correction

FF/FR .... Fast Forward/Fast Reverse
FFT .... Fast Fourier Transform
FGS .... Fine Granularity Scalability
FIR .... Finite Impulse Response
FIT .... FAP Interpolation Table
FLC .... Fixed Length Code
FPGA .... Field Programmable Array
fps .... frames per second
FR .... Full Rate
FS .... Frame Store
FTP .... File Transfer Protocol
G .... Green
GA .... General Audio
GDDS .... Group-Decision Support System
GFA .... Gateway Foreign Agent
GII .... Global Information Infrastructure
GMC .... Global Motion Compensation
GOB .... Group of Blocks
GOP .... Group of Pictures
GOV .... Group of Video Planes
GSM .... Global System for Mobile
GSTN .... General Switched Telephone Network
GW .... Gateway
HDTV .... High Definition Television
HFC .... Hybrid Fiber Coax
HILN .... Harmonic and Individual Lines plus Noise
HL .... High Level
HMIHY ...... How May I Help You
HMM .... Hidden Markov Model
HP .... High Profile
HP .... High Pass
HPNA .... Home Phoneline Network Alliance
HR .... Half-Rate
HSCSD .... High-speed Circuit-Switched Data
HSV .... Hue Saturation Value
HTML .... Hypertext Markup Language

HTTP	Hypertext Transport Protocol
HVS	Human Visual System
HVXC	Harmonic Vector Excitation Coding
IAB	Internet Architecture Board
IAD	Integrated Access Device
IANA	Internet Assigned Number Authority
ICMP	Internet Control Message Protocol
ICT	Irreversible Component Transformation
IDB	Interactive Data Broadcast
IDCT	Inverse DCT
IEC	International Electrotechnical Commission
IEEE	Institute of Electrical and Electronics Engineers
IESG	Internet Engineering Steering Group
IETF	Internet Engineering Task Force
ILP	Integrated Level Processing
IMP	Intelligent Multimedia Processing
IMT	International Mobile Telecommunication
IN	Intelligent Network
I/O	Input/Output
IP	Internet Protocol
IPA	International Phonetic Alphabet
IP-H	IP-based header plus Extensions headers
IPI	Intellectual Property Identification
IPMP	Intellectual Property Management and Protection
IPN	Integrated Packet Network
IPPV	Impulse Pay-Per-View
IPR	Intellectual Property Rights
IROS	Internet Radio Operating System
ISD	Independent Segment Decoding
ISDN	Integrated Services Digital Network
ISI	Intersymbol Interference
ISM	Industrial, Scientific, Medical
ISO	International Organization for Standardization
ISOC	Internet Society
ISP	Internet Service Provider
ISP POP	Internet Service Provider's Point of Presence
ISWG	IETF Integrated Services WG
ITU	International Telecommunications Union
ITU-T	ITU–Telecommunication sector
ITV	Interactive Television
IVB	Interactive Video Broadcast
I-VOP	Intra VOP
IZT	Isolated Zero Tree
JND	Just Noticeable Distortion
JPEG	Joint Photographic Experts Group
JSC	Joint Source Channel Coding
JTC	Joint ISO/IEC Technical Committee
KBD	Kaiser-Bessel Derived
KLT	Karhunen-Loeve Transform
LAN	Local Area Network
LC	Low Complexity
LD-CELP	Low Delay CELP
LFE	Low Frequency Enhancement
LL	Low Level
LLC	Logical Link Control
LMDS	Local Multipoint Distribution Service
LMS	Least Mean Square
LNB	Low Noise Block
LOAS	Low Overhead Audio Stream
LOD	Level of Detail
LOT	Lapped Orthogonal Transform
LPAS	Linear Prediction Analysis by Synthesis
LP	Linear Prediction
LP	Lowpass
LPC	Linear Predictive Coding
LSP	Line Spectral Pair
LTP	Long-Term Prediction

List of Acronyms

LZW ....Lempel-Ziv-Welch
MA ....Multiple Access
MAC............Media Access Control
MAD ....Mean Absolute Difference
MAN ....Metropolitan Area Network
MB...............Macroblock
MBS ....Mobile Broadband System
MC...............Motion Compensation
MCP ....Multimedia Car Platform
MCU............Multipoint Control Unit
MD ....Multidimensional
MDC............Multiple Description Coding
MDCT ....Modified Discrete Cosine Transform
MDS ....Multipoint Distribution Service
MDS ....Multimedia Description Schemes
ME...............Motion Estimation
MEMO ....Multimedia Environment for Mobiles
MF-TDMA ..Multifrequency Time Division Multiple Access
MH ....Multihypothesis
MHBP ....Multihypothesis Block Pattern
MHP ....Multimedia Home Platform
MIDI............Musical Instrument Digital Interface
MIME..........Multipurpose Internet Mail Extension
MIPS ....Millions of Instructions per Second
M-JPEG.......Motion-JPEG
ML...............Main Level
MLP.............Multilayer Perceptron
MM..............Multimedia
MMDS.........Multichannel MDS
MMDS.........Multimedia Description Scheme
MMSP ....Multimedia Signal Processing
MMX ....Multimedia Extension
MOP ....Mesh Object Plane
MOPS ....Mega Operations Per Second
MOS ....Mean Opinion Score

MOT ....Multimedia Object Transfer
MOTIVATE .Mobile Television and Innovative Receivers
MP ....Main Profile
MPE............Multipulse Excitation or Multi-Protocol Encapsulation
MPEG.........Moving Pictures Experts Group
MPEG-J......MPEG–Java
MPLS..........Multiprotocol Label Switching
MPTS..........Multiprogram TS
MPU ....Multimedia Processor Unit
MQ..............Multiple Quantization
MRI ....Magnetic Resonance Imaging
MRC ....Mixed Raster Content
MSB............Most Significant Bit
MSE............Mean Square Error
MSS ....Mobile Support Station
MSTVQ......Multistage Tree-Structured Vector Quantization
MTU ....Maximum Transmission Unit
MV..............Motion Vector
MVPD ....Multichannel Video Program Distribution
NADIB ....Narrow Band Audio Digital Broadcasting
NAPT..........Network Address and Port Translation
NAT............Network Address Translation
NBC............Nonbackward Compatible
NC ....Noiseless Coding
NGN ....Next Generation Network
NISDN........Narrow-band ISDN
NLIVQ........Nonlinear Interpolative Vector Quantization
NM..............Nuclear Medicine
NMR...........Noise-to-Mask Ratio
NN ....Neural Network
NNTP..........Network News Transfer Protocol
NTI ....Null Timing Information

NTP	Network Time Protocol	PSTN	Public Switched Telephone Network
NVoD	Near Video on Demand	PVC	Permanent Virtual Connection
OBMC	Overlapped Block Motion Compensation	P-VOP	Predicted–VOP
		PVR	Packet Voice Receiver
OCFD	Optimum Coding in the Frequency Domain	PVT	Packet Voice Transmitter
		PW	Perceptual Weighted
OCI	Object Content Information	Q	Quantization
OCR	Object Clock Reference	QCIF	Quarter CIF
OD	Object Descriptor	QMF	Quadrature Mirror Filter
OS	Operating System	QoS	Quality of Service
OSI	Open System Interconnection	QP	Quantization Parameter
OSPF	Open Shortest Path First	QPSK	Quadrature Phase-Shift Keying
PAR	Pixel Aspect Ratio	R	Red
PC	Personal Computer	RAM	Random Access Memory
PCA	Principal Component Analysis	RAS	Registration, Admission, and Status
PCF	Picture Clock Frequency	RCPC	Rate-Compatible Punctured Convolution
PCM	Pulse Code Modulation		
PCR	Program Clock Reference	RCST	Return Channel Satellite Terminal
PCS	Personal Communication System	RCT	Reversible Component Transformation
PCU	Processor Complexity Units		
PDA	Personal Digital Assistant	RCU	RAM Complexity Units
PDC	Personal Digital Cellular	R-D	Rate-Distortion
PDF	Probability Density Function	R&D	Research and Development
PDM	Packet Division Multiple	RFC	Request for Comments
PDU	Protocol Data Unit	RGB	Red, Green and Blue
PE	Perceptual Entropy	RIP	Routing Information Protocol
PES	Packetized Elementary Stream	RISC	Reduced Instruction Set Computer
PFGS	Progressive FGS	RL	Run-Length
PID	Packet Identifier	RMOA	Real-Time Multimedia over ATM
PLL	Phase-Locked Loop	RMSE	Root MSE
PNG	Portable Network Graphics	ROI	Region of Interest
PNS	Perceptual Noise Substitution	ROM	Read-Only Memory
POCS	Projection onto Convex Sets	RPE	Regular Pulse Excitation
POP	Point of Presence	RPM	Return Path Multiplexer
POTS	Plain Old Telephone Service	RS	Read-Solomon
PPD	Proposed Package Description	RSVP	Resource Reservation Protocol
PPP	Point-to-point Protocol	RTCP	Real Time Control Protocol
PSI	Program-Specific Information	RTE	Run Time Engine
PSNR	Peak Signal-to-noise Ratio	RTFD	Recommended Technical Framework

List of Acronyms

Document	
RTI	Real Time Interface
RTP	Real-Time Transport Protocol
RTSP	Real-Time Streaming Protocol
RTT	Round-Trip Time
RVLC	Reversible Variable Length Coding
SA	Structured Audio
SA-DCT	Shape Adaptive – DCT
SAM	Split-and-Merge
SAOL	Structured Audio Orchestra Language
SAP	Session Announcement Protocol
SAR	Segmentation and Reassembly
SBBP	Switched-Batch Bernoulli Process
SBC	Subband coding
SCN	Switched Circuit Network
SDC	Single Description Coding
SDH	Synchronous Digital Hierarchy
SDL	Specification and Description Language
SDM	System Decoder Model
SDP	Session Description Protocol
SDRAM	Synchronous Dynamic Random Access Memory
SDTV	Standard Definition Television
SER	Symbol Error Rate
SFM	Spectral Flatness Measure
SG	Study Groups
SH	Supervisor Host
SHDSL	Single Pair High-Speed DSL
SI	Service Information
SIF	Source Input Format
SIMD	Single Instruction Multiple Data
SIP	Session Initiation Protocol
SIT	Satellite Interactive Terminal
SL	Synchronization Layer
SLI	Spoken Language Interface
SM	Simulation Model
SMG	Statistical Multiplexing Gain
SMIL	Synchronized Multimedia Integration Language
SMM	Streaming Multimedia
SMPTE	Society of Motion Picture and Television Engineers
SMR	Signal-to-Mask Ratio
SMTP	Simple Mail Transfer Protocol
SNHC	Synthetic and Natural Hybrid Coding
SNMP	Simple Network Management Protocol
SNR	Signal-to-Noise Ratio
SONET	Synchronous Optical Network
SP	Signal Processing
SP	Simple Profile
SPIE	Society of Photo-optical and Instrumentation Engineers
SPIHT	Set Partitioning in Hierarchical Trees
SPL	Sound Pressure Level
SPS	SL-Packetized Streams
SPTS	Single Program TS
SQ	Scalar Quantization
SRC	Scalable Rate Control
SRM	Session and Resource Manager
SSR	Scalable Sampling Rate
SSRC	Synchronization Source RC
STB	Set-Top Box
STM	Synchronous Transfer Mode
STU	Subscriber Terminal Unit
SVC	Switched Virtual Connection (Circuit)
SZ	Step Size
Tabs	Absolute threshold
TAM	Technical Issues Associated with MHP
TC	Technical Committee, or Transform Coding
TCP	Transmission Control Protocol
TCQ	Trellis Coded Quantization
TD	Tree-Depth

TDM	Time Division Multiplexing
TDMA	Time Division Multiple Access
TDNN	Time-delayed Neural Network
TMT	True Motion Technique
TNS	Temporal Noise Shaping
ToR	Terms of Reference
TR	Technical Report
TS	Transport Stream
TSP	Transport Stream Packet
TTS	Text-to-Speech
TTSI	Text-to-Speech Interface
TWIN-VQ	Transform Domain-Weighted Interleave VQ
UBR	Unspecified Bit Rate
UDP	User Datagram Protocol
UEP	Unequal Error Protection
UMTS	Universal Mobile Telecommunication System
UNII	Unlicensed National Information Infrastructure
UPG	Usage Parameter Control
URL	Uniform Resource Locator
US	Ultrasound, or United States
UTTCQ	Uniform Threshold TCQ
VAD	Voice Activity Detector
VB	Video Buffer
VBR	Variable Bit Rate
VC	Virtual Connection, Virtual Circuit
VCC	Virtual Circuit Connection
VCIP	Visual Communication and Image Processing
VCR	Video Cassette Recorder
VCV	Vowel-Consonant-Vowel
VHS	Video Home System
VLBR	Very Low Bit Rate
VLC	Variable Length Coding
VLD	Variable Length Decoder
VLSI	Very Large-Scale Integration
VM	Verification Model
VO	Video Object
VoATM	Voice over ATM
VoD	Video on demand
VoDSL	Voice over DSL
VoIP	Voice over IP
VOL	Video Object Layer
VOP	Video Object Plane
VOP	Voice Over Packetts
VP	Virtual Path
VPI/VCI	Virtual Path Identifier/Virtual Connection Identifier
VPN	Virtual Private Network
VQ	Vector Quantization
VRML	Virtual Reality Modeling Language
VS	Video Session
VS	Visual Object Sequence
VSB	Vestigial Side Band
VTC	Visual Texture Coding
W3C	World Wide Web Consortium
WAN	Wide Area Network
WATM	Wireless ATM
WBCS	Wireless Broadband Communication System
WD	Working Draft
WG	Working Group
WLAN	Wireless LAN
WMFTWG	Wireless Multimedia Forum Technical Working Group
WTCQ	Wavelet/Trellis Coded Quantization
WWW	World Wide Web
X3D	Extensible 3D
XM	eXperimental Model
XML	Extensible Markup Language
XMT	Extensible MPEG-4 Textual Format
YUV	Luminance Bandwidth-Chrominace
ZTR	Zero Tree Root

CHAPTER 1

Multimedia Communications

Chapter Overview

The challenge of multimedia communications is to provide services that integrate text, sound, image and video information and to do it in a way that preserves the ease of use and interactivity. At first, the concept of a multimedia communication modeling is described. We present a brief description of elements for multimedia systems. After that, user and network requirements are discussed, together with the packet transfer concept. Taking into account that Asynchronous Transfer Mode (ATM) uses a fixed-length packet and is suitable for high-speed applications, we describe multimedia requirements. Finally, we give an overview of multimedia terminals.

1.1 Introduction

Multimedia communications is the field referring to the representation, storage, retrieval and dissemination of machine-processable information expressed in multiple media, such as text, image, graphics, speech, audio, video, animation, handwriting and data files. With the advent of high-capacity storage devices, powerful and yet economical computer workstations and high-speed Integrated Services Digital Networks (ISDNs), providing a variety of multimedia communications services is becoming not only technically, but also economically, feasible. In addition, the Broadband ISDN (BISDN) has been given special attention as a next generation communication network infrastructure that will be capable of transmitting full motion pictures and high speed data at 150 and 600 MB/s and voice, as well as data, throughout the world [1.1].

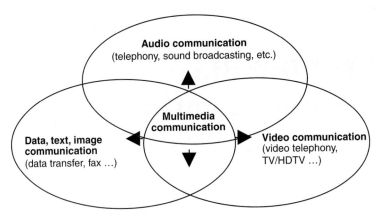

Figure 1.1 Multimedia communication.

Multimedia best suits the human being's complex perception and communicating behavior, as well as the way of acting. Namely, it not only provides communication capabilities and information sharing for people, irrespective of location and time, but it also provides easy and immediate access to widely distributed information banks and information processing centers. Applications in medicine, education, travel, real estate, banking, insurance, administration and publishing are emerging at a fast pace. These applications are characterized by large multimedia documents that must be communicated within very short delays. Computer-controlled cooperative work, whereby a group of users can jointly view, create, edit and discuss multimedia documents, is going to be characteristic of many transactions [1.2]. Some glamorous applications in multimedia processing include distance learning, virtual library access and living books. In distance learning, we learn and interact with instructors remotely across a broadband communication network. Virtual library access means that we instantly have access to all of the published material in the world, in its original form and format, and that we can browse, display, print and even modify the material instantaneously. Living books supplement the written word and the associated pictures with animations, and hyperlinks provide access to supplementary material [1.3, 1.4, 1.5].

Applications that are enabled or enhanced by video are often seen as the primary justification for the development of multimedia networks. Trends toward multimedia communication are represented in Figure 1.1.

Much of the work on packet video has considered a fairly homogenous networking scenario [1.6]. It would be proper if a single type of video service dominated in the networks. However, it is not a valid assumption for the traffic issues. First, video will not constitute a uniform service with easily determined behavior and requirements. Second, video will not share resources with streams of only the same type. This means that multiplexing in the network should be evaluated for a heterogeneous mix of traffic types. In business areas, there is a potential need for various kinds of new communication systems, such as high-speed data networks between geographically distributed LANs, high definition still-picture communication and TV

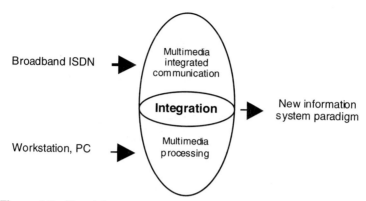

Figure 1.2 New information system paradigm using BISDN and workstations.

conferencing or corporate cable TV services. The new paradigm of the BISDN application system is a result of the integration of multimedia processing by workstations and multimedia communication by BISDN and is shown in Figure 1.2.

It is important to distinguish multimedia material from what is often referred to as multiple-media material. To illustrate the difference, consider using the application of messaging. Today, messaging consists of several types, including electronic mail (email), which is primarily text messaging, voice mail, image mail, video mail, and hand-written mail often transmitted as a facsimile (fax) document. Each of these messaging types is generally a single medium and is associated with a unique delivery mechanism and a unique repository or mailbox. For convenience, most consumers would like to have all messages delivered to a common repository or mailbox. Hence, you have the concept of multiple media being integrated into a single location.

In networked multimedia applications, various entities typically cooperate in order to provide the real-time guarantees to allow data to be presented at the user interface. These requirements are most often defined in terms of Quality of Service (QoS). QoS is defined as the set of parameters that defines the properties of media streams. We distinguish four layers of QoS: user QoS, application QoS, system QoS and network QoS [1.7]. The user QoS parameters describe requirements for the perception of multimedia data at the user interface. The application QoS parameters describe requirements for the application services, possibly specified in terms of media quality (high end-to-end delay) and media relations (like inter-intrastream synchronization). The system QoS parameters describe requirements on the communications services resulting from the application QoS. These may be specified in terms of both quantitative and qualitative criteria. By quantitative criteria, we mean bits per second or task processing time, while multicast, interstream synchronization, error recovery or ordered delivery of data represent qualitative criteria. The network QoS parameters describe requirements on network services, like network load and network performance. This chapter seeks to provide a brief presentation of multimedia communications including multimedia communication models and elements of multimedia systems, as well as the packet transfer concept. User and network requirements are emphasized together with multimedia in the ATM environment. An outline of the multimedia terminals concludes the chapter.

1.2 Multimedia Communication Model

A multimedia communication model is strongly influenced by the manufacturer-dependent solutions for PCs and workstations, including application software on the one hand and the intelligent network concept on the other [1.8, 1.9, 1.10]. A layered model for future multimedia communication comprises five components:

- Partitioning of complex information objects into distinct information types for the purpose of easier communicating, storing and processing. This comprises data, video or audio and takes into account the integration of different information types.
- Standardization of service components per information type, possibly with several levels of quality per information type.
- Creation of platforms at two levels: a network service platform and a multimedia communication platform. The first level hides the transport networks and network building blocks from an application designer's or user's point of view. The second level provides communication support on the basis of information structure and information exchange building blocks for a large number of applications.
- Definition of generic applications for multiple use in various multimedia environments and different branches meeting common widespread needs.
- Specific applications: electronic shopping, teletraining and remote maintenance, based on special information building blocks and using the network service platform and the multimedia communication platform, as well as including generic applications.

With regard to the capability of the available resources in each case, the multimedia communication applications must be scalable in order to run in a constant manner across different network and terminal types and capabilities.

1.3 Elements of Multimedia Systems

Multimedia systems generally use two key communications modes: person-to-person communications and person-to-machine communications. Figure 1.3 presents key elements of multimedia systems. As can be seen, both of these modes have a lot of commonality, as well as some differences.

In the person-to-person mode shown in Figure 1.3(a), there is a user interface that provides the mechanisms for all users to interact with each other, and there is a transport layer that moves the multimedia signal from one user location to some or all other user locations associated with the communications. The user interface creates the multimedia signal and allows users to interact with the multimedia signal in an easy-to-use manner. The transport layer preserves the qual-

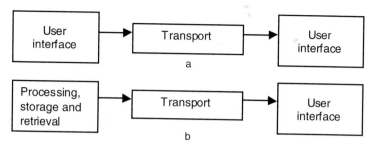

Figure 1.3 Elements of multimedia systems used in (a) person-to-person communications and (b) person-to-machine communications [1.2]. ©1998 IEEE.

ity of the multimedia signals so that all users receive what they perceive to be high-quality signals at each user location.

Examples of applications for the person-to-person mode are teleconferencing, videophones, distance learning and shared workspace scenarios. In the person-to-machine mode, shown in Figure 1.3(b), there is again a user interface for interacting with the machine, along with a transport layer for moving the multimedia signal from the storage location to the user. There is also a mechanism for storage and retrieval of multimedia signals that are either created by the user or requested by the user. The storage and retrieval mechanisms involve browsing and searching to find existing multimedia data. Also, these mechanisms involve storage and archiving in order to move user-created multimedia data to the appropriate place for access by others. Examples of applications for the person-to-machine mode include creation and access of business meeting notes, access of broadcast video and document archives from a digital library or other repositories.

1.4 User Requirements

The user needs a multimedia communication system that prepares and presents the information of interest, allows for the dynamic control of applications and provides a natural interface. From a user's point of view, the most important requirements of multimedia communications are the following:

- Fast preparation and presentation of the different information types of interest, taking into account the capabilities of available terminals and services
- Dynamic control of multimedia applications with respect to connection interactions and quality on demand combined with user-friendly human/machine interfaces
- Intelligent support of users taking into consideration their individual capabilities
- Standardization

User requirements in terms of services are defined by the media, the transmission content and the type of communication, as well as the ability to combine the three. On the other hand, multimedia communication services can be classified as being local (interactive or noninteractive), remote noninteractive or remote interactive and can also as be classified as being for residential, business or mobile use. The context in which multimedia services can be used is shown in Table 1.1.

Table 1.1 Context in which multimedia services can be used.

Local		Remote Noninteractive	Remote Interactive
Residential	Leisure (TV) The arts Teaching Games	Broadcasting	Enhanced telephones Videophones Home shopping Games Remote consultation Video on demand
Mobile	Presentation Demonstration	Broadcasting Remote security Monitoring	Project management Contract negotiation
Business	Multimedia presentation Training Database consultation	Teleinformation Teletraining Telesupervision	Video meeting Video conferencing Distance learning Project management Remote security Monitoring Remote diagnostics

Service usage conditions can be defined by their use, place, independence and degree of urgency. Services can be for private or business use. The terminal and services are usually used in the office, the home, the car or a public place. Independence could be defined by the portability of the terminal and its independence of a given infrastructure as perceived by the user. The degree of independence varies from one type of terminal to another. On the other hand, the degree of urgency, from the user's point of view, determines whether the service should be provided in real time or whether an offline service is sufficient.

A number of key requirements are common to the new multimedia services:

- Instant availability
- Real-time information transfer
- Service always online
- Access their services from any terminal (mobile point of delivery)

Whereas traditional voice services already have these characteristics, data services across the Internet (including voice over data) have typically been limited to basic bit transport with no service guarantees, no guaranteed availability and rather fragmented service interruptions.

With new data service emerging, such as Virtual Private Networks (VPNs) and interconnection service between two network service providers, priorities in the data networking domain have to change. In order to resolve and build robust multimedia networks, it is natural that operators will seek to base their data networks on the proven service delivery capability currently deployed in leading-edge voice networks. This will provide the flexibility, functionality and reliability required to meet the new demands of future users. Also, it will enable operators to offer the sophisticated services currently provided for voice in the multimedia domain.

Multimedia applications have several requirements with respect to the service offered to them by the communication system. These requirements depend on the type of the application and on its usage scenario. For instance, a nonconversational application for the retrieval of audio-visual data has different needs than a conversational application for live audio-visual communication (that is, a conferencing tool). The usage scenario influences the criticality of the demands.

1.5 Network Requirements

From the network point of view, the most important requirements of multimedia communications are the following:

- High speed and changing bit rates
- Several virtual connections using the same access
- Synchronization of different information types
- Suitable standardized services and supplementary service supporting multimedia applications

The requirements of applications regarding the communications services can be divided into traffic and functional requirements. The traffic requirements include transmission bandwidth delay and reliability. They depend on the used kind, number and quality of the data streams. The traffic requirements can be satisfied by the use of resource management mechanisms. They establish a relationship between transmitted data and resources and ensure that the audio-visual data is transmitted in a timely manner. For this, during the transmission of data, the information about the resource needs must be available at all nodes participating in the distributed applications, end systems and centers. Hence, resources must be reserved, and states must be created in these nodes, which basically means that a connection is established. The functional requirements are multicast transmission and the ability to define coordinated sets of unidirectional streams.

Current fixed and mobile networks are built on mature architectures with strong traffic management, configuration capabilities, service platforms and well-defined points of interconnection between the networks of different operators. A key requirement is that the same high-quality network services should exist when building integrated networking platforms for voice, data and multimedia services [1.11].

A future multimedia network must be organized to support heavy traffic flows, a wide variety of service mixes and different traffic patterns, both in terms of routing the traffic efficiently and in terms of scaling for overload. The network must adapt quickly to constantly changing traffic conditions. Reliable security features and firewalls must be in place for interworking between the many operators that will be competing in the market as a result of deregulation.

1.6 Packet Transfer Concept

Today's fiber technology offers a transmission capability that can easily handle high-bit rates like those required for video transmission. This leads to the development of networks, which integrate all types of information services. By basing such a network on packet switching, the services (video, voice and data) can be dealt with in a common format. Packet switching is more flexible than circuit switching in that it can emulate the latter while vastly different bit rates can be multiplexed together. In addition, the network's statistically multiplexing of variable rate services may yield a higher use of the channel capacity than what is obtainable with fixed capacity allocation. Many years ago, most of these arguments were verified in a number of projects [1.10, 1.12, 1.13, 1.14].

The project MAGNET II is an integrated network test bed at Columbia University. It was designed and implemented based on requirements of real-time network mangement and control. Switching is based on the concept of packet switching and the multi-class network model. The requirement made on the resource sharing mechanisms is to guarantee the appropriate quality of service for each traffic class [1.12]. However, bandwidth allocation is one of the most important problems in the management of networks that have guaranteed bandwith policy (ATM [1.12], PARIS [1.14]). Thus, a good bandwidth allocation strategy is crucial for packet switching networks.

Compared to circuit switching, packet switching offers dynamic allocation of bandwidths and switching resources, as well as the elimination of channel structure. Packet networks allow integrated service transport. They can carry voice, video and data using the same hardware protocols. Furthermore, packet communication does not require users to allocate a fixed channel or bandwidth before data transmission. Because users send packets only when necessary, and because many users can send packets over the same shared channel, resources in packet networks are used more efficiently than in circuit switched networks. Video signals are especially well suited for packet transmission. Images usually contain regions of high detail and low detail as well as periods of rapid motion and little motion. Thus, effective video coders should be able to produce data streams with variable bit rates that change with local characteristics [1.15, 1.16]. Packet networks can carry Variable Bit Rate (VBR) signals directly. No buffering or rate control feedback is necessary at the transmitter. If a video coder can specify the order in which a network discards data in case of network congestion, the decoder can often suffer less degradation in picture quality when packet loss occurs. Packet networks that offer prioritization of packets give video coders the ability to protect critical information. However, packet networks also provide some difficulties for video coders.

The ATM networks are based on virtual circuit switching. All fixed size packets of a circuit have the fixed route [1.17]. The tasks of packet video transfer across an asynchronous time division multiplexed network or Internet are to code and transfer digital video signal under quality constraints as shown in Figure 1.4. In Internet Protocol (IP) networks, the packets are of variable length, and there is no pre-established route. Therefore, they may arrive out of order at destination. During transfers in ATM and IP networks, delay and some packet loss are unavoidable [1.18, 1.19, 1.20].

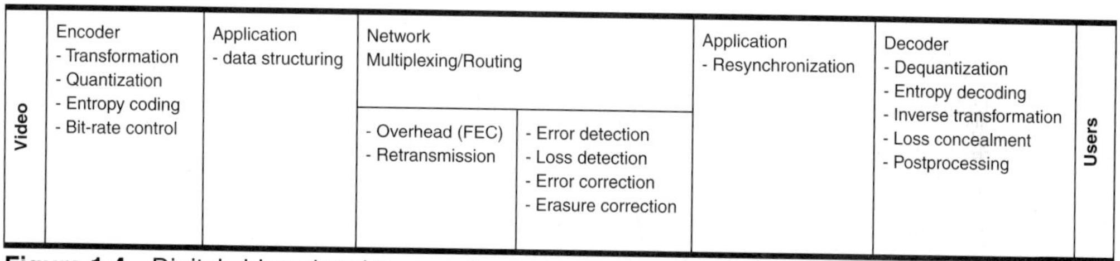

Video	Encoder - Transformation - Quantization - Entropy coding - Bit-rate control	Application - data structuring	Network Multiplexing/Routing		Application - Resynchronization	Decoder - Dequantization - Entropy decoding - Inverse transformation - Loss concealment - Postprocessing	Users
			- Overhead (FEC) - Retransmission	- Error detection - Loss detection - Error correction - Erasure correction			

Figure 1.4 Digital video signal transport.

The generic functions of the network in packet transfer from source to user are routing and multiplexing. Routing provides connectivity and does not depend on the information type used in the transfer. Multiplexing determines much of the transfer quality in the network and is highly dependent on the traffic characteristics, the quality requirements and the user's applications.

Statistical multiplexing with quality guarantees is the best choice for video transfer [1.21]. In order to offer probabilistic guarantees, a network must know its current flow of traffic, based on already accepted connections or measurements of the actual network load. New connections are allowed if they can be guaranteed the quality that they request and their characteristics do not risk the quality of already accepted connections or measurements of the actual network load.

1.7 Multimedia Requirements and ATM Networks

One of the biggest factors in the emergence of multimedia computing and processing was the digitization of the telecommunications network, which was completed by the late 1980s. This brought a new era in communications, where digital representations of both signals and data could interact in a common network, and it led to the concepts behind modern data networks and data protocols like ATM [1.22].

Today the ATM plays a significant role in realizing the flexibility and economy necessary for multimedia communication [1.23]. The most important ATM capabilities for multimedia requirements are the following:

- Constant, variable or burst-oriented bit streams
- Virtual connections or virtual paths through the subscriber access depending on instantaneous needs with the total capacity of about 150 or 600 MB/s
- Uniform bit rate-independent transmission and switching systems

In addition, the ATM concept aims at a universal network that offers all services using one uniform bit rate and information type-independent access based on harmonized protocols. The ATM protocol is based on the concept of designing a packet protocol that would be appropriate for both real-time signals (such as speech, audio and video) and data. Hence, its packet size is small (53 bytes) for low-latency signal processing, and the header size is small (5 bytes) for high efficiency. This facilitates information packet sequencing and synchronization between various information types within one multimedia application. ATM networks are designed to efficiently handle high volume voice, audio, and video traffic, yet still maintain their effectiveness for bursty data. As demand grows, a universal ATM network will be superior both in performance and in cost to the alternative solution where several specialized networks exist side by side [1.24].

The more standard Transmission Control Protocol/Internet Protocol (TCP/IP) used on the Internet uses significantly larger packets (upwards of 1 to 2 KB packets) for greater efficiency in moving large, bursty and data traffic through the IP network.

In spite of these advantages, ATM has some constraints on multimedia communications. Namely, voice packetization and depacketization result in additional delays that may call for echo compensation measures. Nevertheless, the future multimedia communication system will be created offering a large variety of applications based on an efficient universal ATM network.

1.8 Multimedia Terminals

Every major advance in networking has been preceded by an advance in the user interface that has precipitated the acceptance and growth of the networking advance. For example, the invention of the telephone preceded the growth of switch networks, the invention of the television preceded the growth of TV networks and Cable Television (CATV), the radio telephone led to the cellular network, the PC led to the LAN/WAN network and the browser led to the growth of the Internet and the Web. For the multimedia, new smart terminals need to be created in order to facilitate the displaying, accessing, indexing, browsing and searching of multimedia content in a convenient and easy-to-use manner.

For multimedia systems to achieve the vision of the current communications revolution and to become available to everyone, a number of technological issues must be addressed and put into a framework that leads to integration, ease of use, and high quality outputs. Among the issues that must be addressed are the following:

- The basic techniques for compression and coding the various media that constitute multimedia signals, including the signal-processing algorithms, the associated standards and the issues involved with transmission of these media in real communications systems

- The basic techniques for organizing, storing and retrieving multimedia signals, including both downloading and streaming techniques, layering of signals to match characteristics of the network and the display terminal and issues involved with defining a basic QoS for multimedia signal and its constituent components
- The basic techniques for accessing the multimedia signals by providing tools that match the user to the machine
- The basic techniques for searching in order to find multimedia sources that provide the desired information or material, or searching methods, which in essence are based on machine intelligence, provide the interface between the network and the human user and provide methods for searching using text requests, image matching methods and speech queries
- The basic techniques for browsing individual multimedia documents and libraries in order to take advantage of human intelligence to find desired material using text browsing, indexed image browsing, and voice browsing

Multimedia itself denotes the integrated manipulation of at least some information represented as continuous media data, as well as some information encoded as discrete media data (text and graphics). Here, we have the act of capturing, processing, communicating, presenting and/or storing.

Multimedia terminals are needed to retrieve, analyze, store and broadcast the new forms of written, sound and visual content. The architecture of these systems can be defined according to different approaches based on telecommunications data processing and audiovisual technology. By incorporating voice and data as well as still and moving pictures into their communications, business has made functions increasingly sophisticated to improve access to distributed resources and to save valuable time in the decision process. Remote dialog, discussion, information production, maintenance and inspection are now possible from the new multimedia systems at operating costs that are continuing to fall. Existing solutions offer two types of terminals: multifunction office or computer workstations and dedicated equipment, such as enhanced telephone terminals, videophones or shared teleconferencing systems.

Multimedia communication requires powerful terminals: upgraded PCs, desktop workstations or video computers. Today's terminals are enhanced for broadband multimedia applications, for example, PCs by the addition of telecommunication and video-audio capabilities and TV receivers by the addition of intelligence and interactivity. At the same time, High Definition Television (HDTV) is in development, leading the way toward all digital TV. HDTV is a technology driver for memories, image/video processors and flat screens [1.21].

Multimedia terminal equipment also comprises suitable cameras, scanners, printers and mass storage. Special equipment is necessary for editing multimedia information, that is, the creation, alternation and deletion of content and structures. Three-dimensional (3D) display devices and speech recognition systems will further facilitate faster and easier human interaction with multimedia applications or editors.

PCs and workstation architecture are considered for the interconnection of the systems components, based on star configurations and using ATM principles. This could make the integration of all information types easier and could provide the necessary high-bit rates. This concept supports the extension of a PC or workstation into an ATM-oriented desk area network, comprising cameras, printers and other special purpose systems or subsystems offering interfaces to ATM networks.

1.9 Concluding Remarks

The term *multimedia communications* refers to the representation, storage, retrieval and dissemination of computer-processable information expressed in multiple media, such as text, image, graphics, speech, audio, video, animation, handwriting, and data files. Trends toward multimedia communication are audio, video, image, data and text communications.

The key elements of communication models are processing, storage, retrieval, transport and user interface. The user needs a system that prepares and presents the information of interest, allows for the dynamic control of applications and provides a natural interface. From the network point of view, the most important requirements of multimedia communications are high-speed bit rates, synchronization of different information types and suitable standardized services. Multimedia transport based on packet transfer concept has many advantages. The generic functions of network in packet transfer from source to user are routing and multiplexing. Routing provides connectivity and does not depend on information type used in the transfer. Multiplexing determines much of the transfer quality in the network and is highly dependent of the traffic characteristics, the quality requirements and user's applications. The asynchronous transfer mode (ATM) networks are based on virtual circuit switching: All fixed size packets of a circuit have the fixed route. The most important ATM capabilities for multimedia communications are constant, variable or burst-oriented bit streams, virtual connections or virtual paths via the subscriber access depending on instantaneous needs, as well uniform bit rate independent transmission and switching systems.

CHAPTER 2

Audio-Visual Integration

Chapter Overview

In multimedia communication where human speech is involved, audio-visual integration is particularly significant. Not only is it important to consider both verbal and universal information in multimedia communication, but the interaction among different media is also interesting. This chapter explains why multimedia communication is more than simply putting together text, audio, images and video. It reviews a recent trend in multimedia research to exploit the audio-visual interaction and to build the link between audio and video processing. The emphasis is on lip reading, synchronization and tracing, audio-to-visual mapping and bimodal person verification. Some examples cover a broad range of these topics.

2.1 Introduction

Among the possible interactions considering different media types, the interaction between audio and video is the most interesting. A recent trend in multimedia research is to integrate audio and visual processing in order to exploit such interaction. Generally speaking, more interesting research topics can be found when we exploit the interaction among different media types. For example, using speech recognition technology, one can analyze speech waveforms to discover the text that has been spoken. From a sentence of text, a talking-head audio-visual sequence can be generated using computer graphics to animate a facial model and using text-to-speech synthesis to provide synthetic acoustic speech. For person-to-person communication, which is involved in multimedia applications like video telephony and video conferencing, audio and visual interaction is very important. Topics being researched progress from the point of view of audio-visual integration and include automatic lip reading and its use in speech recognition, speech-driven face animation, speech-assisted lip synchronization, facial-feature tracking, audio-visual mapping, bimodal person verification and joint audio-video coding. Some examples cover a broad range, including media interaction, bimodality of human speech, lip reading, lip synchronization, audio-to-visual mapping and bimodal person verification.

2.2 Media Interaction

Integration and interaction among different media types create challenging research topics and new opportunities [2.1].Media interaction is shown in Figure 2.1. As can be seen, media are categorized into three major classes. The first is textual information, the second is audio, including speech and music, and the third represents image and video. The goal of speech recognition is to enable a machine to be able to transcribe spoken inputs literally into individual words, but the goal of spoken language understanding research is to extract meaning from whatever was recognized [2.2, 2.3]. The various Spoken Language Interface (SLI) applications have widely differing requirements for speech recognition and spoken language understanding. Hence, a range of different performance measures on the various systems reflects both the task constraints and the application requirements. Some SLI applications require a speech recognizer to do word-for-word transcription.

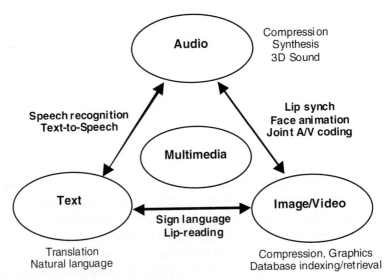

Figure 2.1 Media interaction [2.1]. ©1998 IEEE.

Example 2.1 Sending a textual response to an email message requires having capabilities for voice dictation, entering stock information or ordering from a catalog and entering number sequences or lists of data. For these types of systems, word error rate is an excellent measure of how well the speech recognizer produces a word-for-word transcription of the user's utterance.

The current capabilities in speech recognition and natural language understanding, in terms of word error rates, are summarized in Table 2.1. It can be seen that performance is very good for constrained tasks (digit strings and travel reservations). On the other hand, the word error rate increases rapidly for unconstrained conversational speech. Although methods of adaptation can improve performance by as much as a factor of two, this is still inadequate performance for use in many interesting tasks.

Table 2.1 Word error rates for speech recognition and natural language understanding tasks [2.2].

Corpus	Type	Vocabulary Size	Word Error Rate
Connected digit strings	Spontaneous	10	0.3%
Airline travel information	Spontaneous	2,500	2%
Wall Street Journal	Read text	64,000	8%
Radio (marketplace)	Mixed	64,000	27%
Switchboard	Conversational telephone	10,000	38%
Call home	Conversational telephone	10,000	50%

©1998 IEEE.

For some applications, complete word-for-word speech recognition is not required. Instead, tasks can be accomplished successfully even if the machine detects only certain key words or phrases within the speech. For such systems, the job of the machine is to categorize the user's utterances into one of a relatively small set of categories. The category identified is then mapped to an appropriate action or response [2.2].

Example 2.2 As an example, consider the AT&T system How May I Help You (HMIHY) task, in which the goal is to classify the user's natural language spoken input into one of 15 possible categories, such as billing credit, collect call and so forth. After this initial classification is done, the system transfers the caller to a category-specific subsystem, which uses either another artificial agent or a human operator [2.2]. Concept accuracy is a more appropriate measure of performance for this class of tasks than word accuracy. In the HMIHY task, word accuracy is only about 50%, but concept accuracy approaches 87%.

Another set of applications of speech recognition technology is the so-called spoken language understanding systems where the user is unconstrained in terms of what can be spoken and in what manner, but is highly constrained in terms of the context in which the machine is queried.

Example 2.3 An example of this type of application includes AT&T's CHRONUS system for air-travel information [2.4] and a number of prototype railway information systems. As in

HMIHY (Example 2.2), results show speech-understanding error rates of 6 to 10%, despite recognition error rates of 20 to 23%. These results demonstrate how a powerful language model can achieve high understanding performance despite imperfect Automatic Speech Recognition (ASR) technology.

A good example of using A/V interaction for human speech communication is lip reading, also referred to as speech reading. Human lip reading is widely used by hearing-impaired persons for speech understanding and automated lip reading.

Lip synchronization is one of the most important issues in videotelephony and videoconferencing. A typical situation in videoconferencing equipment is when the frame rate is not adequate for lip synchronization perception. One solution is to extract information from the acoustic signal, which determines the corresponding mouth movements, and then to process the speaker's mouth image accordingly to achieve lip synchronization. It is also possible to warp the acoustic signal to make it sound synchronized with the person's mouth movement. This approach is very useful in non-real-time applications, such as dubbing in a studio.

Researchers have tried to produce visual speech from auditory speech, that is, to generate speech-driven talking heads [2.5, 2.6, 2.7]. The major applications of this technique include human-computer interfaces, computer-aided instruction, cartoon animation, videogames and multimedia telephony for the hearing impaired.

The improvement in multimedia interaction can be obtained by using joint audio-video processing compared to the situation where audio and video are processed independently. Audio-visual interaction is very important in multimedia communication where human speech is involved because of its bimodal nature.

2.3 Bimodality of Human Speech

Due to the bimodality in speech perception audio-visual interaction becomes an important design factor for multimode communication systems, such as videotelephony and video conferencing [2.8, 2.9].

The bimodal nature of human speech perception was demonstrated by McGurk and MacDonald [2.10]. When humans are presented with conflicting audio and visual stimuli, the perceived sound may not exist in either modality.

Example 2.4 When a person "hears" the sound /ba/, but "sees" the speaker saying /ga/, the person may not perceive either /ga/ or /ba/. Instead, what is perceived is something close to /da/. Some other examples of audio-visual combinations are shown in Table 2.2. This shows that the speech that is perceived by a person depends not only on acoustic cues, but also on visual cues such as lip movements.

Psychologists have shown that the reverse McGurk effect also exists, that is, the results of visual speech perception can be affected by the dubbed audio speech [2.11]. The McGurk effect is also robust to a variety of different conditions [2.12]. The same answers are obtained to the conflicting stimuli in cases when there are timing mismatches between the stimuli or even when the face of a male speaker is combined with the voice of female speaker [2.13].

Table 2.2 Examples of McGurk effect [2.1].

Audio +	Visual →	Perceived
ba	ga	da
pa	ga	ta
ma	ga	na

©1998 IEEE.

Speech production is bimodal in nature, together with speech perception. Human speech is produced by the vibration of the vocal cord and the configuration of the vocal tract (articulatory organs, including the nasal cavity, tongue, teeth, velum and lips). These articulatory organs, together with the muscles that generate facial expressions, produce speech. There is an inherent relationship between the acoustic and visible speech because some of these articulators are visible.

The basic unit of acoustic speech is called a *phoneme* [2.2]. Similarly, in the visual domain, the basic unit of mouth movements is called a *viseme* [2.14]. A viseme is the smallest visibly distinguishable unit of speech. Many acoustic sounds are visually ambiguous. These sounds are grouped into the same class that represents a viseme. There is, therefore, a many-to-one mapping between phonemes and visemes.

Example 2.5 The /p/, /b/ and /m/ phonemes are all produced by a closed mouth shape and are visually indistinguishable. Therefore, they form one viseme group. Similarly, /f/ and /v/ both belong to the same viseme group that represents a mouth of which the upper teeth are touching the lower lip.

Example 2.6 Visemes for English consonants can be grouped into nine distinct groups as shown in Table 2.3. A number of visemes are shown in Figure 2.2 [2.1]. The top three are associated with consonants, and the bottom three are associated with vowels.

Table 2.3 Viseme groups for English consonants [2.1]
©1998 IEEE.

Number	Viseme Groups
1.	f, v
2.	th, dh
3.	s, z
4.	sh, zh
5.	p, b, m

Table 2.3 Viseme groups for English consonants [2.1].
©1998 IEEE. (Continued)

Number	Viseme Groups
6.	w
7.	r
8.	g, k, n, t, d, y
9.	l

Example 2.7 Instead of a still image, a viseme can be a sequence of several images that capture the movements of the mouth. This is especially true for some vowels. For example, the viseme /ow/ represents the movement of the mouth from a position close to /o/ to a position close to /w/. Therefore, to illustrate some visemes, we would need to use video sequences. However, most visemes can be approximated by stationary images.

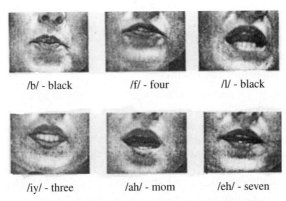

/b/ - black   /f/ - four   /l/ - black

/iy/ - three   /ah/ - mom   /eh/ - seven

Figure 2.2 Example visemes [2.1]. ©1998 IEEE.

Example 2.8 Most of the vowels are distinguishable both in acoustic and in visual modality [2.15]. The sounds /p/, /t/ and /k/ are very similar. The confusion sets in the auditory modality are usually distinguishable in the visual modality. The sounds /p/ and /k/ can be easily distinguished by the visual cue of a closed mouth versus an open mouth.

2.4 Lip Reading

A person skilled in lip reading is able to infer the meaning of spoken sentences by looking at the configuration and the motion of visible articulators of the speaker, such as the tongue, lips, teeth and so forth. Knowledge of the positions of these articulators provides information about the content of the acoustic speech signal. Because all of the articulators are not visible, this information may sometimes be incomplete. By combining the visual content with lexical, syntactic, semantic, and programmatic information, people can learn to understand spoken language by observation of the movements of a speaker's face.

Lip reading performance depends on a number of factors. Viewing conditions may affect the quality of the visual information. Poor lighting may make it difficult to judge the shape of the mouth or to detect the teeth or tongue. Likewise, as the speaker and listener move further apart, it becomes more difficult to view important visual cues. Factors such as viewing angle can also affect recognition. Neely [2.16] found that a frontal view of the speaker led to higher recognition rates than an angled or profile view. Lip reading performance can also be improved through training [2.17].

Example 2.9 Before training, a person could confuse an /s/ or /z/ with a /th/ or /dh/. After training, these confusions can be eliminated. Coarticulation can also affect lip-reading performance. It is the process by which one sound affects the production of neighboring sounds. The place of articulation may not be fixed, but may depend on context [2.18].

Example 2.10 The tongue may be in different positions for the /r/ sound in "art" and "arc." The tongue would be more forward in the word "art." This would affect the visibility of the tongue and thus recognition accuracy.

Syllables with a Vowel-Consonant-Vowel (VCV) context were examined by Benguerel and Pichora-Fuller [2.19]. It was found that consonant recognition depended on the vowels that surrounded it. For example, the middle consonant is more difficult to lip read when the vowel is a /u/ as opposed to an /ae/ or /i/. At the same time, the /u/ was the most recognizable of the vowels. This suggests that the /u/ sound is visually dominant. Its appearance is pronounced, and, because of this, there is a recovery period where neighboring sounds may be masked.

Knowledge of the process by which humans extract and incorporate visual information into speech perception can be useful. Also, it is important to know what information is available in the visual channel, what types of parameters humans use to aid recognition, and what means are used to integrate the acoustic and visual information. The acoustic and visual components of the speech signal are not purely redundant. They are complementary as well. Certain speech characteristics that are usually confusable are acoustically distinct, and those characteristics that are acoustically confusable are visually distinct.

Example 2.11 The /p/ and /k/ phonemes have similar acoustic characteristics, but can easily be differentiated by the closing of the mouth. In contrast, the /p/ and /b/ phonemes have similar visual characteristics, but can be acoustically differentiated by voicing.

The widespread beliefs about how humans integrate acoustic and visual information can be classified into early integration and high integration as shown in Figure 2.3. In early integration, the acoustic and visual parameter sets are combined into a larger parameter set. Recognition occurs by finding the word with the template that matches best to the audio-visual parameter sets. On the other hand, in late integration, the audio is compared against an acoustic template for each word, and the video is compared against a visual template for each word. The resulting audio and visual recognition scores are then combined using a certain mechanism.

The recognition systems that have been designed for automated lip reading vary widely and have been used to help answer a number of questions concerning audio-visual recognition. Many systems have been designed to show that speech recognition is possible using only the visual information. Some researchers have done comparisons on a number of visual feature sets in attempts to find those features that yield the best recognition performance. Next, researchers have attempted to integrate these visual-only recognition systems with acoustic recognition systems in order to enhance the accuracy of the acoustic speech-recognition system. A number of studies have examined strategies of early integration, late integration and other novel means for combining audio and video. Some studies have done more investigations into the resiliency of audio-visual recognition systems to varying levels of acoustic noise. Finally, many of the systems in the open literature have attacked different recognition tasks, implementing speaker-dependent and speaker-independent systems and examining isolated vowels, Consonant-Vowel-Consonant (CVC) syllables, isolated words, connected digits and continuous speech [2.1].

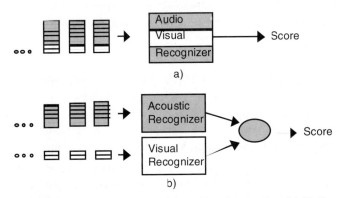

Figure 2.3 Early integration (a) and late integration (b) [2.1]. ©1998 IEEE.

The recognizers that have been designed vary in the acoustic feature sets, visual feature sets, recognition engine and integration strategies. The visual feature sets provide some of the greatest variations between systems. Systems range from performing relatively little processing and having a visual input that consists of a rectangular region of the image to using computer vision techniques to extract a visual feature set to be used for recognition [2.20].

The recognition engine also takes many forms among the various research groups. Some of the earlier recognition systems were based on dynamic time warping. A number of systems have used neural-network architectures, and an increasing number of systems have relied on Hidden Markov Models (HMM) for recognition [2.21].

A number of integration strategies have been proposed. Some of the early systems used either sequential recognition or a rule-based approach. One system converted the visual signal into an acoustic power spectrum and averaged the information sources in this domain. Other systems have employed either early or late integration strategies and compared the differences between the two [2.22, 2.23]. Most of the lip reading systems can be intrusive to the users to some extent. In particular, the user has to remain relatively stationary for the visual analysis system to work well. In Duchnowski et al. [2.24], a modular system was developed to alleviate such constraints. The visual analysis of the system is composed of an automatic face tracker, followed by the lip locator. The capability of face tracking results in a speech recognizer that allows the speaker reasonable freedom of movement. Besides trying to derive the acoustic domain information from visual information, researchers have also tried to produce visual speech from auditory speech, that is, to generate speech-driven talking heads [2.5, 2.6]. In the next section, we study the generation of talking-head images.

2.5 Speech-Driven Talking Heads

The major applications of the talking-heads technique include human-computer interfaces, computer-aided instruction, video games, cartoon animation, and multimedia telephony for the hearing impaired.

Two approaches are used in generating talking-head images: the flipbook method [2.7] and the wireframe model approach, which can be two-dimensional (2D) or (3D) [2.25].

In the flipbook method, a number of mouth images of a person, called key frames, are captured and stored. Each image represents a particular mouth shape, for example a viseme. Then, according to the speech signal, the corresponding mouth images are flipped one by one to the display to form animation. This method results in jerkiness during key-frame transition, especially if the number of key frames is limited. Image warping can be used to create some intermediate frames to make the transition look smoother [2.26]. Image warping is a process whereby one image is distorted through the use of geometric transformations to look like another image. It is useful for producing realistic intermediate images between video frames. To accomplish this, correspondences are made between points in two adjacent video frames. Another similar flipbook approach uses some techniques to enhance the quality of the images [2.1]. Instead of using a small number of mouth images representing the phonemes, this technique uses a long video sequence of a talking person. The sequence contains all the possible mouth shapes of the person. It is analyzed to derive the phonetic transcription. When presented with input audio, matching segments from the original video are found. These segments are concatenated together with an image-processing technique. In order to achieve better synchronization, time warping is also done to these segments.

The wireframe method uses computer-graphics techniques to achieve better realism. A wireframe is composed of a large number of triangular patches that model the shape of a human face. One of the early facial models was developed by Parke [2.27]. Figure 2.4 shows the facial model called Candide developed at Linkoping University [2.28]. A 3D model would contain vertices that correspond to points throughout the head and would allow synthesis of facial images with arbitrary orientations. Because most animation is primarily concerned with the frontal view of the face, and not the back of the head, models are often developed only for a frontal portion of the face.

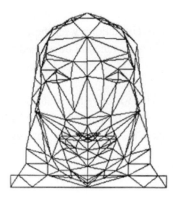

Figure 2.4 The wireframe model Candide [2.1]. ©1998 IEEE.

To synthesize various facial expressions, the Facial Action Coding System (FACS) is often used to generate the required trajectories of the vertices for a large variety of facial expressions and movements. A wireframe model gives only a structural representation of the face. To generate natural looking synthetic faces, a wireframe model must be combined with lighting models that specify how to map the shape and position of the wireframe into intensity when the wireframe is projected onto a 2D image. It is possible to use simple algorithms to synthesize artificial-looking faces together with texture mapping. It is an algorithm that maps pixel values of a 2D image, for example, a photo of a real person, onto patches of a wireframe. The texture from the original image helps create realistic synthetic images.

Generally speaking, there is a duality between the flipbook and the wireframe approaches. The flipbook approach is less computationally intensive, but it requires more data and the right number of images as key frames. The wireframe approach is more computationally intensive, but is needed for texture-mapping purposes. Then, arbitrarily oriented images can be synthesized with the model.

So far, we have discussed only the mechanisms that can be used to create talking heads. Now, we deal with the problem of how to produce the parameters to drive talking heads to make them "say" certain sentences. In [2.5], a 3D wireframe facial model was used to synthesize facial expressions, particularly lip motion. The lip parameters used form an eight-dimensional (8D) vector that includes the position of the upper lip, the position of the chin, and so forth. These parameters are derived by text input or speech input. In the case of text input, a sequence of the 8D feature vector is manually extracted for each phoneme. The input text is then analyzed into a sequence of phonemes, and the corresponding lip feature vectors are used to drive the facial model to produce the lip motion. In the case of speech input, a classifier that derives lip parameters from the input acoustic speech drives the lip feature points [2.2].

A work that focuses on a multimedia telephone for hearing-impaired people is presented by Lavagetto [2.6]. Here, the conversion from speech to lip movements is performed by a number of Time-Delayed Neural Networks (TDNNs). A major advantage of this approach is that TDNNs operate not only the current speech frame, but also its neighbors. Therefore, the estimated lip features can incorporate information from neighboring sounds. This helps model coarticulation effects from speech production.

2.6 Lip Synchronization

One of the most important questions in videotelephony and video conferencing is lip synchronization because human speech perception is bimodal. A typical situation in videoconferencing equipment with bandwidth constraints is when the frame rate is not adequate for lip synchronization perception. The solution can be to extract information from the acoustic signal, which determines the corresponding mouth movement and then process the speaker's mouth image accordingly to achieve lip synchronization.

It is possible to warp the acoustic signal to make it sound synchronized with the person's mouth movement [2.1]. This approach is very useful in non-real-time applications such as dubbing in a studio. In movie production, the dialog is usually recorded in a studio to replace the dialog recorded while filming a scene because the latter has poor quality due to background noises. To ensure lip synchronization, a spectrum analyzer analyzes both the studio audio and the original recording. The results are then input to a processor to find the best time-warping path that is required to modify the time scale of the studio audio to align the original recording. According to the time-warping path, the studio audio is edited pitch synchronously, that is, a period of sound segment being cut out or repeated. Thus, the studio dialog can be made in synchronization with the original lip movement.

One can time warp the video instead of warping the audio. We can warp the image of the speaker to make the lip movement hit the studio dialog. A video codec often skips some frames to meet the bandwidth requirement, which results in a lower frame rate at the decoder. Frame skipping introduces artifacts, such as jerky motion and loss of lip synchronization. To solve this, we can extract information from the speech signal and process the mouth image accordingly to achieve lip synchronization [2.29]. Figure 2.5 represents the block diagram of this approach. In this system, the feature-tracking module analyzes the input frames to find the location and shape of the mouth [2.30]. The audio-to-visual mapping module analyzes the audio signal and produces a sequence of the corresponding mouth shapes that are missing in the low-frame-rate video. Image warping is then applied to the input frames to modify the mouth shape to produce new frames that are to be inserted in the original video. Hence, lip synchronization is achieved in the high-frame-rate output video. Note the interaction between image analysis and speech analysis. The results of image analysis can be used to improve the accuracy of speech recognition, as is done in automatic lip reading. On the other hand, speech information can be used to improve the result of image analysis.

Example 2.12 We can decide whether the mouth is open or closed by speech analysis and then apply different algorithms to locate the lip points. Mouth closeness, for example, during /p/, /b/, /m/ and silence are important perceptual cues for lip synchronization. Therefore, the lip synchronization is good as long as speech analysis and image analysis together detect these closures correctly and image synthesis renders mouth closures precisely.

In addition to the low frame rate, another issue that causes loss of lip synchronization in videoconferencing is the transmission. The transmission delay for video is longer than the audio delay. We can always delay the audio to match the video. The speech-assisted video-processing

24   Chapter 2 • Audio-Visual Integration

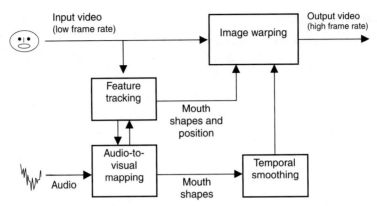

Figure 2.5 Block diagram of information extraction from the speech signal and processing the mouth image to achieve lip synchronization [2.1]. ©1998 IEEE.

technique can solve this problem by warping the mouth image of the speaker to be synchronized with the audio. Therefore, we can actually decrease the overall delay.

The speech-assisted interpolation scheme can be embedded into a video codec. In a typical video codec, such as H.263, some constraint usually exists on the number of frames that can be skipped between two coded frames [2.31]. This is needed in order to prevent too much jerkiness in the motion rendition and also to prevent too much loss of lip synchronization. When a speech-assisted interpolation scheme is in place at the decoder, the encoder can skip more frames than usual. Therefore, more bits can be assigned to each frame that is coded so that image quality and lip synchronization can be improved. For better lip synchronization, frame skipping can be controlled also by the perceptual importance of mouth shapes and ease of speech-assisted interpolation at the decoder. The encoder can avoid skipping frames that are crucial for lip synchronization, for example, frames that contain mouth closures. There are a number of ways to parameterize speech waveforms. For example, linear predictive coefficients and line spectral pairs are often used for coding purposes, and filter-bank outputs are used for recognition purposes. The question is how to analyze the visual signal, that is, the lip movement. Unlike the speech signal, which is one-dimensional (1D), the visual input is a 3D video signal with two spatial dimensions and one temporal dimension. A visual analysis system must convert this sequence of images into a meaningful parameter. We next discuss the enabling technologies for audio-visual research, including lip tracking.

2.7 Lip Tracking

A number ways to parameterize speech waveforms have been developed. As an example, linear predictive coefficients and like-spectral pairs are often used for coding purposes, and filter bank outputs are used for recognition purposes. In connection with this fact, the question often arises as to how to analyze the visual signal, that is, the lip movement. Unlike the speech signal, which is essentially 1D, the visual input is a 3D video signal with two spatial dimensions and one temporal dimension. A visual analysis system must convert this sequence of images into a meaningful parameter. Visual analysis systems can be divided into two major classes. The first classifies the mouth image into one of several categories, for example, into visemes. The other measures parameters or dimensions from the input image, for example, the mouth height and the mouth width. For the first class, vector quantization and neural networks are standard methods for classifying input images into several categories. For these systems, intensity images, Fourier transform coefficients and thresholded binary images are often used as the input. For the second class of image analysis systems, the task is to obtain parameters or dimensions that have some physical significance. For instance, we may want to measure the height between the lips and the width between the corners of the mouth. The next task is to construct a model for the lips and to find the parameters of the model that provide the closest match between the model and the image [2.1].

The system used in Pentajan [2.32] for visual speech recognition took an input image and applied a threshold. The resulting binary images were then analyzed, and parameters, such as the area of the mouth opening, height, width and perimeter length, were extracted to provide an adequate representation of the shape of the mouth. In another system [2.33], the vertical and horizontal projections of both intensity and edge images were used to locate points of interest on the mouth. The distances between these points were successfully used in speech-recognition applications.

The system in Rao and Mersereau [2.30] uses state-embedded deformable templates, a variation of deformable templates that exploits statistical differences in color to track the shape of the lips through successive video frames. Assume that, based on pixel colors, the image can be divided into foreground (pixels within the outer contour of the lips) and background (pixels that are part of the face) regions. The shape of the foreground is modeled by a template composed of two parabolas, as shown in Figure 2.6. This template is specified by the five parameters, a, b, c, d and e. When the parameters change, the shape and position of the template change. This template divides the image into foreground and background regions. Last, we assume that distinct Probability Density Functions (*pdfs*) govern the distribution of pixel colors in the foreground and background. The assumption is valid because the lips and face have different colors.

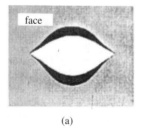
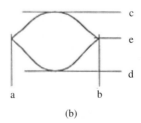

Figure 2.6 Lip tracking: (a) The face image and (b) the template [2.1]. ©1998 IEEE.

The *pdf* for the foreground pixels (the lips and interior of the mouth) and the *pdf* for the background pixels (the face) are first estimated. Based on these two *pdf*s, the joint probability of all pixels in the image can be calculated. Then, the visual analysis system uses a maximization algorithm to find the template that maximizes the joint probability. Sample results from this tracking algorithm are presented in Figure 2.7. Images in the top row show the derived template overlaid on the original image. The bottom images show pixels that are more likely to be part of the foreground.

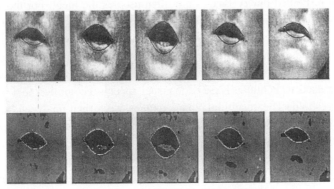

Figure 2.7 Lip tracking results [2.1]. ©1998 IEEE.

It has been suggested that geometric features like mouth shapes could be combined with other image-based features like the Karhunen-Loeve Transform (KLT) of the mouth image. In [2.34], the result of the KLT of the gray-scale mouth image to assist the tracing of mouth shapes is used. The result [2.35] directly combined mouth shapes with KLT for lip reading with color video.

Motivated by applications in surveillance and in human computer interfaces, an increasing amount of research has been done on the analysis of human body motion from video. As an example, human body tracking is presented in Figure 2.8, with a) segmentation of two walking human bodies into heads and legs, b) detection of four independently moving humans in an outdoor scene, and c) display of the heads, bodies, hands and feet of the humans in b) [2.36].

Figure 2.8 Human body tracking [2.36]. ©1998 IEEE.

Based on computer-vision techniques for tracking lip movements of a speaking person, a computer can be trained to understand visual speech [2.37]. One key issue in bimodal speech analysis and synthesis is the establishment of the mapping between acoustic and mouth-shape parameters. It means that, given the acoustic parameters, such as filter-bank coefficients, we need to estimate the corresponding mouth shape and vice versa. Thus, the next section seeks to provide audio-to-visual mapping, or the task of converting acoustic speech to mouth-shape parameters.

2.8 Audio-to-Visual Mapping

The problem for converting acoustic speech to mouth shape parameters can be solved in two different ways. The first one stresses that speech is a linguistic entity. Namely, the speech is first segmented into a sequence of phonemes. Then, each phoneme is mapped to the corresponding viseme. This scheme could be implemented using a complete speech recognizer followed by a lookup table [2.29]. The advantage of this approach is that the acoustic speech signal is explored to the full extent so that all the context information is used, and coarticulations are completely incorporated. Therefore, this approach provides the most precise speech analysis. Also, this approach has a certain amount of computation overhead because we do not really need to recognize the spoken words or sentences in order to achieve audio-to-visual mapping. The construction of the lookup table that maps phonemes to visemes is not trivial. Because a physical relationship exists between the shape of the vocal tract and the sound that is produced, a functional relationship may exist between the speech parameters and the visual parameters set. The conversion problem becomes one of finding the best functional approximation given sets of training data. Many algorithms can be modified to perform this task. These approaches include vector quantization (VQ), neural networks (NNs) and HMMs with Gaussian mixtures [2.5, 2.6, 2.8].

2.8.1 Classification-Based Conversion

An approach to classification-based conversion is given as a block scheme in Figure 2.9. It contains two stages. In the first one, the acoustics must be classified into one of a number of classes. The second stage maps each acoustic class into a corresponding visual output. In the first stage, VQ can be used to divide the acoustic training data into a number of classes. For each acoustic class, the corresponding visual code words are then averaged to produce a visual centroid. Thus, each input acoustic vector would be classified using the optimal acoustic vector quantizer and would then be mapped to the corresponding visual centroid. One shortcoming with this approach is the error that results from averaging visual feature vectors together to form the visual centroids. Another problem invoked by applying the classification-based method is that it does not produce a continuous mapping, but rather produces a distinct number of output levels. A few application examples exist that apply temporal neural models to conversion and/or synchronization including HMMs for audio-to-visual conversation and another example using TDNNs.

Figure 2.9 Block scheme of a classified-based approach.

2.8.2 HMM for Audio-to-Visual Conversion

HMMs have been used in speech recognition for many years. Although the majority of speech-recognition systems train HMMs on acoustic parameter sets, they can also be used to model the visual parameter sets. Recent multimedia results exploit the audio-visual interaction, which includes speech-assisted lip synchronization and joint audio-video coding [2.38]. The goal of speech-driven facial animation is to synthesize realistic video sequence from acoustic speech.

Chen and Rao [2.38] accomplished the audio-to-visual conversion process with HMMs. The correlation between audio and video was exploited for speech-driven facial animation. One problem addressed is that frame skipping due to limited bandwidth commonly introduces artifacts, such as jerky motion and loss of lip synchronization in talking-head video.

Consider estimating a single visual parameter v from the corresponding multidimensional acoustic parameter a. Defining the combined observation to be

$$O = [a, v]^T \tag{2.1}$$

the audio-to-visual conversion process using HMMs can be treated as a missing data problem. More specifically, a continuous-density HMM was trained with a sequence of O for each word in the vocabulary. In the training phase, the Gaussian mixtures in each state of the HMM are modeling the joint distribution of the audio-visual parameters. When presented with a sequence of acoustic vectors that correspond to a particular word, conversion can be made by using the HMM to segment the sequence using the Viterbi algorithm [2.39]. More exactly, when presented with a sequence of acoustic vectors $\{a\}$ that correspond to a particular word, the maximum likelihood estimator for the associated visual parameter vectors $\{v\}$ is equal to the conditional expectation, which can be derived from the optimal state sequence using the Viterbi algorithm of the HMM.

2.8.3 Audio and Visual Integration for Lip-Reading Applications

Due to the maturity of digital-video technology, it is now feasible to incorporate visual information in the speech-understanding process, that is, lip reading. These new approaches offer effective integration of visually derived information into the state-of-the-art speech-recognition systems so as to gain an improved performance in noise without suffering degraded performance on clean speech.

A complete audio-visual lip reading system can be decomposed into three major components [2.40]:

- Audio-visual information preprocessing (explicit feature extraction from audio and visual data)

- Pattern-recognition strategy (hidden Markov modeling, pattern matching with dynamic or linear time warping and various forms of NNs)
- Integration strategy (decision from audio and visual signal recognition)

2.8.4 Audio-Visual Information Preprocessing

Audio information processing has been well discussed in the speech-recognition literature [2.40]. The digitized speech is commonly sampled at 8 KHz. The sampled speech is pre-emphasized and then blocked and Hamming windowed into frames with a fixed-time interval (say, 32 ms long) and with some overlap (say, 16 ms). For each frame, an N-dimensional feature vector is extracted. Two major types of visual features are useful for lip reading: contour-based and area-based features [2.41, 2.42, 2.43]. The active contour models are a good example of contour-based features, which have been applied to object contours found in many image-analysis problems. Principal Component Analysis (PCA) of a gray-level image matrix, which is a typical area-based method, has been successfully used for principal feature extraction in pattern-recognition problems [2.41, 2.42].

2.8.5 Pattern-Recognition Strategies

Most lip-reading systems used similar pattern-recognition strategies as the traditional speech recognition, such as dynamic time warping and HMMs [2.44, 2.45]. NNs can also be used to convert acoustic parameters into visual parameters. In the training phase, input and output patterns are presented to the network. An algorithm called back propagation can be used to train the network weights. The design choice lies in selecting a suitable topology for the network. The number of hidden layers and the number of nodes per layer can be experimentally determined. A single network can be trained to reproduce all the visual parameters, or networks can be trained with each network estimating a single visual parameter [2.46]. NN architectures have also been extensively explored, such as static-feed forward-back propagation networks [2.47], multistage TDNNs [2.48] and the HMM recognizer with NNs for observation probability calculation [2.49]. One project [2.34] combines the acoustic and visual features for effective lip reading. Instead of using NNs as the temporal sequence classifier, the HMM is adopted, and Multilayer Perceptron (MLP) is used to calculate the observation probabilities {P(phone|audio,visual)}. The system combines the ten-order PCA transform coefficients (and/or the deltafeatures) from a gray-level eigen lip matrix (instead of the PCA from the snake points) derived from the video data of the acoustic features from the audio data [2.50]. The discriminatively trained MLP as one of the most popular NN models is used to compute the observation probabilities needed by the Viterbi algorithm. The bimodal hybrid speech recognition system has already been applied to a multispeaker spelling task and to a speaker-independent spontaneous speech-recognition system.

2.8.6 Integration Strategy

The audio and visual features can be combined into one vector before pattern recognition. Then, the decision is solely based on the results of the pattern recognizer. In some lip-reading systems that perform independent visual and audio evaluations, some rule is required to combine the two evaluation scores into a single one. Typical examples include the use of heuristic rules to incorporate knowledge of the relative confusability of phonemes in the evaluation of two modalities. Others used a multiplicative combination of independent evaluation scores for each modality. These postintegration methods possess the advantages of conceptual and implementional simplicity and they also give the user the flexibility to use just one of the subsystems if desired.

Example 2.13 Let us compare the results of the HMM-based method and the NNs based method in considering audio-to-visual mapping. Lip height versus time is observed. Figure 2.10 shows the results of the HMM-based method. The dotted line represents the height variation of the mouth when speaking a particular phrase. The solid line represents the estimation.

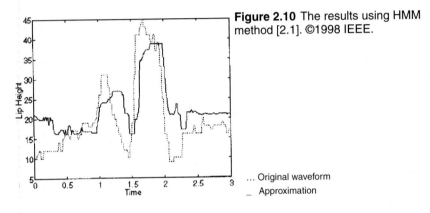

Figure 2.10 The results using HMM method [2.1]. ©1998 IEEE.

... Original waveform
_ Approximation

Figure 2.11 represents the results of the same phrase using the NN approach. As can be seen, the HMM-based approach gives better approximation to the original waveform.

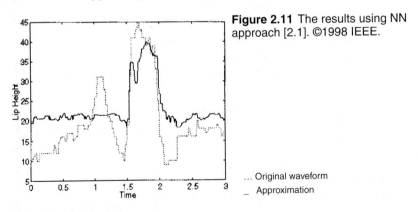

Figure 2.11 The results using NN approach [2.1]. ©1998 IEEE.

... Original waveform
_ Approximation

2.9 Bimodal Person Verification

Audio-visual interaction can also be used for person verification. Existing methods for person verification are mainly based on either face image or voice [2.1]. Using each single modality has certain limitations in both security and robustness. Using still images alone can be ineffective because it is easy to store and use prerecorded images. Image-only person verification can also suffer from image-coding artifacts and variations in lighting conditions. On the other hand, use of voice only for verification is not reliable because it is possible to rearrange phonemes from a pre-recorded speech of a person to synthesize different phrases. In addition, voice-only systems may fail when the acoustic environment is noisy or contains echo, such as in a typical office environment. Joint use of voice and video can solve these problems. By combining these two modalities, we can obtain more secure and more robust person-verification systems.

A number of techniques use lip movement together with acoustic speech to identify or verify a person. PCs with multimedia capabilities (cameras and microphones) make these techniques attractive. During the registration phase, the user says a chosen phrase while the voice and lip movements of the user are recorded into a database. During the verification phase, the user is then asked to read the displayed phrase. The user's voice and video data are then compared with those in the database to verify the user [2.51]. Lip movement has been used mainly for speech recognition and not for speaker verification until recently. Luettin, Thacker and Beet [2.52] showed that lip movement also contained information about a person's identity.

Figure 2.12 demonstrates the time variations of the mouth height of two persons who each say "Hello! How are you?" two times. The lip movements while saying the same phrase vary a great deal from individual to individual, but they stay relatively consistent for the same person. With dynamic time warping [2.2], a technique commonly used in acoustic-based speaker verification, to match the features, the scores of match and no match could differ by a factor of more than 40.

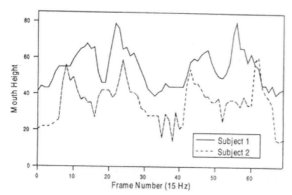

Figure 2.12 Time variation of mouth height [2.1]. ©1998 IEEE.

2.10 Joint Audio-Video Coding

Audio-visual information can be exploited in many ways. The correlation between audio and video can be used to achieve more efficient coding of both audio and video. One example to exploit this correlation is a predictive coding manner. Predictive-coding of video uses information from video frames to help construct an estimate of the current frame. The difference between the original and estimated signals can then be transmitted to allow the receiver to reconstruct the original video frame. This method is useful for removing the temporal redundancy in video. Also, the prediction can be done in a cross-modal manner to explore cross-modal redundancy. The basic idea is that here is information in the acoustic signal that can be used to help predict what the video signal should look like. Because the acoustic data is also transmitted, the receiver is able to reconstruct the video with very little side information. The process of cross-modal prediction is shown in Figure 2.13. This system provides a coding scheme that is scalable to a wide range of bit rates. An acoustic-to-visual mapping module estimates a visual parametric set, such as mouth height and width, given the acoustic data. The image analysis module measures the actual parameter set from the video.

The measured parameter set is compared with the parameter set estimated from the acoustics, and the encoder decides what information must be sent. If the acoustic data lead to a good prediction, no data has to be sent. If the prediction is slightly off, an error signal can be sent. If the prediction is wrong, the measured parameter set can be sent directly. The decision of what information needs to be sent is based on the Rate Distortion (R-D) criteria.

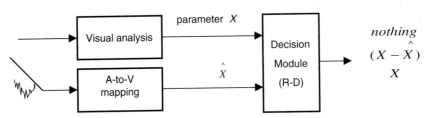

Figure 2.13 Block scheme of the cross-modal predictive coding [2.1]. ©1998 IEEE.

2.11 Concluding Remarks

The joint processing of audio and video provides additional capabilities that are not possible when audio and video are studied separately. Future communication will place a major emphasis on media integration for human communication. Multimedia systems can achieve their potential only when they are truly integrated in three key ways: integration of content, integration of human users and integration with other media systems [2.53]. A prime example of the audio-visual interaction due to the bimodality in speech perception is lip reading. It is not only used by the hearing impaired for speech understanding. In fact, everyone uses lip reading to some extent, in particular in a noisy environment. Based on computer-vision techniques for tracking lip movements of a speaking person, a computer can be trained to understand visual speech. In addition, automatic lip reading has also been used to enhance acoustic speech recognition.

If the frame rate is not adequate for lip-synchronization perception, one solution is to extract the information from the acoustic signal that determines the corresponding mouth movements and then process the speaker's mouth image accordingly to achieve lip synchronization [2.29]. On the other hand, it is also possible to warp the acoustic signal to synchronize with the person's mouth movements. This approach is very useful in non-real-time applications, such as dubbing in a studio. One key issue in bimodal speech analysis and synthesis is the establishment of the mapping between acoustic parameters and mouth shape parameters. In other words, given the acoustic parameters, one needs to estimate the corresponding mouth shape and vice versa. A number of approaches have been proposed for this task that use VQ, NNs and HMMs.

Audio-visual interaction can be exploited in many other ways. One of the characteristic examples is person verification. The correlation between audio and video can be used to achieve more efficient coding of both audio and video [2.5]. Other applications include dubbing of movies, segmentation of image sequences using video and audio signals [2.54], human-computer interfaces and cartoon animation. Also, joint use of audio and video has been applied to multimedia content classification [2.55, 2.56, 2.57, 2.58]. Further discussion of different aspects concerning audio-visual integration in multimedia communication is given in other research [2.39, 2.59, 2.60, 2.61, 2.62, 2.63, 2.64, 2.65, 2.66, 2.67, 2.68, 2.69, 2.70]. Multimedia Signal Processing (MMSP) provides additional capabilities that are not possible when audio and video are studied separately [2.39]. After we break down the artificial boundary between audio/speech and image/video processing, many new research opportunities and applications will arise. Thus, in the next chapter, we deal with the past, present and future of MMSP.

CHAPTER 3

Multimedia Processing in Communications

Chapter Overview

Multimedia has at its very core the field of signal-processing technology. With the exploding growth of the Internet, the field of multimedia processing in communications is becoming more and more exciting. Although multimedia leverages numerous disciplines, signal processing is the most relevant. Some of the basic concepts, such as spectral analysis, sampling theory and partial differential equations, have become the fundamental building blocks for numerous applications and, subsequently, have been applied in such diverse areas as transform coding, display technology and neural networks. The diverse signal-processing algorithms, concepts and applications are interconnected and, in numerous instances, appear in various reincarnated forms.

This chapter is organized as follows. First, we present and analyze digital media and signal processing elements. To address the challenges of multimedia signal processing while providing higher interactivity levels with the media and increased capabilities to access a wide range of applications, multimedia signal-processing methods must allow efficient access to processing and retrieval of multimedia content. Then, we review audio and video coding. During the last decade new digital audio and video applications have emerged for network, wireless, and multimedia computing systems and face such constraints as reduced channel bandwidth, limited storage capacity and low cost. New applications have created a demand for high-quality digital audio and video delivery. In response to this need, considerable research has been devoted to the development of algorithms for perceptually transparent coding of high-fidelity multimedia.

Next, we describe a general framework for image copyright protection through digital watermarking. In particular, we present the main features of an efficient watermarking scheme and discuss robustness issues. The watermarking technique that has been proposed is to hide secret information in the signal so as to discourage unauthorized copying or to attest the origin of the media. Data embedding and watermarking algorithms embed text, binary streams, audio, image or video in a host audio, image or video signal. The embedded data is perceptually inaudible or invisible to maintain the quality of the source data.

We also review the key attributes of neural processing essential to intelligent multimedia processing. The objective is to show why NNs are a core technology for efficient representation for audio-visual information. Also, we will demonstrate how the adaptive NN technology presents a unified solution to a broad spectrum of multimedia applications (image visualization, tracking of moving objects, subject-based retrieval, face-based indexing and browsing and so forth).

Finally, this chapter concludes with a discussion of recent large-scale integration programmable processors designed for multimedia processing, such as real-time compression and decompression of audio and video as well as the next generation of computer graphics. Because the target of these processors is to handle audio and video in real time, the promising capability must be increased compared to that of conventional microprocessors, which were designed to handle mainly texts, figures, tables and photographs. To clarify the advantages of a high-speed multimedia processing capability, we define these chips as multimedia processors. Recent general-purpose microprocessors for workstations and personal computers use special built-in hardware for multimedia processing.

3.1 Introduction

Multimedia signal processing is more than simply putting together text, audio, images and video. It is the integration and interaction among these different media that creates new systems and new research challenges and opportunities. Although multimedia leverages numerous disciplines, signal processing is the most relevant. Some of the basic concepts, such as spectral analysis, sampling theory and partial differential equation theory, have become the fundamental building blocks for numerous applications and, subsequently, have been reinvented in such diverse areas as transform coding, display technology and NNs. The diverse signal-processing algorithms, concepts and applications are interconnected.

The term "multimedia" represents many different concepts. It includes basic elementary components, such as different audio types. These basic components may originate from many diverse sources (individuals or synthetic). For audio, the synthetics may be traditional musical presentation. One may also argue that multimedia is based on the extended visual experience, which includes representation of the real world, as well as its model, through a synthetic representation.

The "multimedia" technologies have dramatically changed and will keep changing. However, it is erroneous to favor advances simply because the final product is based on better technology.

Multimedia consists of {multimedia data}+{set of instructions}. Multimedia data is informally considered as the collection of the three multimedia data, that is, multisource, multitype and multiformat data [3.1]. The interactions among the multimedia components consist of complex relationships without which multimedia could be a simple set of visual, audio and often data [3.2].

We define multimedia signal processing as the representation, interpretation, encoding and decoding of multimedia data using signal-processing tools. The goal of multimedia signal processing is effective and efficient access, manipulation, exchange and storage of multimedia content for various multimedia applications [3.3].

The Technical Committee (TC) on MMSP is the youngest TC in the IEEE Signal Processing (SP) society. It took them a long time to raise some questions like the following:

- What is multimedia signal processing all about?
- What impact has signal processing brought to multimedia technologies?
- Where are the multimedia technologies now?

Multimedia signal-processing technologies will play major roles in the multimedia-network age. Researchers today working in this area have the privilege of selecting the future direction of MMSP technologies, so what they are doing will deeply influence our future society.

3.2 Digital Media

Digital media take advantage of advances in computer-processing techniques and inherit their strength from digital signals. The following distinguishing features make them superior to the analog media:

- *Robustness*—The quality of digital media will not degrade as copies are made. They are most stable and more immune to the noises and errors that occur during processing and transmission. Analog signals suffer from signal-path attenuation and generation loss (as copies are made) and are influenced by the characteristics of the medium itself.
- *Seamless integration*—This involves the integration of different media through digital storage and processing and transmission technologies, regardless of the particular media properties. Therefore, digital media eliminate device dependency in an integrated environment and allow easy data composition of nonlinear editing.
- *Reusability and interchangeability*—With the development of standards for the common exchange formats, digital media have greater potential to be reused and shared by multiple users.
- *Ease of distributed potential*—Thousands of copies may be distributed electronically by a simple command.

Digital image Digital images are captured directly by a digital camera or indirectly by scanning a photograph with a scanner. They are displayed on the screen or printed.

Digital images are composed of a collection of pixels that are arranged as a 2D matrix. This 2D or spatial representation is called the image resolution. Each pixel consists of three components: red (R), green (G) and blue (B). On a screen, each component of a pixel corresponds to a phosphor. A phosphor glows when excited by an electron gun. Various combinations of different RGB intensities produce different colors. The number of bits to represent a pixel is called the color depth, which decides the actual number of colors available to represent a pixel. Color depth is in turn determined by the size of the video buffer in the display circuitry.

The resolution and color depth determine the presentation quality and the size of image storage. The more pixels and the more colors there are means the better the quality and the larger the volume. To reduce the storage requirement, three different approaches can be used:

- *Index color*—This approach reduces the storage size by using a limited number of bits with a color lookup table (or color palette) to represent a pixel. Dithering can be applied to create additional colors by blending colors from the palette. This is a technique taking advantage of the fact that the human brain perceives the media color when two different colors are adjacent to one another. With palette optimization and color dithering, the range of the overall color available is still considerable, and the storage is reduced.
- *Color subsampling*—Humans perceive color as brightness, hue and saturation rather than as RGB components. Human vision is more sensitive to variation in the luminance (or brightness) than in the chrominance (or color difference). To take advantage of such differences in the human eye, light can be separated into the luminance and chrominance components instead of the RGB components. The color subsampling approach shrinks the file size by down-sampling the chrominance components, that is, using less

bits to represent the chrominance components while having the luminance component unchanged.
- *Spatial reduction*—This approach, known as data compression, reduces the size by throwing away the spatial redundancy within the images.

Digital video Video is composed of a series of still-image frames and produces the illusion of movement by quickly displaying frames one after another. The Human Visual System (HVS) accepts anything more than 20 Frames Per Second (fps) as smooth motion. Television and video are usually distinguished. Television is often associated with the concept of broadcast or cable delivery of programs, whereas video allows more user interactivity, such as recording, editing and viewing at a user-selected time.

The biggest challenges posed by digital video are the massive volume of data involved and the need to meet the real-time constraints on retrieval, delivery and display. The solution entails the compromise in the presentation quality and video compression. As for the compromise in the presentation quality, instead of video with full frame, full fidelity and full motion, one may reduce the image size, use less bits to represent colors, or reduce the frame rate. To reduce the massive volume of digital video data, compression techniques with high compression ratios are required. In addition to throwing away the spatial and color similarities of individual images, the temporal redundancies between adjacent video frames are eliminated.

Digital audio Sound waves generate air pressure oscillations that stimulate the human auditory system. The human ear is an example of a transducer. It transforms sound waves to signals recognizable by brain neurons. As with other audio transducers, two important considerations are frequency response and dynamic range. Frequency response refers to the range of frequencies that a medium can reproduce accurately. The frequency range of human hearing is between 20 Hz and 20 KHz. Dynamic range describes the spectrum of the softest to the loudest sound-amplitude levels that a medium can reproduce. Human hearing can accommodate a dynamic range greater than a factor of millions. Sound amplitudes are perceived in logarithmic ratio rather than linearly. Humans perceive sounds across the entire range of 120 dB, the upper limit of which will be painful to humans. Sound waves are characterized in terms of frequency (Hz), amplitude (dB) and phase (degree), whereas frequencies and amplitudes are perceived as pitch and loudness, respectively. Pure tone is a sine wave. Sound waves are additive. In general, sounds are represented by a sum of sine waves. Phase refers to the relative delay between two waveforms. Distortion can result from phase shifts.

Digital audio systems are designed to make use of the range of human hearing. The frequency response of a digital audio system is determined by the sampling rate, which in turn is determined by the Nyquist theorem.

Example 3.1 The sampling rate of Compact Disk (CD) quality audio is 44.1 KHz. Thus, it can accommodate the highest frequency of human hearing, namely, 20 KHz. Telephone quality sound adopts an 8 KHz sampling rate. This can accommodate the most sensitive frequency of human hearing, up to 4 KHz.

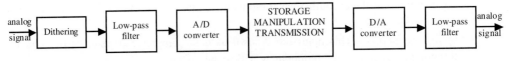

Figure 3.1 Block diagram for digital audio signal processing.

Digital audio aliasing is introduced when one attempts to record frequencies that exceed half the sampling rate. A solution is to use a low-pass filter to eliminate frequencies higher than the Nyquist rate. The quantization interval, or the difference in value between two adjacent quantization levels, is a function of the number of bits per sample and determines the dynamic range. One bit yields 6 dB of dynamic range. For example, 16 bits audio contributes 96 dB of the dynamic range found in CD-grade audio, which is nearly the dynamic range of human hearing. The quantized samples can be encoded in various formats, such as Pulse Code Modulation (PCM), to be stored or transmitted. Quantization noise occurs when the bit number is too small. Dithering, which adds white noise to the input analog signals, may be used to reduce quantization noise. In addition, a low-pass filter can be employed prior to the digital-to-analog (D/A) stage to smooth the stairstep effect resulting from the combination of a low sampling rate and quantization. Figure 3.1 summarizes the basic steps for processing digital audio signals [3.4].

The quality of digital audio is characterized by the sampling rate, the quantization interval and the number of channels. The higher the sampling rate, the more bits per sample and the more channels means the higher the quality of the digital audio and the higher the storage and bandwidth requirements.

Example 3.2 A 44.1 KHz sampling rate, 16-bit quantization and stereo audio reception produce CD-quality audio, but require a bandwidth of 44,100x16x2=1.4 Mb/s. Telephone-quality audio, with a sampling rate of 8 KHz, 8-bit quantization and mono audio reception, needs only a data throughput of 8,000x8x1=64 Kb/s. Digital audio compression or a compromise in quality can be applied to reduce the file size.

Integrated media systems will only achieve their potential if they are truly integrated in three key ways: integration of content, integration with human users and integration with other media systems. First, such systems must successfully combine digital video and audio, text, animation and graphics and knowledge about such information units and their inter-relationships in real time. Second, they must integrate with the individual user by cooperatively interactive multidimensional dynamic interfaces. Third, integrated media systems must connect with other such systems and content-addressable multimedia databases, both logically (information sharing) and physically (information networking, compression and delivery).

3.3 Signal-Processing Elements

Many classical signal-processing procedures have become deeply embedded in the multidimensional fields. A key driver is optimization for representation of multimedia components, as well as the associated storage and delivery requirements. The optimization procedures range from very simple to sophisticated. Some of the principal techniques are the following:

- Nonlinear analog (video and audio) mapping
- Quantization of the analog signal
- Statistical characterization
- Motion representation and models
- 3D representations
- Color processing

A nonlinear analog (video and audio) mapping procedure may be purely analog. Its intention may be the desire to enhance the delivery process. It could also be introduced to mask the limitations of various components of the overall multimedia chain. Typical constraints are introduced by bandwidth limitations and constrained dynamic range in the display terminal.

Quantization of the analog signal is fundamental to any digital representation that has originated in the analog world. The quantization process is an inherently lossy procedure and fundamentally noninvertible. This classical signal-processing element still remains the basic constraint in limiting performance, although not very exciting compared with other multimedia issues [3.5]. Quantization techniques comprise a whole field by themselves. The major relevant issues include uniform and nonuniform techniques and adaptive and nonadaptive procedures [3.6].

Statistical concepts and applications are directly and indirectly strongly embedded in processing components associated with multimedia. This relevant field is part of classical signal processing, and we can only highlight the major categories. A spectral analysis is fundamental to the entire range of image models for filtering and algorithm design. The procedures are critical to both visual and audio data components [3.7, 3.8]. Statistical redundancy is the basic concept upon which the entire field of data compression is based. Mathematical extension of the concept leads to optimum transform for decorrelation. This in turn leads to the entire field of modern transform-coding technology [3.9]. Model-based representations, primarily for compression, are determined from assumed or derived statistical models. The classes of transform-coding algorithms are based on this technology [3.10]. The utility of Fourier transform and its discrete extensions such as Discrete Cosine Transform (DCT), wavelets and others are based on the principle that these transforms asymptotically approach the optimum transform, assuming a reasonable statistical behavior [3.11]. Visual and audio models are fundamental to the relevant multimedia representations, primarily compression procedures. These models are based on fundamental statistical representations of the elementary components, including their evaluation by the human observer [3.12, 3.13].

The models are:

- Implementation of motion detection and associated compensation in subsequent image frames can significantly reduce the required bandwidth. Successful prediction of image segment locations in future frames reduces the required information update to the required motion vectors. Thus, under this condition, the associated update information is dramatically reduced.

- Combining the presence of motion in video segments with the limitations for human visual systems provides additional bandwidth-reduction potentials. Because the human vision deteriorates when observing moving areas, image blur associated with these regions becomes significantly less noticeable. Consequently, additional image compression can be introduced in segments that contain motion, with minimal noticeable effect.

Human vision is basically 3D. Efficient representation of a 3D signal is a major challenge of multimedia. The most common 3D techniques are based on 2D display techniques. The 3D scene is projected onto two dimensions in the rendering phase of the multimedia chain. The proper hierarchy of object elements and behavior maintains the 3D illusion. The relevant processes include shadowing consideration and preserving the proper hidden body behavior. The required processing resources are still significant. A substantial industry produces various processing components, such as chip sets and graphics boards, to develop solutions for many diverse applications including desktop computing. The associated technology is very effective in high-end applications. Virtual reality models using large screens are impressive even though the presentation remains 2D. In 3D representations, the stereo projection is the best known. The same 3D scene is recorded from two slightly different perspectives, essentially replicating our eyes. The two separate recordings are subsequently presented to the eyes separately. Unlike the early stereo film-based recordings, modern techniques are heavily dependent on digital processing, which corrects for camera-projection inaccuracies, resulting in significantly enhanced stereo display.

Projection techniques comprise an effective group to recreate multidimensionality from individual projections through the original object. Although this technology has been used very effectively in medical applications, its utility to multimedia applications is not likely to be useful in the near future. The primary limitations are complexity and lack of easy real-time implementation [3.14].

For efficient representation of color processing, modeling and communication applications, color plays a very important role. The correlation properties among color planes are used in image and video compression algorithms.

3.4 Challenges of Multimedia Information Processing

Novel communications and networking technologies are critical for a multimedia database system to support interactive dynamic interfaces. A truly integrated media system must connect with individual users and content-addressable multimedia databases. This will be a logical connection through computer networks and data transfer.

To advance the technologies of indexing and retrieval of visual information in large archives, multimedia content-based indexing would complement the text-based search. Multimedia systems must successfully combine digital video and audio, text animation, graphics and knowledge about such information units and their interrelationships in real time.

The operations of filtering, sampling, spectrum analysis and signal representation are basic to all of signal processing. Understanding these operations in the multidimensional (mD) case has been a major activity since 1975 [3.15, 3.16, 3.17]. More key results since that time have been directed at the specific applications of image and video processing, medical imaging, and array processing. Unfortunately, there remains considerable cross-fertilization among the application areas.

Algorithms for processing mD signals can be grouped into four categories:

- Separable algorithms that use 1D operators to process the rows and columns of a multi-dimensional array
- Nonseparable algorithms that borrow their derivation from their 1D counterparts
- mD algorithms that are significantly different from their 1D counterparts
- mD algorithms that have no 1D counterparts.

Separable algorithms operate on the rows and columns of an mD signal sequentially. They have been widely used for image processing because they invariably require less computation than nonseparable algorithms. Examples of separable procedures include mD Discrete Fourier Transforms (DFTs), DCTs and Fast Fourier Transform (FFT)-based spectral estimation using the periodogram. In addition, separable Finite Impulse Response (FIR) filters can be used in separable filter banks, wavelet representations for mD signals and decimators and interpolators for changing the sampling rate.

The second category contains algorithms that are uniquely mD in that they cannot be decomposed into a repetition of 1D procedures. These can usually be derived by repeating the corresponding 1D derivation in an mD setting. Upsampling and downsampling are some examples. As in the 1D case, bandlimited multidimensional signals can be sampled on periodic lattices with no loss of information. Most 1D FIR filtering and FFT-based spectrum analysis algorithms also generalize straightforwardly to any mD lattice [3.18]. Convolutions can be implemented efficiently using the mD DFT either on whole arrays or on subarrays. The window method for FIR filter design can be easily extended, and the FFT algorithm can be decomposed into a vector-radix form, which is slightly more efficient than the separable row/column approach for evaluating multidimensional DFTs [3.19, 3.20]. Nonseparable decimators and interpolators have also been derived that may eventually be used in subband image and video coders [3.21]. Another major area of research has been spectral estimation. Most of the modern spectral estimators, such as the maximum entropy method, require a new formulation based on constrained optimization. This is because their 1D counterparts depend on factorization properties of polynomials [3.22]. An interesting case is the maximum likelihood method, where the 2D version was developed first and then adopted to the 1D situation [3.23].

There are also mD algorithms that have no 1D counterparts, especially algorithms that perform inversion and computer imaging. One of these is the operation of recovering an mD distribution from a finite set of its projections, equivalently inverting a discretized Radon transform. This is the mathematical basis of computed tomography and positron emission tomography.

Another imaging method, developed first for geophysical applications, is Fourier integration. Finally, signal recovery methods unlike the 1D case are possible. The mD signals with finite support can be recovered from the amplitudes of their Fourier transforms or from threshold crossings [3.24].

3.4.1 Pre and Postprocessing

In multimedia applications, the equipment used for capturing data, such as the camera, should be cheap, making it affordable for a large number of users. The quality of such equipment drops when compared to their more expensive and professional counterparts. It is mandatory to use a preprocessing step prior to coding in order to enhance the quality of the final pictures and to remove the noise that will affect the performance of compression algorithms. Solutions have been proposed in the field of image processing to enhance the quality of images for various applications [3.25, 3.26]. A more appropriate approach would be to take into account the characteristics of the coding scheme when designing such operators. In addition, pre- and postprocessing operators are extensively used in order to render the input or output images in a more appropriate format for the purpose of coding or display.

Mobile communications is an important class of applications in multimedia. Terminals in such applications are usually subject to different motions, such as tilting and jitter, translating into a global motion in the scene due to the motion of the camera. This component of the motion can be extracted by appropriate methods detecting the global motion in the scene and can be seen as a preprocessing stage. Results reported in the literature show an important improvement of the coding performance when a global motion estimation is used [3.27].

It is normal to expect a certain degree of distortion of the decoded images for very low-bit-rate applications. However, an appropriate coding scheme introduces the distortions in areas that are less annoying to the users. An additional stage could be added to reduce the distortion further due to compression as a postprocessing operator. Solutions were proposed in order to reduce the blocking artifacts appearing at high compression ratios [3.28, 3.29, 3.30, 3.31, 3.32, 3.33]. The same types of approaches have been used in order to improve the quality of decoded signals in other coding schemes, reducing different kinds of artifacts, such as ringing, blurring and mosquito noise [3.34, 3.35].

Recently, advances in postprocessing mechanisms have been studied to improve lip synchronization of head-and-shoulder video coding at a very low bit rate by using the knowledge of decoded audio in order to correct the positions of the lips of the speaker [3.36]. Figure 3.2 shows an example of the block diagram of such a postprocessing operation.

3.4.2 Speech, Audio and Acoustic Processing for Multimedia

The primary advances in speech and audio signal processing that contributed to multimedia applications are in the areas of speech and audio signal compression, speech synthesis, acoustic processing, echo control and network echo cancellation.

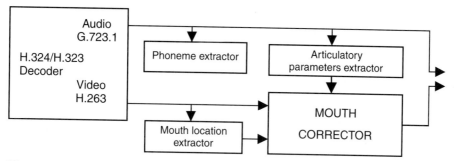

Figure 3.2 Block diagram for audio-assisted head and shoulder video [3.36]. ©1998 IEEE.

Speech and audio signal compression Signal compression techniques aim at efficient digital representation and reconstruction of speech and audio signals for storage and playback as well as transmission in telephony and networking.

Signal-analysis techniques such as Linear Predictive Coding (LPC) [3.37], and all-pole autoregressive modeling [3.38] and Fourier analysis [3.39], played a central role in signal representation. For compression, VQ [3.40, 3.41] marks a major advance. These techniques are built upon rigorous mathematical frameworks that have become part of the important bases of digital signal processing. Incorporation of knowledge and models of psychophysics in hearing have been proven as beneficial for speech and audio processing. Techniques such as noise shaping [3.42] and explicit use of auditory masking in the perceptual audio coder [3.43] have been found very useful. Today, excellent speech quality can be obtained at less than 8 Kb/s, which forms the basis for cellular as well as Internet telephony. The fundamental structure of the Code- Excited Linear Prediction (CELP) coder is ubiquitous in supporting speech coding at 4 to 16 Kb/s, encompassing such standards as G.728 [3.44], G.729 [3.45], G.723.1, IS-54 [3.46], IS-136 [3.47], GSM [3.48] and FS-1016 [3.49]. CD or near-CD-quality stereo audio can be achieved at 64 to 128 Kb/s, less than one twelfth of the original CD rate, and is ready for such applications as Internet audio (streaming and multicasting) and digital radio (digital audio broadcast). Advances in audio-coding standards are supported in MPEG activities.

Speech synthesis The area of speech synthesis includes generation of speech from unlimited text, voice conversion and modification of speech attributes such as time scaling and articulatory mimic [3.50]. Text-to-speech conversion takes text as input and generates human-like speech as output [3.51]. Key problems in this area include conversion of text into a sequence of speech inputs (in terms of phonemes, dyades or syllables), generation of the associated prosodic structure and intonation and methods to concatenate and reconstruct the sound waveform. Voice conversion refers to the technique of changing one person's voice to another, from person A to person B or from male to female and vice versa. It is useful to be able to change the time scale of a signal (to speed up or slow down the speech signal which changes the pitch) or to change the mode of the speech (making it sound happy or sad) [3.52]. Many of these signal-processing techniques have appeared in animation and computer graphics applications.

Acoustic processing and echo control Sound pickup and playback is an important area of multimedia processing. In sound recording, interference, such as ambient noise and reverberation, degrade the quality. The idea of acoustic signal processing and echo control is to allow straightforward high-quality sound pickup and playback in applications, such as a duplex device like a speakerphone, a sound source-tracking apparatus like microphone arrays, teleconferencing systems with stereo input and output, hands-free cellular phones and home theatre with 3D sound.

Signal processing for acoustic echo control includes modeling of reverberation, design of dereverberation algorithms, echo suppression, double-talk detection and adaptive acoustic echo cancellation, which is still a challenging problem in stereo full-duplex communication environments [3.53].

Example 3.3 For typical environments, the system modeling time for reverberation is of the order of 100 ms. This at a sampling rate of 16 KHz translates into a echo-canceling filter of 1600 taps, requiring seconds to converge.

For sound pickup, acoustic processing aims at the design of transducers or transducer arrays to achieve a durable directionality (beam steering and width control) as well as noise resistance. Understanding of near and far-field acoustics is important in achieving the required response in specific applications [3.54]. Various 1D and 2D microphone arrays have been shown in teleconferencing and auditorium applications with good results [3.55].

Network echo cancellation In telephony, both near-end and far-end echo exists due to the hybrid coil that is necessary for two-wire and four-wire conversions. Network echo can be so severe that it hampers telephone conversation. Network echo cancellers were invented to correct the problem in the late 1960s, based on the Least Mean Squares (LMS) adaptive echo cancelation algorithm [3.56]. The network echo delay is of the order of 16 ms, typically requiring a filter with 128 taps at a sampling rate of 8 KHz.

3.4.3 Video Signal Processing

Digital video has many advantages over conventional analog video, including bandwidth compression, robustness against channel noise interactivity and ease of manipulation. Digital-video signals come in many formats. Broadband TV signals are digitized with ITU-R 601 format, which has 30/25 fps, 720 pixels by 488 lines per frame, 2:1 interlaced, 4:3 aspect ratio, and 4:2:2 chroma sample. With the advent of high-definition digital-video, standardization efforts between the TV and PC industries have resulted in the approval of 18 different digital video formats in the United States. Exchange of video signals between TV and PCs requires effective format conversion. Some commonly used interframe/field filters for format conversion, for example, ITU-R 601 to the Source Input Format (SIF) and vice versa and 3:2 pull-down to display 24 Hz motion pictures in 60 Hz format, have been reviewed [3.57]. As for video filters, they can be classified as interframe/field (spatial), motion-adaptive and motion-compensated filters [3.58]. Spatial filters are easiest to implement. However, they do not make use of the high temporal correlation in the video signals. Motion-compensated filters require highly accurate motion estimation between successive views. Other more sophisticated format conversion methods include motion-adaptive field-rate doubling and deinterlacing [3.59] as well as motion compensated frame rate conversion [3.58].

Video signals suffer from several degradations and artifacts. Some of these degradations may be acceptable under certain viewing conditions. However, they become objectionable for freezeframe or printing from video applications. Some filters are adaptive to scene content in that they aim to preserve spatial and temporal edges while removing the noise. Examples of edge-preserving filters include median, weighted median, adaptive linear mean square error and adaptive weighted-averaging filtering [3.58]. Deblocking filters can be classified as those that do require a model of the degradation process (inverse, constrained, least square, and Wiener filtering) and those that do not (contrast adjustment by histogram specification and unsharp masking). Deblocking filters smooth intensity variations across amounts of temporal redundancy. Namely, successive frames generally have large overlaps with each other. Assuming that frames are shifted by subpixel amounts with respect to each other, it is possible to exploit this redundancy to obtain a high-resolution reference image (mosaic) of the regions covered in multiple views [3.60]. High-resolution reconstruction methods employ least-squares estimation, back projection, or projection-autoconvex sets methods based on a simple instantaneous camera model or a more sophisticated camera model including motion blur [3.61].

One of the challenges in digital video processing is to decompose a video sequence into its elementary parts (shots and objects). A video sequence is a collection of shots, a shot is a group of frames and each frame is composed of synthetic or natural visual objects. Thus, temporal segmentation generally refers to finding shot boundaries, spatial segmentation corresponds to extraction of visual objects in each frame and object tracking means establishing correspondences between the boundaries of objects in successive frames.

Temporal segmentation methods edit effects as cuts, dissolves, fades and wipes. Thresholding and clustering using histogram-based similarity methods have been found effective for detection of cuts [3.62] . Detection of special effects with high accuracy requires customized methods in most cases and is a current research topic. Segmentation of objects by means of chroma keying is relatively easy and is commonly employed. However, automatic methods based on color, texture and motion similarity often fail to capture semantically meaningful objects [3.63]. Semiautomatic methods, which aim to help a human operator perform interactive segmentation by tracking boundaries of a manual initial segmentation, are usually required for object-based video editing applications. Object-tracking algorithms, which can be classified as boundary region or model-based tracking methods, can be based on 2D or 3D object representations. Effective motion analysis is an essential part of digital video processing and remains an active research topic.

Storage and archiving of digital video in shared disks and servers in large volumes, browsing of such databases in real time and retrieval across switched and packet networks pose many new challenges, one of which is efficient and effective description of content. The simplest method to index content is by assigning manually or semiautomatically the content to programs, shots and visual objects [3.64]. It is of interest to browse and search for content using compressed data because almost all video data will likely be stored in compressed format [3.65].

Video-indexing systems may employ a frame-based, scene-based or object-based video representation. The basic components of a video-indexing system are temporal segmentation, analysis of indexing features and visual summarization. The temporal-segmentation step extracts shots, scenes and/or video objects. The analysis step computes content-based indexing features for the extracted shots, scenes, or objects. Content-based features may be generic or domain dependent. Commonly used generic indexing features include color histograms, type of camera-motion direction and magnitude of dominant object motion entry and exit instances of objects of interest and shape features for objects [3.66, 3.67]. Domain-dependent feature extraction requires a priori knowledge about the video source, such as new programs, particular sitcoms, sportscasts and particular movies. Content-based browsing can be facilitated by a visual summary of the contents of a program, much like a visual table of contents. Among the proposed visual summarization methods are story boards, visual posters and mosaic-based summaries.

3.4.4 Content-Based Image Retrieval

To address their challenges, multimedia signal-processing methods must allow efficient access to processing and retrieval of content in general, and visual content in particular. This is required across a large range of applications, in medicine, entertainment, consumer industry, broadcasting, journalism, art and e-commerce. Therefore, methods originating from numerous research areas, that is, signal processing, pattern recognition, computer vision, database organization, human-computer interaction and psychology, must contribute to achieving the image-retrieval goal. An example of image retrieval is

```
Given:      A query
Retrieve:   All images that have similar content to that of the query.
```

Image-retrieval methods face several challenges when addressing this goal [3.68]. These challenges, which are summarized in Table 3.1, cannot be addressed by text-based image retrieval systems, which have had an unsatisfactory performance so far. In these systems, the query keywords are matched with keywords that have been associated to each image. Because of difficult automatic selection of the relevant keywords, time consuming and subjective manual annotation is required. Moreover, the vocabulary is limited and must be expanded as new applications emerge.

To improve performance and address these problems, content-based image retrieval methods have been proposed. These methods have generally focused on using low-level features such as color, texture and shape layout, for image retrieval, mainly because such features can be extracted automatically or semiautomatically.

Texture-Based Methods

Statistical and syntactic texture description methods have been proposed. Methods based on spatial frequencies, co-occurrence matrixes and multiresolution methods have been frequently employed for texture description because of their efficiency [3.69]. Methods based on spatial frequencies evaluate the coefficients of the autocorrelation function of the texture. Co-occurrence matrixes identify repeated occurrences of gray level pixel configurations within the tex-

Table 3.1 Image retrieval challenges [3.68].

Challenges	Remarks
Query types	Color based/shape based/color and shape based
Query forms	Quantitative, for example, find all images with 30% amount of red
	Query by example, for example, image region/image/sketch/other examples
Various content	For example, natural scenes/head-and-shoulder images/MRIs
Matching types	Object to object/image to image/object to image
Precision levels	Application specific
	Exact versus similarity-based match
Presentation of results	Application specific

ture. Multiresolution methods describe the texture characteristics at coarse-to-fine resolutions. A major problem that is associated with most texture description methods is their sensitivity to scale, that is, the texture characteristics may disappear at low resolutions or may contain a significant amount of noise at high resolutions [3.70, 3.71, 3.72].

Shape-Based Methods

Describing quantitatively the shape of an object is a difficult task. Several contour-based and region-based shape description methods have been proposed. Chain codes, geometric border representations, Fourier transforms of the boundaries, polygonal representations and deformable (active) models are some of the boundary-based shape methods that have been employed for shape description. Simple scalar region descriptors, moments, region decompositions and region neighborhood graphs are region-based methods that have been proposed for the same task [3.73, 3.74]. Contour-based and region-based methods are developed in either the spatial or transform domains, yielding different properties of the resulting shape descriptors. The main problems that are associated with shape description methods are high sensitivity to scale, difficult shape description of objects and high subjectivity of the retrieved shape results.

Color-Based Methods

Color description methods are generally color histogram based, dominant color based and color moment based [3.75, 3.76]. Description methods that employ color histograms use a quantitative representation of the distribution of color intensities. Description methods that employ dominant colors use a small number of color ranges to construct an approximate representation of color distribution. Description methods that use color moments employ statistical measures of the image characteristics in terms of color.

The performance of these methods typically depends on the color space, quantization, and distance measures employed for evaluation of the retrieved results. The main problem that is associated with histogram-based and dominant-color-based methods is their inability to allow the localization of an object with the image. A solution to address this problem is to apply color segmentation, which allows both image-to-image matching and object localization. The main problem of color-moment-based methods is their complexity, which makes their application to browsing or other image-retrieval functionalities difficult.

Examples of content-based image and video-retrieval systems are included in Table 3.2. Some or all of the limitations of these systems are the following [3.68]:

- Few query types are supported
- Limited set of low-level features
- Difficult access to visual objects
- Results partially match user's expectations
- Limited interactivity with the user
- Limited system interoperability
- Scalability problems

Table 3.2 Examples of content-based image and video-retrieval systems [3.68].

Features	System	Image/Video	Provider
Color and text	WebSeek	I, V	Columbia University
	Picasso	I	University of Florence
	Chabot	I	University of California, Berkeley
	*	I	University of Toronto
Color, texture and shape	QBIC	I	IBM
	PhotoBook	I	MIT
	BlobWorld	I	University of California, Berkeley
	VIR	I, V	Virage
Color, shape and scale	Nefertiti	I	National Research Council of Canada

* No name has been adopted for the corresponding system.

Table 3.2 Examples of content-based image and video-retrieval systems [3.68]. (Continued)

Features	System	Image/Video	Provider
Color, texture, shape and spatial location	NeTra	I	University of California, Santa Barbara
	Digital storyboard	I	Kodak
Color, texture and motion	WebClip	V	Columbia University
	Jacob	I, V	University of Palermo
	*	V	IMAX
N/A	*	V	NASA

* No name has been adopted for the corresponding system.

3.5 Perceptual Coding of Digital Audio Signals

Audio coding is used to obtain compact digital representations of high-fidelity (wideband) audio signals for the purpose of efficient transmission or storage. The central objective in audio coding is to represent the signal with a minimum number of bits while achieving transparent signal reproduction, for example, while generating output audio that cannot be distinguished from the original input, even by a sensitive listener [3.77].

The introduction of the CD brought to the fore all of the advantages of digital audio representation, including high fidelity, dynamic range and robustness. However, these advantages came at the expense of high data rates. Conventional CD and Digital Audio Tape (DAT) systems are typically sampled at 44.1 or 48 KHz, using PCM with a 16-bit sample resolution [3.78]. This results in uncompressed data rates of 705.6/768 Kb/s for monaural channel, or 1.41/1.54 Mb/s for a stereo pair at 44.1/48 KHz, respectively. Although high, these data rates were accommodated successfully in first-generation digital-audio applications, such as CD and DAT [3.79]. Unfortunately, second-generation multimedia applications and wireless systems in particular are often subject to bandwidth or cost constraints that are incompatible with high data rates. Because of the success enjoyed by the first generation, end users have come to expect CD quality audio reproduction from any digital system [3.80]. Therefore, new network and wireless multimedia digital audio systems must reduce data rates without compromising reproduction quality. These and other considerations have motivated considerable research in the area of compression schemes that can satisfy simultaneously the conflicting demands of high compression ratios and transparent reproduction quality for high-fidelity audio signals [3.81].

3.5.1 General Perceptual Audio-Coding Architecture

Although the enormous capacity of new storage media, such as digital versatile disk (DVD), can accommodate lossless audio coding, the research interests are lossy compression schemes

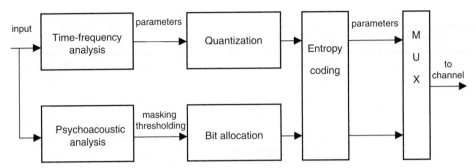

Figure 3.3 Perceptual audio-coder architecture [3.82]. ©1997 IEEE.

[3.82], which seek to exploit the psychoacoustic principles. Lossy schemes offer the advantage of lower bit rates, for example, less than 1 bit per sample relative to lossless schemes (for example, 1 bit per sample).

The lossy compression systems achieve coding gain by exploiting both perceptual irrelevancies and statistical redundancies. All of these algorithms are based on the architecture shown in Figure 3.3.

The coders typically segment input signals into quasistationary frames ranging from 2 to 50 ms in duration. A time-frequency analysis section then decomposes each analysis frame. The time-frequency analysis approximates the temporal and spectral analysis properties of the human auditory system. It transforms input audio into a set of parameters, which can be quantized and encoded according to a perceptual distortion metric. Depending on the overall system objectives, the time-frequency analysis section may contain the following:

- Unitary transform
- Time-invariant bank of uniform bandpass filters
- Time-varying, critically sampled bank of nonuniform bandpass filters
- Hybrid transform/filterbank signal analyzer
- Harmonic/sinusoidal analyzer
- Source-system analysis (LPC/multipulse excitation).

The choice of time-frequency analysis methodology always involves a fundamental tradeoff between time and frequency resolution requirements. Perceptual distortion control is achieved by a psychoacoustic signal analysis section that estimates signal masking power based on psychoacoustic principles. The psychoacoustic model delivers masking thresholds that quantify the maximum amount of distortion that can be injected at each point of the time-frequency plane during quantization and encoding of the time-frequency parameters, without introducing audible artifacts in the reconstructed signal. Therefore, the psychoacoustic model allows the quantization and encoding section to exploit perceptual irrelevancies in the time-frequency parameter set. The quantization and encoding section can also exploit statistical redundancies through classical techniques, such as Differential Pulse Code Modulation (DPCM) or Adaptive

DPCM (ADPCM). Quantization might be uniform or pdf optimized (Lloyd-Max quantizer). It might be performed on either scalar or vector quantities. After a quantized compact parametric has been formed, remaining redundancies are removed through Run-Length (RL) and entropy-coding techniques, for example Huffman, arithmetic, and Lempel-Ziv-Welch (LZW). Because the psychoacoustic distortion control model is signal adaptive, most algorithms are inherently variable rate. Fixed-channel rate requirements are usually satisfied through buffer feedback schemes, which often introduce encoding delays. The study of Perceptual Entropy (PE) suggests that transparent coding is possible in the neighborhood of 2 bits per sample for most high- fidelity audio sources [3.83]. Regardless of design details, all perceptual audio coders seek to achieve transparent quality at low rates with tractable complexity and manageable delay.

3.5.2 Review of Psychoacoustic Fundamentals

Audio-coding algorithms must rely upon generalized receiver models to optimize coding efficiency. In the case of audio, the receiver is ultimately the human ear, and sound perception is affected by its masking properties. The field of psychoacoustics has made significant progress toward characterizing human auditory perception and the time-frequency analysis capabilities of the inner ear [3.84, 3.85, 3.86]. Most current audio coders achieve compression by exploiting the fact that irrelevant signal information is not detectable by even a well-trained or sensitive listener. Irrelevant information is identified during signal analysis by incorporating into the coder several psychoacoustics principles, including:

- Absolute threshold of hearing
- Critical band frequency analysis
- Simultaneous masking and the spread of masking
- Temporal masking

Combining these psychoacoustic notions with basic properties of signal quantization has also led to the development of PE, a quantitative estimate of the fundamental limit of transparent audio-signal compression [3.87].

Absolute Threshold of Hearing

The absolute threshold of hearing is characterized by the amount of energy needed in a pure tone such that it can be detected by a listener in a noiseless environment. The frequency dependence of this threshold was quantified when test results for a range of listeners were reported [3.88]. The quiet threshold is well approximated by the nonlinear function

$$T_q(f) = 3.64(f/1000)^{-0.8} - 6.5e^{-0.6(f/1000-3.3)^2} + 10^{-3}(f/1000)^4 \quad \text{[dB SPL]} \quad (3.1)$$

where SPL is the sound pressure level.

This is representative of a young listener with acute hearing. When applied to signal compression, $T_q(f)$ can be interpreted as a maximum allowable energy level for coding distortions introduced in the frequency domain. The absolute threshold of hearing is shown in Figure 3.4. Algorithm designers have no a priori knowledge regarding actual playback levels. Therefore, the

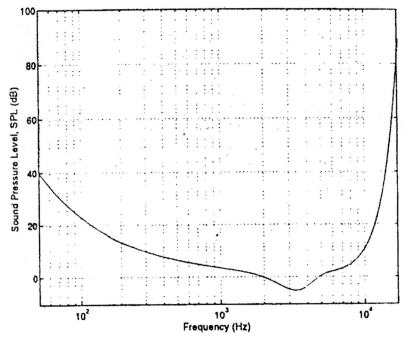

Figure 3.4 Absolute threshold of hearing [3.85].

Sound Pressure Level (SPL) curve is often referenced to the coding systems by equating the lowest point on the curve (that is, 4 KHz) to the energy in +/-1 bit of signal amplitude. Such a practice is common in algorithms that use the absolute threshold of hearing.

Critical Band Frequency Analysis

Using the absolute threshold of hearing to shape the coding-distortion spectrum represents the first step toward perceptual coding. Consider how the ear actually does spectral analysis. It is evident that a frequency-to-place transformation takes place in the inner ear along the basilar membrane. Distinct regions in the cochlea, each with a set of neural receptors, are tuned to different frequency bands. In the experimental sense, critical bandwidth can be loosely defined as the bandwidth at which subjective responses change abruptly. For example, the perceived loudness of a narrowband noise source at a constant sound-pressure level remains constant even as the bandwidth is increased up to the critical bandwidth. The loudness then begins to increase.

Example 3.4 Critical band-measurement methods are presented in Figure 3.5. In a different experiment (Figure 3.5a), the detection threshold for a narrow-band noise source between two masking tones remains constant as long as the frequency separation between the tones remains within a critical bandwidth. Beyond the bandwidth, the threshold rapidly decreases (Figure 3.5c). A similar notched-noise experiment can be constructed with measure and mask roles reversed (Figure 3.5b, and d).

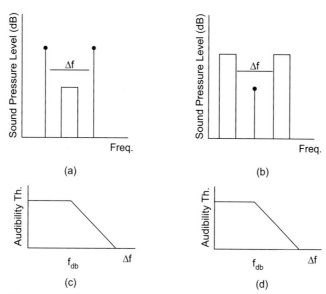

Figure 3.5 Control band-measurement methods [3.82]. ©1997 IEEE.

Simultaneous Masking and the Spread of Masking

Masking refers to a process where one sound is rendered inaudible because of the presence of another sound. Simultaneous masking refers to a frequency-domain phenomenon that has been observed within critical bands (in-band). For the purposes of shaping coding distortions, it is convenient to distinguish between two types of simultaneous masking: tone-masking noise and noise-masking tone [3.85]. In the first case, a tone occurring at the center of a critical band masks noise of any subcritical bandwidth or shape, provided the noise spectrum is below a predictable threshold directly related to the strength of the masking tone. The second masking type follows the same pattern with the roles of masker and mask reversed. A simplified explanation of the mechanism for both masking phenomena is as follows. The presence of a strong noise or tone masker creates an excitation of sufficient strength on the basilar membrane at the critical band location to block transmission of a weaker signal effectively. Interband masking has also been observed. It means that a masker centered with one critical band has some predictable effect on detection thresholds in other critical bands. This effect, also known as the spread of masking, is often modeled in coding applications by an approximately triangular spreading function that has slopes of +25 and -10 dB per bark. A convenient analytical expression is given by

$$SF_{dB}(x) = 15.81 + 7.5(x+0.474) - 175\sqrt{1+(x+0.474)^2} \quad [dB] \quad (3.2)$$

where x has units of barks, and basilar spreading function $SF_{dB}(x)$ is expressed in dB [3.82]. After critical band analysis is done and the spread of masking has been accounted for, masking thresholds in psychoacoustic coders are often established by the decibel [dB] relations

$$TH_N = E_T - 14.5 - B \quad (3.3)$$

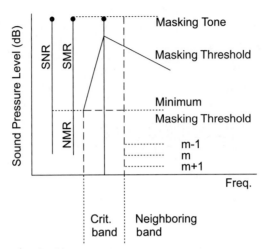

Figure 3.6 Schematic representation of simultaneous masking [3.82]. ©1997 IEEE.

$$TH_T = E_N - K \tag{3.4}$$

where TH_N and TH_T, are the noise and tone-masking thresholds, respectively, due to tone-masking noise and noise-masking tone. E_N and E_T are the critical band noise and tone-masker energy levels, while B is the critical band number [3.89]. Depending upon the algorithms, the parameter K has typically been set between 3 and 5 dB. Masking thresholds are commonly referred to in the literature as functions of Just Noticeable Distortion (JND). One psychoacoustic coding scenario might involve first classifying masking signals as either noise or tone, next computing appropriate thresholds and then using this information to shape the noise spectrum beneath JND. The absolute threshold (T_{abs}) of hearing is also considered when shaping the noise spectra. Also, MAX(JND, T_{abs}) is most often used as the permissible distortion threshold.

Schematic representation of simultaneous masking is shown in Figure 3.6. Consider the case of a single masking tone occurring at the center of a critical band. All levels are given in terms of dB SPL. A hypothetical masking tone occurs at some masking level. This generates an excitation along the basilar membrane, which is modeled by a spreading function and a corresponding masking threshold. For the band under consideration, the minimum masking threshold denotes the spreading function in-band minimum. Under the assumption, the masker is quantized using an m-bit uniform scalar quantizer, and noise might be introduced at the level m. Signal-to-Noise Ratio (SNR) and Noise-to-Mask Ratio (NMR) denote the log distances from the minimum masking threshold to the masker and noise levels, respectively.

Temporal Masking

In the context of audio-signal analysis, abrupt signal transients create pre-and postmasking regions in a time during which a listener will not perceive signals beneath the elevated audibility

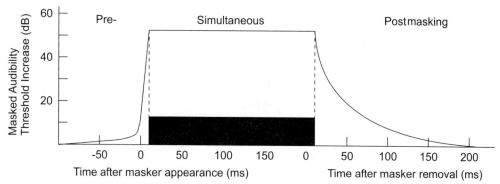

Figure 3.7 Schematic representation of temporal masking properties of the human ear [3.82]. ©1997 IEEE.

thresholds produced by a masker. Schematic representation of temporal masking properties of the human ear is shown in Figure 3.7. Absolute audibility thresholds for masked sounds are artificially increased prior to, during and following the occurrence of a masking signal. Premasking tends to last only about 5 ms. Postmasking will extend anywhere from 50 to 300 ms, depending upon the strength and duration of the masker [3.86]. Temporal masking has been used in several audio-coding algorithms [3.82]. In particular, premasking has been exploited in conjunction with adaptive block-size transform coding to compensate for pre-echo distortions.

PE

This is a measure of perceptually relevant information contained in any audio record. Expressed in bits per sample, PE represents a theoretical limit on the compressibility of a particular signal. PE measurements are reported in Johnston [3.87], who suggests that a wide variety of CD quality audio material can be transparently compressed at approximately 2.1 bits per sample.

The PE estimation process is accomplished as follows. The signal is first windowed and transformed to the frequency domain. A masking threshold is then obtained using perceptual rules. Finally, a determination is made on the number of bits required to quantize the spectrum without injecting perceptible noise. The PE measurements are obtained by constructing a PE histogram over many frames and then choosing a worst-case value as the actual measurement.

Masking thresholds are obtained by performing critical band analysis, masking a determination of the noise or toneline nature of the signal, applying thresholding rules for the signal quality and then accounting for the absolute hearing threshold. First, real and imaginary components are converted to power spectral components.

$$P(w) = [\text{Re}(\omega)]^2 + [\text{Im}(\omega)]^2 \qquad (3.5)$$

Then, a discrete bark spectrum is formed by summing the energy in each critical band

$$B_i = \sum_{\omega=bl}^{bh} P(\omega) \qquad (3.6)$$

where the summation limits are the critical band boundaries (*bl*-bandlow, *bh*-bandhigh). The range of the index, i, is sampling rate dependent, and in particular for $i \in \{1, 25\}$ CD-quality signals. A basilar spreading function (3.2) is then convolved with the discrete bark spectrum

$$C_i = B_i * SF_i \qquad (3.7)$$

to account for interband masking. An estimation of the toneline or noiselike quality for C_i is then obtained using the spectral flatness measure (SFM)

$$SFM = \frac{M_g}{M_a} \qquad (3.8)$$

where M_g and M_a correspond to geometric and arithmetic means of the power-spectral-density components for each band. The SFM has the property that it is bounded by 0 and 1. Values close to 1 will occur if the spectrum is flat in a particular band, indicating a decorrelated (noisy) band. Values close to zero will occur if the spectrum in a particular band is nearly sinusoidal. A coefficient of tonality, α, is next derived from the SFM on a dB scale, that is

$$\alpha = \min\left(\frac{SFM_{dB}}{-60}, 1\right) \qquad (3.9)$$

This is used to weigh the thresholding rules (3.3) and (3.4), with $K = 5.5$ as follows for each band to form an offset.

$$O_i = \alpha(14.5 + i) + (1 - \alpha)5.5 \quad [dB] \qquad (3.10)$$

A set of JND estimates in the frequency power domain are then formed by subtracting the offsets from the bark spectral components.

$$T_i = 10^{\log_{10} C_i - O_i / 10} \qquad (3.11)$$

These estimates are scaled by a correction factor to simulate deconvolution of the spreading function. Then each T_i is checked against the absolute threshold of hearing and replaced by $\max(T_i, T_{ABS}(i))$. As previously noted, the absolute threshold is referenced to the energy in a 4 KHz sinusoid of +/-1 bit amplitude. By applying uniform quantization principles to the signal and associated set of JND estimates, it is possible to estimate a lower bound on the number of bits required to achieve transparent coding. The perceptual entropy in bits per sample is represented by [3.82]

$$PE = \sum_{i=1}^{25} \sum_{\omega=bl_i}^{bh_i} \log_2\left\{2\left|\operatorname{int}\left(\frac{\operatorname{Re}(\omega)}{\sqrt{6T_i / k_i}}\right)\right| + 1\right\} + \log_2\left\{2\left|\operatorname{int}\left(\frac{\operatorname{Im}(\omega)}{\sqrt{6T_i / k_i}}\right)\right| + 1\right\} \qquad (3.12)$$

where i is the index of critical band, bl_i and bh_i are the upper and lower bounds of band i, k_i is the number of transform components in band i, T_i is the masking threshold in band i (3.11), while *int* denotes rounding to the nearest integer. If 0 occurs in the log, we assign 0 for the result.

3.6 Transform Audio Coders

Transform coding for high-fidelity audio makes use of unitary transforms for the time/frequency analysis. These algorithms typically achieve high-resolution spectral estimates at the expense of adequate temporal resolution. Many transform-coding algorithms for wideband and high-fidelity audio have been proposed in the last 15 years. The individual algorithms have been proposed by several groups [3.77, 3.78, 3.90, 3.91, 3.92, 3.93]. Much of this work was motivated by standardization activities, and the International Organization for Standardization/International Electrotechnical Commission (ISO/IEC) eventually clustered these proposals into a single candidate algorithm, Adaptive Spectral Entropy Coding (ASPEC) of high-quality music signals, which competed successfully for inclusion in the ISO/IEC MPEG audio-coding standards [3.94].

The algorithms that were eventually clustered into the ASPEC proposal submitted to ISO/IEC for MPEG audio came from researchers in both the United States and Europe. Novel transform-coding algorithms are not associated with ASPEC. Here, we will report some proposals for transform audio coding, like the following:

- Optimum coding in the frequency domain
- Perceptual transform coder
- Hybrid coder
- Transform coding that exploits DFT interblock redundancy
- ASPEC
- Differential perceptual audio coding
- DFT noise substitution
- DCT with vector quantization
- Modified Discrete Cosine Transform (MDCT)
- MDCT with vector quantization

3.6.1 Optimum Coding in the Frequency Domain

The 132 Kbps algorithm known as OCFD was proposed by Brandenburg [3.77]. It is in some respects similar to the well-known Adaptive Transform Coding (ATC) for speech. Block scheme of the OCFD is shown in Figure 3.8. The input signal is first buffered in 512 sample blocks and transformed to the frequency domain using the DCT. Next, transform coefficients are quantized and entropy coded. A single quantizer is used for all transform coefficients. Adaptive quantization and entropy coding work together in an iterative procedure to achieve a fixed bit rate. The initial quantizer step size is derived from the SFM given by formula (3.8). In the inner loop, the quantizer step size is iteratively increased, and a new entropy-coded bit stream is formed at each update until the desired bit rate is achieved. Increasing the step size at each update produces fewer levels, which in turn reduces the bit rate. Using a second iterative procedure, psychoacoustic masking is introduced after the inner loop is done. First, critical bound analysis is applied. Then, a masking function is applied which combines a flat 6 dB masking threshold with an interband masking threshold leading to an estimate of JND for each critical band. If after inner loop quantization and entropy coding the measured distortion exceeds JND in at least one critical band, quantization step sizes are adjusted in the out-of-tolerance critical bands only. The outer loop repeats until JND criteria are satisfied or a maximum loop cannot be reached. Entropy-coded transform coefficients are then

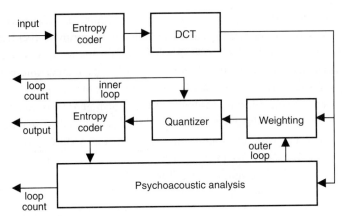

Figure 3.8 Optimum coding in the frequency domain [3.77]. ©1987 IEEE.

transmitted to the receiver along with side information, which includes the log-encoded SFM, the number of quantizer updates during the inner loop and the number of step-size reductions that occurred for each critical band in the outer loop. This side information is sufficient to decode the transform coefficients and to perform reconstruction at the receiver.

3.6.2 Perceptual Transform Coder

The idea behind the perceptual transform coder is to estimate the amount of quantization noise that can be inaudibly injected into each transform domain subband using PE estimates. The coder is memoryless and works as follows [3.78]. The signal is first windowed into overlapping (1/16) segments and transformed using a 2,048-point FFT. Next, the PE procedure is used to estimate JND thresholds for each critical band. Then, an iterative quantization loop adapts a set of 128 subband quantizers to satisfy the JND thresholds until the fixed rate is achieved. Finally, quantization and bit packing are performed. Quantized transform coefficients are transmitted to the receiver along with appropriate side information. Quantization of subbands consists of 8 sample blocks of complex-valued transform coefficients. The quantizer adaptation loop fast initializes the $j \in [1,128]$ subband quantizers (1,024 unique FFT coefficients /8 coefficients per subband) with k_j levels and step sizes of T_i as follows:

$$k_j = 2 * \text{int}\left(\frac{P_j}{T_i} + 1\right) \quad (3.13)$$

where T_i are the quantized critical band JND thresholds, P_j is the quantized magnitude of the largest real or imaginary part of transform coefficients in the j-th subband, while int(.) is the nearest integer rounding function. The block scheme of the perceptual transform coder is shown in Figure 3.9. The adaptation process involves repeated application of two steps. First, bit packing is attempted using the current quantizer set. Although many bit-packing techniques are possible, one simple scenario involves sorting quantizers in k_j order and then filling 64-bit words with encoded transform coefficients according to the sorted results. After bit packing, T_i are adjusted by a carefully controlled scale factor, and the adaptation cycle repeats. Quantizer adaptation halts as soon as the packed data length satisfies the desired bit rate. Both P_j and the modified T_i are quantized on a dB scale using 8-bit nonuniform quantizers with a 170-dB dynamic range. These parameters are transmitted as side information and are used at the receiver to recover quantization levels (and thus implicit bit allocations) for each subband. They are in turn used to decode quantized transform coefficients.

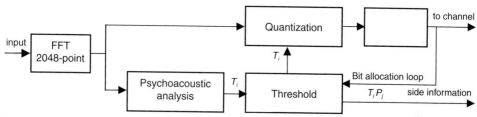

Figure 3.9 Perceptual transform coder.

3.6.3 Hybrid Coder

A hybrid coder is both a subband and transform coder. The idea behind the hybrid coder is to improve time and frequency resolution to OCFD and perceptual transform coder by constructing a filter bank that more closely resembles the human ear. This is accomplished at the encoder by first splitting the input signal into four octave-width subbands using a Quadrature Mirror Filter (QMF) filterbank. The decimated output sequence from each subband is then followed by one or more transforms to achieve the desired time-frequency resolution. Filterbank structure is shown in Figure 3.10 [3.95].

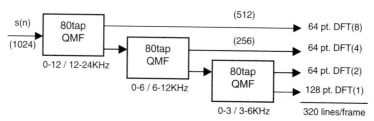

Figure 3.10 Filterbank structure [3.95].

Both DFT and MDCT transforms were investigated. Given the tiling of the time-frequency plane shown in Figure 3.11, frequency resolution at low frequency (24 KHz) is well matched to the ear, while the time resolution at high frequencies (2.7 ms) is sufficient for pre-echo control [3.95]. The quantization and coding schemes of the hybrid coder combine elements from both perceptual transform coder and OCFD. Masking thresholds are estimated using the perceptual

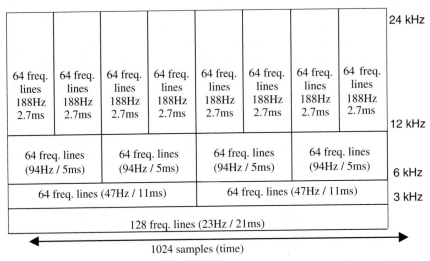

Figure 3.11 Time-frequency tiling [3.95].

transform coder approach for eight time slices in each frequency subband. The hybrid coder employs a quantization and coding scheme borrowed from OCFD.

As far as quality, the hybrid coder without any explicit pre-echo control mechanisms was reported to achieve quality better than or equal to OCFD at 64 Kb/s. The only disadvantage noted by the authors was increased complexity.

3.6.4 Transform Coding Using DFT Interblock Redundancy

A DFT-based audio-coding system that introduced a scheme to DFT interblock redundancy was proposed by Mahieux et al. [3.92]. Nearly transparent quality for 15 KHz audio at 96 Kbps, except for some highly harmonic signals, was reported. The encoder applies first-order backward-adaptive predictors (across time) to DFT magnitudes and differential pulse components. Then it quantizes separately the prediction residuals. Magnitudes and differential phase residuals are quantized using an adaptive nonuniform pdf-optimized quantizer designed for a Laplacian distribution and an adaptive uniform quantizer, respectively. The backward-adaptive quantizers are reinitialized during transients. Bits are allocated during step-size adaptation to shape quantization noise such that a psychoacoustic noise threshold is satisfied for each block. The use of linear prediction is justified because it exploits magnitude and different phase-time redundancy, which tends to be large during periods when the audio signal is quasistationary, especially for signal harmonics. For example, quasistationarity might occur during a sustained note.

3.6.5 ASPEC

ASPEC was claimed to produce better quality than any of the individual coders at 64 Kbs. The structure of ASPEC combines elements from all of its predecessors. Like OCFD, and transform coding using DFT interblock redundancy, ASPEC uses the MDCT for time-frequency mapping. The masking model is similar to that used in the perceptual transform coder and the hybrid coder, including the sophisticated tonality estimation scheme at lower bit rates. The quantization and coding procedures use the pair of nested loops, which is the block differential coding scheme developed in transform coding using DFT interblock redundancy. Moreover, long runs of masked coefficients are run length and Huffman encoded. Quantized scale factors and transform coefficients are Huffman coded also. Pre-echoes are controlled using a dynamic window-switching mechanism. Pre-echoes occur when a signal with a sharp attack begins near the end of a transform block immediately following a region of low energy. ASPEC offers several modes for different quality levels, ranging from 64 to 192 Kbps per channel. A real-time ASPEC implementation for coding one channel at 64 Kbps was realized on a pair of 33 MHz Motorola DSP56001 devices.

3.6.6 Differential Perceptual Audio Coder

DPAC makes use of a scheme for exploiting long-term correlations [3.96]. DPAC works as follows. Input audio is transformed using modified DCT (see Section 3.6.9). A two-state classifier then labels each new frame of transform coefficients as either a reference frame or a simple frame. Reference frames contain significant audible differences from the previous frame. The classifier labels nonreference frames as simple. Reference frames are scalar quantized and encoded using psychoacoustic bit-allocation strategies similar to perceptual audio coder. However, simple frames are subjected to coefficient substitution. The magnitude differences of coefficients with respect to the previous reference frame below an experimentally optimized threshold, those coefficients are replaced at the decoder by the corresponding reference frame coefficients. The encoder then replaces subthreshold coefficients with zeros, thus saving transmission bits. Unlike the interframe predictive coding schemes, the DPAC coefficient substitution system is advantageous in that the simple frame bit allocation will always be less than or equal to the bit allocation that would be required if the frame was coded as a reference frame. Superthreshold simple frame coefficients are coded in the same way as reference frame coefficients. DPAC performance can be evaluated for frame classifiers using different selection criteria.

Under the Euclidean criterion, test frames satisfying the inequality

$$\left[\frac{\mathbf{s}_d^T \mathbf{s}_d}{\mathbf{s}_r^T \mathbf{s}_r}\right]^{1/2} \leq \lambda \qquad (3.14)$$

are classified as simple, where the vectors $\mathbf{s}_r$ and $\mathbf{s}_i$, respectively, contain reference and test frame time-domain samples, while the difference vector, $\mathbf{s}_d$ is defined as

$$\mathbf{s}_d = \mathbf{s}_r - \mathbf{s}_t \qquad (3.15)$$

Under the PE criterion (3.12), a test frame is labeled as simple, if it satisfies the inequality

$$\frac{PE_s}{PE_R} \leq \lambda \qquad (3.16)$$

where PE_s corresponds to the PE of the simple (coefficient-substituted) version of the test frame, and PE_R corresponds to the PE of the unmodified test frame. λ is the decision threshold.

Finally, under the SFM criterion (3.8), a test frame is labeled as simple if it satisfies the inequality

$$abs\left(10\log_{10}\frac{SFM_T}{SFM_R}\right) \leq \lambda \tag{3.17}$$

where SFM_T corresponds to the test frame SFM, and SFM_R corresponds to the SFM of the previous reference frame. The decision threshold λ was experimentally optimized for all three criteria. Best performance was obtained while encoding source material using the PE criterion. As far as overall performance is concerned, NMR measurements were compared between DPAC and perceptual transform coding algorithm at 64, 88 and 128 Kbps. Despite an average drop of 30 to 35% in PE measured at the DPAC coefficient-substitution stage output relative to the coefficient substitution input, comparative NMR studies indicated that DPAC outperforms PTC only below 88 Kbps and then only for certain types of source material such as pop or jazz music. The desirable PE reduction led to an undesirable drop in reconstruction quality.

3.6.7 DFT Noise Substitution

In this method, noise-like spectral regions are identified as follows [3.97]. First, LMS adaptive Linear Predictors (LP) are applied to the output channels of a multiband QMF analysis filterbank that has as input, the original audio $s(n)$. A predicted signal, $\hat{s}(n)$, is obtained by passing the LP output sequences through the QMF synthesis filterbank. Prediction is done in subbands rather than across the entire spectrum to prevent classification errors that could result if high-energy noise subbands are allowed to dominate predictor adaptation, resulting in misinterpretation of low-energy tonal subbands as noisy. Next, the DFT is used to obtain magnitude $\left(S(k), \hat{S}(k)\right)$ and phase components $\left(\Theta(k), \hat{\Theta}(k)\right)$ of the input, $s(n)$, and prediction, $\hat{s}(n)$, respectively. Then, tonality, $T(k)$, is estimated as a function of the magnitude and phase predictability, for example,

$$T(k) = \alpha\left|\frac{S(k) - \hat{S}(k)}{S(k)}\right| + \beta\left|\frac{\Theta(k) - \hat{\Theta}(k)}{\Theta(k)}\right| \tag{3.18}$$

where α and β are experimentally determined constants. Noise substitution is applied to contagious blocks of transform coefficient bins for which $T(k)$ is very small. The 15% average bit-rate savings realized using this method in conjunction with transform coding is offset to a large extent by a significant complexity increase due to the additions of the adaptive linear predictors and a multiband analysis-synthesis QMF filterbank.

3.6.8 DCT With Vector Quantization

After computing the DCT on 512 audio sample blocks, the algorithm uses Multistage Tree-structured VQ (MSTVQ) scheme for quantization of normalized vectors, with each vector containing 4 DCT components. Bit allocation and vector normalization are derived at both the encoder and decoder from a sampled power-spectral envelope that consists of 29 groups of transform coefficients. A simplified masking module assumes that each sample of the power-spectral envelope represents a single masker [3.79]. Masking is assumed to be additive, as in the ASPEC algorithm. Thresholds are computed as a fixed offset from a masking level. To achieve high quality, a strong correlation between SFM and the amount of offset was observed. Two-segment scalar quantizers that are piecewise linear on a dB scale are used to encode the power-spectral envelope. Quadratic interpolation is used to restore full resolution to the subsampled envelope.

In another approach to quantization of transform coefficients, Constrained-Storage Vector Quantization (CS-VQ) techniques are combined with the MSTVQ from the original coder, allowing the new coder to handle peak NMR requirements without impractical codebook storage requirements [3.98]. In fact, CS-MSTVQ enabled quantization of 1,274 coefficient vectors using only 4 unique quantizers. Power-spectral envelope quantization is enhanced by extending its resolution to 127 samples. The samples are then encoded using a two-stage process. The first stage applies Nonlinear Interpolative Vector Quantization (NLIVQ), a dimensionality reduction process that represents the 127-element power spectral envelope vector using only a 12-dimensional feature power-spectral envelope. Unstructured VQ is applied to the feature power envelope. Then a full-resolution quantized envelope is obtained from the unstructured VQ index in the corresponding interpolation codebook. In the second stage, segments of the envelope residual are encoded using a set of 8-, 9- and 10-element quantizers.

3.6.9 MDCT

The MDCT offers the advantage of overlapping time windows while managing to preserve critical sampling. The analysis window must be carefully designed such that the time domain aliasing introduced by 50% overlap and 2:1 decimation will cancel in the inverse transformation [3.99]. The MDCT analysis expression is

$$X(k) = \sum_{n=0}^{2N-1} h(n)\, x(n) \cos\left[\frac{\pi}{2N}(2k+1)(2n+1+N)\right] \qquad (3.19)$$

where $k = 0, 1, ..., 2N-1$. The analysis window must satisfy

$$h^2(N-1-n) + h^2(n) = 2, \quad 0 \le n < N \qquad (3.20)$$

$$h^2(N+n) + h^2(2N-1-n) = 2, \quad 0 \le n < N \qquad (3.21)$$

An example analysis window that produces the desired time-domain aliasing cancellation is given by

$$h(n) = \pm\sqrt{2}\,\sin\left[\left(n+\frac{1}{2}\right)\frac{\pi}{2N}\right] \qquad (3.22)$$

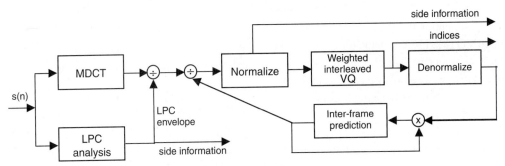

Figure 3.12 TWIN-VQ encoder [3.100]. ©1995 IEEE.

The development of FFT-based fast algorithms for the MDCT has made it viable for real-time applications [3.100, 3.101, 3.102].

3.6.10 MDCT with VQ

Transform Domain-Weighted Interleave Vector Quantization (TWIN-VQ), an MDCT-based coder that involves transform coefficients VQ, was developed by Iwakani et al. [3.100]. This algorithm exploits LPC analysis, spectral interframe redundancy, and interleaved VQ. A TWIN-VQ encoder is presented in Figure 3.12.

At the encoder, each frame of MDCT coefficients is first divided by the corresponding elements of the LPC spectral envelope, resulting in a spectrally flattened quotient (residual) sequence. This procedure flattens the MDCT envelope, but does not affect the fine structure. Therefore, the next step divides the first step residual by a predicted fine-structure envelope. This predicted fine-structure envelope is computed as a weighted sum of three previously quantized fine structure envelopes, that is, using backward prediction. Interleave VQ is applied to the normalized second step residual. The interleave VQ vectors are structured in the following way. Each N-sample normalized second step residual vector is split into K subvectors, each containing N/K coefficients. Second step residuals from N-sample vector are interleaved in the K subvectors such that subvector i contains elements $i+nK$, where $n = 0,1,...,(N/K)-1$. Perceptual weighting is also incorporated by weighting each subvector by a nonlinearly transformed version of its corresponding LPC envelope component prior to the codebook search. VQ indexes are transmitted to the receiver. Side information consists of VQ normalization coefficients and the LPC-encoded envelope. Enhancements to the weighted interleaving scheme and LPC envelope representation are reported by Moriya et al. [3.101]. This has enabled real-time implementation of stereo decoders on Pentium and power PC platforms.

3.7 Audio Subband Coders

Like the transform coders, subband coders also exploit signal redundancy and psychoacoustic irrelevance in the frequency domain [3.102]. Instead of unitary transforms, these coders rely upon frequency-domain representations of the signal obtained from banks of bandpass filters.

The audible frequency spectrum (20Hz to 20KHz) is divided into frequency subbands using a bank of bandpass filters. The output of each filter is then sampled and encoded. At the receiver, the signals are demultiplexed, decoded, demodulated and then summed to reconstruct the signal [3.103, 3.104, 3.105]. Audio subband coders realize coding gains by efficiently quantizing and encoding the decimated output sequences from perfect reconstruction filterbanks. Efficient quantization methods usually rely upon psychoacoustically controlled dynamic bit allocation rules, which allocate bits to subbands in such a way that the reconstructed output signal is free of audible quantization noise or other artifacts [3.106]. In a generic subband audio coder, the output signal is first split into several uniform or nonuniform subbands using some critically sampled, perfect reconstruction filterbanks. Nonideal reconstruction properties in the presence of quantization noise are compensated for by using subband filters that have very good attenuation. This requires high-order filters. Then, decimated output sequences from the filterbank are normalized and quantized for short, 2 to 10 ms blocks. Psychoacoustic signal analysis is used to allocate an appropriate number of bits for the quantization of each subband. The usual approach is to allocate a just-sufficient number of bits to mask quantization noise in each block while simultaneously satisfying some bit-rate constraint. Because masking thresholds and hence, bit-allocation requirements, are time varying, buffering is often introduced to match the coder output to a fixed rate. The encoder sends to the decoder quantized subband output samples, normalization scale factors for each block of samples and bit-allocation side information. Bit allocation may be transmitted as explicit side information, or it may be implicitly represented by some parameters, such as the scale-factor magnitudes. The decoder uses side information and scale-factors in conjunction with an inverse filterbank to reconstruct a coded version of the original input. Numerous subband coding algorithms for high-fidelity audio have appeared in the literature [3.106, 3.107, 3.108, 3.109, 3.110].

3.7.1 Wavelet Decompositions

In-depth technical information regarding wavelets is available in many references [3.111]. It is useful to summarize some basic wavelet characteristics. Wavelets are a family of basis functions in the space of square integrable signals. A finite energy signal can be represented as a weighted sum of the translates and dilates of a single wavelet. Continuous-time wavelet-signal analysis can be extended to discrete time and square summable sequences. Under certain assumptions, the Discrete Wavelet Transform (DWT) acts as an orthonormal linear transform $T: R^N \to R^N$. For a finite support wavelet of length K, the associated transformation matrix, Q, is fully determined by a set of coefficients $\{c_k\}$ for $0 \leq k \leq K-1$. This transformation matrix has an associated filter bank interpolation. One application of the transform matrix, Q, to an $N \times 1$ signal vector, x, generates an $N \times 1$ vector of wavelet-domain transform coefficients, y. The $N \times 1$ vector y can be separated into two $(N/2) \times 1$ vectors of approximation and detail coefficients, y_{lp} and y_{hp}, respectively. The spectral content of the signal x captured in y_{lp} and y_{hp} corresponds to the frequency subbands realized in 2:1 decimated-output sequences from a QMF filterbank that obeys the power complementary condition, that is,

$$|H_{lp}(\Theta)|^2 + |H_{lp}(\Theta + \pi)|^2 = 1 \qquad (3.23)$$

where $H_{lp}(\Theta)$ is the frequency response of the lowpass (LP) filter.

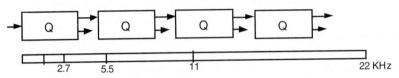

Figure 3.13 Wavelet decomposition [3.82]. ©1997 IEEE.

Successive applications of the DWT can be interpreted as passing input data through a cascade of banks of perfect reconstruction LP and highpass (HP) filters followed by 2:1 decimation. In effect, the forward/inverse transform matrixes of a particular wavelet are associated with a corresponding QMF analysis/synthesis filterbank. The usual wavelet decomposition implements an octave-band filterbank structure shown in Figure 3.13. Frequency subbands associated with the coefficients from each stage are schematically represented for an audio signal sampled at 44.1 KHz.

Wavelet packet representations decompose both the detail and approximation coefficients at each stage of the tree as shown in Figure 3.14. Frequency subbands associated with the coefficients from each stage are schematically represented for an audio signal sampled at 44.1 KHz. A filterbank interpolation of wavelet transforms is attractive in the context of audio coding algorithms for at least two reasons. First, wavelet or wavelet packet decompositions can be tree structured as necessary (unbalanced trees are possible) to decompose input audio into a set of frequency subbands tailored to some application. For example, it is possible to approximate the critical band auditory filterbank using a wavelet packet approach. Second, many K-coefficient finite support wavelets are associated with a single magnitude frequency-response QMF pair. Therefore, a specific subband decomposition can be realized while retaining the freedom to choose a wavelet basis, which is in some sense optimal.

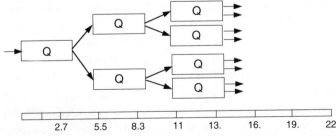

Figure 3.14 Wavelet packet decomposition [3.82]. ©1997 IEEE.

3.7.2 DWT-based Subband Coders

The basic idea behind DWT-based subband coders is to quantize and encode efficiently the coefficient sequences associated with each stage of the wavelet decomposition tree. Irrelevancy is exploited by transforming frequency-domain masking thresholds to the wavelet domain and shaping wavelet-domain quantization noise such that it does not exceed the masking threshold. Wavelet-based subband algorithms also exploit statistical signal redundancies through differential, run-length, and entropy coding schemes. The next few subsections concentrate on DWT-based subband coders developed in [3.112, 3.113, 3.114], including the hybrid sinusoidal/wavelet transform algorithm [3.115]. Other studies of DWT-based audio coding schemes concerned with low-complexity, low-delay, combined wavelet/multipulse LPC coding, and combined scalar/vector quantization of transform coefficients were reported, respectively in [3.116, 3.117, 3.118, 3.119, 3.120].

3.8 Speech Coder Attributes

Speech-coding attributes can be divided into four categories: bit rate, delay, complexity and quality [3.121]. It is possible to relax requirements for the less important attributes so that the more important requirements can be met.

Bit rate is the attribute that most often comes to mind first when discussing speech coders. The range of bit rates that has been standardized is from 2.4 Kb/s for secure telephony to 64 Kb/s for network applications. The coders standardized by the ITU are of primary interest [3.122]. The 64 Kb/s G.711 PCM coder is used in digital telephony networks and switches throughout the world. The 32 Kb/s G.726 Adaptive Differential (AD) PCM coder is used for circuit-manipulation equipment to increase effectively the capacity of undersea cables and satellite links. The 64/56/48 Kb/s G.722 and 16 Kb/s G.728 coders are used in video teleconferencing across ISDN or frame relay connections. The G.723.1 and G.729 coders have been standardized for low-bit-rate multimedia applications across telephony modems.

The codec delay can have a large impact on individual coder suitability for a particular application. Speech coders for real-time conversations cannot have too much delay, or they will quickly become unsuitable for network applications. For multimedia storage applications with only one-way transmission of speech, the coder can have virtually unlimited delay and still be suitable for the application. Psychologists who have studied conversational dynamics know that, if the one-way delay of a conversation is greater than 300 ms, the conversation will become more like a half-duplex, or a push-to-talk experience, rather than an ordinary conversation. In contrast, a speech or audio file of 300 ms or more before starting will be virtually imperceptible to the user. Thus, a conversation is an application that is the most sensitive to coder delay, while one involving speech storage is the least-delay-sensitive application. The components of the total system delay include the frame size, the look-ahead, other algorithmic delay, multiplexing delay, processing delay for computation and transmission delay. The algorithm used for the speech coder will determine the frame size and the look-ahead. Its complexity will have an impact on the processing delay. The network connection determines the multiplexing and transmission delays.

Recent multimedia speech coders have been implemented on the host Central Processing unit (CPU) of personal computers and workstations. The measures of complexity for a Digital Signal Processor (DSP) and a CPU are somewhat different due to the nature of these two systems. At the heart of complexity is the row number of computational instructions required to implement the speech coder. DSP chips from different vendors have different architectures and consequently different efficiencies in implementing the same coder. The measure used to indicate the computational complexity is the number of instructions per second required for implementation. This is usually expressed in Millions of Instructions per Second (MIPS). The numbers given in Table 3.3 are DSP MIPS for the ITU-T standards-based coders.

Table 3.3 ITU-T speech-coding standards [3.121]. ©1998 IEEE.

Standard	Bit Rate	Frame Size/Look-Ahead	Complexity
G.711 PCM	64 Kb/s	0 / 0	0 MIPS
G.726, G.721, G.723, G.727, ADPCM	16, 24, 32, 40 Kb/s	0.125 ms / 0	2 MIPS
G.722 Wideband coder	48, 56, 64 Kb/s	0.125 / 1.5 ms	5 MIPS
G.728 LD-CELP	16 Kb/s	0.625 ms / 0	30 MIPS

©1998 IEEE.

The ideal speech coder has a low bit rate, high perceived quality, low signal delay and low complexity. No ideal coder as yet exists with all these attributes. Real coders make tradeoffs among these attributes, for example, trading off higher quality for increased bit rate, increased delay or increased complexity. Figure 3.15 shows a plot of speech quality as measured subjectively in terms of Mean Opinion Scores (MOSs) for a range of telephone bandwidth coders spanning bit rates from 64 Kb/s down to 2.4 Kb/s. Curves of quality based on measurements made in 1980 and 1990 are shown. The MOS subjective test of speech quality uses a five-point rating scale with the following attributes:

  5—Excellent quality, no noticeable impairments

  4—Good quality, only very slight impairments

  3—Fair quality, noticeable but acceptable impairments

  2—Poor quality, strong impairments

  1—Bad quality, highly degraded speech

As can be seen, the telephone bandwidth coders maintain a uniformly high MOS for bit rates ranging from 64 Kb/s down to about 8 Kb/s, but fall steadily for bit rates below 8 Kb/s.

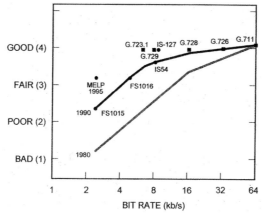

Figure 3.15 Subjective quality of various speech coders versus bit rate [3.121]. ©1998 IEEE.

3.9 CD Audio Coding for Multimedia Applications

A key aspect of multimedia systems is the ability to provide CD quality audio across telecommunications networks. Because uncompressed CD audio requires 1.4 Mb/s for transmission [3.121], high-quality coding of CD audio is essential for its practical use in any multimedia application. The state of the art in CD audio coding has improved dramatically in the course of the last decade. This is due to the relentless exploitation of unknown and well-understood properties of the human auditory system. Almost all modern CD audio coders use a quantitative model of the human auditory system to drive the signal quantization so that the resulting distortion is imperceptible. Those features of the audio signal that are determined not to be audible are discarded. The amount of quantization noise that is inaudible can also be calculated. This combination of discarding inaudible features and quantizing the remaining features so that the quantization noise will be inaudible is well known as perceptual coding. It has made its greatest impact to date in the field of audio coding and has been extended to speech, image and video coding. Remember that perceptual coding is a lossy coding method, that is, the output signal is not the same as the input signal. The imperceptible information removed by the perceptual coder is called the irrelevancy. In practice, most perceptual coders attempt to remove both irrelevancy and redundancy in order to make a coder that provides the lowest bit rate possible for a given quality. Most perceptual coders have lower SNR than source coders, but have better subjective quality for an equivalent bit rate.

3.10 Image Coding

Image coding involves compressing and coding a wide range of still images, including bilevel (fax) images, photographs and document images containing text, handwriting, graphics, and so forth. There are a number of important multimedia applications. These include the following:

- Slideshow graphics for applications such as ordering from catalogs, shopping from home, viewing real estate, and so forth. For this class of applications, it is essential that the images be presented at a variety of resolutions, including low resolution for fast browsing and searching and high resolution for detailed inspection. This type of application demands a 3D model that allows the user to view the image from different frames of reference.
- Creation, display, editing, access and browsing of banking images (electronic checks) and forms (insurance forms, medical forms, and so forth). For this application, it is essential that the user be able to interact with the system that creates and displays the images and that multimedia attachments be easily accessible by the user.
- Medical applications where sophisticated high-resolution images need to be stored, indexed, accessed and transmitted from site to site on demand. Examples of such images include medical X-rays, NMR, Electroencephalogram (EEG) and Electrocardiogram (EKG) scans, and so forth.

In order to compress and code image and video signals, it is essential to take advantage of any observable redundancy in the signal. The obvious forms of signal redundancy in most image and video signals include spatial redundancy and temporal redundancy [3.123].

Spatial redundancy takes a variety of different forms in an image, including correlations in the background image, correlations across an image and spatial correlations that occur in the spectral domain. A variety of techniques have been devised to compensate for spatial redundancy in image and video sequences [3.121]. Temporal redundancy takes two different forms in video sequences. The most obvious form is redundancy from repeated objects in consecutive frames of a video sequence. Such objects can move horizontally, vertically or any combination of directions, and they can disappear from the image as they move out of view [3.124].

The second basic principle of image coding is to take advantage of the HVS, which is used to view the coded image and video sequences in the same way that we take advantage of the human hearing system for listening to speech and audio signals. By understanding the perceptual masking properties of the HVS and their sensitivity to various types of distortion as a function of image intensity, texture and motion, we can develop a profile of the signal levels that provide JND in the image and video signals. By creating this JND profile for each image to be coded, it is possible to create image quantization schemes that hide the quantization noise under the JND profile and thereby make the distortion become perceptually invisible.

To illustrate the potential coding gains that can be achieved using a perceptual coding distortion measure, consider the set of curves of PE (distortion) of a black-and-white still image versus the image-coding rate measured in bits/sample or equivalently bits/pixel as shown in Figure 3.16. The upper curve (uniform quantization) shows that, in theory, it would take more than 8 bits/pixel to achieve low (essentially zero) distortion. By using proper compounding methods, this zero-distortion point can be reached with just 8 bits/pixel in practice as shown by the second curve. Taking into account noiseless coding methods (Huffman coding or arithmetic coding), the

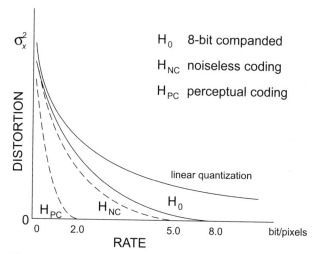

Figure 3.16 PE versus the coding rate [3.89]. ©1993 IEEE.

Noiseless Coding (NC) threshold can be reduced by a factor of nearly 2 to 1, down to 5 bits/pixel, as shown by the third curve. Last, by exploiting the perceptual model, the perceptually lossless coding threshold can be reduced by another factor of nearly three, down to around 2 bits/pixel or below depending on the image as shown by the dotted curve.

A block diagram of a generic image coding algorithm based on perceptual-coding principles is shown in Figure 3.17. The first step is to perform a short-term spectral analysis of the input image in order to determine a profile of image intensity and image texture. This short-term profile is then fed into the JND estimation box, which converts the measured image characteristics into a JND profile. It is then used as the input to an adaptive coder, which provides either a constant bit-rate (variable quality) output or a constant quality (variable bit rate) output signal.

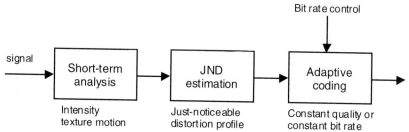

Figure 3.17 Image-coding algorithm based on perceptual coding [3.89]. ©1993 IEEE.

3.11 Video Coding

Video signals differ from image signals. The most important difference is that video signals have an associated frame rate of anywhere from 15 to 60 frames/s, which provides the illusion of motion in the displayed signal. A second difference is the frame size. Video signals may be as small as Quarter Common Intermediate Format (QCIF), 176x144 pixels, and as large as HDTV (1920x1080 pixels), whereas still images are sized primarily to fit PC color monitors (640x480 pixels or 124/768 pixels). A third difference between images and video is the ability to exploit temporal masking as well as spectral masking in designing compression methods for video. One can also take advantage of the fact that objects in video sequences tend to move in predictable patterns and can therefore be motion compensated from frame to frame if we can reliably detect both the object and its motion trajectory over time.

Some major initiatives in video coding have led to a range of video standards. These include the following:

- Video coding for video teleconferencing, which has led to ITU standards H.261 and for ISDN video conferencing [3.125], H.263 for POTS video conferencing [3.126, 3.127], and H.262 for ATM/broadband video conferencing, and digital TV.
- Video coding for storing movies on CD Read-Only Memory (ROM), on the order of 1.2 Mb/s allocated to video coding and 256 Kb/s allocated to audio coding, which led to the initial ISO MPEG-1 standard [3.128].
- Video coding for storing broadcast video on digital video disk (DVD), on the order of 2-15 Mb/s allocated to video and audio coding, which led to the ISO MPEG-2 standard [3.129, 3.130, 3.131, 3.132].
- Video coding for low bit-rate video telephony over POTS networks, with as little as 10 Kb/s allocated to video and as little as 5.3 Kb/s allocated to voice coding, which led to H.324 standard [3.57].
- MPEG-4 audio-video coding (both synthetic and natural). Also object/content-based coding with user interactivity. Designed for Internet and mobile communications. A Very Low Bit Rate (VLBR) core coder is to be compatible with H.263.
- Video coding for advanced HDTV, with 15 to 4000 Mb/s allocated to the video coding.
- MPEG-7 Multimedia content description interface. To facilitate truly integrated multimedia search engines based on descriptors, description schemes and description definition language.

3.11.1 TC and Subband Coding (SBC)

TC and SBC refer to compression systems where the signal decomposition is implemented using an analysis filterbank. At the receiver, the signal is reassembled by a synthesis filterbank. By TC, we usually mean that the linear transform is block based. When transform coding is interpreted as an SBC technique, the impulse responses of the analysis and synthesis filters are at most as long as the subsampling factor employed in the subbands. Thus, the image can be subdivided into blocks that are processed in an independent manner. On the other hand, general SBC allows the impulse responses to overlap and thus includes transform coding as a special case.

One of the most important tasks of the transform is to pack the energy of the signal into as few transform coefficients as possible. The DCT yields nearly optimal energy concentration [3.133]. Almost all image transform coders today employ the block-wise DCT, usually with a block size of 8x8 pixels. The transform is followed by quantization and entropy coding. Typically, the transform coefficients are run-length encoded. That is, successive zeros along a zigzag path are grouped with the first nonzero amplitude into a joint symbol, which is then Huffman coded. The Lapped Orthogonal Transform (LOT) could be substituted for the DCT to avoid some of the typical blocking artifacts that become visible with coarse quantization.

The full potential of SBC is unleashed when nonuniform band splitting is used to build multiresolution representations of an image. Besides excellent compression, multiresolution coders provide the successive approximation feature. As higher-frequency components are added, higher-resolution, better quality images are obtained. Moreover, multiresolution techniques fit naturally into joint source-channel coding schemes. Subband coders with octave band decomposition are often referred to as DWT coders, or wavelet coders [3.134].

At present, many state-of-the-art multiresolution image coders draw on the ideas introduced by Shapiro in his Embedded Zero-Tree Wavelet (EZW) algorithm [3.135]. The algorithm employs a data structure called zero-tree, where one assumes that, if a coefficient at a low frequency is zero, it is highly likely that all the coefficients at the same spatial location at all higher frequencies will also be zero. Thus, when encountering a zero-tree root, one can discard the whole tree of coefficients in higher frequency bands. Moreover, the algorithm uses successive approximation quantization, which allows termination of encoding or decoding at any point. These ideas have produced a new class of algorithms aimed at exploiting both frequency and spatial phenomena [3.136]. Research has shown that wavelet coders can produce superior results, but transform coders employing a block-wise DCT are still dominant. After years of use, DCT coders are very well understood and many improvements have been made, for example, in the area of fast algorithms or by imposing perceptual criteria.

3.11.2 Predictive Coding

Except when used with SBC or TC, predictive coders do not decompose the image into independent components. Instead, both the coder and decoder calculate a prediction value for the current signal sample. Then, the prediction error, rather than the signal sample itself, is transmitted. This principle can be used for both lossy and lossless image coding. The predictors calculate linear combinations of previous image samples because general nonlinear predictors, addressed by combinations of, say, 8-bit pixels, would often require enormous lookup tables for the same performance. For lossy predictive coding, DPCM has been used since the early days of image coding. Intraframe DPCM exploits dependencies within a frame of a video sequence. Typically, pixels are encoded in line-scan order, and previous samples in the current line and samples from the previous line are combined for prediction. Today, this simple scheme has been displaced by vastly superior transform/SBC schemes. In fact, lossy predictive intraframe coding is alive and well in the form of predictive closed-loop pyramid coders that feed back the quantization error before encoding the next higher-resolution layer. It has been shown that closed-loop pyramid coders even outperform the equivalent open-loop over complete pyramid representations when combined with scalar quantizers [3.137].

For interframe coding where statistical dependencies between successive frames of a video sequence are exploited, DPCM is the dominating scheme at present and for the foreseeable future. For example, other than spatio-temporal SBC, interframe DPCM avoids undesirable delay due to buffering of one or several frames. It is straightforward to incorporate motion adaptation and motion compensation into a temporal prediction loop and to combine motion-compensated prediction with other schemes for encoding the prediction error.

3.11.3 Motion-Compensated Video Coding

Motion compensation is a key element in most video coders. All modern video compression coders, such as those standardized in the ITU-T Rec. H.261 [3.125] and H.263 [3.126] or in the ISO MPEG standards [3.138], are motion-compensated hybrid coders. Motion-compensated hybrid coders estimate the displacement from frame to frame and transmit the motion vector field as side information in addition to the motion-compensated prediction error image. The prediction error image is encoded with an intraframe source encoder that exploits statistical dependencies between adjacent samples. The intraframe encoder is an 8x8 DCT coder in all current video coding standards, but other schemes, such as SBCs or VQs, can be used as well.

Figure 3.18 shows a block diagram of a motion-compensated image coder. The key idea is to combine TC in the form of the DCT of 8x8 pixel blocks with predictive coding in the form of differential PCM in order to reduce storage and computation of the compressed image and at the same time to give a high degree of compression and adaptability. Because motion compensation is difficult to perform in the transform domain, the first step in the interframe coder is to create a motion-compensated prediction error using macroblocks of 16x16 pixels. This computation requires only a single frame store in the receiver. The resulting error signal is transformed using DCT, followed by an adaptive quantizer, entropy encoded using a Variable Length Coder (VLC) and buffered for transmission across a fixed-rate channel.

The way that a motion estimator works is illustrated in Figure 3.19. A 16x16 pixel macroblock in the current frame is compared with a set of macroblocks in the previous frame to determine the one that best predicts the current macroblock. The set of macroblocks includes those within a limited region of the current macroblock. When the best matching macroblock is found, a motion vector is determined that specifies the reference macroblock and the direction of the motion of that macroblock.

Motion compensation works well for low spatial frequency components in the video signal. For high spatial frequency components, even a small inaccuracy of the motion compensation will render the prediction ineffective. Hence, it is important to use spatially a lowpass filter for the prediction signal by a loop filter. This loop filter is explicitly needed for integer-pixel

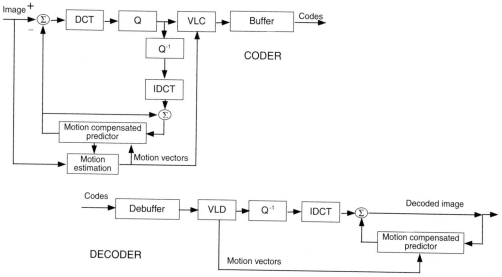

Figure 3.18 Motion-compensated coder and decoder for interframe coding [3.121]. ©1998 IEEE.

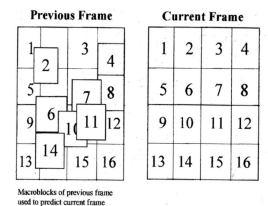

Macroblocks of previous frame used to predict current frame

Figure 3.19 Illustration of motion-compensated coding of interframe macroblocks [3.121]. ©1998 IEEE.

accurate motion compensation. For subpixel accurate motion compensation, it can be incorporated into the interpolation kernel required to calculate signal samples between the original sampling positions. The loop filter also improves prediction by acting as a noise-reduction filter. Prediction can be further improved by combining multiple independently motion-compensated signals. Examples are the bidirectionally predicted B-frames in MPEG [3.138] or overlapped block motion compensation [3.139] that has also been incorporated in the ITU-T Recommendation H.263 [3.126, 3.127].

Motion-compensated hybrid coding can theoretically outperform an optimum interframe coder by at most 0.8 bit/pixel in moving areas of an image if motion compensation is performed with only integer-pixel accuracy [3.140]. For half-pixel accuracy, this gain can be up to 1.3 bits/pixel. In addition, in nonmoving areas, or other parts of the image that can be predicted perfectly, no prediction error signal has to be transmitted, and these areas can simply be repeated from a frame store, a technique often referred to as conditional replenishment.

At low bit rates, motion compensation is severely constrained by the limited bit rate available to transmit the motion vector field as side information. Therefore, rate constrained estimation and a rate-efficient representation of the motion vector field are very important [3.141]. For simplicity, most practical video-coding schemes still employ blockwise constant motion compensation. More advanced schemes interpolate between motion vectors, employ arbitrarily shaped regions or use triangular meshes for representing a smooth motion vector field.

3.12 Watermarking

Data transmitted through a network may be protected from unauthorized receivers by applying techniques based on cryptography [3.142]. Only people who possess the appropriate private key can decrypt the received data using a public algorithm implemented either in hardware or in software. Fast implementation of encryption-decryption algorithms is highly desirable. Data-content manipulation can be performed for various legal or illegal purposes (compression, noise removal or malicious data modification). The modified product is not authentic with respect to the original one.

The technology of multimedia services grows rapidly, and distributed access to such services through computer networks is a matter of urgency. However, network access does not protect the copyright of digital products that can be reproduced and used illegally. An efficient way to solve this problem is to use watermarks [3.143, 3.144]. A watermark is a secret code described by a digital signal carrying information about the copyright property of the product. The watermark is embedded in the digital data such that it is perceptually not visible. The copyright holder is the only person who can show the existence of his own watermark and to prove the origin of the product.

Reproduction of digital products is easy and inexpensive. In a network environment, like the Web, retransmission of copies all throughout the world is easy. The problem of protecting the intellectual property of digital products has been treated in the last few years with the introduction of the notion of watermarks.

The following requirements should be satisfied by a watermarking algorithm:

- Alterations introduced in the image should be perceptually invisible.
- A watermark must be undetectable and not removable by an attacker.
- A sufficient number of watermarks in the same image, detectable by their own key, can be produced.
- The detection of the watermark should not require information from the original image.
- A watermark should be robust, as much as possible, against attacks and image processing, which preserves desired quality for the image.

Watermarks slightly modify the digital data to embed nonperceptible encoded copyright information. Digital data embedding has many applications. Foremost is passive and active copyright protection. Digital watermarking has been proposed as a means to identify the owner or distributor of digital data. Data embedding also provides a mechanism for embedding important control, descriptive or reference information in a given signal. A most interesting application of data embedding is providing different access levels to the data [3.145]. Most data-embedding algorithms can extract the hidden data from the host signal with no reference to the original signal [3.146].

The first problem that all data-embedding and watermarking schemes need to address is that of inserting data in the digital signal without deteriorating its perceptual quality. We must be able to retrieve the data from the edited host signal. Because the data insertion and data recovery procedures are intimately related, the insertion scheme must take into account the requirement of the data-embedding applications. Data insertion is possible because the digital medium is ultimately consumed by a human. The human hearing and visual systems are imperfect detectors. Audio and visual signals must have a minimum intensity or contrast level before they can be detected by a human. These minimum levels depend on the spatial, temporal and frequency characteristics of the human auditory and visual systems. Most signal-coding techniques exploit the characteristics of the human auditory and visual systems directly or indirectly. Likewise, all data-embedding techniques exploit the characteristics of the human auditory and visual systems implicitly or explicitly. A diagram of a data-embedding algorithm is shown in Figure 3.20. The information is embedded into the signal using the embedding algorithm and a key. The dashed lines indicate that the algorithm may directly exploit perceptual analysis to embed information. In fact, embedding data would not be possible without the limitations of the human visual and auditory systems.

Data embedding and watermarking algorithms embed text, binary streams, audio, image or video in a host audio, image or video signal. The embedded data are perceptually inaudible or invisible to maintain the quality of the source data. The embedded data can add features to the host multimedia signal, for example, multilingual soundtracks in a movie, or they can provide copyright protection.

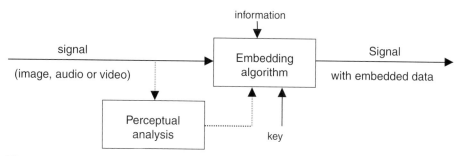

Figure 3.20 Block diagram of a data-embedding algorithm [3.146]. ©1998 IEEE.

3.12.1 Watermarking Techniques

Different watermarking techniques have been proposed by various authors in the last few years. The proposed algorithms can be classified into two main classes on the basis of the use of the original image during the detection phase: the algorithms that do not require the original image (blind scheme) [3.147, 3.148, 3.149] and the algorithms where the original image is the input in the detection algorithms along with the watermarked image (nonblind scheme) [3.150, 3.151, 3.152, 3.153]. Detectors of the second type have the advantage of detecting the watermarks in images that have been extensively modified in various ways.

Watermarking embedding can be done either in the spatial domain or in an appropriate transform domain, like a DCT domain [3.148, 3.152, 3.153], a wavelet transform domain [3.150, 3.151] or a Fourier transform domain [3.154]. In certain algorithms, the imposed changes take into account the local image characteristics and the properties of the human visual system (perceptual masking) in order to obtain watermarks that are guaranteed to be invisible [3.148, 3.150, 3.155].

The DCT-based watermarking method has been developed for image watermarking that could survive several kinds of image processings and lossy compression [3.156]. In order to extend the watermarking techniques into video sequences, the concept of temporal prediction exploited in MPEG is considered. For intraframe, the same techniques of image watermarking are applied, but for non-intraframe, the residual mask, which is used in image watermarking to obtain the spatially neighboring relationship, is extended into the temporal domain according to the type of predictive coding. In considering the JPEG-like coding technique, a DCT-based watermarking method is developed to provide an invisible watermark and also to survive the lossy compression.

The human eyes are more sensitive to noise in a lower frequency range than its higher frequency counterpart, but the energy of most natural images is concentrated in the lower frequency range. The quantization applied in lossy compression reflects the human visual system, which is less sensitive to quantization noise at higher frequencies. Therefore, to embed the watermark invisibly and to survive the lossy data compression, a reasonable trade-off is to embed the watermark into the middle-frequency range of the image. To prevent an expert from extracting the hidden information directly from the transform domain, the watermarks are embedded by modifying the relationship of the neighboring blocks of midfrequency coefficients of the original image instead of embedding by an additive operation.

Example 3.5 The original image is divided into 8x8 blocks of pixels, and the 2D DCT is applied independently to each block. Then, the coefficients of the midfrequency range from the DCT coefficients are selected. A 2D subblock mask is used in order to compute the residual pattern from the chosen midfrequency coefficients.

Let the digital watermark be a binary image. A fast 2D pseudorandom number-traversing method is used to permute the watermark so as to disperse its spatial relationship. In addition to the pixel-based permutation, a block-based permutation according to the variances of both the image and watermark is also used. Although the watermark is embedded into the mid-frequency coefficients, for those blocks with little variances, the modification of DCT coefficients introduces quite visible artifacts. In this image-dependent permutation, both variances of the image blocks and watermark blocks are sorted and mapped according to importance of the invisibility. After the residual pattern is obtained for each marked pixel of the permuted watermark, the DCT coefficients are modified according to the residual mask, so that the corresponding polarity of residual value is reversed. Finally, inverse DCT of the associated results is applied to obtain the watermarked image.

Example 3.6 The extraction of a watermark requires the original image, watermarked image and also the digital watermark. At first, both the original image and the watermarked images are DCT transformed. Then, we make use of the chosen midfrequency coefficients and the residual mask to obtain the residual values. Perform the EXCLUSIVE-OR operation on these two residual patterns to obtain a permuted binary signal. Reverse both the block and the pixel-based permutations to get the extracted watermark.

A video sequence is divided into a series of Group of Pictures (GOP). Each GOP contains an interframe (I-frame), forward-predicted frame (P-frame) and bidirectional predicted/interpolated frame (B-frame). P-frame is encoded relative to intraframe or another P-frame. B-frame is derived from two other frames, one before and one after. These non-intraframes are derived from other reference frames by motion-compensation that uses the estimated motion vectors to construct the images. In order to insert the watermark into such kind of motion-compensated images, the residual patterns of neighboring blocks are extended into the temporal domain and other parts of the image. Watermarking techniques can be applied directly into non-intraframes.

For a forward-predicted P-frame, the residual mask is designed between the P-frame and its reference I- or P-frame, that is, the watermarks are embedded by modifying the temporal relationship between the current P-frame and its reference frame. For a bidirectionally predicted or interpolated B-frame, the residual mask is designed between the current B-frame and its past and future reference frames. The polarity of the residual pattern is reversed to embed the watermark.

3.12.2 Main Features of Watermarking

Watermarks are digital signals that are superimposed on a digital image causing alternations to the original data. A particular watermark belongs exclusively to one owner who is the only person that can proceed to a trustworthily detection of the personal watermark and, thus, prove the ownership of the watermark from the digital data. Watermarks should possess the following features [3.157]:

- *Perceptual invisibility*—The modification caused by the watermark embedding should not degrade the perceived image quality. However, even hardly visible differences may become apparent when the original image is directly compared to the watermarked one.
- *Trustworthily detection*—Watermarks should constitute a sufficient and trustworthily part of ownership of a particular product. Detection of a false alarm should be

extremely rare. Watermark signals should be characterized by great complexity. This is necessary in order to be able to produce an extensive set of sufficiently well distinguishable watermarks. An enormous set of watermarks prevents the recovery of a particular watermark by trial-and-error procedure.

- *Associated key*—Watermarks should be associated with an identification number called watermark key. The key is used to cast, detect and remove a watermark. Subsequently, the key should be private and should exclusively characterize the legal owner. Any digital signal, extracted from a digital image, is assumed to be a valid watermark if and only if it is associated to a key using a well established algorithm.
- *Automated detection/search*—Watermarks should combine easily with a search procedure that scans any publicly accessible domain in a network environment for illegal deposition of an owner's product.
- *Statistical invisibility*—Watermarks should not be recovered using statistical methods. For example, the possession of a great number of digital products, watermarked with the same key, should not disclose the watermark by applying statistical methods. Therefore, watermarks should be image dependent.
- *Multiple watermarkings*—We should be able to embed a sufficient number of different watermarks in the same images. This feature seems necessary because we cannot prevent someone from watermarking an already watermarked image. It is also convenient when the copyright property is transferred from one owner to another.
- *Robustness*—A watermark that is of some practical use should be robust to image modifications up to a certain degree. The most common image manipulations are compression, filtering, color quantization/color-brightness modifications, geometric distortions and format change. A digital image can undergo a great deal of different modifications that may deliberately affect the embedded watermark. Obviously, a watermark that is to be used as a means of copyright protection should be detectable up to the point that the host image quality remains within acceptable limits.

Example 3.7 A visible watermark must be obvious to any person with normal or corrected vision (including the color blind), be flexible enough that it can be made as obstructive or unobstructive as desired, have bold features that form a recognizable image, allow the features of the unmarked image to appear in the marked image, and be very difficult, if not impossible, to remove.

To fulfill these criteria, we begin with the construction of a mask corresponding to the watermark. The mask determines which pixels in an image will remain unchanged and which will have their intensity altered. The mask is then resized, and the masking purpose and the location at which the watermark will be placed are chosen. Last, the intensity in the pixels specified by the mask is altered. We use a mathematical model of the intensity in an image

$$\hat{Y}_{m,n} = Y_{m,n} + C \times \Delta L \tag{3.24}$$

where $Y_{m,n}$ and $\hat{Y}_{m,n}$ represent the intensity of the (m,n)th pixel in the original and marked images, respectively, the constant C is a function that reflects various properties of the specific image and watermark mask, and L is the intensity, that is, the amount of the light received by the eye, regardless of color [3.158]. The appearance of the watermark is controlled by varying the intensity L by ΔL. If the same value of ΔL were used to alter all the pixels that fall under the mask, the watermark could be easily removed by a hostile party. To render robustness to the mask, randomness is introduced by using $2R_{m,n}\Delta L$ in place of ΔL, where $R_{m,n} \in [0,1]$ is a discrete random variable that satisfies

$$\lim_{M\to\infty}\lim_{N\to\infty}\frac{2}{MN}\sum_{m=1}^{M}\sum_{n=1}^{N}R_{m,n}\Delta L = \Delta L$$

(if ΔL is truly randomly distributed).

A watermark needs to have bold features because the introduction of the random variable $R_{m,n}$, depending on its values, can make fine details of the mark less discernible.

3.12.3 Application Domains

Because a generic watermarking algorithm cannot fit for a variety of applications, there is need for standardization of watermarking technology on an application-by-application basis. Table 3.4 [3.157] shows six different application domains and their requirements in terms of robustness and resilience:

Table 3.4 Categorization of applications by different robustness and resilience and the need of detection in every decoder [3.157].

Application Domains	Unintentional Attacks			Intentional Attacks		Every Decoder	High Capacity	Application Examples
	AT1	AT2	AT3	AT4	AT5			
A1.	Yes	Yes	Maybe	No	No	Yes	Yes	Value-added metadata
A2.	Yes	Yes	Yes	Yes	Yes	Yes	No	Copy protection
A3.	Yes	Yes	Yes	Yes	Yes	No	No	Ownership/fingerprint
A4.	Yes	No	No	No	Some	Yes	No	Authentication
A5.	Yes	Yes	No	No	Yes	Yes	Yes	Broadcasting
A6.	Yes	Yes	Maybe	Maybe	Yes	No	Yes	Secret communication

© 2000 ISO/IEC.

- **A1. Carrying value-added metadata**—This kind of application uses an embedded watermark to carry some additional information, for example, hyperlinks, annotation and content-based indexing information, which is to survive common content-preserving transformations. The watermarks should be detectable when the content is used not only in MPEG-4, but also in MPEG-1 or MPEG-2 if possible. In general, these applications often require higher capacity watermarks compared to other applications. On the other hand, in many cases, such watermarks are not subject to purely malicious attacks.
- **A2. Copy protection and conditional access**—In this kind of application, an embedded Intellectual Property Rights (IPR) system uses an embedded watermark to control Intellectual Property Management and Protection (IPMP)-related issues, for example, view options and copy options, in compliant decoders (for example, in the second generation of DVD). Every compliant decoder must be able to trigger protection or royalty collection mechanisms at the time when the contents are decoded. In general, such systems are predicated on the fact that unauthorized people should not be able to forge, remove or invalidate the watermarks by any means.
- **A3. Ownership assertion, recipient tracing**—In this kind of application, content owners use embedded watermarks to establish ownership or to determine the origin of unauthorized duplication. Normally, consumers do not need to know the original ownership when playing back the content. This kind of information is thus required only when prosecuting for IPR infringement. Again, such systems' effectiveness depends on the inability of unauthorized users to counterfeit, remove or invalidate watermarks.
- **A4. Authentication and verification**—Applications can use fragile or semi-fragile (for example, robust authentication) watermarks for verification of authenticity of source and content, verification of integrity of content and so forth. If contents are altered, watermarks should disappear or change to indicate the extent and possibly location of changes, depending on the precise scheme. In general, the viewer would like to know the verification results when playing back the contents. Although concerns of people removing or invalidating watermarks are less rampant in such applications, preventing forgery and counterfeiting becomes a particularly important issue to address.
- **A5. Broadcast monitoring**—We embed watermarks into contents and monitor where and when the contents are played. For example, advertisers may watermark their commercial. Then, upon verifying transmission of the commercial, the advertisers will pay a certain amount of money to whomever the advertisement is distributed. Every broadcast monitoring decoder that receives the commercial should report the usage. In general, detecting watermarks from heavily degraded content is less of an issue in such applications. However, depending on the precise application, watermark removal, invalidation and/or forgery can be significant concerns. In the broadcast-monitoring case, for example, counterfeiting should be intractable for the system to be effective.
- **A6. Secret communication or steganography**—This application, that is, data hiding in its most general sense, uses an embedded watermark to carry some secret information.

Often, these applications may need higher capacity watermarks than many other applications. Maintaining secrecy of the message so that unauthorized users cannot extract the watermark may often be an overriding concern in these kinds of applications. Removal and counterfeiting may or may not be an issue depending on the application.

The watermarking application environment is similar to that of compression. The intended information recipients (users of decompression or watermark extraction) need to know which method is used by the information provider (compression or watermark insertion). For those application domains that require every decoder to detect the watermarks, there is a need for standardization (A1, A2, A4 and A5). For those application domains that do not require every decoder to detect the watermarks, there is no rush for standardization (A3 and A6).

The watermark robustness testing conditions are defined as follows:

- AT1 Basic attacks—Lossy compression, frame dropping and temporal rescaling, that is, frame-rate changes
- AT2 Simple attacks—Blurring, median filtering, noise addition, gamma correction, and sharpening
- AT3 Normal attacks—Translation, cropping and scaling
- AT4 Enhanced attacks—Aspect ratio change and random geometric perturbations, for example, Stirmark, and local permutation of pixels
- AT5 Advanced attacks—Delete/insert watermarks, single-document watermark estimation attacks and multiple-document statistical attacks

3.13 Organization, Storage and Retrieval Issues

Once multimedia material is compressed and coded, it needs to be sent across a network to one or more end-users. To ensure that communication does not break down in transmission, we need to ensure the method and speed of delivery of the material using either a streaming implementation or a full download. We have to provide mechanisms for different resolutions of the receiving system and a guaranteed QoS so that the received multimedia signal has essentially the quality that is expected and is being paid for [3.121].

3.13.1 Streaming Issues for Speech and Audio

Streaming refers to the transmission of multimedia signals for real-time delivery without waiting for the entire file to arrive at the user terminal. Streaming can be either narrowcast (from the server to just one client) or broadcast (one transmission stream to multiple clients). The key point is that the real-time information is flowing solely from one source in both cases. There are four main elements to a streaming system:

- The compressed (coded) information content, for example, audio, video, speech, multimedia, data, and so forth

- The content, which is stored on a server
- The server, which is connected to the Internet and/or possibly other networks (POTS, ISDN, ATM or frame relay)
- The clients

Each of these elements can cause impairments. There are well-established methods for minimizing the degradations to the signal that result from these impairments. To set the scenario, assume that a user has requested a real-time audio or video stream from the Internet. Further, we assume that the client is not directly connected to the stream on the Internet, but instead accesses it through an Internet Service Provider (ISP). The access could be through a modem on a POTS line, ISDN, a corporate LAN running on IP or could even include ATM or frame relay in the access link. Although the backbone of the Internet is a high-speed data network, there are a number of potential bottlenecks within the path from the server that streams the data to the final client that receives the real-time stream. Heavy traffic causes congestion and results in variable delays and possibly dropping packets. Thus, the two manifestations of network congestion on the packet stream that represents the real-time signal are highly variable delays and lost packets. The degree to which these two problems occur determines the QoS that can be provided. The solutions to these problems must focus on these two issues: delayed and lost packets [3.121].

One potential way of addressing these problems is through the compression scheme. Using compression to represent the real-time signal provides an excellent mechanism for dealing with lost packets. Because the signal has been analyzed and reduced to its component parts as part of the compression scheme, this often makes handling lost and delayed packets practical because parts of the lost signal are often highly predictable. Although the coding algorithms attempt to remove all redundancy in the signal, some redundancy still remains, even for the most advanced coders. We can also use concepts of statistical predictability to extrapolate some of the component parts for missing packets. Hence, for both speech and audio, the best extrapolation policy for handling lost and delayed packets seems to be to assume stationarity of the missing signal. The degree to which this assumption is valid makes the strategy more or less viable in practice. If the signal is in a transitional state, holding its statistical properties constant will cause an obvious and highly perceptual degradation. However, shutting of the signal (playing silence in place of the signal) would lead to an even more obvious degradation most of the time. There is also the problem that lost packets cause the decoder state to lose synchronization with the encoder state. Generally, forward adaptive coders can resynchronize the encoder and decoder faster than backward adaptive coders. Hence, forward-adaptive coders are preferred for a highly congested data network with streaming speech signals.

If the speech or audio was encoded or uncompressed, an analysis of the immediate past material could be used to determine the amount of prediction that could be applied during periods of lost or delayed packets. To the extent that this occurs, the QoS may become unacceptable.

A technique that can be used with speech and also to some extent with audio is to change the buffer size to match the network delay characteristics. This method can also be used to account for any long term differences in the clock rates of the server and the client.

The client's software needs must be adapted to the network so that it can respond to missing packets. Obviously, the client must take some action to deal with the variation in the time of receipt of packets due to the variable delay of packets. Most commonly, the delay is accounted for by creating a buffer at the client to smooth over the variations in delay. For information retrieval, this is generally acceptable. It means that the start of playback will be delayed in proportion to the size of the buffer, but, for audio distribution, the extra delay will probably not be noticed.

The server's function in a streaming transaction is to transmit the information (the coded speech or audio) at an average rate designed to maintain real-time decoding of the compressed material. If the server is working too slowly (that is, heavily overloaded), the real-time signal received by the client will have gaps caused by the decoder running out of bit stream to decode. If the server is transmitting it quickly (that is, under loaded conditions), the bit stream will build up in the decoder buffers, eventually overfilling them and causing a loss of signal because of the buffer overflow. The server must serve multiple clients and must respond to changes in the network due to variable traffic and congestion. Thus, if the server is requested to reduce the amount of traffic it is generating on the network, it will be expected to do so. If it has too many clients, it can also cause interruptions in the regular transmission of packets. In general, because it is serving many clients, the server will not transmit packets at regular intervals. Namely, there will necessarily be some jitter in its transmission instants. Additionally, it will be more efficient to transmit packets of larger size.

The transmitted streaming information is now sent across the Internet and into the user's network. The principal problem is congestion at a few nodes or across a few links. The congestion results in both variable delays and lost packets. If the rate of packet loss is too high, real-time streaming is just not viable. The extrapolation techniques may be adequate to cover up losses of a few percent, but will not suffice when the loss rate is within the focus of percent range. Similarly, if the delay variation is too great, the longest delayed packets will be treated as lost packets.

3.13.2 Streaming Issues for Video

Video has a much higher data rate than speech or audio, and in a great deal of the time video requires only one-way streaming (movies, shows, documentaries, video clips, and so forth) and can therefore tolerate long delays in the streaming network. A successful streaming application requires a well-designed system that takes into account each of these elements [3.159]. Streaming in data networks is implemented as part of the application-layer protocols of the transmission, that is, it uses User Datagram Protocol (UDP) and TCP at the transport layer. Because of the known shortcomings of TCP, most streaming implementations are based on the inherently unreliable UDP. Thus, whenever there is network congestion, packets are dropped. Also, because delays can be large (on the order of seconds) and often unpredictable on the Internet, some packets may arrive after their nominal presentation time, effectively turning them into lost packets. The extent of the losses is a function of the network congestion, which is highly correlated with the time of day and the distance (in terms of the number of the routers) between the client and the multimedia source [3.160]. The practical techniques that have evolved for improving the performance of streaming-based real time signal delivery can be classified into four broad areas:

- Client-side buffer management, which determines how much data needs to be buffered both prior to the start of the streaming playback and during the playback. It also determines a strategy for changing the buffer size as a function of the network congestion and delay and the load on the media server.
- Error-resilient transmission techniques, which increase client-side resilience to packet losses through intelligent transport techniques, such as using higher priority for transmitting more important parts (headers, and so forth) of a stream and/or establishing appropriate retransmission mechanisms where possible.
- Error-resilient coding techniques, which use source and perhaps combined source and channel-coding techniques that have built-in resilience to packet losses.
- Media control mechanisms, which use efficient implementations of Video Cassette Recorder (VCR)-type controls when serving multiple clients.

None of these techniques is sufficient to guarantee high-quality streaming, but, in combination, they serve to reduce the problems to manageable levels for most practical systems.

3.14 Signal Processing for Networked Multimedia

Real-time transmission of multimedia data across packet networks poses several interesting problems for signal-processing research. Although the range of these problems covers a large variety of topics, two groups attract the most attention. The first group concerns adapting the signal-compression techniques to address the special requirements imposed by the packet networks, including accommodating for packet losses, delays and jitter; providing capability for multipoint and coping with the heterogeneous nature of today's networks. The second group of problems is related to protecting the IPR associated with the transmitted multimedia data. The increasing availability of high bandwidth networking makes it extremely easy to duplicate and disseminate digital information illegally. Unless mechanisms can be established to protect the rights of the content providers, commercial use of networked multimedia will remain extremely limited.

Adapting signal compression techniques to networked applications may require some changes in the fundamental approach to this problem. The goal of classical signal compression is to achieve the highest possible compression ratio. The compression and transmission aspects have generally been treated as separate issues. The first problem with this approach is that the resulting compression algorithms usually do not address the needs of networked transmission.

Example 3.8 In a networked multimedia multicast, several receivers may be connected to the network with bandwidths that may range from very low, for example, a 28.8 Kbs modem, to very high, for example, a 150 Mbps optical link. Using a compression rate that satisfies the high bandwidth, receivers will cut off the low bandwidth ones completely.

On the other hand, using the lowest bandwidth will not be acceptable for the high bandwidth receivers. The alternate approach, simulcasting several data rates, requires about a two-fold increase in the main network bandwidth. A compression algorithm designed with this application requirement in mind must provide for easy extraction of data streams having different rates from a single compressed stream. Such compression algorithms are also useful for adapting to the changing network conditions caused by congestion [3.161, 3.162]. The second difficulty in separately designing the compression and transmission components is caused by the fact that a successful compression algorithm removes all the redundancy. Hence, the compressed data must be delivered error free. As an example, if a single bit of a compressed picture is lost, the entire picture may become undecodable. Effective error-concealment techniques must be present in the receiver to minimize the visual impact of any errors. A straightforward approach to provide robustness is to insert pointers into the compressed data to make the partially received data usable [3.163]. This approach works when the compressed data have natural boundaries (blocks in an image) and its error resilience increases as the size of the independently decodable data segments gets smaller. Because the added pointers (restart markers in JPEG images and slice start codes in MPEG video) introduce overhead, using smaller segments reduces the compression efficiency. A compression method designed to generate an output with independently decodable segments may perform better than this.

Another consideration in designing compression techniques for network use is to identify the impact of losing different portions of a compressed stream. For example, losing a header that contains the quantization information may render a large segment of data useless, but the same size loss in the data portion may only destroy a short segment that could be concealed at the receiver. Considering the relative ease of providing error-free transmission for shorter data segments, it is preferable to have the important parts of a compressed stream concentrated into a short and identifiable segment.

3.15 NNs for Multimedia Processing

Future multimedia technologies will need to handle information with an increasing level of intelligence, that is, automatic recognition and interpretation of multimodal signals. The main attribute of neural processing is its adaptive learning capability, which enables machines to be taught to interpret possible variations of some object or pattern, for example, scale, orientation and perspective [3.163, 3.164]. Moreover, we are able to approximate accurately unknown systems based on sparse sets of noisy data. Some neural models have also effectively incorporated statistical signal processing like expectation-maximization, Gaussian mixture and optimization techniques. In addition, spatial/temporal neural structures and hierarchical models are promising for multirate, multiresolution multimedia processing. NNs have thus received increasing attention in many multimedia applications, such as the following:

- Human perception—Facial expression and emotional categorization, human color perception and multimedia data visualization [3.165, 3.166, 3.167, 3.168]
- Computer-human communication—Face recognition, lip-reading analysis and human-human and computer-human communications
- Multimodal representation and information retrieval—Hyperlinking of multimedia objects, queries and search of multimedia information; 3D object representation and motion tracking, image sequence generation and animation [3.169, 3.170, 3.171, 3.172, 3.173, 3.174, 3.175, 3.176].

A major impact may be achieved by integrating adaptive neural processing into the state-of-the-art multimedia technologies. A complete multimedia system consists of many information processing stages, for which neural processing offers an efficient and unified core technologies.

Neural processing and Intelligent Multimedia Processing (IMP) share the following characteristics:

- A universal data-processing engine for multimodal signals
- Multimodality: multiple sensor/data sources
- Unsupervised clustering and/or supervised learning by example mechanisms

Future IMP applications include speech recognition/understanding, character recognition, texture classification, image/video segmentation, face-object detection/recognition, tracking 3D objects and analysis of facial expressions and gestures.

3.15.1 NNs for Optimal Visualization

For some image-processing applications (medical), a display that maximizes diagnostic information would be very desirable. NNs have been successfully applied for optimal visualization so that information can be more noticeably displayed. Note that raw data may contain more bits than what can be displayed in an ordinary computer monitor.

Example 3.9 A magnetic resonance image contains 12-bit data, but most monitors only have an 8-bit display. To map 12-bit data to an 8-bit display, the appearance of the image depends upon a proper selection of window width/center, which is a typical representation of image dynamic range in the medical field. An NN-based system is used to estimate window width/center parameters for optimal display.

To reduce the input dimension of NN, a feature vector of an input image is first extracted through PCA transformations. Then a competitive layer (unsupervised) NN is applied to label the feature vector into several possible classes with their confidence measure. For each class, both nonlinear and linear adaptive estimators are used to best calibrate window width/center. A nonlinear estimator is vulnerable to drastic and very unreasonable failures, but is very efficient in reaching local optimum. To alleviate such concern, a safety net is provided through a linear estimator.

A final data fusion scheme outputs the optimal window/center parameters by combining the results from all possible classes with appropriate weighting of the confidence measures [3.68].

3.15.2 Neural Techniques for Motion Estimation

Neural techniques for motion estimation have been under investigation. A motion estimation algorithm based on the Expectation Maximization (EM) technique was proposed by Fan, Nomazi and Penafiel [3.177]. First, the motion field is represented by a model characterized by a series of motion coefficients. Smoothness of motion is imposed on the assumption. Then the EM-based iterative algorithm is adopted to estimate the image motion coefficients from noisy measurements.

A feature true motion technique (TMT) for object-based motion tracking was proposed by Chen, Lin and King [3.178]. Based on a neighborhood relaxation neural model, it can effectively find true motion vectors of the prominent features of an object. By prominent feature, we mean the following:

- Any region of an object contains a good number of blocks that have motion vectors that exhibit certain consistency.
- Only true motion vectors for a few blocks per region are needed.

Therefore, at the outset, it would disqualify some reference blocks that are doomed unreliable to track. The method adopts a multicandidate prescreening to provide some robustness in selecting motion candidates. Furthermore, assuming that the true motion field is piecewise continuous, the method calculates the motion of a feature block after consulting all of its neighboring blocks' motions. This precaution allows a singular and erroneous motion vector to be corrected by its surrounding motion vectors, yielding an effect very much like median filtering. The tracker has also found useful application in motion-based video segmentation [3.177].

Example 3.10 One Foreman example is shown in Figure 3.21. Two frames of Foreman sequence are represented. Motion vectors found by the original full-search block-matching algorithm are shown also. Finally, we have motion vectors obtained by the neural method through neighborhood relaxation.

3.15.3 NN Application to Face Detection and Recognition

NNs have been recognized as an established and mature tool for many pattern-classification problems. Particularly, they have been successfully applied to face-recognition applications. By combining face information with other biometric features such as speech, feature fusion should not only enhance accuracy, but also provide some fault tolerance, that is, it could tolerate temporary failure of one of the bimodal channels. For many visual monitoring and surveillance applications, it is important to determine human eye positions from an image or an image sequence containing a human face. After the human eye positions are determined, all of other important facial features, such as positions of the nose and mouth, can easily be determined. The basic

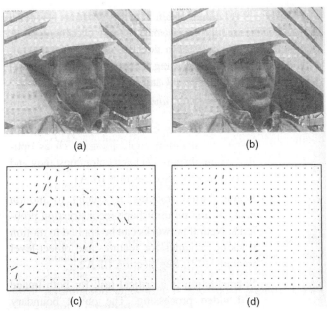

Figure 3.21 Two frames of the Foreman sequence (a, b). Motion vectors found by the original full-search block-matching algorithm (c). Motion vectors obtained by the neural method through neighborhood relaxation (d) [3.179]. ©1998 IEEE.

facial geometry information, such as the distance between eyes, nose and mouth size, can further be extracted. This geometry information can then be used for a variety of tasks, such as to recognize a face from a given face database.

There are many successful NN examples for face detection and recognition. Eigen-face subspace is used to determine the classes of the face patterns in [3.42]. Eigen-face and Fisher-face recognition algorithms were studied and compared in [3.68]. Cox et al. proposed mixture-distance VQ network for face recognition and reached a 95% success rate on a large database (685 persons) [3.180]. In Lin, Chan and King [3.181], NNs have been successfully applied to find such patterns with specific applications to detecting human faces and locating eyes in the faces.

3.15.4 Personal Authentication by Fusing Image and Speech

The fusion network has been applied to person recognition. The nonlinear fusion network proposed by Huang et al. [3.182] is based on the McCulloch-Pits NN where information from image and speech channels is combined as in Figure 3.22. The combined use of two channels yields a performance that is much improved over using either channel alone [3.182]. Noisy face images, each containing a 64x64, 8-bit grayscale image, serve as one source of information for classification. The images were decomposed into 13 channels by 4-level biorthogonal kernels. In

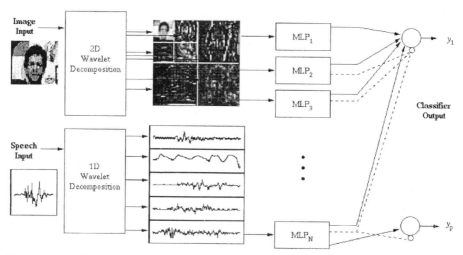

Figure 3.22 The nonlinear fusion network [3.182]. ©1997 IEEE.

speech channel, we have a noisy segment of speech, consisting of the spoken name of the same person. Speech segments digitized at 8 KHz were decomposed into eight channels using a length-eight wavelet kernel.

3.15.5 Subject-Based Retrieval for Image and Video Databases

A NN-based tagging algorithm was proposed for subject-based retrieval for image and video data bases [3.183]. Object classification for tagging is performed offline using Decision-Based NN (DBNN). A hierarchical multiresolution approach is used, which helps out the search space for looking for a feature in an image. The classification is performed in two phases. In the first phase, color is used, and in the second, texture features are applied to refine the classification, both using DBNN.

Compared to most of the other existing content-based retrieval systems, which only support similarity-based retrieval, the subject-based indexing system supports subject-based retrieval, allowing the system to retrieve by visual object. The difference between subject- and similarity-based retrieval lies in the necessity for visual object recognition. NNs provide a natural effective technology for intelligent information processing.

The tagging procedure includes four steps. In the first step, each image is cut into 25 equal-size blocks. Each block may contain single or multiple objects. In the second step, color information is employed for an initial classification, where each block is classified into one of the following families in the color space: black, gray, white, red, yellow, green, cyan, blue and magenta. In the next step, texture features are applied to refine the classification using DBNN if the result of color classification is a nonsingleton set of subject categories. Each block may be further classified into one of the following categories: sky, foliage, fleshtone, blacktop, white-object, ground, light, wood, unknown, and unsure. Last, an image tag generated from the lookup table using the object-recognition results is saved in the tag database. The experimental results on the Web-based information show that this model is very efficient for a large film or television, program-oriented digital video database.

3.15.6 Face-Based Video Indexing and Browsing

A video indexing and browsing scheme based on human faces is proposed by two studies [3.181, 3.184]. The scheme is implemented by applying face detection and recognition techniques [3.185]. In many video applications, browsing through a large amount of video material to find the relevant clips is an important task. The video database indexed by human faces provides users the facility to acquire video clips about the person of interest, efficiently. A probabilistic DBNN face-based video browsing system is shown in Figure 3.23. A scene-change algorithm divides the video sequences into several shots. A face detector examines all the representative frames to see if they contain human faces. If they do, the face detector passes the frame to a face recognizer to find out whose face it is. The scheme contains three steps. The first step of our face-based video browser is to segment the video sequence by applying a scene-change-detection algorithm. Scene-change detection gives an introduction of when a new shot starts and ends. Each segment created by scene change detection can be considered as a story unit of this sequence. After video sequence segmentation, a probabilistic DBNN face detector is invoked to find the segments (shots) that might possibly contain human faces. From every video shot, we take its representative frame (Rframe) and feed it into face detector. Those representative frames from which the detector gives high face detection confidence scores are annotated and serve as the indexes for browsing. This scheme can be very helpful to algorithms for constructing hierarchies of video shots for video-browsing purposes.

3.16 Multimedia Processors

In the area of multimedia processors, the pipelining used in Reduced Instruction Set Computer (RISC) chips has been a key advance. These chips are used as host CPUs for PCs and workstations with operating systems. On the other hand, DSPs have achieved their high performance by incorporating hardware function units, such as multiply accumulators, Arithmetic and Logic Units (ALUs) and counters, which are controlled by parallel operations with moderate clock frequencies. These processors are used for speech compression for mobile phones, voice-band modems, and facsimile machines, as well as for the acceleration of sound and still-picture image processing on PCs.

Multimedia processing is the driving force in the evolution of both microprocessors and DSPs. The introduction of digital audio and video was the starting point of multimedia because it enabled audio and video as well as text, figures and tables to be used in a digital form in a computer and to be handled in the same manner. However, digital audio and video require a tremendous amount of information bandwidth unless compression techniques are used. Also, the amount of audio and video data for a given application is highly dependent on the required quality and can vary across a wide range. For example, HDTV (1920x1080 pixels with 60 fields/s) is

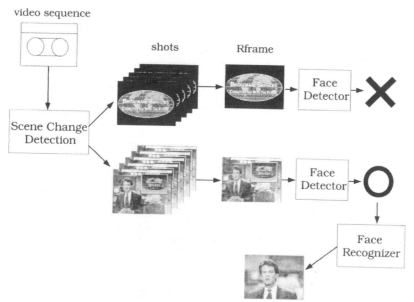

Figure 3.23 Probabilistic DBNN [3.181]. ©1996 IEEE.

expected to be compressed into around 20 Mb/s, but an H.263 video-phone terminal using sub-quarter-common intermediate format (128x96 pixels) with 7.5 frames/s is expected to be compressed into 1 to 20 Kb/s. Compression techniques call for a large amount of processing, but this also depends on the desired quality and information throughput. The required processing rate for compression ranges from 1,000 Mega Operations Per Second (MOPS) to more than 1 tera operations per second.

The wide variety of demands for processing multimedia has led to the software implementations of such compression techniques on microprocessors and DSPs to create an affordable multimedia environment. Microprocessors and programmable DSP chips offer powerful processing capabilities that enable real-time video and audio compression and decompression. Because a wide variety of applications is available to users, careful selection of these chips is essential to ensure flexibility in the independent application areas. This is because the chips' basic architectures differ significantly, and their respective advantages are highly related to their architectures.

3.16.1 Image-Processing Hardware and Software

We are now seeing PC systems that have the capability to acquire and process digital images as part of their display systems through the use of media coprocessors [3.186] and/or that are built into the instruction set of the microprocessor [3.187]. Another driving force in hardware and software is the fact that digital images and video are now part of many commercial and consumer applications.

In the early 1980s, Texas Instruments introduced the TMS320 digital signal processor for applications in speech and audio [3.188]. Others including AT&T/Lucent, Motorola, and Analog Devices have also developed DSP architectures. These DSPs have the capability of implementing real-time digital video processing. Other developments include Chromatic's Mpact [3.186] media processor, which is capable of accelerating several multimedia functions simultaneously [3.189].

The new microprocessors that are being developed have the capability of real-time image-processing operations. As clock speeds exceed 300 MHz, many real-time image-processing operations can now be performed on these processors without hardware acceleration. The most visible use is Intel's Multimedia Extension (MMX) processors [3.187, 3.190]. In MMX, 57 special instructions are added to the Intel processors that allow speedup in the execution of DSP operations.

The development of image-processing software has trailed behind the development of faster processors. Quite recently, few tools have been widely available for image-processing software designed for specific imaging applications. Software tools that have appeared in recent years include Mathematica [3.191], MATLAB's Image Processing Toolbox [3.192], Lab-View [3.193], and NIH Image [3.194].

The image and video-processing field will continue to benefit from the trends in computer technology. DSPs and microprocessors will be able to do real-time image processing. New software tools will be developed that will allow one to use the new microprocessors/PCs that will be arriving soon on everyone's desk.

3.16.2 Multimedia Processors' Classification

With the goal of improving multimedia-processing capabilities a wide variety of processors have been derived from microprocessors and DSPs. The multimedia processors can be classified in terms of their structure into the five categories as shown in Figure 3.24. These categories are

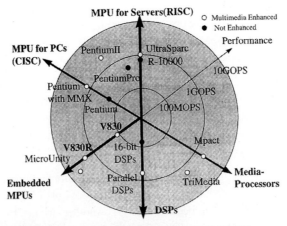

Figure 3.24 Multimedia Processor unit (MPU) classification [3.195]. ©1998 IEEE.

RISC microprocessors for workstations and servers, Complex Instruction Set Computer (CISC) microprocessors for PCs, embedded microprocessors, low power consumption DSPs and DSPs for PC acceleration, which are also called media processors [3.195]. This figure reflects certain trends in multimedia processor features and development. First, microprocessors for servers and PCs have been enhanced to handle multimedia processing while maintaining compatibility with their previous generations. To allow compatibility with the huge amount of existing software, there have been many restrictions on their enhancement. In other words, the heavy burden of hardware increase resulted. This is why these processors use very high clock frequencies and consume a lot of power.

New architectures with aggressive performance enhancements are being introduced for embedded microprocessors and DSPs, even though some of them have the functionality of a standalone CPU with high-level language support.

Multimedia processors target MPEG-2 decoding software. Thus, the performance and functionality for MPEG-2 video decoding are key issues in the design of the multimedia processor architectures. Decoding and playback of the compressed bitstream start with variable-length decoding, followed by inverse quantization to DCT coefficients from the compressed bitstream. Then an inverse DCT operation produces the prediction error signal. This signal retrieves the decoded frame with the addition of a motion-prediction signal. This signal is calculated by pixels' interpolation using one or two previously decoded frames (Figure 3.18). The decoded frame is transformed into a display format and transferred to the video Random Access Memory (RAM) and video output buffer. This transformation to display formats includes a Luminance Bandwidth Chrominance (YUV) to red-green-blue transform as well as dithering. This decompression process is carried out by a square image block called the macroblock (16x16 pixels with color components) or the block (8x8 pixels). Table 3.5 shows the major parameters for MPEG-2 MP@ML [3.196].

MPEG decoding for multimedia processors requires that the following five important functions be included in the system architectures:

- Bit-manipulation function, which parses and selects bit strings in serial bit streams. Variable-length encoding and decoding belong to this category.
- Arithmetic operations, which consist of multiplication, add/subtract, and other specific arithmetic operations, such as the sum of the absolute difference for motion estimation. Different word lengths are also desirable to improve hardware efficiency in handling many different media, such as video and audio data. Parallel processing units are also important for efficient Inverse DCT (IDCT) processing, which requires a lot of multiplications due to the nature of 2D IDCT algorithms.
- Memory access to a large memory space, which provides a video frame buffer that usually cannot reside in a processor on chip memory. The frequent access to the frame buffer for motion compensation requires a high-bandwidth memory interface.
- Stream data Input/Output (I/O) for media streams such as video and audio as well as compressed bitstreams. The I/O functionality is also needed for compressed bit streams

Table 3.5 MPEG-2 MP@ML parameters [3.196].

Parameter	MPEG-2 MP@ML
Horizontal size (pixels)	720
Vertical size (lines)	480
Frames/s	30
Display bits/s	15.55 M
Compressed bits/s	4-15 M
Number of macroblocks/s	40,500
Number of blocks/s	243,000

©1994 ISO/IEC.

- for storage media, such as hard disks; bit streams for storage media, such as hard disks and compact disks and for the communication networks.
- Real-time task switching, which supports hard real-time deadlines. This requires sample-by-sample and frame-by-frame time constraints. One example is switching between different types of simultaneous media processing to synchronize video and audio decoding.

In what follows, we discuss how the different processors are enhanced in terms of five key functions.

3.16.3 General Purpose Microprocessors

At present, high-end general-purpose microprocessors can issue two to four instructions per cycle by using superscalar control, which enables more than one floating-point instruction or several multimedia instructions to be issued at one time [3.197]. This control mechanism has two types of issuing mechanisms. One is the in-order-issue control, which issues instructions in the order that they are stored in the program memory. The other is the out-of-order issue control, where the issue order depends on the data priority rather than the storage order. This is effective for microprocessors that operate above 200 MHz and have long pipeline latency instructions where out-of-order control can maximize the high-speed pipelined ALU performance.

An example of an out-of-order superscalar microprocessor is shown in Figure 3.25. Implementation of out-of-order control requires several additional hardware functional units, such as a reorder buffer that controls the instruction issue and completion and a reservation station that reorders the actual instruction issues for the execution units and renamed register files, as well as

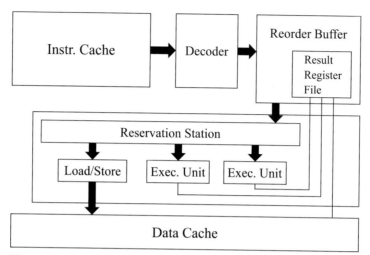

Figure 3.25 Superscalar microprocessor [3.195]. ©1998 IEEE.

control circuits for these units. These components take up a large part of the silicon area and contribute to the power dissipation.

In out-of-order control, the number of registers can actually be increased by register renaming. This improves the processing performance by reducing the number of load and store operations of intermediate calculation results to the memory as well as by reducing the processor stall cycles due to data dependencies. On the other hand, image processing has inherent parallelism in pixels and macroblocks. Although a large amount of hardware must be implemented for superscalar control, the issue of two to four parallel instructions does not fully take advantage of the parallelism in image processing. General-purpose microprocessor architecture related to media processing is illustrated in Figure 3.26. The arithmetic operations of microprocessors have a word-length problem. Microprocessor word lengths have been increased to 32 or 64 bits. The word lengths needed for multimedia processing are 8, 16 or 24 bits, which are much shorter than the word lengths of current state-of-the art microprocessors. This provides extravagant margins if we handle multimedia data with arithmetic instructions on microprocessors.

The memory access to a large memory space in microprocessors is a problem. Namely, in image processing, there is frequent access to large video frames that cannot reside in the first- or second-level cache. Therefore, we cannot expect the high-cache bit rate usually assumed in general-purpose applications. Moreover, a difference in the data locality affects the memory access performance. The cache mechanism is designed to use the 1D locality of consecutive addresses, but image processing has the 2D locality of access. In media processing for audio and video, processing of each audio sample or video frame should be completed within a predetermined sample or frame-interval time. This requires the predictability of the execution time, which is not easy to achieve in microprocessors. Data-dependent operations and memory access using cache mechanisms make it difficult to predict the processing cycles in the micro-

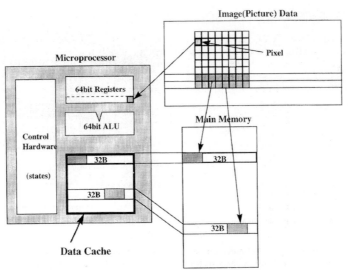

Figure 3.26 General-purpose microprocessor architecture [3.195]. ©1998 IEEE.

processors. Task switching with less overhead is also required. However, internal state introduced by out-of-order superscalar processors makes it difficult to achieve fast task switching.

The multimedia processing capability of recent microprocessors has been improved by multimedia extensions. These microprocessors enhance the arithmetic performance by dividing the long-word arithmetic unit (for example, 64 bits) to execute two to eight operations in parallel by using Single Instruction Multiple Data (SIMD)-type multimedia instructions as shown in Figure 3.27.

These instructions are implemented in either an integer data path or a floating point data path (coprocessor) in multiprocessors [3.197, 3.198, 3.199, 3.200]. The cache-miss penalties by off-chips main memory access are going to be a serious problem in microprocessor-based multimedia processing because the memory access latencies of microprocessors tend to be longer due to their higher clock frequencies. For most multimedia applications, the address for the memory access can be calculated beforehand.

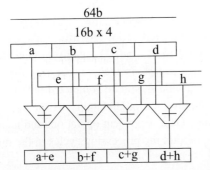

Figure 3.27 SIMD [3.195]. ©1998 IEEE.

3.16.4 Microprocessors for Embedded Applications

Target applications of this class of microprocessors include applications that originally used DSPs. In addition to these applications, multimedia applications, such as Internet terminals, set-top boxes, car navigations and personal digital assistants will be targets for these microprocessors.

A class of embedded RISC processors are inexpensive and consume little power. They do not employ complicated control mechanisms such as out-of-order controls. The arithmetic performance of embedded microprocessors can be enhanced by using a hardware multiply accumulator such as that shown in Figure 3.28. The requirements for real-time processing and the large

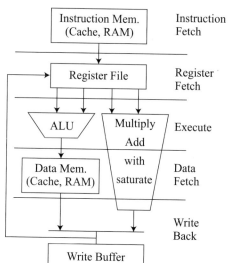

Figure 3.28 Data path for the arithmetic performance of embedded microprocessors [3.195]. ©1998 IEEE.

memory space are met by having both caches and buffers. Because microprocessors are normally used in low-cost systems that do not have a second-level cache, the cache-miss penalty is likely to be heavy. To present this, an internal RAM that is guaranteed not to cause a cache miss has been developed.

Another class of embedded microprocessors includes more sophisticated chips, such as the MicroUnit Mediaprocessor [3.201]. The processor has special memory interfaces as well as a stream I/O input/output interface with accompanying chips as shown in Figure 3.29. High-bandwidth memory access is realized by using special memory interfaces such as Synchronous Dynamic RAM (SDRAM) and ram bus dynamic random access memory [3.202]. The processor also has the capability for general-purpose microprocessors, such as virtual memory and memory management for standalone use. The arithmetic performance of the processor has been enhanced by using SIMD-type multimedia instructions with long word size. A large register file

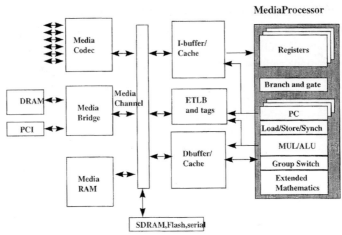

Figure 3.29 MicroUnit Media processor [3.195]. ©1998 IEEE.

(128x32 bits) also helps to improve the arithmetic performance of the processor. A memory-mapped I/O avoids coherence and latency problems in a Division Multiple Access (DMA)-based I/O system. Analog interfaces for audio and video are implemented on a separate chip [3.203]. With the advanced features described here and a higher clock frequency (1 GHz), this processor should be able to handle broadband media. This will result in higher power consumption.

3.17 Concluding Remarks

MMSP technologies play major roles in the multimedia network age. At this point, it is important to consider what the major issues are for MMSP technologies. Engineers have been concentrating their research on computer models that have functions in communicating with human beings.

The human communication mechanism can be explained by using a two-layer model. The first layer, which is the surface, enables communications based on language usage. This means that the logical information involved in human communications comes from this layer. However, this logical information is only a part of the whole information that constitutes speech. Other information, like information on emotions or senses, is also included.

The interaction layer is the second layer and is a deeper layer. It controls interactive behaviors in human communications. This layer also controls very basic behaviors like turning our face toward the direction of a sound or closing our eyes upon suddenly sensing strong light. These functions are considered to be important keys to humanlike behavior in daily life.

Algorithms for processing m-dimensional signals can be grouped into four categories:

- The separable algorithms that use 1D operators to process the rows and columns of a multidimensional array
- The nonseparable algorithms that borrow their derivation from their 1D counterpart

- The MD algorithms that are significantly different from their 1D counterparts
- The MD algorithms that have no 1D counterparts

Separable algorithms operate sequentially on the rows and columns of an MD signal. They have been widely used for image processing since the 1960s, because they require less computation than nonseparable algorithms. Examples include MD DFT, DCT, and FFT-based spatial estimation. In addition, separable FIR filters can be used in separable filter banks, wavelet representations for MD signals and decimations and interpolators for changing the sampling rate.

Algorithms that are MD and cannot be decomposed into a repetition of 1D procedures are uniquely MD in that they cannot be decomposed into a repetition of 1D procedures. They are a straightforward generalization of 1D techniques. They can usually be derived by repeating the corresponding 1D derivation in an MD setting. Sampling and downsampling are one example. As in the 1D case, band-limited MD signals can be sampled on periodic lattices with no loss of information. Most 1D filtering and FFT-based spectrum analysis algorithms generalize straightforwardly to any MD lattice. The window method for FIR filter design can be easily extended, and FFT algorithms can be decomposed into a vector-radix form, which is slightly more efficient than the separable row/column approach for evaluating MD DFTs.

The MD algorithms that have no 1D counterpart are algorithms that perform inversion and computer imaging. One of these is the operation of recovering an MD distribution from a finite set of its projections, equivalently inverting a discretized Radon transform. This is the mathematical basis of computer tomography. Another imaging method, developed first for geophysical applications, is Fourier migration. This is an efficient algorithm for image formation. Finally, signal recovery methods unlike the 1D case are possible. The MD signals with finite support can be recovered from the amplitudes of their Fourier transforms or from threshold crossing.

Adapting signal compression to networked applications may require some changes in the fundamental approach to this problem. The compression and transmission aspects have generally been treated as separate issues. The first problem with this approach is that the resulting compression algorithms usually do not address the needs of networked transmission. A successful compression algorithm removes all the redundancy, and, hence, the compressed data must be delivered error free. Another consideration in designing compression techniques for network use is to identify the impact of losing different portions of a compressed stream. It is preferable to have the important parts of the compressed stream concentrate into a short and identifiable segment.

Signal-processing techniques can be valuable for hiding a watermark (or identifying information) in the media. Watermarks can play a number of roles. First, a watermark can mark or identify the original owner of the content, such as the image creator. Second, it can identify the recipient of an authorized single-user copy. Third, a watermark can be used to identify when an image has been appreciably modified. An appropriate solution for the watermarking problem requires understanding of both the signal coding and networking or security issues.

Multimedia processors that realize multimedia processing through the use of software include those for bit manipulation, arithmetic operations, memory access, stream data I/O and real-time switching. The programmable processors for multimedia processing are classified into media-enhanced microprocessors (CISC or RISC), embedded microprocessors, DSPs and media processors.

Many critical research topics remain yet to be solved. From the commercial system perspective, there are many promising application-driven research problems. These include analysis of multimodal scene-change detection, facial expressions and gestures, fusion of gesture/emotion and speech/audio signals; automatic captioning for the hearing impaired or second-language television audiences; multimedia telephone and interactive multimedia services for audio, speech, image and video contents.

From a long-term research perspective, there is a need to establish a fundamental and coherent theoretical ground for intelligent multimedia technologies. A powerful preprocessing technique capable of yielding salient object-based video representation would provide a healthy footing for online, object-oriented visual indexing. This suggests that a synergistic balance and interaction between representation and indexing must be carefully investigated. Another fundamental research subject needing our immediate attention is modeling and evaluation of perceptual quality in multimodal human communication. For a content-based visual query, incorporating user feedback in the interactive search process will be also a challenging but rewarding topic.

CHAPTER 4

Distributed Multimedia Systems

Chapter Overview

In this chapter, we outline the issues concerning Distributed Multimedia Systems (DMS). We give an overview of DMS: main features, resource management, networking and multimedia operating system. Next, we identify the applications (interactive television, telecooperation and hypermedia) and survey the important enabling technologies. These topics will continue to be of great interest in the near future.

4.1 Introduction

A DMS is an integrated communication, computing and information system that enables the processing, management, delivery and presentation of synchronized multimedia information that the quality of service guarantees [4.1, 4.2]. It integrates and manages the information communication and computing subsystems to realize multimedia applications. Such a system enhances human communications by exploiting both visual and aural senses and provides the ultimate flexibility in work and entertainment by allowing you to collaborate with remote participants, view movies on demand and access online digital libraries from the desktop. DMS will create an electronic world. Technological advances in computers, high-speed networks, data compression and consumer electronics—coupled with the availability of multimedia resource mechanism, and manipulation functions; the development of the relevant standards and the convergence of the computer, telecommunications, and digital TV industries—are accelerating the realization of such systems. An example of DMS is a number of multimedia PCs and/or workstations interconnected with continuous media servers usings the Internet that allow users to

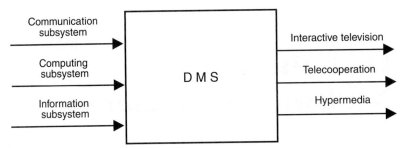

Figure 4.1 Block scheme of a summarized DMS.

retrieve, browse and manipulate video or audio. Besides constraints on bit error rates, packet-loss probabilities and delivery delays required in a point-to-point information delivery system, additional constraints are introduced in a DMS, such as the synchronization among multiple media streams from distributed sources to achieve a meaningful presentation. Figure 4.1 summarizes a DMS.

The inputs of the system consist of the factors that drive a DMS from concepts to reality, and the outputs consist of a wide range of distributed multimedia applications. The system inputs are a collection of the enabling technologies of the communication subsystem (for transmission), the computing subsystem (for processing) and information subsystem (for storage). The communication subsystem consists of the transmission medium and transport protocols. It connects the users with distributed multimedia resources and delivers multimedia materials with QoS guarantees, such as real-time delivery for video or audio data and error-free delivery for text data. The computing subsystem consists of a multimedia platform, operating system (OS), presentation and authoring tools and multimedia manipulation software. It allows users to manipulate the multimedia data. An authoring tool is specialized software that allows a producer or designer to design and assemble multimedia elements for a multimedia presentation. The information subsystem consists of the multimedia servers, information archives and multimedia database systems.

The outputs of the system can be broadly classified into three different types of distributed multimedia applications: Interactive Television (ITV), telecooperation and hypermedia. ITV allows subscribers to access video programs and interact with them. Services include home shopping, interactive video games (which can be classified as hypermedia applications), financial transactions, Video on Demand (VoD), news on demand and so forth. Telecooperation overcomes time and location restrictions and allows remote participants to join a group activity. Services include remote learning, telecommuting, teleservicing, teleoperation, multimedia emails, videophone, desktop conferencing, electronic meeting rooms, joint editing and group drawing. A hypermedia document is a multimedia document with links to other multimedia documents and it allows users to browse multimedia information in a consequential manner. Services include digital libraries, electronic encyclopedias, multimedia magazines, multimedia documents, information kiosks, computer-aided learning tools and the Web.

4.2 Main Features of a DMS

The main features of a DMS can be summarized as follows [4.3]:

- *Technology integration*—Integrates information, communication and computing systems to form a unified digital processing environment.
- *Multimedia integration*—Accommodates discrete data as well as continuous data in an integrated environment.
- *Real-time performance*—Requires the storage systems processing systems and transmission systems to have real-time performance. Hence, huge storage volume, high network transmission rate and high CPU processing rate are required.
- *Systemwide QoS support*—Supports diverse QoS requirements on an end-to-end basis along the data path from the sender, through the transport network and to the receiver.
- *Interactivity*—Requires duplex communication between the user and the system and allows each user to control the information.
- *Multimedia synchronization support*—Presents the playback continuity of media frames within a single continuous media stream, and temporal relationships among multiple related data objects.
- *Standardization support*—Allows interoperability despite heterogeneity in the information content, presentation format, user interfaces, network protocols and consumer electronics.

According to the different requirements imposed upon the information, communication and computing subsystems, distributed multimedia applications may be broadly classified into ITV, telecooperation and hypermedia. ITV requires a very high transmission rate and QoS guarantees. It demands point-to-point, switched connections as well as good customer services and excellent management for information sharing, billing and security. Telecooperation, such as videophone and desktop conferencing, allows lower picture quality and therefore has a lower bandwidth requirement. It requires powerful multimedia database systems rather than just continuous media servers. Sharing information among groups is the key to effective collaboration. Hypermedia systems may be treated as an application of database systems because they provide flexible access to multimedia information and a novel method to structure and manage data. Hypermedia applications require point-to-point and switched services.

4.3 Resource Management of DMS

A DMS integrates and manages the information, communication and computing subsystems. The resource management ensures end-to-end QoS guarantees. Such guarantees have the following characteristics:

- Systemwide resource management and admission control to ensure desired QoS level.

- Quantitative specification (packet loss probability and delay jitter) rather than qualitative description as in the Internet TCP/IP. This gives the flexibility to accommodate a wide range of applications with diverse QoS requirements [4.4].
- Dynamic management, which means that QoS is dynamically adjusted rather than statistically maintained throughout the lifetime of the connection.

In the context of DMSs, QoS is defined as the quantitative description of whether the services provided by the system satisfy the application needs and is expressed as a set of parameter-value pairs [4.5]. Packet-loss probability (10^{-3}) and packet delay (10^{-6}s) are examples of such parameter-value pairs. These QoS parameters are negotiated between the users and the service providers. QoS requirements can be mapped into desired resources in system components. Then, they can be managed by integrated resource managers to maintain the QoS commitment according to the negotiated service contracts. The system supports three levels of QoS commitment: deterministic (guarantees that the performance is the same as the negotiated service contract), statistical (guarantees the performance with some probability) and best effort (does not offer service guarantees). The goal of systemwide resource management is the coordination among system components to achieve end-to-end QoS guarantees. The major functions include the following:

- Negotiate, control and manage the service contracts of the users.
- Reserve, allocate, manage, adopt and release system resources according to the negotiated values.

After the service contract has been negotiated, it will be preserved throughout the lifetime of the connection. It is also possible, through proper notification and renegotiations, to tune the QoS level dynamically [4.6]. Admission control protects and maintains the performance of existing users in the system. The principle of admission control is that new requests can be accepted so long as the performance guarantees of existing connections are not violated [4.7].

4.4 Networking

The network transports multimedia traffic to satisfy QoS guarantees on an end-to-end basis. The transmission media may be wired (coaxial cables or fiber optics) or wireless (radio channels or satellite channels). The transport protocols may provide connection-oriented or connectionless services and best effort, statistical or deterministic performance guarantees. The network can be a LAN, a Metropolitan Area Network (MAN) or a WAN. A LAN (Ethernet, token ring or token bus) may cover an area within a building, a campus or an organization. A MAN (for example, a Fiber Distributed Data Interface ([FDDI]), may cover a metropolitan area, such as a small city. A WAN (TCP/IP, ATM) is a nationwide or an international network.

Multimedia traffic has diverse characteristics and various QoS requirements. Discrete media traffic (for example, file transfer or image retrieval) requires error-free services, but is tolerant of delay. Continuous media traffic (for example, video or audio playback) requires real-time, high-speed transmission and is connection oriented. It is sensitive to delay and delay jitter or the packet delay variations between consecutive packets. On the other hand, it is tolerant of occasional packet losses. In addition, the network has to support application-specific requirements. For example, video conferencing needs multicast service for group distribution, but ITV requires switched point-to-point services and asymmetric bandwidth for the downstream (video server to user) and the upstream (user to video server) directions.

The Internet has rapidly evolved into a significant network infrastructure for communications. It runs the IP with its best effort delivery service and enjoys a large user base. Another promising technique, ATM, is rapidly appearing in the market. It allows bandwidth on demand and guaranteed QoS and is expected to be the best candidate for high-quality media delivery. In what follows, we will examine the Internet effort on the support of multimedia transmission, the properties of ATM that are especially suited for distributed multimedia applications and the issues of integrating ATM networks with the emerging integrated services Internet.

4.4.1 IP Networking

IP networking is a booming market for telecom and datacom service providers and equipment vendors. Service providers are quickly defining and bringing to market differentiated IP services, including voice transport, VPNs, application/policy prioritization, multimedia and transport LANs.

Internet refers to the global network to which a large percentage of existing networks are now interconnected by routers or gateways, running the TCP/IP protocol suite. As the key to success for the Internet, IP provides datagram delivery (best effort and connectionless) service and leaves the reliability issues (delay, out-of-order delivery, packet loss and misdelivery) to the end systems. It uses a global addressing scheme for a vast range of services. In addition to data, each datagram carries routing information in the header to independently forward to the destination.

Integrated management architecture leverages common data and human resources across applications such as email, and voice transport across an IP-based infrastructure demands low latency and jitter. The Round-Trip Time (RTT), which is the time required by a network to travel from the source to the destination and back, including the time to process the message and generate a reply, should be less than 250 ms. The servicing of these packets is done based on a priority scheduling scheme.

The primitive service model of the Internet provides only point-to-point and best-effort services. Such services are suited for traditional applications, such as file transfer and remote login. It performs well for real-time media traffic (for example video and audio) only under lightly loaded networks. UDP and Real-Time Transport Protocol (RTP) are typically used to transfer a Voice over IP (VoIP) packet. UDP is a connectionless transport layer protocol in the TCP/IP protocol stack. It is a simple protocol that exchanges datagrams without acknowledgment or guarantee of delivery, requiring that error processing and retransmission function be handled by other protocols. RTP is a protocol designed to provide end-to-end network transport functions for applications transmitting real-time data, such as voice, video or simulation data across multicast or multicast network services. RTP provides services such as stamping and delivery and monitoring of real-time applications.

The support of IP multicast, resource reservation and higher level RTPs are the major development efforts on the Internet toward multimedia applications. We will examine these three issues.

IP Multicast

Multicast refers to the ability to send a single packet to multiple destinations. Applications that require multicast include teleconferencing, email multicast, remote learning, and group communication. Traditional protocols like TCP or UDP provide only unicast transmission. The provision of multicast with unicast services requires the delivery of replicated unicast packets to each recipient. To avoid sending multiple unicast copies, thereby increasing network use, multicast transport service is required. With IP multicast, each data source sends a packet to a multicast address and lets the network forward a copy of the packet to each group of hosts. Figure 4.2 shows unicast and multicast services for multipoint communication. As it can be noted, there is a difference between the two transport services.

Multicast routing protocols provide a mechanism that enables routers to propagate multicast datagrams in order to minimize the number of excess copies transmitted to any particular subnetwork. In addition, they must be flexible enough to allow participants to join and leave the system without affecting others. The most important thing is to determine which hosts on which subnetwork will participate in the group. Two important multicast routing protocols are link state and distance vector multicast. They are extensions from the corresponding unicast routing algorithms.

In a link state protocol, such as Open Shortest Path First (OSPF), each router monitors the state (or metric) of its adjacent links and sends this information to all other routers whenever the state changes. The metric may be the reliability, delay, cast, and so forth of the link. Each router has a picture of the entire network. If router A wants to multicast to a group of users, it needs to identify the subset of routers. This is accomplished by adding the identifications of the groups that have members on a particular link to the state of the link. Then it computes the shortest distance tree routed at A and spanning this subset of routers based on the link-state metrics. The multicast is then performed on this shortest distance tree.

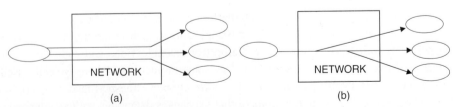

Figure 4.2 Unicast (a) and multicast (b) services for multipoint communication.

In a distance vector multicast protocol, each router maintains a routing table with entries (destination, cost, and next node) indicating for each destination the least cost to reach it and the next node on the shortest path to it. For each destination that it can reach, a node sends the destination cost to its directly connected neighbors, informing them of the cost of reaching the destination if this node is used as a relay. Each node will update its routing table if it finds that a neighbor can provide a shorter path to the destination. To support multicasting, each node will also send the identification of the groups that can be reached using this node to each of its directly connected neighbors. This information can be included in the routing table, and each router then knows across which links it should forward multicast packets for each group. To avoid sending duplicate multicast packets when multiple routers are connected to a router or link, we can designate the router as the parent router or link. The identification of this parent is a function of the source of the multicast. The router with the shortest path to the service is selected as the parent. Only the parent will forward a multicast packet.

Resource Reservation Protocol (RSVP)

This protocol attempts to provide guaranteed QoS for heterogeneous receivers across the Internet with multipoint-to-multipoint communications. Figure 4.3 summarizes the RSVP architecture. To reserve resources at a router, the RSVP block communicates with the admission control and policy control modules. Admission control determines if these are sufficient resources to satisfy the QoS of the new request while guaranteeing the QoS requirements for existing connections. Policy control determines if the user has the administrative permission for resource reservation. If both check out, the RSVP block sets parameters in the packet classifier, which determine the QoS for each packet, and the packet scheduler, which orders packet transmissions.

RSVP is receiver initiated. Two types of messages are used to reserve resources, that is, PATH and RESV. Each data source sends a PATH message, which contains a flow specification (for example, bandwidth) to the destination multicast address. When a router receives a PATH message, it records the relevant information (for example, IP multicast address, flow specification, source identification, and so forth). As a result, not only are the receivers informed of the

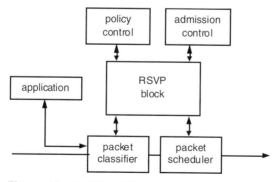

Figure 4.3 RSVP architecture.

flow specification of the data traffic, but the intermediate nodes also obtain the PATH state information. Based on information obtained from PATH messages and from higher layer protocols, each receiver can determine its QoS requirements and can initiate a RESV message to reserve specific resources along the reserved route of the PATH message. Multiple receivers may send RESV messages toward the same data source. RSVP scales well and accommodates heterogeneous receiver requirements. Resources will be received along the route from the source toward the receiver. For reliability and robustness, RSVP periodically updates the soft states at intermediate routers to allow for users joining and learning a group [4.8].

Three reservation styles have been defined: fixed filter, wildcard filter and shared explicit. Fixed filter allows explicit sender selection and distinct reservation and is suitable for applications, such as video conferencing, in which each videostream has its own RSVP flow. Wildcard sender selection includes any sender. Explicit sender selection refers to the reservation made only for senders explicitly listed in the RSVP message. Distinct reservation means that each sender has its own RSVP flow, and shared reservation uses the same RSVP flow for multiple senders. Wildcard filter allows wildcard sender selection and shared reservation. It is suitable for applications, such as audio conferencing, in which participants typically take turns speaking. Shared explicit allows explicit sender selection and shared reservation.

RTP

One of the IPs that can be used in conjunction with reservation models at the network layer is RTP [4.9]. RTP is an end-to-end protocol for the transport of real-time data. An important application type supported by RTP is multiparty conferencing because of its support for synchronization, framing, encryption, timing and service identification. RTP has its companion RTP Control Protocol (RTCP), which is used to interchange QoS and failure information between the QoS monitor applications in the end systems.

RTP does not define any kind of QoS itself and does not provide reordering or retransmission of lost packets. However, it provides a sequence number that enables the application using RTP to initiate such steps. RTP is directly used on top of UDP/IP. The RTP stack provides the information necessary to make educated guesses about the behavior of the datastream based on the RTP's knowledge of the data format. In addition to the base RTP specification, a number of companion documents exist that provide encapsulations for various continuous media formats, such as Motion Joint Photographic Experts Group (M-JPEG) or MPEG. Hence, RTP itself provides no real QoS support. It relies on other appropriate protocols and mechanisms.

4.4.2 Integrated Management Architecture for IP-Based Networks

Ensuring profitability from multimedia services requires a comprehensive service management architecture that enables service providers to plan carefully, provide quickly, operate efficiently and bill accurately for these services [4.10]. Figure 4.4 illustrates the typical structure of an IP network including core, edge and access subnetworks. The backbone technology is based on Multiprotocol Label Switching (MPLS), such as core and edge-label switch routers. MPLS networks integrate IP routing protocols, which allow efficient support of services such as IP

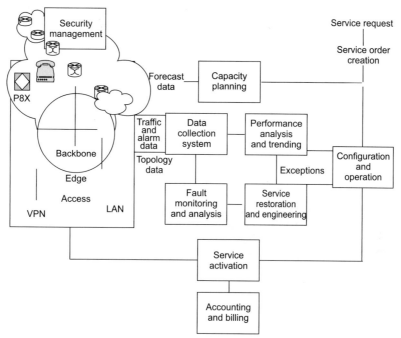

Figure 4.4 Typical structure of an IP network [4.10]. ©2000 IEEE.

multicast, IP class of service and IP VPNs. Customer sites are connected to the edge network, which is typically owned and controlled by a service provider. The extended functions include service restoration, traffic engineering, data collection, service activation and network planning.

In what follows, we will discuss an extended framework of the FCAPS functions for IP-based networks. The FCAPS acronym is used to refer to integrated management of various types of networks. It refers to the Open System Interconnection (OSI) for five functional areas: performance management, fault management, configuration management, security management and accounting and billing management [4.11].

Capacity planning provides a long-term view of network demands and requirements. It computes network element growth rates and generates a long-term capacity expansion plan. The network administrator who wants to do hypothetical studies typically carries out the capacity planning function to determine the required capacity as well as optimal equipment locations based on forecast or expected demands.

The data collection function collects and forwards data on a regular basis to the appropriate module. For example, alarms and related fault-statistics data are forwarded to the fault module to provide comprehensive diagnostic capabilities. Traffic-statistics data is forwarded to the performance module for data analysis. Traffic-statistics as well as network topology data may be forwarded to the planning module [4.12]. Estimated factors can then be used to estimate the future forecasting traffic loads. Finally, traffic statistics can also be used as the basis to observe traffic loads and to estimate the load for use in network engineering.

Performance Management

This is the process of converting IP traffic measurements into meaningful performance measures. It can be divided into real-time and long-term management. The real-time performance management process is a mechanism to guarantee that enough bandwidth is reserved for time-sensitive IP voice traffic, while other applications sharing the same link get their share without interfacing with the mission traffic. Another example is constant monitoring of high-priority customer services as well as customers who have been complaining about the performance of their services. Long-term performance management supports studies that monitor the ability of the existing IP networks to meet service objectives. The purpose is to identify situations where corrective planning is necessary. This is needed when objectives are not being satisfied and, where possible, to provide early warning of potential service degradation so that a corrective plan can be formulated before service is affected.

Traffic measurements are collected, validated by data-collection systems and then stored in batch mode in a database. Examples of IP performance traffic measurements include the following:

- Number of packets received per interface
- Number of packets transmitted per interface
- Number of packets dropped due to mild and severe congestion per interface (wild and severe congestion states are defined by the network administrator for each service)
- Number of packets dropped due to protocol errors
- Amount of time a network element is in a mild or a severe congestion state
- Number of times a network element enters a mild or severe congestion state

The performance management process then converts the validated measurements into meaningful network element loads (use, packet loss ratio, delay, jitter, and so forth). Next, it calculates statistics to characterize the load for traffic engineering purposes (for example, average peak values or average busy season). The process then computes network element performance measures (route-delay, end-to-end packet loss, average and peak packet loss, and so forth) based on the characteristic engineering loads. Finally, the performance management process compares the calculated performance results for the short and long terms with the service objectives to identify service or performance.

Fault Management

This process is similar to the real-time performance process except that it uses the collected alarms and fault statistics to detect and correct problems by pointing and correlating faults through the system. It simplifies the service provider's ability to monitor customer services by providing the status of the subscribed services.

One of the most challenging functions of the management process for IP networks is traffic engineering. It represents the action that the network should consider in order to relieve a potential servicing problem before the service is affected. This may include rerouting, load balancing and congestion control. Traffic engineering is also on for network dimensioning and planning as well as for capacity expansions. It is an optimization process that involves a set of algorithms that determine the required network resources (capacity) to meet a specific set of performance objectives.

The development of appropriate models for traffic engineering depends primarily on clear understanding of quality and grade-of-service requirements and the statistical characteristics of the traffic. Several traffic models and network dimensioning methods for packet networks have been proposed. Generally speaking, the models can be divided into two categories: those that exhibit long-range dependence (the fractional Brownian motion model on the on/off model with heavy-tailed distributions for the on/off duration) and Markovian models that exhibit only short-range dependence (on/off models with exponential on/off distributions, Markov-modulated Poisson process or Gaussian autoregressive models, which typically have exponentially decaying correlation functions). The on/off model has been proposed to model VoIP calls with alternating active periods and silent periods. The parameters of the on/off models can be estimated from actual traffic traces or by using typical default values. Finally, traffic engineering methods depend on the function of the network element. For example, traffic techniques for IP edge routers include packet classification, admission control and configuration management. Congestion management and congestion avoidance are typical considerations of backbone routers or switches.

Configuration Management

It deals with the physical and geographical interconnections of various IP network elements, such as routers, switches, multiplexers and lines. It includes the procedure for initializing, operating, setting and modifying the set of parameters that control the day-to-day operation of the networks. Configuration management also deals with service provisioning, user profile management and collection of operational data, which is the basis for recognizing changes in the state of the network. The main functions of configuration management are creation, deletion and modification of network elements and network resources. This includes the action of setting up an IP network or extending an already existing network, setting various parameters, defining threshold values, allocating names to managed IP objects and taking out existing network elements.

Security Management

This process includes authentication, authorization and other essential secure communication issues. Authentication establishes the identity of both the sender and the receiver of information. Integrity checking of confidential information is often done if the identity of the sending or receiving party is not properly established. Authorization establishes what a user is allowed to do after the user is identified. Authorization usually follows any authentication procedures. Issues related to authentication and authorization include the robustness of the methods used in verifying an entity's identity, the establishment of trusted domains to define authorization boundaries and the requirement of namespace uniqueness.

Accounting and Billing Management

This process deals with the generation and processing functions of end-user usage information [4.13]. This includes measuring the subscribers and possibly the network resources for auditing purposes and managing call detail information generated during the associated call processing. The records created in the application servers are of growing importance in IP networks. Such records are contents and services delivered by the network. Billing data collection and systems between the IP architecture and the billing platforms may aggregate usage-related data and usage detail records. The access usage detail can then be transferred to a billing system to render invoices to the subscribers that use IP services. Fraud detection and subscriber-related profile information, such as authorization to charge, are also a function of accounting and billing management.

4.4.3 ATM

The development of ATM is motivated by the merger of the computer networks and the telecommunications approaches toward multimedia communication. ATM technology was selected to support BISDN. Two groups work on the standardization of ATM: the telecommunications sector of the ITU-T, which is the international standardization organization for telecommunication, and the ATM Forum, which is a consortium of industrial and research organizations. Although originally designed for WANs, it is also suited for LANs and MANs. Therefore, it is a perfect vehicle for their seamless integration. ATM uses small, fixed-size cells (53 bytes, 5 for header and 48 for payload) and simple communication protocols to reduce per-cell processing and to speed up switching.

The ATM performs the following functions:

- Asynchronously multiplexes small packets called cells going from a number of information services to various destinations in a constantly flowing train if there is a seat for the specified destination with the required QoS
- Switches cells during transport if necessary
- Lets cells jumps off the train at the destination

The ATM uses a connection-oriented operation. It establishes a sequence of switches so that a connection is made from the source to the destination. Such a connection is called a Virtual Circuit Connection (VCC). The switches can be established to perform simplex, duplex, multicast and broadcast communications. A Virtual Connection (VC) is a connection between a switching node and the next node. Thus, a VCC consists of services of VCs. There are two kinds of VCs [4.14]:

- A Permanent VC (PVC) for a leased line
- A Switched VC (SVC) for a dynamically established connection

To simplify the management of VCs, a number of VCs with the same starting and ending nodes is grouped together as a virtual path (VP). To identify a VP as a VC, a number is used as the identifier and is labeled VP Identifier/VC Identifier (VPI/VCI).

Networking

Example 4.1 In Figure 4.5 VC1, VC2 and VC3 are grouped into VP1, VC4 and VC5 are grouped into VP2, and VP3 contains only one, namely, VC6.

A connection or call is assigned a VC. The end node uses the VC1 of the cell to direct it to the corresponding terminal. At the transit nodes, the VP1 of the cell provides enough information to direct this cell to the corresponding path in the network.

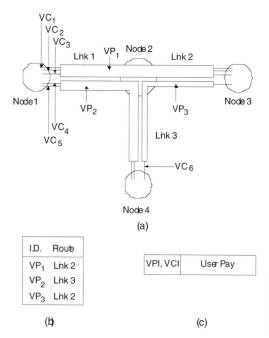

Figure 4.5 The ATM VP concept: a) VP, VCs, links and nodes. b) Routing table at node 2. c) VP, VC identifiers in a cell [4.2]. ©1997 IEEE.

The ATM network has a layered structure allowing multimedia traffic to be mixed in the network. It includes the upper layer, the ATM Adaption Layer (AAL), the ATM layer and the physical layer. Figure 4.6 shows the relationship between the ATM layer structure and the OSI protocol stacks.

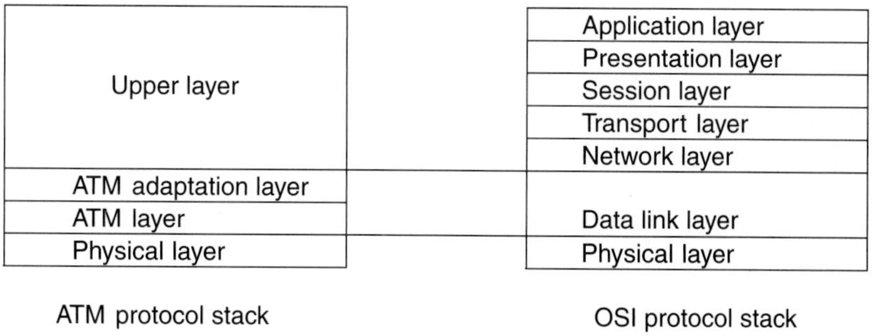

Upper layer
ATM adaptation layer
ATM layer
Physical layer

ATM protocol stack

Application layer
Presentation layer
Session layer
Transport layer
Network layer
Data link layer
Physical layer

OSI protocol stack

Figure 4.6 ATM protocol stack versus OSI protocol stacks [4.2]. ©1997 IEEE.

The upper layer includes the higher layer protocol, such as TCP, RTP and Xpress Transport Protocol [4.15].

The AAL layer adapts the upper layer protocols to the ATM format. It inserts or extracts user information as 48-byte payloads. The AAL layer consists of two sublayers: the Convergence Sublayer (CS) and the Segmentation and Reassemble (SAR) sublayer. The CS converges different types of user traffic and encapsulates/decapsulates data flow to and from the SAR sublayer. The SAR sublayer in the sender segments data into a series of 48-byte cells, and, in the receiver, it reassembles cell sequences back to the original data.

The ATM layer adds or strips the 5-byte header to or from the payload. For the sender, it takes 48-byte data from the AAL and adds the 5-byte header that contains routing information to ensure that the cell is sent on the right connection. For the receiver, it strips the 5-byte header and passes the 48-byte payload to the corresponding AAL.

The physical layer defines the electrical characteristics and network interfaces and places ATM cells into the transmission medium.

Figure 4.7 shows the rate of the ATM layered structure in the ATM network.

The ATM is suitable for multimedia communication because it provides a guaranteed QoS. ATM does not prevent cells from being lost, yet it guarantees that the cell order is always maintained in a connection. QoS is conceptually negotiated between three entities: the calling party (initiator of the connection), the network and the called party. The calling party requires a connection to the called party with a SETUP message, in which it provides its QoS requirements to the network and to the called party.

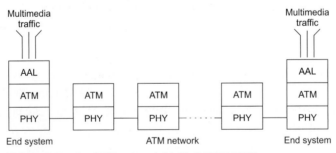

Figure 4.7 An ATM network [4.2]. ©1997 IEEE.

The QoS parameters supported for ATM connections differ slightly from those that are considered in networks that are based on variable length packets. The ATM parameters are the following:

- **Sustainable rate**—The minimum number of cells per second that must be supported by the network for the entire length of a connection
- **Peak rate**—The number of cells that must be expected at each node in the network in rapid succession (one burst)

- **Maximum burst length**—The length of an interval in which at most one burst must be expected by a network node
- **Cell loss ratio**—The maximum rate of lost or corrupted cells that an application can accept for a connection
- **Maximum end-to-end delay**—The restriction on the sum of all waiting times that each cell can stand in the queues between the sender and the receiver of the cell
- **Maximum cell delay variation**—The maximum difference in end-to-end transmission time that two cells of a connection can experience

To adapt to different characteristics of the traffic, ATM provides five types of adaptation. Type 1 is for circuit emulation at a Constant Bit Rate (CBR) service for isochronous data. Type 2 is for VBR connection-oriented service for isochronous data. Type 3 is for connection-oriented data service. Type 4 supports the services such as connectionless data communications. This provides the features equivalent to AAL Type 3 features. Type 5 is for LAN emulation and all other possible traffic. The ATM Forum also defines Available Bit Rate (ABR), which guarantees a minimum rate, but the delay may vary. The Unspecified Bit Rate (UBR) is defined, too. It is similar to ABR, but does not guarantee a minimum rate, and cells may be lost due to congestion.

4.4.4 Integration of IP and ATM

The strength of IP is its large installed base. The Internet is a packet-switched network and basically provides connectionless and best-effort transport services. The major problem is that the best-effort delivery of messages does not support the various QoS required in a true integrated service network. In addition, the existing Internet has very limited bandwidth, although there are plans to upgrade it to higher bandwidth. The development of IP multicast, resource reservation protocols and real-time transport protocols allows the Internet to provide limited forms of multimedia services. Even with the upgraded bandwidth, the Internet will be inadequate for many multimedia applications, for example, VoD. On the other hand, the strength of ATM lies in the possibility of setting up connections with a class of service that matches the requirements of the application involved. The planned bit rates and bit error rates also meet the requirements of real-time transmission of audio and videostreams. ATM works best in an environment where everyone uses ATM. The global interworking protocol will be the one that supports QoS guarantees and may include both ATM and IP technologies.

Although the Internet community is sponsoring ATM as the promising high-speed bearer technology of subnetworks, the vision of the Internet is still focusing on the protocol suite to be executed excessively among hosts at network edges and routers in subnetwork boundaries. The perspective is to consider ATM as a high-speed data pipe technology and to guarantee QoS by means of RSVP [4.16, 4.17, 4.18].

Cooperation between RSVP and ATM signaling protocols to guarantee QoS is an attractive perspective for building novel broadband networks to carry Internet traffic across public infrastructures and across enterprise Internets. The underlying subnetwork layer protocols are

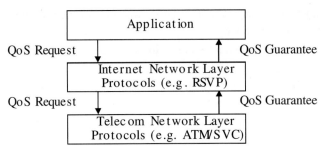

Figure 4.8 Application support for QoS [4.18]. ©1997 IEEE.

cooperating with the Internet network layers protocols to provide QoS. Application support for QoS is shown in Figure 4.8.

A number of proposals have been introduced to integrate IP and ATM protocols. Examples are IP switching [4.19] and tag switching [4.20] schemes. The former addresses a radical substitution of conventional routes and switching technologies to build high-speed IP backbones. The latter suggests a smoother overlay approach to be implemented across conventional routers and switching technology.

IP switching nodes dynamically shift between store-and-forward and cut-through switching [4.19]. IP switches base their operation on the traffic flow concept. A traffic flow is a sequence of IP packets sent from a particular service to a particular destination, sharing the same protocol type, the same type of service, and other characteristics as determined by examing information in the packet header. Prior to cut-through, an IP switch acts like a router with store-and-forward routing of IP datagrams.

Tag switching enables the forwarding of IP packets directly across a network of tag-switching-compliant routers [4.20]. A tag-switching network is made of two types of nodes: tag-edge routers and tag switches. The former are routers at the edge of the tag-switching network that apply tags and perform Internet network layer functions. The information can be carried in data units in two ways:

- As a part of the ATM layer header
- As a part of the Internet network layer header

Tag switching is independent of the routing protocols implied.

4.4.5 Real-Time Multimedia over ATM (RMOA)

The communication industry is currently engaged in extensive work on two flavors of voice over packet transport: Voice Over ATM (VoATM) and VoIP. VoATM is likely to be successful in delivering toll quality with special gateways preparing native Public Switched Telephone Network (PSTN) voice for ATM transport while using the QoS features of ATM technology. VoIP is becoming increasingly important for enterprises that can more easily provide resources to deliver quality to the voice traffic within their corporate networks.

Because ATM is becoming increasingly ubiquitous in the core of service provider networks, the ATM Forum developed an efficient and scalable means to transport native H.323 VoIP traffic over ATM that is not possible with the existing IP over ATM solutions. This effort was carried out within the RMOA working group and defined a new type of gateway called H.323-H.323 gateway [4.21].

ITU-T has created the H.323 standard describing the system components call model and signaling procedures to be used by entities engaged in multimedia communications across a network such as the Internet.

Example 4.2 The H.323 standard does not impose any QoS requirements on the network used to carry the H.323 media streams. We will focus on the scenario depicted in Figure 4.9. Here, an Internet backbone has an ATM core where the devices (routers or switches) at the edges of the core (devices X and Y) are capable of receiving H.323 RTP/UDP/IP packets and sending them across VCs in the ATM core.

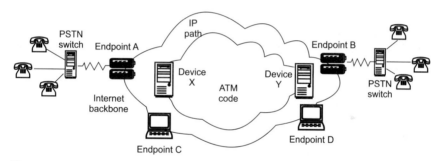

Figure 4.9 The ATM core in an Internet backbone [4.21]. ©2000 IEEE.

The H.323 endpoints in Figure 4.9 are IP hosts, and the H.323 media traffic traverses an access IP path to reach devices X and Y at the edge of ATM core. Two types of H.323 endpoints are illustrated: gateways (endpoints A and B) and terminals (endpoints C and D). H.323 terminals are capable of initiating and receiving H.323 calls, and H.323 gateways do the same on behalf of the other non-H.323 terminals, such as PSTN phones. This particular type of H.323 gateway provides a form of VoIP service.

Example 4.3 Figure 4.10 elaborates on the scenario in Figure 4.9, focusing on an H.323 call between endpoints A and B. It indicates the access IP paths used by the traffic originating at the end points and introduces another standard H.323 entity, the gatekeeper. Here, we have three gatekeepers used in the signaling architecture to break the end-to-end call into three legs. The different legs result from control and media termination forced devices Gateway (GW) X and GW Y at the edge of the ATM core.

An H.323 gatekeeper provides the services of Registration, Admission and Status (RAS) through the RAS channel, which also includes address resolution. Endpoint A will use the RAS

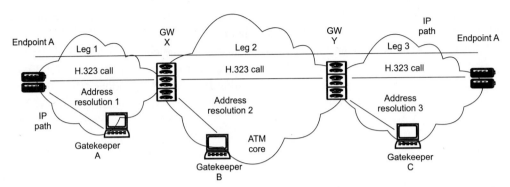

Figure 4.10 The ATM core in an Internet backbone scenario focusing on an H.323 call between endpoints A and B [4.21]. ©2000 IEEE.

channel to gatekeeper A, with which it is registered, to place calls to endpoint B. Usually, endpoint A will know an alias address for endpoint B. Gatekeeper A will translate this address to an IP address and a transport protocol port number. Three gatekeepers are illustrating a scenario with three zones where endpoints A and B belong to different zones. As such, address resolution 1 provided by gatekeeper A tells endpoint A to use GW X as a gateway for calls to endpoint B.

H.323 includes two other ITU-T standards: H.225.0 and H.245. These define control messages to be exchanged among H.323 endpoints.

The H.225.0 signaling establishes the H.323 call. This implies the opening of a call-control channel, subsequently used to open media channels through H.245 procedures. Similar to H.225.0, the H.245 channel is defined by a set of transport addresses used for the exchange of H.245 control messages. The initial H.225.0 messages carry these transport addresses that are replaced with local addresses for the gateways as the H.225.0 messages are relayed across adjacent legs of a call. These addresses' replacements force the H.245 messages to be addressed to the H.323-H.323 gateways, breaking the end-to-end H.245 channel into three legs as well.

4.5 Multimedia Operating Systems

OSs manage computer resources, for example, CPU, memory, I/O devices, and so forth. They also hide the physical characteristics of the underlying hardware and provide an efficient and convenient environment for end-users. A multimedia OS extends the functionalities of OSs to accommodate multimedia data manipulation to provide an environment for real-time support. The major concerns include real-time support while simultaneously running traditional applications efficiently. The major concerns include real-time processing and QoS-based resource-management. A multimedia OS may be developed as an extension of a traditional OS or constructed using the microkernel architecture [4.22]. It should provide CPU management, memory management, I/O management and file system management.

CPU Management

Real-time processing can be achieved through efficient real-time scheduling. In the context of continuous media, a deadline can be the acceptable playback time of each frame. Therefore, it is a soft deadline and appears periodically. The challenges of multimedia scheduling are due to two conflicting goals: non-real-time processes and real-time processes. Non-real-time processes should not suffer from the execution of real-time processes, because multimedia applications equally depend on discrete and continuous media data. Real-time processes should be allowed to pre-empt non-real-time processes or other real-time processes with lower priorities.

The most important real-time scheduling approaches include Earliest Deadline First (EDF) and rate monitoring scheduling [4.23]. With EDF, each task is preemptive and is assigned a priority according to the deadline. The highest priority is assigned to the job with the earliest deadline, and tasks are executed in a priority order. When a new task arrives, the scheduler recomputes the priorities of all pending tasks and then reorganizes such that the order of the task being executed is preempted and the new task gets served immediately. The interrupted process is resumed later from the interruption point. Otherwise, the new task will be put in an appropriate position.

With rate-monotonic scheduling, each task is pre-empted and is assigned a priority according to the request rate. The highest priority is assigned to the job with the highest rate. In contrast to EDF, such assignments are performed only at the connection establishment time and are maintained through the lifetime of the connection. For preemptive periodic tasks, rate-monotonic scheduling is optimal in the sense that no other-static algorithm can schedule a task that the rate-monotonic algorithm cannot also schedule [4.24].

Comparing these two algorithms, EDF is more dynamic. It has to be executed frequently and thus incurs higher scheduling overhead. The advantage is that it can achieve processor utilization up to 100%. On the other hand, a rate-monotonic algorithm is static because the priority assignment is only calculated once. Because the priorities are assigned according to the request rate, more context switches occur in rate-monotonic scheduling than EDF. The worst-case upper bound of the process use is about 69% even though, on the average, the use is suitable for continuous media applications because it has no scheduling overhead and is optimal for periodic jobs.

Memory Management

The memory manager allocates memory to processes. Continuous media data is typically very large in size and requires stringent timing requirements. One solution is to avoid swapping and to lock continuous media data in memory during the process [4.24]. This approach, however, may affect resource use. Other important practical implementation techniques include using scatter buffers and passing pointers. With scatter buffers or scatter loading, the address space of a process is loaded into possibly discontinuous regions of memory. This tends to be more space efficient than loading into a single continuous region, but may result in fragmentation. With passing pointers, objects are passed by reference rather than having to pass the objects themselves. This may result again in more efficient usage of memory space.

IO Management

The main function of the I/O subsystem is to transfer multimedia information between the main memory and the network adapter or multimedia peripherals (camera, loudspeaker, CD-ROM drive, microphone, disk, keyboard, monitor, and so forth). The important issues include device management, interrupt latency, and real-time transmission. Device management integrates all hardware components and provides a uniform interface for the control and management of these devices. Multimedia applications are I/O intensive. The continuous media frames will not frequently interrupt the kernel and lower the system throughput. There are three strategies to alleviate this problem: changing the internal structure of the kernel to make it highly preemptive, including a set of safe pre-emption points to the existing kernel or converting the current kernel to a user program and running on top of a microkernel [4.25]. Real-time I/O is necessary for most multimedia applications to ensure the continuity of each stream and the synchronization of multiple related streams. With advances in networking and communication technologies, it is possible to achieve network bandwidth well above a gigabit per second. The network I/O becomes the bottleneck that limits overall system performance. Therefore, the focus is to improve I/O system throughput.

File System Management

File management is responsible for making efficient use of the storage capacity and for providing file access and control of stored data to the users. Multimedia file systems demand additional real-time requirements to handle continuous media streams.

4.6 Distributed Multimedia Servers

Servers are an integral part of the multimedia environment such as digital libraries, in-house training systems, VOD or near-VOD services and so forth. In a typical environment, a multimedia server is connected through an interconnection to clients who request information on demand. The servers not only store and provide information to clients, but they perform various management operations as well (billing, accounting, encryption and so forth). One of the most important elements of a multimedia service environment is a videostream. Videostream elements require high I/O bandwidth at the server for their delivery and large amounts of memory for their storage [4.26].

A number of issues make the design of the video server difficult. First, a video server needs to provide video services simultaneously to multiple clients, to guarantee the QoS to multiple clients and to guarantee the QoS to each client. Second, a video server needs to manage system resources, including CPU, disk and memory, and it needs to schedule network activity to get the maximum use from the resources, while not overloading the system. Third, a video server needs to be able to support a variety of operations, such as playback, fast forward, slow forward, pause, resume, indexing and scrolling. Finally, a user watching a video may change from one service to another service (for example, from playback to fast-forward or from playback to slow forward). A video server should support these dynamic service changes while efficiently using system resources [4.27].

In general, servers can be classified into two large categories: centralized and distributed. In centralized servers, a high-end system stores servers and manages the video streams, and in a distributed server, a collection of workstations or PCs may constitute the server. A lot of effort has been concentrated on distributed server environments because it is not clear which type of server provides the most cost-effective solution [4.28, 4.29, 4.30].

In order to describe precisely the activities between a client and a server, we can divide them into three levels: session, transaction and services. A session is a connection between a client and a server. Its lifetime is defined as the duration from the login to the logout of a video client. Within a session, a user can execute two types of transactions, query or video. A query transaction asks for metadata of the video server. A video transaction is a sequence of video-related activities and is encapsulated by the open and close of a video file. The major difficulties between a query transaction and a video transaction are that the latter involves time-critical, video-related activities, whereas the former does not. After a video transaction is established, a user can initiate a video service, terminate a video service, or switch from one video service to another where a video service is a playback, fast forward, or slow forward operation on the chosen video [4.31].

The most important parameters in a multimedia server are its I/O bandwidth requirements and its storage requirements. The I/O bandwidth designates how many clients can be simultaneously served. On the other hand, the available amount of storage determines the number of videostreams that can be stored in the server. In a centralized server, one can easily observe that no client request for a videostream can be blocked as long as I/O bandwidth is available. This is not necessarily true for a distributed server [4.32].

In distributed multimedia servers where a client requests different videostreams that may have different probabilities, placement of videostreams is an important parameter because it may result in an unbalanced request to the system's stations, and thus to high blocking probabilities of requests.

Example 4.4 Assume that a client request arrives at the server for a stream that is stored on a station with I/O bandwidth that is allocated to other served streams. Although aggregate I/O bandwidth may be available in the system, the request has to be blocked. One solution to resolve such problems is to provide copies of videostreams in multiple stations so that alternative stations can serve a request. This demonstrates that placement and replication of videostreams is an important issue in distributed servers.

4.6.1 Multimedia Packing

Multimedia packing is a method that achieves load and storage balancing in the distributed multimedia server with two basic operations:

- Placement of videostreams in server stations and replication of a small subset of videostreams
- Weighted scheduling of client request for the replicated videostreams

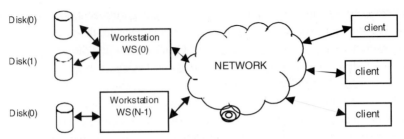

Figure 4.11 Distributed server configuration environment.

This results in the probability of a video request being the same for all stations in the distributed server. When client requests are uniformly routed to disks (or servers, assuming that a server corresponds to one disk), the blocking probability of a client's request is minimized [4.30]. In addition to traffic load balancing, multimedia packing achieves storage balancing, because it results in approximately an equal number of videostreams stored in all stations. The number of videostreams stored on any two workstations differs by a number less than or equal to two.

Consider a distributed server configuration in an environment as shown in Figure 4.11. A distributed video server is employed to deliver video services to clients. The server is composed of N stations WS(0),...,WS(N-1) storing a total of M different videostreams $VS_0,..., VS_{M-1}$ and is connected to a total of C clients through a network. Here, we have $M \gg N$. Clients generate requests for videostreams to the server. Videostreams are requested with different probabilities: videostream VS_i is required with probability P_i. Thus, we have $\sum_{i=0}^{M-1} P_i = 1$. Given a placement of videostreams to the workstations, the probability that a client's request is routed to a certain workstation WS(i) is equal to the cumulative probability SP_i of the videostreams stored on WS(i). Multimedia packing achieves a cumulative probability $SP_i = 1/N$ for $0 \leq i \leq (N-1)$. Multimedia packing also achieves storage balancing by resulting in a configuration with $|NS_i - NS_j| \leq 2$ for all i,j. Here NS_i is the number of streams stored on workstation WS(i).

4.7 Distributed Multimedia Applications

Multimedia integration and real-time networking create a wide range of opportunities for multimedia applications. Distributed multimedia applications have several requirements with respect to the service achieved for them by the communication system. These requirements depend on the type of application and on its usage scenario. Furthermore, the requirements of applications regarding the communication services can be divided into traffic and functional requirements. The traffic requirements include transmission bandwidth, delay and reliability. They depend on the quality of the datastreams. The functional requirements are multicast transmission and the ability to define coordinated sets of unidirectional streams. On the other hand, the reliability requirements are sometimes lower than for traditional communication applications, for example, if a fault-tolerant data-encoding scheme is used. Furthermore, retransmissions, which are traditionally used for the provisioning of reliability, increase the end-to-end delay and are often worse than lost data for multimedia applications.

The traffic requirements can be satisfied by the use of resource management mechanisms. They establish a relationship between transmitted data and resources and ensure that the audio-visual data are transmitted in a timely manner. For this, during the transmission of data, the information about the resource needs must be available at all nodes participating in the distributed applications, that is, end systems and routers. Hence, a resource must be reserved, and a state must be created in these nodes, which means that a connection is established. This connection should then be used for the transmission of data.

For various multimedia applications, multiple receivers are interested in receiving the same data. For instance, in a talk distributed using the network, all listeners must receive the same data. Sending each person a single copy wastes resources because, for parts of the path from the sender to the receivers, the same nodes are traversed. Thus, multicast should be used, which provides for the transmission of a single copy of data to multiple receivers, as shown in Figure 4.2. In addition to reduced network load, multicast also lowers the processing load of the sender. Multicast must not be limited to a single sender. In conferencing scenarios, it is usual to have several senders who normally do not use the resources at the same time.

The delivery of audio-visual data to large receiver groups, such as the distribution of Internet Engineering Task Force (IETF) meetings across the multicast backbone, must also take into account that the resource capabilities and the participations can vary widely from high-speed network links and fast workstations to low-end personal computers connected using relatively narrow band links. Therefore, support for heterogeneous systems must be provided (heterogeneous is with respect to networks as well as to end-system capabilities).

According to the different requirements imposed upon the information, communication and computing subsystems and distributed multimedia applications may be classified into three types: ITV, telecooperation and hypermedia.

4.7.1 ITV

ITV requires a very high transmission rate and stringent QoS guarantees. It is therefore difficult to provide such broadband services across a low-speed network, such as the current Internet, due to its low bandwidth and best-effort-only services. ITV typically demands point-to-point switched connections, good customer services and excellent management for information sharing, billing and security. The bandwidth requirement is asymmetric in that the bandwidth of a downstream channel that carries video programs from the server to the user is much higher than that of the upstream channel from the user to the server [4.33].

The four main components of ITV systems are a home terminal commonly known as the Set-Top Box (STB) or Customer Premises Equipment (CPE), an access network, a network-based server and a powerful user-friendly interface. The STB (or CPE) typically takes the form of a box sitting on the top of the TV set. This STB connects to both the television and an external communication network using a subscriber drop or loop. When interactive services are offered, the complexity and cost of the STB increases. The new STB must provide both the analog functions and the digital functions of audio and video decompression, demodulation to recover the digital feeds, decryption, an upstream modem for communicating consumer control requests back to the program source and a user-friendly interface [4.33]. ITV servers are a collection of computing, storage and communications equipment that implements interactive video services. A service may require that more than one server be implemented, or a server may implement more than one service. All subsystems of the server communicate with one another using local high-bandwidth interconnect and a switch. This architecture can be used to scale capabilities of the server incrementally, and it provides isolation between the various subsystems. Two key components of the server technology are logical organization of the multimedia samples in a file system or database and techniques by which media components can be continuously recorded or played back from the server. The multimedia database server must ensure that the recording and presentation follow a real-time data rate. User interface designs for ITV are more involved than those for standard TV, owing to the richness of the types of possible interactions with the consumers. They will vary among different applications. An important ITV service is VoD.

VoD

VoD provides electronic video-rental services across the broad band network [4.34]. Customers are allowed to select programs from remote massive video archives, view them at the time they want without leaving the comfort of their homes and interact with the programs. A VoD system that satisfies requirements, like any video, or any VCR or like user interaction, is called a true VoD system and is said to provide true VoD services. Otherwise, it is called a near VoD system. One way to allow true VoD services is to have a dedicated videostream for each customer. This is not only expensive, but is wasteful of the system resources because, if multiple users are viewing the same video, the system has to deliver multiple identical copies at the same time. To reduce this cost, batching may be used. This allows multiple users accessing the same video to be served by the same videostream. Although batching complicates the provision of user interactions, it increases the system capability in peaks of the number of customers that the system can support [4.35, 4.36, 4.37].

Example 4.5 In a VoD system, multiple identical copies (streams) of the same video program are broadcast every five minutes. A user is served by one of the streams, and user interactions are simulated by jumping to a different stream. Not all user interactions can be simulated in this fashion. Even for those that can be simulated, the effect is not exactly what the user requires. For example, one cannot issue fast forward because there is no stream in the fast-forward mode. Also, one may pause for 5, 10, 15 and so forth minutes, but not for seven minutes from the original stream.

A protocol called Split-and-Merge (SAM) has been proposed to provide true VoD services while fully exploiting the benefit of batching, thereby reducing the per-user video delivery cost.

To implement the complete set of interactive services, a VoD system contains many components, including the video server, transport network, Subscriber Terminal Unit (STU) and ser-

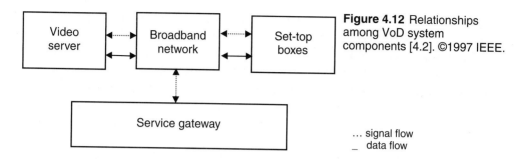

Figure 4.12 Relationships among VoD system components [4.2]. ©1997 IEEE.

... signal flow
— data flow

vice gateway. The VoD system architecture and the relationship among its components is shown in Figure 4.12.

A video server consists of massive storage and media controllers (Figure 4.13). The video server stores a large number of digitized videos and serves a considerable number of simultaneous video requests to the same or to different videos on demand. The storage media consist of magnetic disks, optical disks, and magnetic tape and are usually organized hierarchically for cost-effectiveness. Under this configuration, popular videos are stored in the disks. Less popular ones are stored in tape devices with terabyte capacity and are retrieved as necessary to the disk drive for processing. The video server may be located at the local or regional switch of the network provider or at remote information archives. The basic functions supported by video servers include request handling, random access, and user interactions, in addition to admission control and QoS guarantees. The networked video jukebox and the video library system are two examples of video server prototypes [4.38, 4.39].

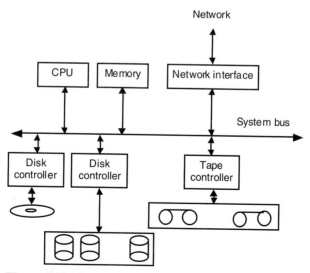

Figure 4.13 Video server architecture [4.2]. ©1997 IEEE.

The transport network delivers video programs from the video server to the customers. The network must have a very high data rate and must satisfy the real-time delivery constraints of video traffic. It consists of two major components: the backbone network with high-speed switches and the local access network. The backbone network links the remote video server at geographically dispersed locations and the regional, national or international information archives. The trend is toward a Synchronous Optical Network (SONET) backbone with ATM switching because of the low error rate, high data transfer rate, bandwidth on demand and seamless services.

The STU (or set-top box) along with the television monitor and the infrared controller (that is, remote control) serves as the bridge between the subscribers and the system. The major functions of STU include receiving the incoming videostreams; demodulating; demultiplexing and decoding the signals; performing the necessary signal conversion, such as D/A transformation for playback on the TV monitor; and sending outgoing control messages. STU must accommodate the heterogeneity of technologies and formats from various controls to services [4.40]. The usefulness of an STU is in its adaptation to the diversity of access networks, service providers, applications and user interfaces.

A service gateway component may be integrated with an access node or may be a separate element in the network. The main functions performed by the service gateway include the following:

- Directory services to provide menu browsing and program scheduling
- Mapping from service identity to corresponding location and program provider
- Controlling, coordinating and signaling for multimedia session establishment, maintenance and disconnection
- System management, including operation management, fault management, configuration, resource management and performance management
- Subscriber profile maintenance and billing
- Secure communication to prevent unauthorized access, including authentication, encryption and scrambling

Video server placement is an important design issue in VoD systems. The alternatives include centralized video servers, hierarchical video servers and fully replicated distributed video servers. A centralized video server system is relatively simple to manage. All requests are sent to and served at one site. Hierarchical server placement exploits the user access pattern and the nonuniform popularity of videos. The distributed server system distributes the video copies to many switches located closer to the users, thus alleviating the congestion in the network and the bottleneck due to the central server, but at the expense of higher cost [4.41, 4.42]. However, managing distributed servers is more complex. One has to decide which video and how many copies to maintain at each distributed server. In addition, due to varying rates of requirements, the video offerings at each distributed server need to be changed periodically. Which alternative is highly preferable depends on the tradeoff between storage and communication costs, the application needs, the underlying infrastructure and other factors. In Li et al. [4.34], a performance model that may be used to evaluate the requirements of network bandwidth and server storage is proposed. Hence, we obtain the tradeoff between communication and storage costs for various placement alternatives.

4.7.2 Telecooperation

Telecooperation, also known as Computer-Supported Cooperative Work (CSCW) refers to a system that provides an electronic shared workspace to geographically dispersed users with communication, collaboration and coordination supports [4.43, 4.44, 4.45]. Group communication provides an electronic channel for the users to exchange messages either synchronously or asynchronously. It allows individuals to cooperate regardless of time and location constraints. Sharing information among groups is the key to effective collaboration.

Telecooperation requires multicast, multipoint and multiservice network support for group distribution. In contrast to the strict requirement on the video quality of ITV, telecooperation, such as videophone and desktop conferencing, allows lower picture quality and therefore has a lower bandwidth requirement. It is possible to provide such services with the development of real-time transport protocols across the Internet. Telecooperation requires powerful multimedia database systems rather than continuous media servers with the support of visual query and content-based indexing and retrieval.

CSCW may be classified into four different interactions, which are shown in Figure 4.14:

- Centralized synchronous
- Distributed synchronous
- Centralized asynchronous
- Distributed asynchronous

Synchronous and asynchronous refer to the time dimension, while centralized and distributed refer to the space dimension. Synchronous exchanges demand real-time communication, but distributed interactions require broadcast or multicast support for group distribution.

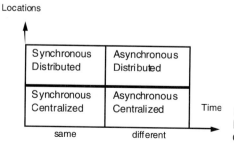

Figure 4.14 Classification of interactions by time and space for CSCW [4.2]. ©1997 IEEE.

The centralized synchronous mode requires face-to-face interactions. Examples are applications in the meeting room. The distributed synchronous mode provides real-time interaction in groups dispersed at different locations. Examples include network chat, real-time joint editing, multimedia conferencing and videophone. This type of application poses the greatest challenge in the design of group collaboration systems. The centralized asynchronous mode refers to those activities held at the same place, but at different times. News groups and electronic bulletin boards are such examples. The distributed asynchronous mode allows the exchange of messages within the group asynchronously. Electronic mail and voice mail are examples.

Telecooperation Infrastructure

A telecooperation infrastructure provides a robust framework to facilitate group work and to share information. It consists of a network model, a system model and a communication protocol model [4.46]. The telecooperation network model defines the functional elements and their relationships in the telecooperation system. It includes a multicast, multipoint and multiservice network and a collection of group collaboration agents. The group multicast, multipoint and multiservice network connects and distributes multimedia materials to remote participants. Figure 4.15 contains the block scheme of the network model. The collaboration agent includes the hardware and software that provide the necessary facilities and functionalities for cooperation and management of group work.

The telecooperation system model consists of five major modules as shown in Figure 4.16: cooperation control, application sharing, conferencing, interface and database. The cooperation control module administers a dynamic set of participants during cooperation sessions. The major functions include access control, group dynamic control and floor control. Access control validates membership in a group activity, registers a session, initiates or closes a session, and modifies the membership from one session to another. Group dynamic control allows participants to add in or drop out of a session dynamically. Floor control allows only one participant to own the floor at a time, that is, only one user can interact with programs at a time during the session. Floor control policies have two variations: centralized control and distributed control [4.47]. The

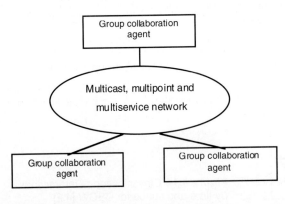

Figure 4.15 The telecooperation network model [4.2]. ©1997 IEEE.

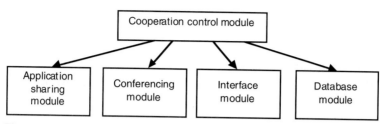

Figure 4.16 The telecooperation system model [4.2]. ©1997 IEEE.

application-sharing module handles the shared activities of the participants. It provides outputs to all participants with appropriate views while coordinating and controlling the inputs to satisfy the cooperation.

The application share module has two possible implementations: centralized or replicated. In the centralized approach (that is, client-server architecture), shared applications are run in a dedicated server. This control server processes all user requests and distributes the results to all local machines for display. In the replicated approach, every local machine keeps and runs a copy of the shared applications, and the input and processing events of the floor holder are distributed to all other sites [4.48]. This approach is more tolerant of machine failures and more scalable for large and heterogeneous user environments. It also facilitates the support of different views at the local displays. The major characteristics with the replicated approach are the maintenance of data consistency and cooperation control among the participants. The conferencing module supports asynchronous messaging and real-time audio-visual communications for distributed, multiparty collaboration. For video and audio transmissions, full-duplex communication channels and real-time transport services are required among the participants. The interface module is concerned with the display of the shared data. It supports private and shared windows for the participants to share only part of the information and allows the shared information to be presented to different users in different ways. The database module stores shared data and knowledge. Concurrency control is required to resolve the conflicts in the shared data and shared operations between participants. In contrast to database concurrency control (for example, locking shared data for exclusive access), group concurrency control needs to consider real-time and cooperation issues [4.49, 4.50].

The communication protocol model provides the protocols to exchange information with groups. Two kinds of protocols are required: user presentation and group work management. User-presentation protocols are concerned with client interactions, such as opening, closing or dynamically joining and leaving a telecooperation session. The group-work management protocols perform the communication between clients and servers, such as registering active sessions and inquiring about the current status of cooperative work.

Telecooperative Applications

Of the telecooperative applications, three are important:

- Multimedia email

- Collaborative authorship applications
- Multimedia conferencing

Multimedia email is an electronic messaging system that allows all moments to exchange multimedia messages asynchronously. Email is the most widely used service on the Internet. Unlike the older Simple Mail Transfer Protocol (SMTP) standard, which understands only ASCII characters, Multipurpose Internet Mail Extensions (MIME) specifies a set of encoding rules and header extensions that describe new types of contents (image, video or audio) and the structure of multimedia messages to embed multimedia information.

Collaborative authorship applications mean the activity of collaborative editing and composing a multimedia document by a group of people. This type of application can be either synchronous or asynchronous. Each member works on a part, and the completed product is the combination of all individual parts [4.51, 4.52].

Multimedia conferencing supports participants with distributed multiparty synchronous collaboration to simulate face-to-face interactions or real-time telephone conversations. The service may range from point-to-point videophone to multipoint conferencing. In videophone, the hardware requirements include a telephone equipped with a video camera, which transmits low-quality images at low frame rates through existing phone lines. In desktop conferencing, the hardware requirements include desktop PCs or workstations equipped with audio-visual facilities. The electronic meeting room, or Group-Decision Support System (GDSS), uses several networked workstations, large computer-controlled public displays and audio and video equipment to simulate a face-to-face meeting electronically [4.53, 4.54].

A video conferencing system uses both intraframes and interframes. The intraframe is only sent for the first picture or after a change of scene. Intraframe does not have motion estimation for the DCT [4.55], quantization, zigzag scan, and variable length and Huffman coding are used for each macroblock. The inverse quantization and the IDCT for the quantized frame form a referenced frame. The input frame uses motion estimation by comparing the input frame to the referenced neighbor frame (the reconstructed neighbor picture) to find the motion vectors. If the difference between the input block and the referenced block is below the threshold, no information need be sent. Otherwise, the difference is transformed by DCT, quantized, zigzag scanned and coded using variable length and Huffman coding. A referenced frame can be generated by reconstructing the frame using inverse quantization and the IDCT on the quantized difference and by adding the motion-compensated picture to this difference. A loop filter that removes the high-frequency noise can be used to improve the visual effects.

There are two approaches to implementing CSCW systems: collaboration transparency and collaboration awareness [4.48, 4.56]. Collaboration transparency performs multiuser cooperation on existing single-user applications using an application-sharing system. One key feature of collaboration-transparent applications is that all the users have to use the same application-sharing system and usually only one participant is able to access or control the shared windows [4.43]. Collaboration awareness performs multiuser access using the development of a new special-purpose application explicitly to handle collaboration. This approach embeds collaborative facilities within the applications and usually allows multiple users simultaneous access to the shared windows.

Telemedicine

Although a telemedicine concept is very simple (we acquire medical data from appropriate devices and transfer it to other centers), its realization is very difficult due to very hard technical requirements, particularly to transmit store and search for an extremely large number of large files, as medical images [4.57]. Within the last 10 years, various image-processing, transmission and archiving systems have been developed for medical applications. These have been focused in the areas of radiology and pathology. They are now finding their way into such areas as cardiology, neurology, orthopedics and surgery. The medical images, acquired from different imaging devices—Computer Tomography (CT), Magnetic Resonance Imaging (MRI), Nuclear Medicine (NM) imaging, Ultrasound (US) imaging, different radiology images, images from digital microscopes, and so forth—are in different formats having different spatial and level resolutions. The large volume of image data requires image compression. Different approaches are derived for ensuring the required level of QoS with cost-effectiveness [4.58]. Videostreams are usually compressed before being transferred across a network. Experimental studies with the lossy compression algorithm, such as JPEG or wavelet-based transforms, confirm that compression ratios of 10:1 or 20:1 produce no perceptible differences in the quality of the medical image. At the user level, a perceived QoS is defined as the percentage of diagnosis producing the same results as those carried out with original images.

4.7.3 Hypermedia Applications

Hypermedia applications are retrieval services and require point-to-point or multipoint-to-point and switched services. They also require user interfaces, powerful authoring and presentation tools. Some applications are particularly suited for hypermedia, such as encyclopedias, dictionaries and other reference books [4.59]. They are composed of a number of independent units that are seldom accessed sequentially, but rather by selection and cross-reference with other entries. On the Internet, even for technical papers and reports that are considered more suitable for linear reading in sequence, there is an increasing tendency to include Hypertext Markup Language (HTML). The terms "hypertext" and "hypermedia" are usually distinguished. The former refers to a system with text-only information, and the latter refers to multimedia data.

A hypermedia system may be tracked as an application of database systems because it provides flexible access to multimedia information and a novel method to structure and manage data. A hypermedia system allows the user more freedom in assimilating and exploring information, as the conventional database has well-defined structures and manipulation languages for data processing [4.60].

Basic Features of a Hypermedia System

The basic features of a hypermedia system are the following [4.61]:

- Information may be divided into several smaller units or nodes. Each node may contain single media data, such as text, source code, graphics, video, audio or animation, or

combinations of two or more media types. The contents in different nodes may have different authors.
- The links that interconnect the units of information can be bidirectional to facilitate backward traversals. Node contents are displayed by activating links.
- The system consists of a network of nodes and links, which may be distributed across multiple sites and remote servers in a computer network.
- Linear reading is used to move up and down among the document pages at a node. Nonsequential browsing allows users to jump back and forth in the information space and to navigate the hypermedia network.
- With authoring tools, users can build their own information structure for various purposes, through creation, manipulation and linkage of shared information units.
- It is necessary to maintain a database system to manage the shared multimedia data. In addition to the database functions, such as data-integrity guarantee, query processing, concurrency control, failure recovery and security mechanisms, support of rich modeling, document hierarchy and hypertext link should be included.

Due to proprietary storage mechanisms and document formats, most current hypermedia systems are not amenable to interoperability, and it is difficult to integrate the materials created in different systems. To make hypermedia systems completely open and interchangeable, various hypermedia models and document architectures have been developed [4.62, 4.63, 4.64].

The Web

A nice property of the Web is that it shields the implementation details of both the formats and the access protocols and presents a uniform interface to users. The Web is evolving toward a universal information hyperspace with a distributed, collaborative infrastructure. In practice, the Web is a vast collection of digitized information documents stored in computers connected to the worldwide Internet. It is accessible through Web browsers, such as Netscape and Microsoft Explorer, and runs on a request-and-response paradigm known as Hypertext Transport Protocol (HTTP) to allow online service to information located in remote sites. HTML is the standard document representation format for Web pages.

The Web documents created by HTML are static in nature. The author of a Web page determines its contents. After it is stored on the Web server, each browser requesting the document obtains the same response. In contrast, in dynamic documents, the contents are generated by application programs dynamically during runtime, depending on the current input from the user. Two approaches may be used to create such dynamic documents. The first is exemplified by the Common Gateway Interface (CGI), which is a widely used technology. As soon as a browser requests a document, the Web server invokes a CGI program that creates a document according to the input information. The server then forwards this document to the client browser. The other approach allows the dynamics of documents even after they have been loaded into the client browsers. A popular example of this approach is Java. Whenever a browser requests a document, the server sends a copy of a computer program that the browser must run locally. As a result, the display of the document can change continuously and can also interact with user inputs, alleviating the burden of the servers. For example, one may use this approach to display animation of one's home page. There is no central authority on the Web. Anyone can create a Web document and reference other documents. The Uniform Resource Locator (URL) is the specification scheme to locate a document unambiguously on the Web.

Example 4.6 In a URL such as *commsci.usc.edu/faculty/li.html, http* indicates the access method and *commsci.usc.edu* is the machine name. Other methods to retrieve information from the Web include File Transfer Protocol (FTP) and Telnet.

4.8 Concluding Remarks

A DMS is an integrated communication, computing, and information system that enables the processing, management, delivery and presentation of synchronized multimedia information. Such a system includes discrete media data (text, data and images) and continuous media data (video or audio). It enhances human communications by exploiting both visual and aural senses and provides the ultimate flexibility in work and entertainment by allowing one to collaborate with participants, to view video movies on demand, to access online digital libraries from the desktop, and so forth. Solutions for DMS issues have to be developed and have to provide a complete multimedia communication infrastructure, which is needed to support distributed multimedia applications. More user control and interactivity are desired. Faster processors and hardware, higher network bandwidth and data compression ratios, as well as improvements in a variety of related technologies, are necessary. Standardization is also important to accommodate the heterogeneity of techniques and to provide probability of applications. For the applications to be commercially viable, cost is an important consideration.

CHAPTER 5

Multimedia Communication Standards

Chapter Overview

Multimedia communication standards have to rely on compromises between what is theoretically possible and what is technologically feasible. Standards can only be successful in the marketplace if the cost performance ratio is well balanced. This is specifically true in the field of audio-video coding where a large variety of innovative coding algorithms exist, but may be too complex for implementation.

In this chapter, we discuss MPEG-1, MPEG-2, MPEG-4, MPEG-4 VTC, JPEG2000, MPEG-7, MPEG-21, ITU-T and Internet standards. MPEG-1 is targeted at CD-ROM with applications at a bit rate of about 1.5 Mb/s. It has also proved useful for computer-generated multimedia where transmission bandwidth and storage capacity are limited or expensive. MPEG-2 addresses high-quality coding for all digital multimedia transmissions at data rates of 2 to 50 Mb/s. It can produce the video quality needed for multimedia entertainment piped to the home and for more demanding business and scientific applications. The scope and potential of the MPEG-4 standard is discussed in the context of audio-visual multimedia communication environments. We show that this standard provides tools and algorithms for coding both natural and synthetic audio and video, as well as provisions to represent the audio-visual data at the user terminal in a highly flexible manner. JPEG-2000 not only provides rate distortion and subjective image quality performance superior to existing standards, but also provides functionalities that current standards can either not address efficiently or not address at all. The objective of the MPEG-7 standardization process is to facilitate the browsing and retrieval of multimedia. We discuss audiovisual content presentation issues from the MPEG-21 multimedia framework. We discuss the ITU-T standardization process in multimedia communications from the video and speech coding, as well from the multimedia multiplex and synchronization points of view (H.32x, H.26x, H.22x). The Internet standardization process concludes the chapter.

5.1 Introduction

In a broad sense, multimedia is assumed to be a general framework for interaction with information available from different sources. With the digital revolution, it became possible to exploit a well-known concept further: the more that is known about the content means the better can be its representation, processing, and so forth, in terms of efficiency and allowed functionalities. After we break down the boundary between speech research and image research, a large number of new techniques and applications can be developed [5.1].

A multimedia standard is expected to provide support for a large number of applications. These applications translate into a specific set of requirements that may be very different from one another. One theme common to most applications is the need for supporting interactivity with different kinds of data. Communications mean standards, but the production of standards for multimedia communications is beset by the problem that the many industries having a stake in multimedia communications have radically different approaches to standardization. Standards play a major role in the multimedia revolution because they provide interoperability between hardware and software provided by multiple vendors.

Example 5.1 Although, in practice, one could use a nonstandard coder, this would lead to a closed architecture system, which would discourage the widespread use of the system in which the coder was embedded. As a result, all of the major vendors of both terminals and software for multimedia communications have embraced the concept of standardization so that their various products will operate at a basic level.

The success of the MPEG [2.63] is based on a number of concurrent elements. MPEG appeared at a time when the coding algorithms of audio and video were reaching asymptotic performance. By relying on the support in terms of technical expertise, of all industries interested in digital audio and video applications, MPEG contributed to the practical acceptance of the audio-visual representation layer, independent of the delivery system. A last element of success has been the focus on the decoder instead of the traditional encoder-decoder approach.

Therefore, MPEG could provide the standard solution to the major players who were considering the use of digital coding of audio and video for innovative mass-market products and could allow a faster achievement of a critical mass than would have been possible without it. The different industries have been diverging, but multimedia communications necessarily need some convergence zone that can only be achieved by standardization in key areas. Putting every stakeholder together and producing communication standards accepted by all is a big task. After the great success of the MPEG-1 and MPEG-2 standards, which opened the digital frontiers to audiovisual information and allowed the deployment of high performance services, MPEG is striking again with the emerging MPEG-4 standard. The MPEG-4 standard is the acknowledgement by MPEG, the leading standardization body in audiovisual representation technology, that the data models underpinning MPEG-1 and MPEG-2 were limited and could not fulfill new needs of emerging multimedia applications, such as hyperlinking, interaction and natural and synthetic data integration. MPEG-4 is the answer to the requirements coming from the new ways in which audio-visual information is nowadays produced, delivered and consumed. To reach this target, MPEG-4 follows an object-based representation approach where an audiovisual scene is coded as a composition of objects, natural as well as synthetic, which provides the first powerful hybrid playground. The objective of MPEG-4 is thus to provide an audiovisual representation standard supporting new ways of communication access and interaction with digital audiovisual data, and it offers a common technical solution to various services. It also extends to layered coding (scalabilities), multiview (stereoscopic video), shape/texture/motion coding of objects and animation. Its role extends to the Internet, Web TV, large databases (storage, retrieval and transmission) and mobile networks [5.2]. MPEG-4 Version 1 became an international standard in February 1999, and Version 2 became a standard in November 1999. Version 2 with extended functionalities is backward compatible with Version 1.

Multimedia databases on the market today allow searching for pictures using characteristics, such as color, texture and information about the shape of objects in an image. MPEG started a new work item to provide a solution to the problem of facilitating multimedia search engines. One of the members of the MPEG family (called Multimedia Content Description Interface) is MPEG-7 [5.3, 5.4, 5.5]. It extends the limited current search capabilities to include more information types, such as video, image, audio, graphics and animation. In other words, MPEG-7 specifies a standardized description of various types of multimedia information. This description is associated with the content itself and allows fast and efficient searching for multimedia that is of interest to users. The description can be attached to any kind of multimedia material, no matter what the format of the description is. Stored material that has this information attached to it can be indexed, searched and retrieved.

When the scope of new work has been sufficiently clarified, MPEG usually makes open requests for proposals. So far proposals have been requested for the following:

- MPEG-1 Audio and Video (July 1989)
- MPEG-2 Audio and Video (July 1991)

- MPEG-4 Audio and Video (July 1995)
- MPEG-7 and MPEG-21

In the original ITU-T work plan, the goal was to define a near-term recommendation in 1996, followed by a long-term recommendation several years later. The near-term recommendation is referred to as H.263. The long-term recommendation H.26L (previously called H.263L) is scheduled for standardization in the year 2002 and may adopt a completely new compression algorithm. After H.263 was completed, it became apparent that incremental changes could be made to H.263 that could visibly improve its compression performance. Thus, ITU-T decided in 1996 that a revision to H.263 would be created that incorporated these incremental improvements. This is H.263 plus with several new features. Hence, the name H.263+ (now called H.263 Version 2). H.263+ contains approximately 12 new features that do not exist in H.263. These include new coding modes that improve compression efficiency, support for scalable bit streams, several new features to support packet networks [5.6] and error-prone environments, added functionality and support for a variety of video formats.

5.2 MPEG Approach to Multimedia Standardization

MPEG was established in January 1988 with the mandate to develop standards for the coded representation of moving pictures, audio and their combination. It operates in the framework of the Joint ISO/IEC Technical Committee (JTC 1) on Information Technology under WG11 of SC29.

Starting from its first meeting in May 1988 when 25 experts participated, MPEG has grown to an unusually large committee. Usually some 350 experts from some 200 companies and organizations from about 20 countries take part in MPEG meetings. As a rule, MPEG meets three times a year (in March, July and November), but meets more frequently when the workload so demands.

Depending on the nature of the standard, documents of different nature may be produced. For audio and video coding standards, the first document is called a *Verification Model* (VM). In MPEG-1 and MPEG-2, this was called Simulation and Test Model, respectively. The VM describes, in some sort of programming language, the operation of the encoder and the decoder. The VM is used to carry out simulations to optimize the performance of the coding scheme. When MPEG has reached sufficient confidence in the stability of the standard under development, a *Working Draft* (WD) is produced. This is already in the form of a standard, but is kept internal to MPEG for revision. At the planned time, the WD has become sufficiently solid and becomes *Committee Draft* (CD).

A WD usually undergoes several revisions before moving to the CD stage. A key role is played by core experiments where different technical options are studied by at least two different partners. Each revision involves a large number of experts who draw the committee's attention to possible errors contained in the document.

MPEG Approach to Multimedia Standardization

A list of work items in MPEG multimedia standardization is as follows:

- *ISO/IEC IS 11172*—Coding of moving pictures and associated audio at up to about 1.5 Mb/s (MPEG-1)
  - Part 1 Systems
  - Part 2 Video
  - Part 3 Audio
  - Part 4 Conformance testing
  - Part 5 Software simulation
- *ISO/IEC IS 13818*—Generic coding of moving pictures and associated audio (MPEG-2)
  - Part 1 Systems
  - Part 2 Video
  - Part 3 Audio
  - Part 4 Conformance testing
  - Part 5 Software simulation
  - Part 6 System extensions—Digital Storage Media-Command and Control (DSM-CC)
  - Part 7 Audio extension—Advanced Audio Coding (AAC)
  - Part 8 VOID—(withdrawn)
  - Part 9 System extension—Real Time Interface (RTI) for system decoders
  - Part 10 Conformance extension for DSM-CC
- *ISO/IEC IS 14496*—Coding of audiovisual objects (MPEG-4)
  - Part 1 Systems
  - Part 2 Visual
  - Part 3 Audio
  - Part 4 Conformance testing
  - Part 5 Reference software
  - Part 6 Delivery Multimedia Integration Framework (DMIF)
  - Part 7 Optimized reference software
  - Part 8 Carriage of MPEG-4 content across IP networks
  - Part 9 Reference hardware description
- *ISO/IEC IS 15938*—Multimedia content description interface (MPEG-7)
  - Part 1 Systems
  - Part 2 Description Definition Language (DDL)
  - Part 3 Visual
  - Part 4 Audio
  - Part 5 Multimedia Description Schemes (MDS)
  - Part 6 Reference software
  - Part 7 Conformance
  - Part 8 Extraction and use of MPEG-7 descriptors

- *ISO/IEC IS 18034*—Multimedia framework (MPEG-21)
    - Part 1 Vision, technologies and strategy
    - Part 2 Digital item declaration
    - Part 3 Digital item identification and description
    - Part 4 Intellectual property management and protection
    - Part 5 Rights expression language
    - Part 6 Rights data dictionary

Although the television paradigm dominated audiovisual communications for many years, the situation now is evolving very quickly in terms of the ways audiovisual content is produced, delivered and consumed [5.7]. Moreover, hardware and software are getting more and more powerful, opening new frontiers to the technologies used and to the functionalities provided.

Producing content today is made very easy. Digital still cameras directly storing images in JPEG format have hit the mass market. Together with the first digital video cameras recording directly in MPEG-1 format, this represents a major step for the acceptance in the consumer market of digital audiovisual acquisition technology. This step transforms every one of us into a potential content producer, capable of creating content that can be easily distributed and published using the Internet. Moreover, more content is being synthetically produced, computer generated and integrated with natural material as a truly hybrid audiovisual content. The various pieces of content, digitally encoded, can be successively reused without the quality losses typical of the previous analog processes.

Although audiovisual information was, until recently, only carried across very few networks, the trend is now toward the generalization of visual information in every single network. Moreover, the increasing mobility in telecommunications is a major trend. Mobile connections will not be limited to voices, and other types of data, including real-time media, will be the next. Because mobile telephones are replaced every two to three years, new mobile devices can finally make the decade-long promise of audiovisual communications turn into reality [5.8]. The explosion of the Web and the acceptance of its interactive mode of operation have clearly shown in the last few years that the traditional television paradigm would no longer suffice for audiovisual services. Users will want to have access to audio and video like they now have access to text and graphics. This requires moving pictures and audio of acceptable quality at low bit-rates on the Web and Web-type interactivity with live content.

Standardization items have been identified well in advance and no MPEG standard has endorsed an industry standard. MPEG standards do not specify complete systems. Therefore, it is possible that industry standards are needed with MPEG standards to make full-fledged products.

Example 5.2 Industries by definition need to make vertically integrated specifications in order to make products that satisfy some needs. Audiovisual decoding may well be a piece of technology that can be shared with other communities, but, in the event industries need to sell a satellite receiver or a video CD player, these require an integrated standard. However, if different industries need the same standard, they quite likely will have different systems in mind. Therefore, only the components of a standard, the tools as they are called in MPEG, can be specified in a joint effort. The implementation of this principle requires the change of the nature of standards from system standards to component standards. Industries will assemble the tool specification from the standards body and build their own product specification.

If tools are the object of standardization, a new process must be devised to produce meaningful standards. The following sequence of steps has been found to be practically implementable and to produce the desired result:

1. Select a number of target applications for which the generic technology is intended to be specified.
2. List the functionalities needed by each application.
3. Break down the functionalities into components of sufficiently reduced complexity so that they can be identified in different applications.
4. Identify the functionality components that are common across the systems of interest.
5. Specify the tools that support the identified functionality components, particularly those common to different applications.
6. Verify that the tools specified can actually be used to assemble the target systems and provide the desired functionalities.

These standardized sets of tools have been called profiles in MPEG-2 Video [5.9]. It is advisable that certain major contributions of tools be specified as normative, making sure that these are not application specific, but functionally specific.

In some environments, it is proper to add those nice little things to a standard that bring a standard nearer to a product specification. This is, for example, the case of industry standards or when standards are used to enforce the concept of guaranteed quality so important to broadcasters and telecommunication operators because of their public service nature. However, in the case when a standard is to be used by multiple industries, only the minimum that is necessary for interoperability can be specified. The profile-level philosophy successfully implemented by MPEG provides a solution: within a single tool one may define different grades called levels in MPEG [5.10].

Example 5.3 When a standard is defined by a single type of industry generally, an agreement exists on where a certain functionality resides in the system. In a multi-industry environment, this is not possible. Take the case of encryption. Depending on our role in the audiovisual distribution chain, you would like to have the encryption function located where it serves your place in the chain best, because encryption is an important value-added function. If the standard endorses our business model, we will adopt the standard. If it does not, we will antagonize it.

After the work is nearing completion, it is important to make sure that it does indeed satisfy the requirements (product specification) originally set. MPEG does that through a process called verification tests, with the scope of ascertaining how well the standard produced meets the specification. We give now a brief account of MPEG multimedia communications standards—some established and some under development—by giving a description of their functionalities and usage.

5.3 MPEG-1 (Coding of Moving Pictures and Associated Audio)

The first standard developed by the group, nicknamed MPEG-1, was the coding of the combined audiovisual signal at a bit rate around 1.5 Mb/s. This was motivated by the prospect, becoming apparent in 1988, of storing video signals on a CD with a quality comparable to VHS cassettes. In 1988, coding of video at such low bit rates had become possible thanks to decades of research in video-coding algorithms. These algorithms, however, had to be applied to subsampled pictures—a single field from a frame and only half of the samples in a line—to show their effectiveness. Also, coding of audio, as separate from speech, allowed reduction by one-sixth of the PCM bit rate, typically 256 Kb/s for a stereo source, with virtual transparency. Encoded audio and video streams, with the constraint of having a common time base, were combined into a single stream by the MPEG systems layer. As previously indicated, MPEG-1, formally known as ISO/IEC 11172, is standardized in five parts [5.11]. The first three parts are Systems, Video and Audio. Two more parts complete the suite of MPEG-1 standards; Conformance Testing, which specifies the methodology for verifying claims of conformance to the standard by manufacturers of equipment and producers of bitstreams, and Software Simulation, a full C-language implementation of the MPEG-1 standard (encoder and decoder).

Part 1 addresses the problem of combining one or more datastreams from the video and audio parts of the MPEG-1 standard with timing information to form a single stream (Figure 5.1). This is an important fuction because, after being combined into a single stream, the data is in a form well suited to digital storage or transmission.

Part 2 specifies a coded representation that can be used for compressing video sequences—both 625-line and 525-line—to bit rates around 1.5 Mb/s. Part 2 was developed to operate from storage media offering a continuous transfer rate of about 1.5 Mb/s. Nevertheless, it can be used more widely than this because the approach taken is generic [5.12].

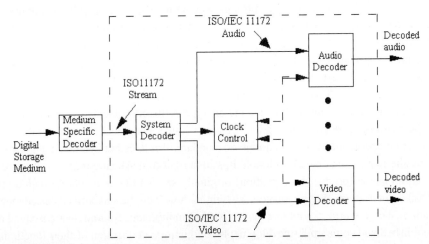

Figure 5.1   ISO/IEC 11172 decoder [5.11]. ©1993 ISO/IEC.

MPEG-1 (Coding of Moving Pictures and Associated Audio)

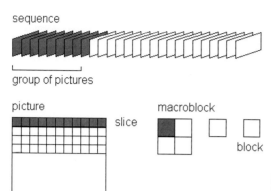

Figure 5.2 Temporal picture structure [5.12]. ©1993 ISO/IEC.

A number of techniques are used to achieve a high compression ratio. The first is to select an appropriate spatial resolution for the signal. The algorithm then uses block-based motion compensation to reduce the temporal redundancy. Motion compensation is used for causal prediction of the current picture from a previous picture, or noncausal prediction of the current picture from a future picture or for interpolative prediction from past and future pictures. The difference signal, the prediction error, is further compressed using DCT to remove spatial correlation and is then quantized. Finally, the motion vectors are combined with the DCT information and coded using variable length codes. Figure 5.2 illustrates a possible combination of the three main types of pictures that are used in the standard.

Part 3 specifies a coded representation that can be used for compressing audio sequences, both mono and stereo as shown in Figure 5.3. Input audio samples are fed into the encoder. The mapping creates a filtered and subsampled representation of the input audio stream. A psychoacoustic model creates a set of data to control the quantizer and coding. The quantizer and coding block create a set of coding symbols from the mapped input samples. The block frame packing assembles the actual bit stream from the output data of the other blocks and adds other information (for example error correction) if necessary [5.13].

Part 4 specifies how tests can be designed to verify whether bit streams and decoders meet the requirements as specified in Parts 1, 2 and 3 of the MPEG-1 standard. These tests can be used by the following:

- Manufacturers of encoders and their customers to verify whether the encoder produces valid bit streams
- Manufacturers of decoders and their customers to verify whether the decoder meets the requirements specified in Parts 1, 2 and 3 of the standard for the claimed decoder capabilities
- Applications to verify whether the characteristics of a given bit stream meet the application requirements, for example, whether the size of the coded picture does not exceed the maximum value allowed for the application [5.14].

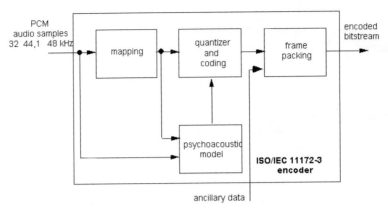

Figure 5.3 Basic structure of the MPEG-1 audio encoder [5.13].
©1993 ISO/IEC.

Part 5, technically not a standard, but a technical report, gives a full software implementation of the first three parts of the MPEG-1 standard [5.15].

Example 5.4 The different layers have been defined because they all have their merits. Basically, the complexity of the encoder and decoder, the encoder delay and the coding efficiency increase when going from Layer I through Layer II to Layer III. Layer I has the lowest complexity and is specifically suitable for applications where the encoder complexity also plays an important role. Layer II requires a more complex encoder and a slightly more complex decoder and is directed toward one-to-many applications, that is, one encoder serves many decoders. Compared to Layer I, Layer II is able to remove more of the signal redundancy and applies the psychoacoustic threshold more efficiently. Layer III is again more complex and is directed toward lower bit-rate applications due to the additional redundancy and irrelevancy extraction from enhanced frequency resolution in its filterbank.

MPEG-1 was also the first signal-processing standard developed and eventually documented using the C-programming language. This facilitated the recognition that what matters in a coding standard is just the syntax used to represent the operations carried out on the signal. Bit rate, frame rate, number of lines, number of pixels per line, and so forth are just parameters with their overriding importance in the analog domain reduced to size in the digital domain.

MPEG-1 not only was a technological achievement, but also contributed to define the highly politicized issues of television standards. MPEG recognized that what matters in a television signal is not the number of lines or the number of fields per second, but the bandwidth of the signal in the analog domain and the number of pixels per second in the digital domain. The result has been the normative definition of the Constrained Parameter Set (CPS) for MPEG-1, where there is no reference to television standards. In terms of memory requirement, what matters is the total number of pixels in a frame, and, in terms of processing requirements, what matters is the number of macroblocks, that is, number of coded 16x16 pixels. Elements of CPS are given in Table 5.1.

Table 5.1 MPEG-1 Video CPS [5.10].

Parameter	Value
Horizontal size	≤ 768
Vertical size	≤ 576
Number of macroblocks / picture	≤ 396
Number of macroblocks / second	≤ 9900
Picture rate	≤ 30 Hz
Interpolated pictures	≤ 2
Bit rate	≤ 1856 Kb/s

©1998 IEEE.

The audio part of the MPEG-1 standard has become the key component for radio broadcasting at CD quality offered by digital audio broadcasting, which is actively being deployed in several countries. MPEG-1 Audio Layer II and, more recently, Layer III have become the standard form for music distribution on the Web. The full MPEG-1 standard (Audio-Video-Systems) is the standard format for distribution of video material across the Web.

MPEG-1 provided the first concrete opportunity for the microelectronics industry to invest in digital video technology. MPEG-1 decoder chips are produced by multiple sources, some of which incorporate the electronics needed to read bits from a CD. Several suppliers exist for MPEG-1 encoder chips. One consumer-electronics manufacturer has already put its digital video camera on the market. This is made up of an optical part, audio and video sensors, a single chip for audio-video systems encoding and a hard disk for 20 minutes of recording. These devices and the growing number of personal computers are creating the conditions for the popularization of multimedia contents production. Detailed information on MPEG-1 Audio can be found in ISO/IEC IS13818-3 (MPEG-1) and Brandenburg et al. [5.16, 5.17].

The MPEG-1 video algorithm was primarily targeted for multimedia CD-ROM applications, requiring additional functionality supported by both encoder and decoder. Important features provided by MPEG-1 include frame-based random access of video, fast forward/fast reverse (FF/FR) searches through compressed bit streams, reverse playback of video and editability of the compressed bit stream.

5.3.1 The Basic MPEG-1 Interframe Coding Scheme

The basic MPEG-1 video-compression technique is based on a macroblock structure, motion compensation and the conditional replenishment of macroblocks. As outlined in Figure 5.4a, the

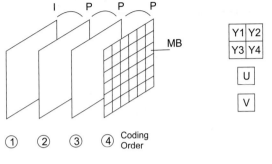

Figure 5.4 Illustration of I-pictures (I) and P-pictures (P) in a video sequence [5.12]. ©1993 ISO/IEC.

MPEG-1 coding algorithm encodes the first frame in a video sequence in intraframe coding mode (I-picture). Each subsequent frame is coded using interframe prediction (P-pictures), which means that only data from the nearest previously coded I- or P-frame is used for prediction. The MPEG-1 algorithm processes the frames of a video sequence as block based. Each color input frame in a video sequence is partitioned into nonoverlapping macroblocks as depicted in Figure 5.4b. Each macroblock contains blocks of data from both luminance and cosited chrominance bands: four luminance blocks (Y1, Y2, Y3 and Y4) and two chrominance blocks (U and V), each with size 8x8 pels. Thus, the sampling ratio between Y:U:V luminance and chrominance pels is 4:1:1. P-pictures are coded using motion-compensated prediction based on the nearest previous frame (I or P). Each frame is divided into disjoint macroblocks. With each macroblock, information related to four luminance blocks (Y1, Y2, Y3 and Y4) and two chrominance blocks (U and V) is coded. Each block contains 8x8 pels.

The block diagram of the basic hybrid DPCM/DCT MPEG-1 encoder and decoder structure is depicted in Figure 5.5. The first frame in a video sequence (I-picture) is encoded in INTRA mode without reference to any past or future frames. At the encoder, the DCT is applied to each 8x8 luminance and chrominance block, and, after output of the DCT, each of the 64 DCT coefficients is uniformly quantized (Q). The quantizer Step Size (SZ) used to quantize the DCT-coefficients within a macroblock is transmitted to the receiver. After quantization, the lowest DCT coefficient (DC coefficient) is treated differently from the remaining coefficients (AC coefficients). The DC coefficient corresponds to the average intensity of the component block and is encoded using a differential DC prediction method. The nonzero quantized values of the remaining DCT coefficients and their locations are then zig-zag scanned and run-length entropy coded using VLC tables.

The concept of zig-zag scanning of the coefficients is outlined in Figure 5.6. The scanning of the quantized DCT-domain 2D signal followed by variable-length code-word assignment for the coefficients serves as a mapping of the 2D image signal into a 1D bit stream. The nonzero AC coefficient quantized values (length) are detected along the scan line as well as the distance (run) between two consecutive nonzero coefficients. Each consecutive (run, length) pair is

MPEG-1 (Coding of Moving Pictures and Associated Audio) 151

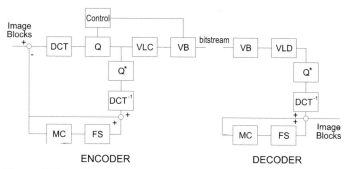

Figure 5.5   Block diagram of a basic DPCM/DCT encoder and decoder structure [5.12]. ©1993 ISO/IEC.

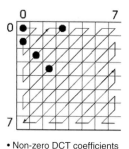

Figure 5.6   Zig-zag scanning of the quantized DCT coefficients [5.12]. ©1993 ISO/IEC.

• Non-zero DCT coefficients

encoded by transmitting only one VLC codeword. The purpose of zig-zag scanning is to trace the low-frequency DCT coefficients (containing most energy) before tracing the high-frequency coefficients.

Example 5.5 Zig-zag scanning of the quantized DCT coefficients in an 8x8 block is presented in Figure 5.6. Only the nonzero quantized DCT coefficients are indicated. The possible locations of nonzero DCT coefficients are indicated. The zig-zag scan attempts to trace the DCT-coefficients according to their significance. The lowest DCT coefficients (0,0) contain most of the energies within the blocks. The energy is concentrated around the low-frequency DCT coefficients.

The decoder performs the reverse operations, first extracting and decoding (VLD) the variable-length coded words from the bit stream to obtain locations and quantized values of the non-zero DCT coefficients for each block. With the reconstruction (Q*) of all nonzero DCT coefficients belonging to one block and subsequent inverse DCT (IDCT), the quantized block pixel values are obtained. By processing the entire bit stream, all image blocks are decoded and reconstructed.

For coding P-pictures, the previously I- or P-picture frame $N-1$ is stored in a Frame Store (FS) in both encoder and decoder. Motion compensation (MC) is performed on a macroblock basis. Only one motion vector is estimated between frame N and frame $N-1$ for a particular Macroblock to be encoded. These motion vectors are coded and transmitted to the receiver. The motion-compensated prediction error is calculated by subtracting each pel in a Macroblock with its motion-shifted counterpart in the previous frame. An *8x8* DCT is then applied to each of the *8x8* blocks contained in the macroblock followed by Q of the DCT coefficients with subsequent run-length coding and entropy coding (VLC). A Video Buffer (VB) is needed to ensure that a constant target bit rate output is produced by the encoder. The quantization SZ can be adjusted for each macroblock in a frame to achieve a given target bit rate and to avoid buffer overflow and underflow.

The decoder uses the reverse process to reproduce a macroblock of frame *N* at the receiver. After decoding the variable length words contained in the video decoder buffer, the pixel values of the prediction error are reconstructed (Q^* and DCT^{-1}-operations). The motion-compensated pixels from the previous frame *N-1* contained in the FS are added to the prediction errors to recover the particular macroblock of frame *N*.

5.3.2 Conditional Replenishment

An essential feature supported by the MPEG-1 coding algorithm is the possibility of updating macroblock information at the decoder only if needed, for example, if the content of the macroblock has changed in comparison to the content of the same macroblock in the previous frame (conditional macroblock replenishment). The key for efficient coding of video sequences at lower bit rates is the selection of appropriate prediction modes to achieve conditional replenishment. The MPEG standard allows three different macroblock coding types (MB types):

- *Skipped MB*—Prediction from previous frame with a zero motion vector is used. No information about the macroblock is coded or transmitted to the receiver.
- *Inter MB*—Motion-compensated prediction from the previous frame is used. The macroblock type, the MB address and, if required, the motion vector, the DCT coefficients and quantization SZ are transmitted.
- *Intra MB*—No prediction is used from the previous frame (intraframe coding only). Only the macroblock type, the MB address and the DCT coefficients and quantization SZ are transmitted to the receiver.

5.3.3 Specific Storage Media Functionalities

For accessing video from storage media, the MPEG-1 video compression algorithm was designed to support important functionalities, such as random access and FF/FR playback functionalities. To incorporate the requirement for storage media and to explore the significant advantages of motion compensation and motion interpolation further, the concept of B-pictures (bidirectional predicted/bidirectional interpolated pictures) was introduced in MPEG-1. This concept is depicted in Figure 5.7 for a group of consecutive pictures in a video sequence. Three types of pictures are considered. Intrapictures (I-pictures) are coded without reference to other pictures contained in the video sequence. I-pictures allow access points for random access and FF/FR functionality in the bit stream, but achieve only low compression. Interframe-predicted pictures (P-pictures) are coded with reference to the nearest previously coded

MPEG-1 (Coding of Moving Pictures and Associated Audio)

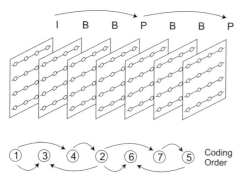

Figure 5.7 I-pictures (I), P-pictures (P) and B-pictures (B) used in a MPEG-1 video sequence [5.12]. ©1993 ISO/IEC.

I-picture or P-picture, usually incorporating motion compensation to increase coding efficiency. Because P-pictures are usually used as reference for prediction for future or past frames, they provide no suitable access points for random access functionality or editability. Bidirectional predicted/interpolated pictures (B-pictures) require both past and future frames as references. To achieve high compression, motion-compensation can be employed based on the nearest past and future P-pictures or I-pictures. B-pictures themselves are never used as references. B-pictures can be coded using motion-compensated prediction based on the two nearest already coded frames (either I-picture or P-picture). The arrangement of the picture-coding types within the video sequence is flexible to suit the needs of diverse applications. The direction for prediction is indicated in Figure 5.7.

The user can arrange the picture types in a video sequence with a high degree of flexibility to suit diverse application requirements. As a general rule, a video sequence coded using I-pictures only (I I I I I I . . .) allows the highest degree of random access, FF/FR and editability, but achieves only low compression. A sequence coded with a regular I-picture update and no B-pictures (that is I P P P P P P I P P P P . . .) achieves moderate compression and a certain degree of random access and FF/FR functionality. Incorporation of all three pictures types, as depicted in Figure 5.7 (I B B P B B P B B I B B P . . .), may achieve high compression and reasonable random access and FF/FR functionality but also increases the coding delay significantly. This delay may not be tolerable for videotelephony or videoconferencing applications.

5.3.4 Rate Control

An important feature supported by the MPEG-1 encoding algorithm is the possibility of tailoring the bit rate (and thus the quality of the reconstructed video) to specific applications requirements by adjusting the quantizer SZ in Figure 5.5 for quantizing the DCT coefficients.

Coarse quantization of the DCT coefficients enables the storage or transmission of video with high compression ratios, but, depending on the level of quantization, may result in significant coding artifacts. The MPEG-1 standard allows the encoder to select different quantizer values for each coded macroblock. This enables a high degree of flexibility to allocate bits in images where needed and to improve image quality. Furthermore, it allows the generation of both constant and variable bit rates for storage or real-time transmission of the compressed video.

Compressed video information is inherently variable in nature. This is caused in general by the variable content of successive video frames. To store or transmit video at a constant bit rate, it is therefore necessary to buffer the variable bit stream generated in the encoder in a video buffer (VB) as depicted in Figure 5.5. The input into the encoder VB is variable over time, and the output is a constant bit stream. At the decoder, the VB input bit stream is constant, and the output used for decoding is variable. MPEG encoders and decoders implement buffers of the same size to avoid reconstruction errors.

A rate-control algorithm at the encoder adjusts the quantizer SZ depending on the video content and activity to ensure that the VB will never overflow. At the same time, it targets to keep the buffers as full as possible to maximize image quality. In theory, overflow of buffers can always be avoided by using a large enough VB. However, besides the possibly undesirable costs for the implementation of large buffers, there may be additional disadvantages for applications requiring low-delay between encoder and decoder, such as for the real-time transmission of conversational video. If the encoder bitstream is smoothed using a VB to generate a constant bit rate output, a delay is introduced between the encoding process and the time the video can be reconstructed at the decoder. Usually the larger the buffer means the larger the delay introduced.

MPEG has defined a minimum VB size that needs to be supported by all decoder implementations. This value is identical to the maximum value of the VB size that an encoder can use to generate a bit stream. However, to reduce delay or encoder complexity, it is possible to choose a virtual buffer size value at the encoder smaller than the minimum VB size that needs to be supported by the decoder. This virtual buffer size value is transmitted to the decoder before sending the video bit stream.

The rate-control algorithm used to compress video is not part of the MPEG-1 standard, and it is thus left to the implementers to develop efficient strategies. It is worth emphasizing that the efficiency of the rate-control algorithms selected by manufacturers to compress video at a given bit rate heavily impacts the visible quality of the video reconstructed at the decoder.

5.4 MPEG-2 (Generic Coding of Moving Pictures and Associated Audio)

The MPEG-2 family of standards outlines the compression technologies and bit-stream syntax that enable transmission of audio and video in broadband networks. These standards also describe the aspects needed to multiplex programs, enable clock synchronization and set up logical network links carrying video and audio content. MPEG-2 is, in many cases, associated only with video compression, which is certainly one of the most important parts of its functionality [5.18, 5.19, 5.20, 5.21]. However, the MPEG-2 standards include more than just pure video. In total, MPEG-2 has different parts, which cover the different aspects of digital video and audio delivery and representation [5.22]. Table 5.2 lists the different MPEG-2 parts.

Table 5.2 Parts of the MPEG-2 standards.

ISO/IEC 13818 MPEG-2	Description
13818-1	Systems
13818-2	Video
13818-3	Audio
13818-4	Compliance
13818-5	Software simulation
13818-6	DSM-CC
13818-9	RTI for system decoders
13818-10	DSM reference script format

Basically, MPEG-2 can be seen as a superset of the MPEG-1 coding standard and was designed to be backward compatible to MPEG-1. Every MPEG-2 compatible decoder can decode a valid MPEG-1 bit stream. Many video-coding algorithms were integrated into a single syntax to meet the diverse application requirements. New coding features were added by MPEG-2 to achieve sufficient functionality and quality, so prediction modes were developed to support efficient coding of interlaced video. In addition, scalable video-coding extensions were introduced to provide additional functionalities, such as embedded coding of digital TV and HDTV and graceful quality degradation in the presence of transmission errors.

For comparison, typical MPEG-1 and MPEG-2 coding parameters are shown in Table 5.3. However, implementation of the full syntax may not be practical for most applications. MPEG-2 has introduced the concept of profiles and levels to stipulate conformance for equipment not supporting the full implementation. Profiles and levels provide means for defining subsets of the syntax and thus the decoder capabilities required to decode a particular bit stream. As a general rule, each profile defines a new set of algorithms added as a superset to the algorithms in the profile below. A level specifies the range of the parameters that are supported by the implementation (that is, image size, frame rate and bit rates). The MPEG-2 core algorithm at the Main profile features nonscalable coding of both progressive and interlaced video sources. It is expected that most MPEG-2 implementations will at least conform to the Main Profile at the Main level, which supports nonscalable coding of digital video with approximately digital TV parameters: a maximum sample density of 720 pixels per line and 576 lines per frame, a maximum frame rate of 30 frames per second and a maximum bit rate of 15 Mb/s.

The upper bound of parameters at each level of a profile is given in Table 5.4. The MPEG-2 algorithm defined in the Main Profile is a straightforward extension of the MPEG-1 coding

Table 5.3 MPEG-1 and MPEG-2 coding parameters.

Parameter	MPEG-1	MPEG-2
Standardized	1992	1994
Main application	Digital video on CD-ROM	Digital TV (and HDTV)
Spatial resolution	SIF format (1/4 TV) 288x360pixels	TV (4xTV) 576x720 (1152x1440)
Temporal resolution	25/30 frames/s	50/60 fields/s (100/120 fields/s)
Bit rate	1.5 Mb/s	4 Mb/s (20 Mb/s)
Quality	Comparable to VHS	Comparable to NTSC/PAL for TV
Compression ratio over PCM	20-30	30-40

scheme to accommodate coding of interlaced video while retaining the full range of functionality provided by MPEG-1. Because it is identical to the MPEG-1 standard, the MPEG-2 coding algorithm is based on the general hybrid DPCM/DCT coding scheme, incorporating a macroblock structure, motion compensation and coding needs for conditional replenishment of macroblocks. The concept of I-picture, P-picture and B-picture is fully retained in MPEG-2 to achieve motion prediction and to assist random access functionality. In what follows we focus on the most essential parts of MPEG-2.

Table 5.4 Upper bound of parameters at each level of profile.

Level	Parameters
High	1,920 pixels/line
	1,152 lines/frame
	60 frames/s
	80 Mb/s
High 1440	1,440 pixels/line
	1,152 lines/frame
	60 frames/s
	60 Mb/s

Table 5.4 Upper bound of parameters at each level of profile. (Continued)

Level	Parameters
Main	720 pixels/line
	576 lines/frame
	30 frames/s
	15 Mb/s
Low	352 pixels/line
	288 lines/frame
	30 frames/s
	4 Mb/s

5.4.1 MPEG-2 Video

The main goal of the MPEG-2 Video standard is to define a format that can be used to describe a coded video bit stream. This video bit stream is the output of an encoding process, which significantly compresses the video information. MPEG-2 does not specify the encoding process. It only defines the resulting bit stream. When MPEG-2 was developed, one of the requirements was to make it flexible enough to handle a range of video applications, like broadcast (satellite) services, cable TV distribution and interactive television services, subject to flexible equipment capabilities, network bandwidth constraints and picture qualities. The MPEG-2 group managed to make the standard very generic by providing a set of tools that can be combined in different ways. The MPEG-2 Video standard consists of the following parts [5.23]:

- *Basic definitions*—Basic objects such as pictures and frames are defined.
- *MPEG-2 Video syntax*—Different syntax elements are derived.
- *Semantic description for the video stream syntax*—Semantic description is given for all syntax elements.
- *Video-decoding process*—Video decoding processes are described, including decoding in interlaced and progressive modes.
- *Scalability extensions*—Different variations of scalability of MPEG-2 Video are described, and the decoding for each mode is explained.
- *Profiles and levels*—Different profiles and levels, which are used to define subsets of MPEG-2 Video, are described.
- *Annexes*—Annexes provide variable length coding tables, tables that define profile and level constraints, and the DCT function. They also contain some information sections.

MPEG-2 Video—The Basics

MPEG-2 deals with a number of basic objects that are used to structure video information. Basic objects in MPEG-2 are shown in Figure 5.8.

The video sequence represents a number of video pictures or group of video pictures. A video sequence contains only a few pictures and not a whole movie.

A frame contains all the color and brightness information that is needed to display a picture. The color and brightness information is organized into three matrices, which contain the luminance and chrominance values. Figure 5.9 shows these matrixes for a 4:4:4 and 4:2:2 sampled frame.

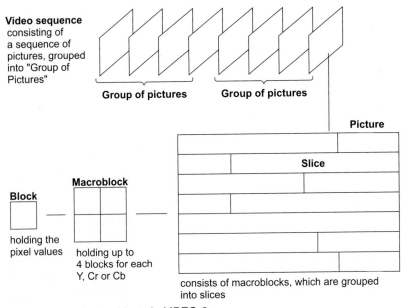

Figure 5.8 Basic objects in MPEG-2.

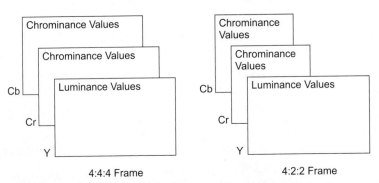

Figure 5.9 Matrixes forming a 4:4:4 and a 4:2:2 frame.

Each picture is divided into a number of blocks, which are grouped into macroblocks. Each block contains eight lines, with each line holding eight samples of luminance or chrominance pixel values from a frame. This gives 64 chrominance or luminance pixel values defining a block. Four blocks with luminance values, plus a number of blocks with chrominance values, form the luminance and chrominance information of a macroblock. The number of chrominance blocks in a macroblock depends on the sampling format used to digitize the video material. A 4:2:0 macroblock holds four blocks of luminance and two blocks of chrominance information. A 4:4:4 macroblock holds four blocks of luminance and eight blocks of chrominance information.

There are three picture types defined in MPEG-2 Video [5.24, 5.25]. Intracoded pictures (I-pictures) are pictures that are coded in such a way that they can be decoded without knowing anything about other pictures in the video sequence. In a video sequence or group of pictures, the first picture is always an I-picture and provides bootstrap information for the following pictures. Predictive-coded pictures (P-pictures) are decoded by using information from another picture, which was decoded earlier. The information that can be used from the previous picture is determined by motion estimation and is coded in what are called intermacroblocks. A P-picture consists of intracoded macroblocks and predictive-coded macroblocks. The latter are always combined with a motion vector indicating which macroblock to use from a previous picture. A P-picture requires 30 to 50% of the number of bits needed for an I-picture. Bidirectionally coded pictures (B-pictures) also use information from other pictures. Like P-pictures, they can use information provided by a picture that occurred previously. A B-picture can also use information from a picture coming in the future. As in P-pictures, picture information that cannot be found in previous or future pictures is intracoded. B-pictures require approximately 50% of the number of bits needed for a P-picture.

Example 5.6 Figure 5.10 represents an example to encode a picture as a B-picture. The plane that is hidden by the cloud in picture #1 starts to appear in picture #2. If this picture would be coded as a B-picture, the clouds could be borrowed from picture #1, and the front part of the plane could be taken from picture #4. As for picture #4, it would be coded as a P-picture, using the clouds from picture #1. Only the plane would actually be coded in the P-picture.

Besides the picture reordering, B-pictures also require more memory in the decoder because an additional frame needs to be stored for later reference. This makes B-pictures quite a complex feature to implement in MPEG-2 Video. Because of the complexity of B-pictures, MPEG-2 defines subsets (profiles/levels) where B-pictures are not allowed. Sequences of pictures are grouped together to form GOPs. This can be done to support random access or editing functions. A typical, widely used GOP is the sequence IBBPBBPBBPBB [5.24]. All the B- and P-pictures of this GOP can be decoded by accessing only the I-picture or P-pictures, all belonging to this GOP. To support editing, the GOP structure contains a time-stamp. The time stamp format is actually defined by the Society of Motion Picture and Television Engineers (SMPTE) and corresponds to the time code that is also used in video studio equipment. There are two types of time stamps. The first type is usually called a reference time stamp. Reference time stamps are to be found in the Packetized Elementary Stream (PES) syntax, in the program syn-

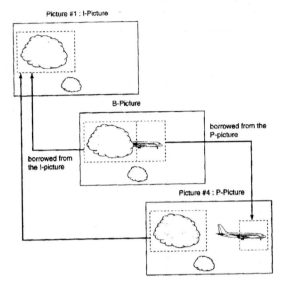

Figure 5.10 Use of a B-picture.

tax and in the transport syntax. The second type of time stamp is called Decoding Time Stamp (DTS). They indicate the exact moment where a video frame or an audio frame has to be decoded or presented to the user, respectively.

Slices are elements to support random access within a picture. A slice is a series of macroblocks. The slice contains information about where to display the contained macroblocks. In case of transmission errors and loss of picture information, the information in a slice can be used to continue the display process within a picture. Not all macroblocks of a picture must be included in slices. From a data compression point of view, slices are not really necessary. They are not coded to have resynchronization points within the picture.

MPEG-2 has introduced the concept of frame pictures and field pictures, along with particular frame prediction and field prediction modes to accommodate coding of progressive and interlaced video. For interlaced sequences, it is assumed that the coder input consists of a series of odd (top) and even (bottom) fields that are separated in time by a field period. Two fields of a frame may be coded separately (field pictures). In this case, each field is separated into adjacent nonoverlapping macroblocks, and the DCT is applied on a field basis. Alternatively, two fields may be coded together as a frame (frame pictures) similar to conventional coding of progressive video sequences. Here, consecutive lines of top and bottom fields are simply merged to form a frame. Notice that both frame pictures and field pictures can be used in a single video sequence.

New motion-compensated field-prediction modes were introduced by MPEG-2 to encode field pictures, and frame pictures efficiently. The concept of field pictures and possible field prediction are illustrated in Figure 5.11 for an interlaced video sequence, which in this figure is assumed to contain only three field pictures and no B-pictures. The top fields and the bottom fields are coded separately [5.24]. Each bottom field is coded using motion-compensated interfield prediction based on the previously coded top field. The top fields are coded using motion-

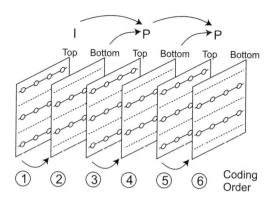

Figure 5.11 The concept of field pictures and possible field prediction.

compensated interfield prediction based on the previously coded bottom field. This concept can be extended to incorporate B-pictures. Generally, the interfield prediction from the decoded field in the same picture is preferred if no motion occurs between fields. An indication about which reference fields are used for prediction is transmitted with the bit stream. Within a field picture, all predictions are field predictions.

Frame prediction forms a prediction for a frame picture based on one or more previously decoded frames. In a frame picture, either field or frame predictions may be used and the particular prediction mode preferred can be selected on a macroblock-by-macroblock basis. As for data compression, it is achieved by combining three techniques:

- Removing picture information that is invisible to the human eye
- Using variable length coding tables
- Using motion estimation

Because of its internal structure, the human eye is quite insensitive to high frequencies in color changes. The idea is, therefore, to represent the picture information in such a way that this characteristic of the eye is used. MPEG-2 uses a method, which is based on DCT, to approximate the original chrominance and luminance information in each block. Instead of using the real color values for each block, a set of frequency coefficients is calculated. This set describes the color transitions in the block. By dividing the resulting coefficients by a certain value, some of them can become zero after rounding. This is the step where picture information is lost. This process is called quantization, and the factors are provided by a quantization matrix. MPEG-2 defines default quantization matrixes, but also allows user-defined quantization matrixes. Quantization is also controlled by a scale factor, which allows the user to adjust the quantization level and the compression ratio. The scale factor is provided for each slice and can optionally be redefined for each macroblock. By having the scale factor in the quantization process, it becomes possible to generate constant bit rate videostreams, which fit into the constraints that might be given by a certain network architecture.

MPEG-2 defines a number of tables with codes to be used for specific patterns in a coefficient data sequence. The trick is to use very short codes with only a few bits for patterns that occur very often in the sequence. The quantization process results in a number of coefficients where certain coefficients equal zero (for example, 2, 0, 0, 1, 0, 0, 1). MPEG-2 Video is coding this sequence of coefficient data by an assigned code for a specific coefficient data pattern. The interpolation of this code returns two values. One value specifies the number of leading zeros in front of a nonzero coefficient. MPEG-2 Video uses the term "run" for this value. The other value is the actual coefficient, which is called "level" in MPEG-2 Video.

Example 5.7 The variable length codes and the corresponding run and level values can be seen in Figure 5.12. Based on this table, the sequence could be represented by the code sequence 0100, 0101 and 0101. The variable length coding tables use between 2 and 13 bits to encode run-level combinations. For the uncovered combinations, MPEG-2 Video defines an escape-coding mechanism. The level is then coded with an actual value.

Coefficient data sequence	Variable length coding table			Coded sequence
	Variable length code	Run	Level	
2, 0, 0, 1, 0, 0, 1	011	1	1	0100, 0101, 0101
	0100	0	2	
	0101	2	1	
	00101	0	3	
	00111	3	1	
	00110	4	1	
	000110	1	2	

Figure 5.12 Variable length coding.

The motion-estimation process uses the macroblocks as basic units for comparison. For each macroblock, the encoder is searching the previous picture (in the case of a P-picture) or the previous and the future pictures (in the case of a B-picture) for a macroblock that matches or closely matches the current macroblock. If such a macroblock is found, the difference between this macroblock and the current macroblock is calculated. The resulting difference is first DCT coded on an 8x8 block basis and then variable length coded, together with the motion vector of the macroblock. At decoding time, the motion vector is used to identify the macroblock in the previous or future picture. The identified macroblock will then be combined with the decoded difference and written into the display or picture buffer. In the optimal case, the current macroblock is found at the same place in the previous picture. This would result in a zero motion vector together with a null difference. MPEG-2 would skip the coding of this macroblock. At decoding time, the previously displayed macroblock would stay on the screen.

MPEG-2 Video Syntax

Generally speaking, a syntax specifies the structure of a bitstream, such as how different parameters, tags, and so forth, are mapped and laid on the bitstream. For multiplexing purposes, it is important for the syntax to provide patterns that can be recognized with an extremely high degree of confidence. These patterns are called synchronization patterns. In addition, an indication of time and of the bit rate of the bit stream may also be provided.

The application requirements that MPEG-2 has addressed made it necessary to develop a formal syntax that supported all requirements. Some syntax elements control the appearance of other syntax elements. Many syntax elements are optional and are only present in the bitstream if a flag indicates it. A flag is mostly located in the syntax structure header [5.22]. In that way, the amount of data that has to be transmitted is further reduced. Instead of transmitting void values, like zeros or special codes, some elements are simply not present in the bit stream. The MPEG-2 standard uses a C-like pseudocode to describe the syntax.

MPEG-2 Video Scalability

The scalability is achieved by the MPEG-2 Video syntax. Video information can be separated into different information streams, which are complementary. Different applications can be realized by combining different streams of information. MPEG-2 uses the term "layer" for the different information streams. The intention of scalable coding is to provide interoperability between different services and to support receivers flexibly with different display capabilities. Receivers either not capable or willing to reconstruct the full resolution video can decode subsets of the layered bit stream to display video at lower spatial or temporal resolution or with lower quality. Another important purpose of scalable coding is to provide a layered video bit stream that is amenable for prioritized transmission. The main challenge here is to deliver video signals reliably in the presence of channel errors, such as cell loss in ATM-based transmission networks or cochannel interference in terrestrial digital broadcasting.

Example 5.8 Flexible supporting multiple resolutions is of particular interest for interworking between HDTV and Standard Definition Television (SDTV). In this case, it is important for the HDTV receiver to be compatible with the SDTV product. Compatibility can be achieved by means of scalable coding of the HDTV source, and the wasteful transmission of two independent bit streams to the HDTV and SDTV receivers can be avoided. Other important applications for scalable coding include video database browsing and multiresolution playback of video in a multimedia environment.

Scalability can be applied for different aspects of video presentation. Although some applications are constricted to low implementation complexity, others call for very high coding efficiency. As a consequence, MPEG-2 has standardized several scalable coding schemes.

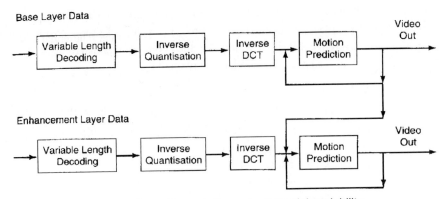

Figure 5.13 The decoding process in the case of spatial scalability.

Spatial scalability has been developed to support displays with different spatial resolutions at the receiver. Lower spatial resolution video can be reconstructed from the base layer. With spatial scalability, the enhancement layer and the base layer data are combined after the IDCT step. The process mainly affected by spatial scalability is motion compensation, which can now use motion vectors from the enhancement layer data or from the base layer data. Decoding flow in the case of spatial scalability is illustrated in Figure 5.13. This functionality is useful for many applications, including embedded coding for HDTV/TV systems, which allows a migration from a digital TV service to higher spatial resolution HDTV services [5.26, 5.27]. The algorithm is based on a classical pyramidal approach for progressive image coding [5.20, 5.28]. Spatial scalability can flexibly support a wide range of spatial resolutions, but adds considerable implementation complexity to the coding scheme.

Temporal scalability defines the possibility to handle different picture rates in a single video stream. This tool was developed with an aim simpler than spatial scalability [5.24]. Namely, stereoscopic video can be supported with a layered bit stream suitable for receivers with stereoscopic display capabilities. The base layer, which provides the basic video picture, can be combined with the enhancement layer to achieve higher frame rates. The enhancement layer uses the base layer to generate final video pictures. Layering is achieved by providing a prediction of one of the images of the stereoscopic video in the enhancement layer based on coded images from the opposite view transmitted in the base layer. Possible usage is in the support of different generations of decoder equipment with different transmission qualities. As in the case of spatial scalability, the enhancement happens after the IDCT step and mainly affects the motion compensation process.

SNR scalability allows for the handling of at least two different video qualities. The video information provided by the base layer can be improved by one or more enhancement layers carrying additional information. However, the base and enhancement layers have the same spatial video resolution [5.22]. If the base layer can be protected from the transmission errors, a version of the video with gracefully reduced quality can be obtained by decoding the base layer bit-stream. The algorithm used to achieve graceful degradation is based on a frequency (DCT domain) scalability technique. At the base layer, the DCT coefficients are coarsely quantized and transmitted to achieve moderate image quality at a reduced bit rate. The enhancement layer encodes and transmits the difference between the nonquantized DCT coefficients and the quantized coefficients from the base layer with finer quantization SZ. At the decoder, the highest quality video signal is reconstructed by decoding both the base layer and the higher layer bit streams. The main application of the SNR scalability is error concealment. In this case, the base layer would carry the most critical information while using a quite robust transport channel in a network. The enhancement layer, bearing the less critical information, could be transported across a transport channel with a lower quality of service. The enhancement process in the case of the SNR scalability happens after the inverse quantization process. The enhancement layer contains mainly DCT coefficients, which are added to the one provided by the base layer. By doing this, the picture quality is refined. Decoding flow in the case of SNR scalability is shown in Figure 5.14. It is also possible to use this method in order to obtain video with lower spatial resolution at the receiver. If the decoder selects the lowest NxN DCT coefficients from the base layer bit stream, nonstandard IDCTs of size NxN can be used to reconstruct the video at a reduced spatial resolution [5.29, 5.30]. However, depending on the encoder and decoder implementation, the lowest layer downscaled video may be subject to drift [5.31].

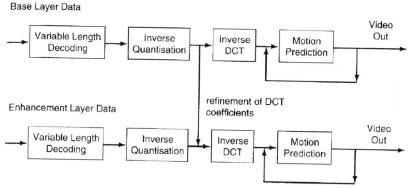

Figure 5.14 The data flow in the decoder for SNR scalability.

Data partitioning is intended to assist with error concealment in the presence of transmission or channel errors in ATM, terrestrial broadcast or magnetic recording environments. Because the tool can be entirely used as a postprocessing and preprocessing tool to any single layer coding scheme, it has not been formally standardized with MPEG-2, but is referenced in the informative Annex of the MPEG-2 Draft International Standard (DIS) document [5.27]. The algorithm is similar to the SNR scalability tool, based on the separation of DCT coefficients, and is implemented with very low complexity compared to the other scalable coding schemes. To provide error protection, the coded DCT coefficients in the bit stream are simply separated and transmitted in two layers with different error likelihoods. In data partitioning, the complete video bit stream is split into parts with relatively higher or lower importance. The most important syntax elements are transmitted on a higher performance channel, and the less important elements are transmitted on a channel with lower performance. A special syntax element, called priority breakpoint, is used to define which parts of the video bit stream syntax are put into which partition.

MPEG-2 Video: Profiles and Levels

Because of the great range of applications, the MPEG-2 standard became quite complex. However, an application might not need the full range of the MPEG-2 Video feature set. If it had to support the complete specification, MPEG-2 equipment would be expensive. Therefore, this standard defines profiles and levels to define subsets of MPEG-2 Video. Profiles and levels defined by MPEG-2 Video are shown in Table 5.5.

Table 5.5 Profiles and levels for MPEG-2 Video.

Profiles	Algorithms	Levels
Simple Profile (SP)	Includes all functionalities provided by the Main profile, but does not support B-picture prediction modes and 4:2:0 YUV-representation.	Low Level (LL)
Main Profile (MP)	Supports functionality for coding-interlaced video, random access, B-picture prediction modes and 4:2:0 YUV-representation with nonscaling coding algorithm.	Main Level (ML)
SNR Scalable Profile (SNR)	Supports all functionalities provided by the MP plus an algorithm for SNR scalable coding (two layers allowed) and 4:2:0 YUV-representation.	High 1440 Level (H14)
Spatial Scalable Profile (Spatial)	Supports all functionalities provided by the SNR Scalable Profile plus an algorithm for spatial scalable coding (two layers allowed) and 4:2:0 YUV-representation.	High Level (HL)
High Profile (HP)	Supports all functionalities provided by the Spatial Scalable Profile plus the provision to support three layers with the SNR and Spatial scalable coding modes and 4:2:2 YUV-representation for improved quality requirements.	

A profile is described as a well-defined subset of the video syntax. Certain syntaxes defined by MPEG-2 Video are not valid and cannot be decoded if the decoder only supports a low profile. Neither the Simple nor the Main profile supports any kind of scalability. However, the lower profiles are always a subset of the higher profiles. A decoder supporting the spatial profile is required to support spatial and SNR scalabilities.

Table 5.6   MPEG-2 Main level at Main profile values.

Parameter	Main level at Main profile value
Samples/line	720
Lines/frame	576
Frames/sec	30
Luminance samples/sec	1,036,8000
Max. video data rate in Mb/s	15
Max. size of decoder buffer (bits)	1,835,008

A level defines values for certain parameters in the video bit stream. The levels describe the number of samples per line, the number of lines per frame and the number of frames per second. Profiles and levels are combined to define exactly which selection or subset from the MPEG-2 Video toolkit is used. The combination Main level at Main profile defines a sufficient subset of the MPEG-2 Video functionality. Table 5.6 shows some of the Main level–Main profile values. Profiles and levels are organized hierarchically, and MPEG-2 Video defines a forward compatibility between different profiles and levels.

5.4.2 MPEG-2 Audio

MPEG-2 Audio is a backward-compatible multichannel extension of the MPEG-1 Audio standard. Namely, the audio part of the MPEG-2 standard is to a great extent based on the MPEG-1 audio part, and a great deal of compatibility exists. The compatibility aspect is valid in two senses [5.16, 5.22]:

- Existing MPEG-1 equipment can make a partial decoding of MPEG-2 signals by extracting the MPEG-1 compatible part (backward compatibility).
- MPEG-2 equipment can decode MPEG-1 signals (forward compatibility).

The fact that the core of the MPEG-2 bit stream is an MPEG-1 bit stream enables fully compatible decoding with an MPEG-1 decoder. In addition, the need to transfer two separate bit streams, called simulcast (one for two-channel stereo and another one for the multichannel audio program), is avoided. This incurs some cost in coding efficiency for the multichannel audio signal compared to AAC, which is a Nonbackward Compatible (NBC) coding algorithm [5.16].

Any combinations of MPEG-1 and MPEG-2 audio and video can be handled by the system as specified by the MPEG Systems standards [5.11, 5.16]. For example, MPEG-2 Audio can be used with MPEG-1 Video. Also, MPEG-1 Audio can be used with MPEG-2 Video without any restrictions.

MPEG-2, as well as MPEG-1, audio compression describes three degrees of compression: layers 1, 2 and 3. These layers represent a family of coding algorithms. Basically, the complexity of the encoder and decoder, the encoder/decoder delay and the coding efficiency increases when going from layer 1 through layer 2 to layer 3. Layer 1 has the lowest complexity and is specifically suitable for applications where the encoder complexity also plays an important role. Layer 2 requires a more complex encoder and a slightly more complex decoder and is directed toward one-to-many applications, that is, one encoder serves many decoders. Compared to layer 1, layer 2 is able to remove more of the signal redundancy and to apply the psychoacoustic threshold more efficiently. Layer 3 is more complex and is directed toward lower bitrate applications due to the additional redundancy and irrelevancy extraction from enhanced frequency resolution in its filterbank. The main functional modules of the lower layers are also used by the higher layers. The subband filter of layer 1 is also used by layer 2 and layer 3. Layer 2 adds a more efficient coding of side information. Layer 3 adds a frequency transform in all the subbands. The three layers have been defined to be compatible in a hierarchical way. The level of compression, the demands for processing power and the sound quality all increase proportionally with the layer number. The required transmission bandwidth decreases with the layer number. Layer 1 has the lowest compression rate, about four times. It demands the smallest amount of processing power and has the lowest delay, realistically below 50 ms. Layer 1 also has the highest requirements for transmission bandwidth, with the highest possible level of compression, going to 448 Kbps in stereo, with the lowest possible level of compression. The sound quality of the layer 1 signal is furthermore inferior to what can be obtained by layers 2 and 3. Layer 3 is intended to yield the best sound quality of the three layers, but it achieves a compression ratio, 10:1. On the other hand, the processing time is more than three times longer. Central characteristics of the three layers of MPEG-2 audio coding are presented in Table 5.7. These three layers are compatible in the sense that a layer N decoder can decode layer N, as well as all the layers below. For example, a layer 3 decoder can decode layers 1, 2 and 3 bit streams, but a layer 2 decoder can only decode bit streams from layers 1 and 2.

The overall audio compression and encoding processes for all three layers are shown in Figure 5.15. In the case of signals with more than one channel, each channel is treated separately. The filter bank used in MPEG-2 audio coding can be one of the following types: polyphase or a hybrid polyphase and MDCT. Regardless of the type, the time domain samples are here con-

Table 5.7 Main characteristics of the three layers of MPEG-2 audio coding.

Layer	Approximate compression ratio	Target bit rate	Realistic delay	Theoretical minimum delay
1	4:1	192 Kb/s	<50 ms	19 ms
2	6:1	128 Kb/s	100 ms	35 ms
3	10:1	64 Kb/s	150 ms	58 ms

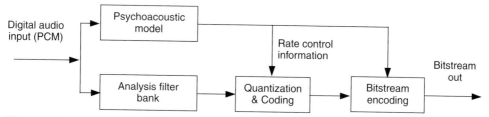

Figure 5.15 The overall audio compression and encoding processes for three layers.

verted into the same number of frequency domain samples. The output of the filter bank is a number of subbands of equal bandwidth. In layers 1 and 2, a filter bank yielding 32 subbands, each containing 12 or 36 frequency domain samples, respectively, is used. In layer 3, the number of subbands can be either 192 or 576. In parallel with the filter bank, the psychoacoustic model process calculates the Signal-to-Mask Ratio (SMR) for each subband. The primary psychoacoustic effect that a perceptual audio coder uses is called auditory masking, where parts of a signal are not audible due to the function of the human auditory system. The parts of the signal that are masked are commonly called irrelevant, as opposed to parts of the signal that are removed by a source coder (lossless or lossy), which are termed redundant.

In order to remove this irrelevance, the encoder has a psychoacoustic model. This psychoacoustic model analyzes the input signals within consecutive time blocks. It then determines for each block the spectral components of the input signals within consecutive time blocks and determines for each block the spectral components of the input audio signal by applying a frequency transform. Then it models the masking properties of the human auditory system and estimates the just-noticeable noise level, sometimes called the threshold of masking.

In parallel, the input signal is fed through a time-to-frequency mapping, resulting in spectrum components for subsequent coding. In its quantization and coding stages, the encoder tries to allocate the available number of data bits in a way that meets both the bit rate and masking requirements. The information on how the bits are distributed across the spectrum is contained in the bit stream as side information.

The principal function of the psychoacoustic model is to calculate a new bit allocation for the frequency samples in the subbands. The new bit allocation is aimed at efficiently allocating the available bits to each of the subbands. If a given subband does not have power, no bits are allocated. The principle applied is the fact that frequencies with higher power make nearby frequencies of lower power inaudible to human hearing. The new bit allocation is calculated separately for each subband.

To obtain the SMR, it is necessary to first make a time-to-frequency domain compression of the original audio signal. This works in parallel with the filter bank process. This conversion is done by the FFT technique, which allows a time-to-frequency domain transformation with a better spectral resolution than that of the polyphase filter bank. On the basis of frequency domain data, the maximum power in each subband is found, the tonal and nontonal (noiselike) parts of the audio signal are determined, the absolute masking threshold is identified, and, finally, the masking thresholds of all the individual subbands are calculated. A global masking threshold is calculated by adding all the individual masking thresholds with the absolute threshold. Now it is possible to calculate for each subband the difference between the actual signal and the masking threshold and to obtain the SMR.

The bit or noise allocation process uses the output samples from the filter bank and the SMR information from the psychoacoustic model to determine the amount of quantization noise that is tolerable to each subband. The higher the allowable quantization noise means the lower the number of necessary bits to represent each sample. In layer 1, a scale factor is applied for each subband (containing 12 audio samples). In layer 2, each subband contains 36 audio samples, which are split into 3 groups of 12 samples. Each of the three groups can have separate scale factors. The scale factors are calculated separately for each subband. The scale factor, which will be transmitted along with the audio samples to the decoder, expresses a certain factor that the resolution steps of the audio samples will have to be multiplied by the decoder. It is possible to express small, as well as large, amplitude steps with a relatively small number of bits. For layers 1 and 2, the number of bits used to represent the audio samples in each subband is variable when the requirements of the bit rate and those of the psychoacoustic model are to be combined. In layer 3, this is done somewhat differently because of noise injected in the subbands is a variable. In both cases, the encoder starts an iterative process of increasing the accuracy of the subband quantization, to the limit possible within the specified bit rate.

In the bit stream formatting process, the subband frequency samples are joined together with the scale factor information in the audio data field of the audio frame. The audio frame contains header, error check and ancillary data fields. For layer 3, Huffman coding of the quantized frequency samples is applied at this step. It means that, instead of fixed-length PCM for each frequency sample, as used in layers 1 and 2, a variable-length code is used. Huffman coding represents the most common bit combinations in the data flow with the shortest codes, and it represents the most uncommon with the longest codes. Thus, further reduction in the bit rate can be obtained.

The decoder is much less complex. It does not require a psychoacoustic model and bit allocation procedure. Its only task is to reconstruct an audio signal from the coded spectral components and associated side information.

With MPEG-2, it is possible to use only half the sampling rate in MPEG-1 and still obtain a very good sound quality. This is especially interesting for applications such as commentary channels, multilingual channels and multimedia. The sampling rate available in MPEG-2 allows the sampling of the time domain signal to be done with 16, 22.05 or 24 KHz samples per second for all three layers. This gives an upper frequency limit of 7.5, 10.3 and 11.25 KHz, respectively. To allow transmission of more realistic stereoscopic representation, MPEG-2 supports audio channels, which together can convey a surround stereo image. The five channels are left channel (L), right channel (R), center channel (C), left rear surround channel (LS) and right rear surround channel (RS). This is also called 3/2 stereo because it makes use of three front loudspeakers and two loudspeakers in the rear. Surround setup in 3/2 stereo is shown Figure 5.16. In addition, a Low Frequency Enhancement (LFE) channel is available for a subwoofer signal in

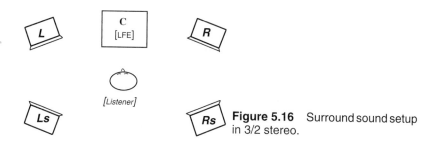

Figure 5.16 Surround sound setup in 3/2 stereo.

the range from 15 to 120 Hz. This channel is mainly used for special effects. An alternative configuration of the five-channel system of MPEG-2 Audio is the application of multilingual/commentary channels, accompanying a specific program with, for example, bilingual comments or sound tracks. The MPEG-2 specifications allow for up to seven multilingual/commentary channels per program. As the surround sound signal consists of five channels, a great deal of redundancy often exists among these five channels, so, in many cases, the same piece of audio information may appear in two or more of the five surround channels with different delays. The MPEG-2 Audio compression can use this redundancy to achieve a higher compression ratio.

The audio part of the MPEG-2 standard has focused on the compatibility with MPEG-1 Audio. However, the compatibility has some drawbacks. For example, the used techniques have not been able to eliminate completely the fact that there is a trade-off between maintaining compatibility and achieving optimal sound quality at a given bit rate. Listening tests have proved the backward compatible coding techniques to be somewhat inferior to other nonbackward compatible techniques under the condition that the same number of bits per second were available.

In the framework of MPEG-2, work was going on in the field of AAC, also known as NBC coding. The MPEG-2 AAC standard is the state-of-the art audio standard that provides very high audio quality at a rate of 64 Kb/s/channel for multichannel generation. It provides a capability of up to 48 main audio channels, 16 low frequency effects channels, 16 overdub/multilingual channels and 16 data streams. Up to 16 programs can be described, each consisting of any number of the audio and data elements. The AAC standard has three profiles called main profile (AAC), Low Complexity (LC) profile and Scalable Sampling Rate (SSR) profile. The main profile is intended for use when processing and especially memory are not a premium. The LC profile is intended for use when cycles and memory use are constrained, and SSR profile is intended for use when a scalable decoder is required. The main and LC profiles have been tested at 320 Kb/s for five-channel audio programs, and both have demonstrated better quality than competing audio-coding algorithms running at 640 Kb/s for the five-channel program.

5.4.3 MPEG-2 Systems

The MPEG-2 Systems standard is an ISO/IEC standard that defines the syntax and semantics of bit streams in which digital audio and visual data are multiplexed. Such bit streams are said to be MPEG-2 Systems compliant. However, this specification does not mandate how equipment that produces, transmits or decodes such bit streams should be designed. As a result, the specification can be used in a diverse array of environments, including local storage, broadcast (terrestrial and satellite) and interactive environments [5.23]. This standard was industry driven and complemented the MPEG-2 activities in audio and video coding. The consumer TV industry actively participated in the definition of MPEG-2 Systems, to ensure that a low-complexity receiver could be built at a reasonable cost.

The MPEG-2 Systems standard enables the widest interoperability in digital video and audio applications and services. The video and audio part of the MPEG-2 standard defines the format with which audio or video information is presented. However, to use this data in a complete video delivery chain, some additional requirements have to be addressed. These requirements result from the applications in which the audio and video data is used. They are also related to the technology that is used to deliver the data.

Example 5.9 Let us take the standard TV broadcasting application. In TV broadcasting, there is a need to transport different programs to the consumer, who can freely choose among them. In other words, at some point, different audio-video streams have to be multiplexed together and have to be delivered together to the consumer. This multiplexing is usually done somewhere in TV broadcasting networks, like satellite or cable distribution systems. In the case of a satellite distribution system, different programs delivered by different broadcasting stations are multiplexed together at some satellite uplink station. The collection of programs, sometimes also called bouquet, is transmitted to the satellite, which then sends it down to earth in a Direct To Home (DTH) broadcasting system.

MPEG-2 Systems are using data structures that are commonly referred to as packets in the data communication world. Packets always consist of a packet header and the packet payload and can be of fixed or variable size. The basic idea behind a packet concept is to create a flexible mechanism to transport any kind of data. Usually the packet header contains the information that is needed to process the data in the packet payload. Depending on the application scenario where the packets are used, it makes sense to use variable or fixed-sized packets. For example, in a network environment, it is useful to have fixed-sized packets that are relatively short. This helps to optimize the network equipment that is processing the packets because the length of the packet is always the same. If a part of the packet is corrupted for some reason (loss of data in the network), only that information is lost. Figure 5.17 shows the scope of the MPEG-2 Systems specifications.

MPEG-2 Systems provide a two-layer multiplexing approach. The first layer is dedicated to ensuring tight synchronization between video and audio. It is a common way for presenting all the different materials that require synchronization (video, audio and private data). This layer is called PES. The second layer is dependent on the intended communication medium. The specification for error-free environments, such as local storage, is called MPEG-2 Program Stream, and the specification addressing error-prone environment is called MPEG-2 Transport Stream.

In MPEG-2, the output bit stream of an audio-video encoder or the private data bit stream is called the elementary stream. In the case of audio or video, this elementary stream can be organized into access units. An access unit is a picture in the case of a video elementary stream or an audio frame, in the case of an audio elementary stream.

MPEG-2 (Generic Coding of Moving Pictures and Associated Audio)

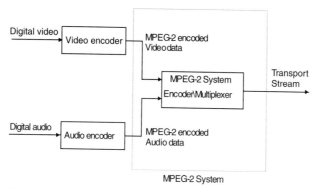

Figure 5.17 Scope of the MPEG-2 Systems standard in relation to the video and audio parts and the broadband equipment.

An elementary stream is then converted into a PES, which consists of PES packets. Each PES packet consists of the PES packet payload (which is a variable-sized part of the elementary stream) and a PES packet header. By having the size of the payload variable, the payload of the PES packets can be an exact access unit of the elementary stream.

A PES packet is a way to packetize the elementary streams uniformly. Embedded in a PES packet, elementary streams may be synchronized with time stamps. They are not protected. The PES packets may be of variable length, which allows them also to be of fixed length. PES packets may be rather long packets. However because elementary streams are continuous streams, it is also possible to know that a PES packet is finished when the next PES packet arrives. Sometimes the length is not relevant (for video PES packets).

The PES packet is mapped into the MPEG-2 Transport Stream Packet (TSP), also consisting of a header and a payload part. Consecutive transport packets form the MPEG-2 transport stream. The functionality of the MPEG-2 Systems layer processor is summarized in Figure 5.18. An MPEG-2 Systems layer processor could extract the PES packets out of the transport stream and put them into program stream packets. MPEG-2 transport streams carry transport packets. These packets carry two types of information: the compressed material and the associated signaling tables. A transport packet is identified by its Packet Identifier (PID). Each PID is assigned to carry either one particular compressed material (and only this material) or one particular sig-

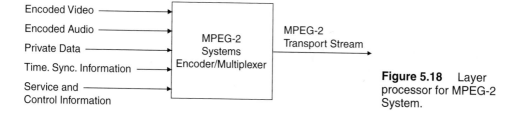

Figure 5.18 Layer processor for MPEG-2 System.

naling table. The compressed material consists of elementary streams, which may be built from video, audio or data material. These elementary streams may be tightly synchronized (because it is usually necessary for digital TV programs or for digital radio programs) or not synchronized (in the case of programs offering downloading of software or games, as an example).

The associated signaling tables consist of the description of the elementary streams, which are combined to build programs, and in the description of those programs. Tables are carried in sections. The signaling tables are called Program Specific Information (PSI).

Transport packets are 188 bytes long because MPEG-2 wanted these packets to be carried across ATM. At that time, according to the AAL, ATM cells are supposed to have a payload of 47 bytes (4x47 bytes = 188 bytes). A transport packet of 188 bytes maps exactly into the payload of four ATM cells. An ATM cell has 48 bytes of payload, but one byte of the payload is used for the overhead information of the AAL.

MPEG-2 Systems distinguish between two kinds of transport streams: Single Program Transport Streams (SPTS) and Multiprogram Transport Streams (MPTS). The SPTS contains different PES, which all share a common time base. The different PES could carry video, different audio and perhaps data information. All would be used with the same time base. An application for this is a movie transmitted in different languages. MPTS is multiples of a number of SPTS. The MPEG-2 System hierarchy based on different transport stream variations is presented in Figure 5.19.

The MPEG-2 transport packet consists of 4 bytes of header information, a variable length adaptation field and the payload, containing the PES packets. An MPEG-2 transport packet header is shown in Figure 5.20. One of the most important fields in the header is the PID. The PID is used to identify transport packets that carry PES data from the same elementary stream. It also defines

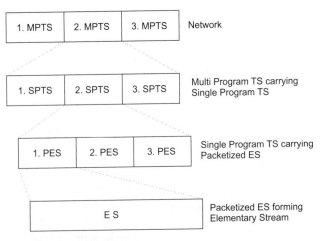

Figure 5.19 MPEG-2 Systems hierarchy based on the different transport streams.

MPEG-2 (Generic Coding of Moving Pictures and Associated Audio)

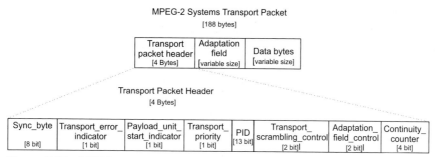

Figure 5.20 MPEG-2 transport packet header.

the type of data that is transported by the packet payload. Besides the PID, the transport packet header contains several control fields, which are used to identify the appearance of other fields in the header and also to provide the information about the payload of the transport packet. A very important field is the adaptation field. It is an optional field in the transport stream packet header, which contains additional information that is used for clock recovery and splicing functions. One of the most important fields in the adaptation field is the Program Clock Reference (PCR) field. This field contains time stamp information that is used by the decoder to synchronize its clock to the encoder clock. The adaptation field also has a section to transport private data, which is not defined by the MPEG-2 standard.

5.4.4 MPEG-2 DSM-CC

MPEG-2 DSM-CC is the specification of a set of protocols that provides the control functions and operations specific to managing MPEG-1 and MPEG-2 bit streams. These protocols may be used to support applications in both standalone and heterogeneous network environments [5.32]. In the DSM-CC model, a stream is sourced by a server and delivered to a client. Both the server and the client are considered to be users of the DSM-CC network. DSM-CC defines a logical entity called the Session and Resource Manager (SRM), which provides a logically centralized management of the DSM-CC sessions and resources. The DSM-CC reference model is shown in Figure 5.21.

To a user, DSM-CC allows the delivery of multimedia across a guaranteed end-to-end QoS irrespective of the transport technology that the user is using. Thus, it allows the end-users the choice of the transport technology and media within the locality that the service is provided and that best suits their budget.

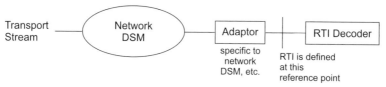

Figure 5.21 DSM-CC reference model [5.25]. ©1995 ISO/IEC.

The DSM-CC protocols supplement other networking protocols, like BISDN signaling or transport protocols, in order to cover all requirements of video networks. DSM-CC signaling assumes that the network links between the different entities are already established.

After an initial link has been set up between the two entities in the video delivery network, DSM-CC provides the functionality to continue the setup of an application session. Because this session setup happens at the interface between network and user equipment, DSM-CC defines a DSM-CC user to network protocol. After the application session has been set up, further logical links are established between a video server and a set-top box. One logical link might be used for user data (like MPEG-2-coded video), and another logical link might be used to control what is happening on the user data link.

DSM-CC defines a set of services to manipulate a videostream in the server, which can be used by the client. Because these services are only relevant between two user entities, that is, the server and the client, the DSM-CC standard refers to them as the DSM-CC user-to-user interface [5.33]. Links between server and client are shown in Figure 5.22, and Figure 5.23 illustrates the two different interfaces that DSM-CC addresses.

The user-network interface has much in common with OSI layer 3 signaling protocols. The user-user part of DSM-CC is application layer oriented and uses an object-oriented approach.

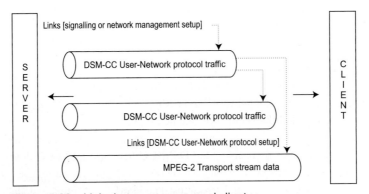

Figure 5.22 Links between server and client.

5.5 MPEG-4—Coding of Audiovisual Objects

Multimedia communication is the possibility to communicate audiovisual information with the following characteristics:

- Is natural, synthetic or both
- Is real time and non-real time
- Supports different functionalities responding to user's needs
- Flows to and from different sources simultaneously
- Does not require the user to bother with the specifics of the communications channel, but uses a technology that is aware of it
- Gives users the possibility to interact with the different information elements

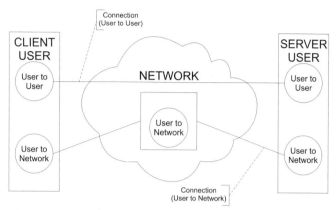

Figure 5.23 DSM-CC user-to-network and user-to-user interfaces.

- Lets the users present the results of their interaction with content in a way that suits their needs [5.9]

MPEG-4 is the MPEG project that started in July 1993 and was developed to provide enabling technology for the previous seven items. It has reached Working Draft level in November 1996, Committee Draft level in November 1997, and International Standard level in the beginning of 1999. To reach its own target, MPEG-4 follows an object-based representation approach where an audiovisual scene is coded as a composition of objects, natural as well as synthetic, providing the first powerful hybrid playground. Thus, the objective of MPEG-4 is to provide an audiovisual representation standard supporting new ways of communication, access and interaction with digital audiovisual data and offering a common technical solution to various service paradigms—telecommunications, broadcast and interactive—among which the borders are disappearing. MPEG-4 will supply an answer to the emerging needs of application fields such as video on the Internet multimedia broadcasting; content-based audiovisual database access; games; audiovisual home editing; advanced audiovisual communications, notably across mobile networks; teleshopping; and remote monitoring and control [5.34].

The fully backward compatible extensions under the title of MPEG-4 Version 2 were frozen at the end of 1999 to acquire the formal International Standard Status in early 2000. Some work on extensions in specific domains is still in progress. MPEG-4 builds on the proven success of three fields [5.35, 5.36]:

- Digital television
- Interactive graphics applications (synthetic content)
- Interactive multimedia (distribution of and access to content on the Web)

5.5.1 Overview of MPEG-4: Motivations, Achievement, Process and Requirements

MPEG-1 and MPEG-2 standards have given rise to widely adopted commercial products and services, such as Video-CD, DVD, digital television and digital audio broadcasting. The aim of the MPEG-4 standard is to define an audiovisual coding standard to address the emerging needs of the communications, interactive and broadcasting service models, as well as the needs of the mixed service models resulting from their technological convergence. The convergence of the three separate application areas, communications, computing and TV/film/entertainment, was evident in the mutual cross-fertilization with functionalities characteristic of each one of these application areas such as user interactivity, sythentic and natural hybrid coding, and Web-based services emerging more and more.

Although audiovisual information, notably the visual part, was until recently only carried across a few networks, the trend is now toward the generalization of visual information on every single network. New mobile devices can finally make the decade-long promise of audiovisual communications turn into reality. The explosion of the Web and the acceptance of its interactive mode of operation have clearly shown that the traditional television paradigm would no longer suffice for audiovisual services. Users want to have access to audio and video like they now have access to text and graphics. This requires moving pictures and audio of acceptable quality at low bit rates on the Web and providing Web-type interactivity with live content.

The MPEG-4 standard provides a set of technologies to satisfy the needs of authors, service providers and end-users. For authors, MPEG-4 enables the production of content that has greater reusability and that has greater flexibility than is possible today with individual technologies, such as digital television, animated graphics, Web pages and their extensions. For network services providers, MPEG-4 offers transparent information, which can be interpreted and translated into the appropriate native signaling messages of each network with the help of relevant standards bodies. The foregoing excludes QoS considerations, for which MPEG-4 provides generic QoS descriptors for different MPEG-4 media. For end-users, MPEG-4 brings higher levels of interaction with content within the limits set by the author. It also brings multimedia to new networks, including those employing a relatively low bit rate, and mobile networks. An MPEG-4 application document exists on the MPEG home page and describes many end-user applications, including interactive multimedia broadcast and mobile communications. MPEG-4 achieves these goals by providing standardized ways to do the following:

- Represent units of aural, visual or audiovisual content called media objects. These media objects can be of natural or synthetic origin; this means they could be recorded with a camera or microphone or generated with a computer.
- Describe the composition of these objects to create compound media objects that form audiovisual scenes.
- Multiplex and synchronize the data associated with media objects so that it can be transported across network channels providing a QoS appropriate for the nature of the specific media objects.

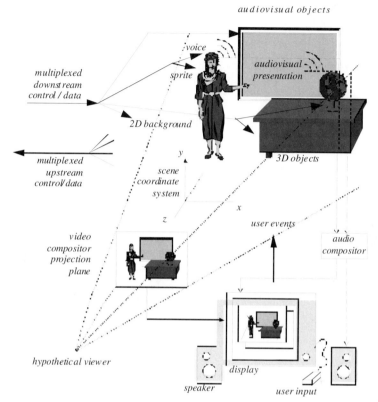

Figure 5.24 An example of an MPEG-4 audiovisual scene [5.36]. ©2001 ISO/IEC.

- Interact with the audiovisual scene generated at the receiver's end.

An audiovisual scene is depicted in Figure 5.24. The figure contains compound media objects that group primitive media objects together. Primitive media objects correspond to leaves in the descriptive tree, and compound media objects encompass entire subtrees.

A major difference from previous audiovisual standards on the basis of new functionalities, is the object-based audiovisual representation model that underpins MPEG-4. The MPEG-4 object-based architecture is shown in Figure 5.25. An object-based scene is built using individual objects that have relationships in space and time and that offer a number of advantages. First, different object types may have different suitable coded representations. A synthetic moving head is clearly best represented using animation parameters, but video benefits from a smart representation of pixel values. Second, it allows harmonious integration of data into one scene. Third, interacting with the objects and hyperlinking from them is now feasible. There are more advantages, such as selective spending of bits, easy sense of content without transcoding, providing sophisticated schemes for scalable content on the Internet, and so forth.

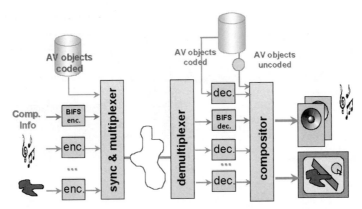

Figure 5.25 The MPEG-4 object-based architecture [5.8]. ©2000 Elsevier.

MPEG-4 audiovisual scenes are composed of several media objects organized in a hierarchical fashion. At the leaves of the hierarchy, we can find primitive media objects like still images (a fixed background), video objects (a talking person without the background), audio objects (the voice associated with that person), and so forth.

MPEG-4 standardizes a number of primitive media objects and is capable of representing both natural and synthetic content types, which can be 2D or 3D. MPEG-4 also defines the coded representation of an object, such as text and graphics, talking synthetic heads and associated text used to synthesize the speech and to animate the head and synthetic sound. A media object in its coded form consists of descriptive elements that allow handling the object in an audiovisual scene. Each media object can be represented in its coded form, independent of its surroundings or background.

The applications that benefit from what MPEG-4 brings are found in different environments [5.37]. Although MPEG-4 is a rather big standard, it is structured in a way that solutions are available at the measure of the needs. The task of implementers is to extract from the MPEG-4 standard the technological solutions adequate to their needs.

Media Objects

Media objects may need streaming data, which is conveyed in one or more elementary streams. An object descriptor identifies all streams associated with one media object. Each stream itself is characterized by a set of descriptors for configuration information, for example, to determine the required decoder resources and the precision of encoded timing information. Furthermore, the descriptors may carry hints to the QoS that it requests for transmission (maximum bit rate, bit error rate, priority and so forth.).

Synchronization of elementary streams is achieved through time stamping of individual access units within elementary streams. The synchronization layer manages the identification of such access units and the time stamping. Independent of the media type, this layer allows identification of the type of access units (video or audio frames or scene-description commands) in

MPEG-4—Coding of Audiovisual Objects

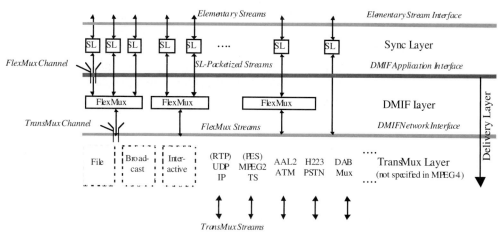

Figure 5.26 The MPEG-4 Systems layer model [5.36]. ©2001 ISO/IEC.

elementary streams and recovery of the media objects or scene-description time base. It also enables synchronization among them. The synchronized delivery of streaming information from source to destination, exploiting different QoS as available from the network, is specified in terms of the synchronization layer and a delivery layer containing a two-layer multiplexer as depicted in Figure 5.26. The first multiplexing layer is managed according to the DMIF. This multiplex may be embodied by the MPEG-defined FlexMux tool, which allows grouping of Elementary Streams (ESs) with a low multiplexing overhead. Multiplexing at this layer may be used. An example is grouping ES with similar QoS requirements, reducing the number of network connections or the end-to-end delay. The transport multiplexing (TransMux) layer models the layer that offers transport services matching the requested QoS. Only the interface to this layer is specified by MPEG-4, and the concrete mapping of the data packets and control signaling must be done in collaboration with the bodies that have jurisdiction over the respective transport protocol. Use of the FlexMux multiplexing tool is optional, and this layer may be empty of the underlying TransMux instance providing all the required functionality. However, the synchronization layer is always present. With regard to the MPEG-4 Systems layer model it is possible to do the following [5.38]:

- Identify access units, transport timestamps and clock reference information and identify data loss
- Optionally interleave data from different ESs into FlexMux streams
- Convey control information
- Indicate the required QoS for each ES and FlexMux stream
- Translate such QoS requirements into actual network resources
- Associate ESs to media objects
- Convey the mapping of elementary streams to FlexMux and TransMux channels

In general, the user observes a scene that is composed following the design of the scene's author. Depending on the degree of freedom allowed by the author, the user has the possibility to interact with the scene. It is also important to have the possibility to identify the intellectual property in MPEG-4 media objects. To support this, MPEG has worked with representatives of different creative industries in the definition of syntax and tools. MPEG-4 incorporates identification of the intellectual property by storing unique identifiers that are issued by international numbering systems. These numbers can be applied to identify a current rights holder of a media object.

MPEG-4 Version 1

The MPEG-4 requirements have been addressed by the six parts of the finalized MPEG-4 Version 1 standard.

Part 1: Systems—Specifies scene description, multiplexing, timing identification, synchronization and recovery mechanisms, buffer management and management and protection of intellectual property [5.39].

Part 2: Visual—Specifies the coded representation of natural and synthetic visual objects (computer-generated scenes). This will, for example, allow the virtual presence of videoconferencing participants. The tools and algorithms of the MPEG-4 Visual standard will support bit rates (typically between 5 Kb/s and 10 Mb/s), formats (progressive as well as interlaced video) and resolutions (typically from sub-QCIF to beyond TV). Also, compression efficiency, content-based functionalities; scalability of textures, images and video; shape and alpha channel coding; robustness in error-prone environments and face animation will be supported [5.40].

Part 3: Audio—Specifies the coded representation of natural and synthetic audio objects. It facilitates a wide variety of applications that could range from intelligible speech to high-quality multichannel audio and from natural sounds to synthesized sounds. In particular, it supports the highly efficient representation of audio objects, consisting of speech signals, synthesized speech, general audio signals and synthesized audio and bounded-complexity synthetic audio [5.35].

Part 4: Conformance testing—Defines conformance conditions for bit streams and devices. This part is used to test MPEG-4 implementations [5.41].

Part 5: Reference software—Includes software corresponding to most parts of MPEG-4 (normative and non-normative). It can be used for implementing compliant products [5.42].

Part 6: DMIF—Defines session protocol for the management of multimedia streaming over generic delivery technologies [5.43].

Parts 1 through 3 and 6 specify the core MPEG-4 technology, and Parts 4 and 5 are supporting parts. Parts 1, 2 and 3 are delivery independent and leave to Part 6 (DMIF) the task of dealing with the delivery layer. Although the various MPEG-4 parts are rather independent and thus can be used by themselves, they were developed to achieve maximum benefit results when they are used together. In October 1998, the first set of MPEG-4 standards was frozen. Work on MPEG-4 continued for Version 2, which adds tools to the MPEG-4 standard in the form of new profiles.

MPEG-4 Version 2

Version 2 is a backward compatible extension of Version 1. Figure 5.27 depicts the relationship between the two versions. MPEG-4 Version 2 is formally seen as amendments to the various parts of Version 1. The Systems layer of Version 2 is backward compatible with Version 1. In the areas of audio and visual, Version 2 adds profiles to Version 1 [5.35].

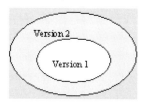

Figure 5.27 Relationship between MPEG-4 Versions 1 and 2.

Version 2 of the MPEG-4 systems extends Version 1 to cover issues like extended Binary Format for Scene Description (BIFS) functionalities and Java programming language (MPEG-J) support. Version 2 also specifies a file format to store MPEG-4 content. The version 2 BIFS (advanced BIFS) includes the following functionalities: advanced environment modeling in interactive virtual scenes; body animation of either a default body model present at the decoder or of a downloadable body model; chroma keying, which is used to generate a shape mask and a transparency value for an image or a video sequence; inclusion of hierarchical 3D meshes to BIFS scenes and associating interactive commands to media nodes. As opposed to the parametric system offered by MPEG-4 Version 1, MPEG-J is a programmatic system that specifies an Application Programming Interface (API) for interoperation of MPEG-4 media players with Java programming language code. Location of interfaces in the architecture of an MPEG-J enabled MPEG-4 system is shown in Figure 5.28. The lower half of this drawing depicts the parametric MPEG-4 system player, also referred to as the presentation engine. The MPEG-J subsystem controlling the representation engine, also referred to as the application engine, is depicted in the upper half of the drawing. The Java programming language application is delivered as a separate elementary stream to the MPEG-4 terminal. For a scene-graph API, the objective is to provide access to the scene graph: to inspect the graph, to alter nodes and their field and to add and remove nodes within the graph. The resource manager API is used for regulation of performance. It provides a centralized facility for managing resources. The terminal capability API is used when program execution is contingent upon the terminal configuration and its capabilities, both static (they do not change during execution) and dynamic. The media decoder API allows the control of the decoders that are present in the terminal. The network API provides a way to interact with the network and is compliant to the MPEG-4 DMIF application interface. Complex applications and enhanced interactivity are possible with these basic packages.

MPEG-4 Visual Version 2 adds technology in the following areas:

- Increased flexibility in object-based scalable coding
- Improved coding efficiency

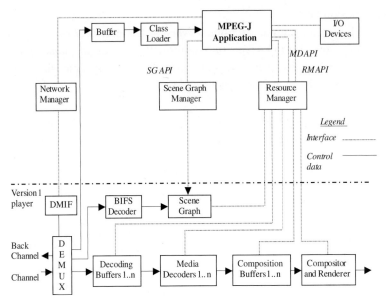

Figure 5.28 Location of interfaces in the architecture of an MPEG-J enabled MPEG-4 system [5.36]. ©2001 ISO/IEC.

- Improved temporal resolution stability with the low buffering delay
- Improved error robustness
- Added coding of multiple views. Intermediate views or stereoscopic views are supported based on the efficient coding of multiple images or video sequences. A particular example is the coding of stereoscopic images or video by redundancy reduction of information contained between the images of different views.

This version improves the motion estimation and compensation of rectangular and arbitrary-shaped objects and significantly improves the texture coding of arbitrary-shaped objects. In the area of motion estimation and compensation, two new technologies are introduced: Global Motion Compensation (GMC) and quarter-pel motion compensation. By GMC, we mean a method based on global motion estimation, image warping, motion trajectory coding and texture coding for prediction errors. On the other hand, quarter-pel motion compensation enhances the precision of the motion compensation scheme, at the cost of only small syntactical and computational overhead. An accurate motion description leads to a smaller prediction error and, hence, to better visual quality. In the area of texture coding, the Shape-Adaptive DCT (SA-DCT) improves the coding efficiency of arbitrary-shaped objects. The SA-DCT algorithm is based on predefined orthonormal sets of 1D DCT basis functions. Subjective evaluation tests within MPEG have shown that the combination of these techniques can result in a bit stream saving of up to 50% compared with Version 1, depending on the content type and data rate.

Another new technique is Dynamic Resolution Conversion (DRC). This is a way to stabilize the transmission buffering delay by minimizing the jitter of the amount of the coded output bits per Video Object Plane (VOP). Large frame skips are also prevented and the encoder can control the temporal resolution even in highly active scenes. The following three new tools for coding textures and still images are provided in Version 2:

- Wavelet tiling allows an image to be divided into several tiles and each tile to be encoded independently. This means that large images are encoded or decoded with very low memory requirements, and that random access at the decoder is significantly enhanced.
- Scalable shape coding allows encoding of arbitrary-shaped textures and still images in a scalable fashion. Using this tool, a decoder can decode an arbitrary-shaped image at any desired resolution. This tool enables applications to employ object-based, spatial and quality scalabilities at the same time.
- The error resilience tool adds new error-resiliency features. Using packetization and segment marker techniques, it significantly improves the error robustness in applications, such as image transmission across mobile channels and the Internet.

In Version 2, the use of multiple alpha channels to transmit auxiliary components is introduced. The basic idea is that the gray-scale shape is not only used to describe the transparency of the video object, but can be defined in a more general way. A gray-scale shape may represent the following:

- Transparency shape
- Disparity shape for multiview video objects (horizontal and vertical)
- Depth shape (acquired by laser range finder or by disparity analysis)
- Infrared or other secondary texture

All the alpha channels can be encoded by the shape-coding tools, that is, the binary shape-coding tool and the gray-scale shape-coding tool that employs a motion-compensated DCT and that usually has the same shape and resolution as the texture of the video object.

The body is an object capable of providing virtual body models and animations in the form of a set of 3D polygonal meshes ready for rendering. Two sets of parameters are defined for the body: Body Definition Parameter (BDP) set and Body Animation Parameter (BAP) set. The BDP set defines the set of parameters to transform the default body to a customized body with its body surface, body dimensions and, optionally, texture. If correctly interpreted, the BAPs will produce reasonably similar high-level results in terms of body posture and animation on different body models, without the need to initialize or calibrate the model. The human body model should be capable of supporting various applications, from realistic simulation of human motions to network games using simple human-like models. The work on body animation includes the assessment of the emerging standard as applied to hand signs for the hearing impaired.

Version 2 MPEG-4 provides a suite of tools for coding 3D polygonal meshes. Polygonal meshes are widely used as a generic representation of 3D objects. Capabilities for 3D mesh coding include the following:

- Coding of generic 3D polygonal meshes enables the efficient encoding of 3D polygonal meshes. The coded representation is generic enough to support both manifold and nonmanifold meshes.
- Incremental representation enables a decoder to reconstruct a number of faces in a mesh proportional to the number of bits in the bit stream that have been processed.
- Error resilience enables a decoder to recover a mesh partially when subsets of the bit stream are missing or corrupted.
- Level of Detail (LOD) scalability enables a decoder to reconstruct a simplified version of the original mesh containing a reduced number of vertexes from a subset of the bit stream. Such simplified representations are useful to reduce the rendering time of objects that are distant from the viewer, but they also enable less powerful rendering engines to render the object at a reduced quality.

MPEG-4 Audio Version 2 is an extension to MPEG-4 Audio Version 1. It adds new tools and functionalities to the MPEG-4 standard, but none of the existing tools of Version 1 is replaced. The following additional functionalities are provided by MPEG-4 Audio Version 2: error robustness, low-delay audio coding, parametric audio coding, CELP, silence compression, environmental spatialization and back channel audio transparent stream. The error-robustness tools provide improved performance on error-prone transmission channels. They can be distinguished into coded specific resilience tools and a common error-protection tool. The error-resilience tools reduce the perceived deterioration of the decoded audio signal that is caused by corrupted bits in the bit stream. The error resilient bit stream payload syntax is mandatory for all Version 2 object types. The Error Protection (EP) tool provides error protection applicable for a wide range of channel error conditions. The main features of the EP tool are as follows:

- Providing a set of error correcting and detecting codes with wide and small-step scalabilities in performance and in redundancy
- Providing a generic and bandwidth-efficient error protection framework that covers both fixed-length frame bit streams and variable-length frame bit streams
- Providing an Unequal Error Protection (UEP) configuration control with low overhead

MPEG-4 Audio Version 2 coding algorithms provide a classification of each bit stream field according to its error sensitivity. Based on this, the bit stream is divided into several classes, which can be separately protected by the EP tool so that more error-sensitive parts are strongly protected.

While the MPEG-4 general audio coder provides very efficient coding of general audio signals at low bit rates, it has an algorithmic encoding or decoding delay of up to several 100/ms and is thus not well suited for applications requiring low coding delay, such as real-time bidirectional communication. As an example, for a general audio coder operating at a 24 KHz sampling rate and 24 Kb/s, this results in an algorithmic coding delay of about 110 ms plus up to additional 210 ms for the use of the bit reservoir. To enable coding of general audio signals with an algorithmic delay not exceeding 20 ms, MPEG-4 Version 2 specifies a low-delay audio coder, which is derived from MPEG-2 and MPEG-4 AAC. Compared to speech-coding schemes, this coder allows compression of general audio signal types, including music, at a low delay. It operates at up to a 48 KHz sampling rate and uses a frame length of 512 or 480 samples, compared to 1,024 or 960 samples used in standard MPEG-2 or 4 AAC. Also, the size of the window used in the analysis and in the synthesis filter banks is reduced by a factor of two.

The parametric audio coding tools combine very low bit-rate coding of general audio signals with the possibility of modifying the playback speed or pitch during decoding without the need for an effects-processing unit. In combination with the speech and audio coding tools of Version 1, improved overall coding efficiency is expected for applications of object-based coding, allowing selection and/or switching between different coding techniques. Parametric audio coding uses the Harmonic and Individual Lines plus Noise (HILN) technique to code general audio signals at bit rates of 4 Kb/s and higher using a parametric representation of the audio signal. The basic idea of this technique is to decompose the input signal into audio objects, which are described by appropriate source models and represented by model parameters. Object models for sinusoids, harmonic tones, and noise are used in the HILN coder. Verification tests have shown that HILN coding has performance comparable to the other MPEG-4 coding technology operating at similar bit rates while providing the additional capability of an independent audio signal speed or pitch change when decoding. The tests have also shown that the scalable HILN coder provides quality comparable to that of a fixed-rate HILN coder at the same bit rate.

The silence compression reduces the average bit rate thanks to a lower bit-rate compression for silence. In the encoder, a voice activity detector is used to distinguish between regions with normal speech activity and those with silence or background noise. During normal speech activity, the CELP coding in Version 1 is used.

The environmental spatialization tools enable composition of an audio scene with a more natural sound source and sound environment modeling than is possible in Version 1. Both physical and perceptual approaches to spatialization are supported. The physical approach is based on a description of the acoustical properties of the environment (for example, room geometry, material properties and position of sound source) and can be used in applications like 3D virtual reality. The perceptual approach, on the other hand, permits a high-level perceptual description of the audio scene based on parameters similar to those of reverberations like movies. Although the environmental spatialization tools are related to audio, they are part of the BIFS in MPEG-4 System and are referred to as Advanced Audio BIFS. The back channel allows a request of client and/or client terminal to server. With this capability, interactivity can be achieved. In MPEG-4

System, the need for an upstream channel (back channel) is signaled to the client terminal by supplying an appropriate elementary stream descriptor declaring the parameters for that stream. In MPEG-4 Audio, the back channel allows feedback for bit-rate adjustment, scalability and error-protection adaptation.

The MPEG-4 Audio transport stream defines a mechanism to transport MPEG-4 Audio streams without using MPEG-4 Systems and is dedicated to audio-only applications. The transport mechanism uses a two-layer approach, namely a multiplex layer and a synchronization layer. The multiplex layer manages multiplexing of several MPEG-4 Audio payloads and audio-specific configuration information. The synchronization layer specifies a self-synchronized syntax of the MPEG-4 Audio transport stream, which is called Low Overhead Audio Stream (LOAS). The interface format to a transmission layer depends on the conditions of the underlying transmission layer.

Extensions to MPEG-4 Beyond Version 2

MPEG is currently working on a number of extensions to Version 2. In the visual area, the following technologies are in the process of being added: fine-grain scalability, tools for use of MPEG-4 in the studio, 2D and 3D animation coding, and digital cinema. Fine-grain scalability is a tool that allows small quality steps by adding or deleting layers of extra information. It is useful in a number of environments, notably for streaming purposes, but also for dynamic (statistical) multiplexing of pre-encoded content in broadcast environments. For the tools in the studio, care has been taken to preserve some form of compatibility with MPEG-2 profiles. Features such as 2D and 3D animation coding are under study. Digital cinema applications will require truly lossless coding [5.35].

Advanced BIFS provides new nodes to be used in the scene graph for monitoring available media and for managing media, such as sending commands to a server, providing advanced control of media playback, and using the so-called EXTERNPROTO, a node that provides further compatibility with Virtual Reality Modeling Language (VRML), which allows writing macros that define the behavior of objects. Also, it covers advanced compression of BIFS data, and, in particular, optimal compression for mesh and for arrays of data.

The extensible MPEG-4 Textual Format (XMT) is a framework for representing an MPEG-4 scene description using a textual syntax. XMT allows the content authors to exchange the content with other authors, tools or service providers and facilitates interoperability with both the Extensible 3D (X3D) being developed by the Web3D Consortium and the Synchronized Multimedia Integration Language (SMIL) from the Web3C Consortium.

The advanced synchronization model (usually called FlexTime) supports synchronization of objects from multiple sources with possibly different time bases. The FlexTime model specifies timing using a flexible, constraint-based timing model. In this model, media objects can be linked to one another in a time graph using relationship constraints such as CoStart, CoEnd, or Meet. In addition, to allow some flexibility to meet these constraints, each object may have a flexible duration with specific stretch and shrink mode preferences that may be applied.

Profiles in MPEG-4

Profiles exist for various types of media content (audio, visual and graphics) and for scene descriptions. MPEG does not prescribe or advise combinations of these profiles, but care has been taken that good matches exist among the different areas. MPEG-4 provides a large and rich set of tools for the coding of audiovisual objects. In order to allow effective implementations of the standard, subsets of the MPEG-4 Systems, Visual and Audio tool sets have been identified and can be used for specific applications. These subsets called profiles limit the total set a decoder has to implement. For each of these profiles, one or more levels have been set, restricting the computational complexity.

The visual part of the standard provides profiles for the coding of natural or synthetic/natural hybrid visual content. There are five profiles for natural video content:

- The simple visual profile provides efficient, error-resilient coding of rectangular video objects, which are suitable for applications on mobile networks.
- The simple scalable objects in the simple visual profile are useful for applications that provide services at more than one level of quality due to bit-rate or decoder source limitations, such as Internet use and software decoding.
- The core visual profile adds support for coding of arbitrary-shaped and temporally scalable objects to the simple visual profile. It is useful for applications such as those providing relatively simple content interactivity (Internet multimedia applications).
- The main visual profile adds support for coding of interlaced, semitransparent and sprite objects to the core visual profile. It is useful for interactive and entertainment-quality broadcast and DVD applications.
- The N-bit visual profile adds support for coding video objects having pixel-depths ranging from 4 to 12 bits to the core visual profile. It is suitable for use in surveillance applications.

The profiles for synthetic and synthetic/natural hybrid visual content are the following:

- The simple facial animation visual profile provides a simple means to animate a face model, suitable for applications such as audio-video presentation for the hearing impaired.
- The scalable texture visual profile provides spatial scalable coding of still image (texture) objects useful for applications needing multiple scalability levels, such as mapping texture on to objects in games and high-resolution digital still cameras.
- The basic animated 2D texture visual profile provides spatial scalability, SNR scalability, and mesh-based animation for still image (textures) objects and also to simple face object animation.
- The hybrid visual profile combines the ability to decode arbitrary-shaped and temporally scalable natural video objects with the ability to decode several synthetic

and hybrid objects, including simple face and animated still image objects. It is suitable for various content-rich multimedia applications.

Version 2 adds the following profiles for natural video:

- The Advanced Real-Time Simple (ARTS) profile provides advanced error-resilient coding techniques of rectangular video objects using a back channel and improved temporal resolution stability with the low buffering delay. It is suitable for real-time coding applications, such as the videophone, teleconferencing and remote observation.
- The core scalable profile adds support for coding of temporal and spatial scalable arbitrary shaped objects to the core profile. The main functionality of this profile is object-based SNR and spatial/temporal scalability for regions or objects of interest. It is useful for applications, such as the Internet, mobile and broadcast.
- The Advanced Coding Efficiency (ACE) profile improves the coding efficiency for both rectangular and arbitrary-shaped objects. It is suitable for applications such as mobile broadcast reception, the acquisition of image sequences and applications where high coding efficiency is requested.

The Version 2 profiles for synthetic and synthetic/natural hybrid visual content are the following:

- The advanced scalable texture profile supports decoding of arbitrary-shaped texture and still images, including scalable shape coding, wavelet tiling and error resilience. It is useful for applications that require fast random access as well as multiple scalability levels and arbitrary-shaped coding of still objects. Examples are fast content-based still image browsing on the Internet, multimedia-enabled Personal Digital Assistants (PDA) and Internet-ready high-resolution digital still cameras.
- The advanced core profile combines the ability to decode arbitrary-shaped video objects (as in the core visual profile) with the ability to decode arbitrary-shaped scalable still image objects (as in the advanced scalable texture profile). It is suitable for various content-rich multimedia applications such as interactive multimedia streaming across the Internet.
- The simple face and body animation profile is a superset of the simple face animation profile, obviously adding body animation.

Four audio profiles have been defined in MPEG-4 Version 1:

- The speech profiles provide Harmonic Vector Excitation Coding (HVXC), which uses a very low bit-rate parametric speech coder, a CELP narrow band/wide band speech coder and a Text-to-Speech Interface (TTSI).

- The synthetic profile provides score-driven synthesis and a TTSI to generate sound and speech at very low bit rates.
- The scalable profile, a superset of the speech profile, is suitable for scalable coding of speech and music for networks, such as Internet and Narrow Band Audio Digital Broadcasting (NADIB). The bit rates range from 6 Kb/s to 24 Kb/s, with bandwidth between 3.5 KHz and 9 KHz.
- The main profile is a rich superset of all other profiles, containing tools for natural and synthetic audio.

Another four profiles were added in MPEG-4 Version 2:

- The high quality audio profile contains the CELP speech coder and the low complexity AAC coder including long-term prediction. Optionally, the new Error-Resilient (ER) bit stream syntax may be used.
- The low-delay audio profile contains the HVXC and CELP speech coders, the low-delay AAC coder and the TTSI.
- The natural audio profile contains all natural audio coding tools available in MPEG-4, but not synthetic ones.
- The mobile audio internetworking profile contains the low-delay and scalable AAC object types, including twin VQ and Bit-Sliced Arithmetic Coding (BSAC).

Graphics profiles define which graphical and textural elements can be used in a scene. These profiles are defined in the Systems part of the standard [5.35]:

- The simple 2D graphics profile provides for only those graphics elements of the BIFS tool that are necessary to place one or more visual objects in a scene.
- The complete 2D graphics profile provides 2D graphics functionalities and support features, such as arbitrary 2D graphics and text, possibly in conjunction with visual objects.
- The complete graphics profile provides advanced graphical elements, such as elevation grids, and allows creating content with sophisticated lighting.

Scene description profiles allow audiovisual scenes with audio-only 2D, 3D or mixed 2D/3D content. The 3D profile is called VRML because it optimizes interworking with VRML material [5.35]:

- The audio-scene graph profile provides for a set of BIFS scene graph elements for use in audio-only applications. The audio-scene graph profile supports applications like broadcast radio.
- The simple 2D scene graph profile provides for only those BIFS scene graph elements necessary to place one or more audiovisual objects in a scene. The simple 2D scene

graph profile allows presentation of audiovisual content with potential update of the complete scene, but no interaction capabilities. The simple 2D scene graph profile supports applications like broadcast television.
- The complete 2D scene graph profile provides for all the 2D scene description elements of the BIFS tool. It supports features such as 2D transformations and alpha blending. The complete 2D scene graph profile enables 2D applications that require extensive and customized interactivity.
- The complete scene graph profile provides the complete set of scene graph elements of the BIFS tool. The complete scene graph profile enables applications like dynamic virtual 3D world and games.

The MPEG-J profile includes the following [5.35]:

- The Personal profile (a lightweight package for personal devices) addresses a range of constrained devices, including mobile and portable devices. Examples of such devices are cellular video phones, PDAs and personal gaming devices. This profile includes the following packages of MPEG-J APIs: network, scene and resource.
- The Main profile (includes all the MPEG-J APIs) addresses a range of consumer devices including entertainment devices. Examples of such devices are set-top boxes, computer-based multimedia systems and so forth. It is a superset of the Personal profile. Apart from the packages in the Personal profile, this profile includes the following packages of the MPEG-J APIs: decoder, decoder functionality, section filter and service information.

The Object Descriptor profile includes the following tools [5.8, 5.35]:

- Object Descriptor (OD) tool
- Synchronization Layer (SL) tool
- Object Content Information (OCI) tool
- IPMP tool

Currently, only one profile is defined that includes all these tools. The main reason for defining this profile is not to create a subset of tools, but rather to define levels for them. This applies especially to the SL because MPEG-4 allows multiple time bases to exist. In the context of levels for this profile, restrictions can be defined, for example, to allow only a single time base.

Verification Testing: Checking MPEG's Performance

MPEG carries out verification tests to check whether the standard delivers what it promises. The test results can be found on MPEG's home page [5.44].

Figure 5.29 Subjective impairment scale for audio.

In video, a number of MPEG-4 capabilities have been formally evaluated using subjective tests. Coding efficiency, although not the only MPEG-4 functionality, is an important selling point of MPEG-4 and one that has been tested more thoroughly. Also, error robustness has been put to rigorous tests. Furthermore, scalability tests were done, and, for one specific profile, the temporal resolution stability was examined. Many of these tests address a specific profile.

MPEG-4 audio technology is composed of many coding tools. Verification tests have focused on small sets of coding tools that are appropriate in an application arena and hence can be effectively compared. Because compression is a critical capability in MPEG, the verification tests have, for the most part, compared coding tools operating at similar bit rates. The results of these tests are presented progressing from higher bit rates to lower bit rates. The exception of this is error robustness tools, the performance of which is noted at the end of this section.

The primary purpose of verification tests is to report the subjective quality of a coding tool operating at a specified bit rate. Most audio tests report this on the subjective impairment scale. This is a continuous five-point scale with subjective anchors as shown in Figure 5.29.

The error-robustness tools provide equivalently good error robustness across a wide range of channel error conditions and do so with only a modest overhead in bit rate. Verification test results suggest that the error-robustness tools used with an audio-coding system provide performance in error-prone channels that is nearly as good as the same coding system operating across a clear channel.

MPEG-4 Standardization Process

A standard must provide a minimum set of relevant tools, which, after being assembled according to industry needs, provide the maximum interoperability at a minimum complexity and lowest cost [5.45]. The success of the MPEG standard is mainly bounded by the "one functionality, one tool" principle. MPEG wants to offer the users interoperability and flexibility at the smallest complexity and cost. In order to fulfill these objectives, MPEG follows a development process with the following steps [5.45]:

1. Identify relevant applications using input from MPEG members.
2. Identify the functionalities needed by the applications determined in Step 1.
3. Describe the requirements following from the functionalities determined in Step 2 in such a way that common requirements can be identified for different applications.
4. Identify which requirements are common across the areas of interest and which are not common, but are still relevant.

5. Specify tools that support the requirements. Verify that the tools developed can be used to assemble the target systems and to provide the desired functionalities with an adequate level of performance. This is done by means of the so-called verification tests. The verification tests consist of formal subjective tests aimed at evaluating the quality of either audio or video signals processed using specific MPEG algorithms.

Two working tools play a major role in the development phase that follows the initial competitive phase: the working model and core experiments [5.46]. In MPEG-4, these phases are independent: VMs for the video, audio and synthetic and Natural Hybrid Coding (SNHC) and systems development.

Requirements for MPEG-4

The vision behind the MPEG-4 standard was explained through the eight new or improved functionalities described in the MPEG-4 Proposal Package Description (PPD). These eight functionalities came from an assessment of the functionalities that would be useful in future applications, but were not supported or not well supported by the available coding standards. The eight new or improved MPEG-4 functionalities formed the following three areas, the convergence of which MPEG-4 wanted to address [5.47].

- *Content-based interactivity*—This includes content-based multimedia data access tools, content-based manipulation and bit stream editing, hybrid natural and synthetic data coding, and improved temporal random access.
- *Compression*—This improves coding efficiency and coding of multiple concurrent data streams.
- *Universal access*—This includes robustness in an error-prone environment and content-based scalability.

Requirements for the MPEG-4 video standard are shown in Table 5.8.

Table 5.8 Requirements for the MPEG-4 Video standard [5.48].

Functionality	MPEG-4 Video Requirements
Content-based interactivity	
Content-based manipulation and bit stream editing	This includes support for content-based manipulation and bit stream editing without the need for transcoding.
Hybrid natural and synthetic data coding	This includes support for combining synthetic scenes or objects with natural scenes or objects. This includes the ability for compositing synthetic data with ordinary video, allowing for interactivity.

Table 5.8 Requirements for the MPEG-4 Video standard [5.48]. (Continued)

Functionality	MPEG-4 Video Requirements
Improved temporal random access	This includes provisions for efficient methods to access parts randomly, within a limited time and with fine resolution, for example, video frames or arbitrarily shaped image content from video sequences. This includes conventional random access at very low bit rates.
Compression	
Improved coding efficiency	MPEG-4 Video shall provide subjectively better visual quality at comparable bit rates compared to existing or emerging standards.
Coding of multiple concurrent datastreams	This includes provisions to code multiple views of a scene efficiently. For stereoscopic video applications, MPEG-4 shall allow the ability to exploit redundancy in multiple viewing points of the same scene, permitting joint coding solutions that allow compatibility with normal video as well as the ones without compatibility constraints.
Universal Access	
Robustness in error-prone environments	This includes provisions for error-robustness capabilities to allow access to applications across a variety of wireless and wired networks and storage media. Sufficient error robustness shall be provided for low bit rate applications under severe error conditions (for example, long error bursts).
Content-based scalability	MPEG-4 shall provide the ability to achieve scalability with fine granularity in content, quality (for example, spatial and temporal resolutions) and complexity. In MPEG-4, these scalabilities are especially intended in content-based scaling of visual information.

5.5.2 MPEG-4 Systems

In MPEG-4, in addition to overall architecture, multiplexing and synchronization, the Systems part encompasses scene description interactivity, content description, and programmability. The combination of the exciting new ways of creating interactivity in audiovisual content offered by MPEG-4 Systems and the efficient representation tools provided by the Visual and Audio parts promise to be the foundation of a new way of thinking about audiovisual information. The goal of specifying a standard way for the description and coding of audio-visual objects was the primary motivation behind the development of the tools in the MPEG-4 Systems [5.49, 5.50].

MPEG-4 Systems requirements may be categorized into two groups: traditional MPEG Systems requirements and specific MPEG-4 Systems requirements [5.48]. Traditional MPEG Systems requirements are streaming, synchronization and stream management. By streaming, we mean that the audiovisual data is to be transmitted piece by piece in order to match the delivery of the content to clients with limited network and terminal capabilities. In synchronization, the different components of an audiovisual presentation are closely related in time. For most applications, audio samples with associated video frames have to be presented together to the user at precise instances in time. In stream management, the complete management of streams of audio visual information implies the need for certain mechanisms to allow an application to consume the content.

Specific MPEG-4 Systems requirements represent the ideas central to MPEG-4 and are completely new in MPEG Systems. The foundation of MPEG-4 is the coding of audiovisual objects. As for MPEG-4 terminology, an audiovisual object is the representation of a natural or synthetic object that has an audio and/or visual manifestation. Natural is generally understood to mean representations of the real world that are captured using cameras, microphones and so on, as opposed to synthetically generated material. The advantages of coding audiovisual objects can be summarized as follows:

- Allows interaction with the content
- Improves reusability and coding of the content
- Allows content-based scalability

In order to be able to use these audiovisual objects in a presentation, additional information needs to be transmitted to the clients' terminals. The individual audiovisual objects are only a part of the presentation structure that an author wants delivered to the consumers.

MPEG-4 Systems Architecture

The overall architecture of the MPEG-4 Systems is shown in Figure 5.30. Starting at the bottom, we first encounter the storage/transmission medium. This refers to the lower layers of the delivery infrastructure. Transportation of the MPEG-4 data uses a variety of delivery systems. This includes MPEG-2 TSs, UDP, ATM AAL2, MPEG-4 files or the Digital Audio Broadcasting (DAB) multiplexer. The FlexMux tool can optionally be used on top of the existing transport delivery layer. Regardless of the transport layer used and whether the FlexMux option is used, the delivery layer provides to the MPEG-4 terminal a number of ESs. In order to isolate the design of MPEG-4 from the specifics of the various delivery systems, the concept of the DMIF Application Interface (DAI) was defined [5.51]. This interface defines the process of exchanging information between the terminal and the delivery layer. The DAI defines procedures for initializing an MPEG-4 session and for obtaining access to the various ESs that are contained in it. These streams can contain a variety of different information: audiovisual object data scene

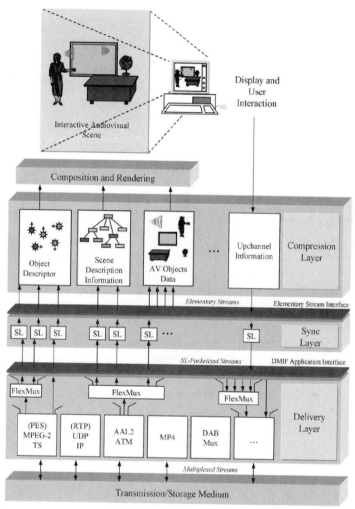

Figure 5.30 MPEG-4 Systems architecture. ©2000 ISO/IEC.

description information, control information in the form of object descriptors and metainformation that describes the content or associated intellectual property rights to it. The SL allows the inclusion of timing, fragmentation and continuity information on associated data packets. Such information is attached to data units that comprise complete presentation units, for example one entire VOP or an audio frame. These are called access units. An SL header contains no packet length indication. This is because it is assumed that the delivery layer that processes SL packets will already make such information available. The SL is the sole mechanism of implementing timing and synchronization mechanisms in MPEG-4. From the SL information, we can recover a time base as well ESs. The streams are sent to their respective decoders that process the data and that produce compensation units (for example, a decoded VOP).

In order for the receiver to know what type of information is contained in each stream, control information in the form of object descriptors is used. These descriptors associate sets of ESs to one audio or visual object, define a scene description stream, or even point to an object descriptor stream. These descriptors are the way with which a terminal can identify the content being delivered to it. The scene description information defines the spatial and temporal position of the various objects, their dynamic behavior and any interactivity features made available to the user. The scene description contains pointers to object descriptors when it refers to a particular audiovisual object. It is possible that an object (in particular synthetic objects like text and simple graphics) may be fully described by the scene description. As a result, it may not be possible to uniquely associate an audiovisual object with just one syntactic component of MPEG-4 Systems. The system's compositor uses the scene description information, together with decoded audiovisual object data, in order to render the final scene that is presented to the user. The scene description tools provide mechanisms to capture user or system events. In particular, they allow the association of events to user operations on desired objects that can, in turn, modify the behavior of the stream. The use of an object-based structure, where composition is performed at the receiver, considerably simplifies the content operation process. Starting from a set of coded audiovisual objects, it is very easy to define a scene description that combines these objects in a meaningful presentation. A similar approach is used in HTML and Web browsers, thus allowing even nonexpert users to create their own content easily. The fact that the content's structure survives the process of coding and distribution also allows for its reuse. For example, content filtering and/or searching applications can be easily implemented using ancillary information carried in object descriptors.

An audiovisual terminal contains three layers: Delivery Layer, Sync Layer and Compression Layer:

- The Delivery Layer encapsulates all the transport and multiplex functionality. It is mostly defined outside the MPEG-4 Systems standard. Some adaptation to (arbitrary) transport layers are defined in the scope of the DMIF, which is published as Part 6 of the MPEG-4 standard. The interface to the Delivery Layer is called the DAI. MPEG-4 Systems just specifies an optional multiplex tool, called FlexMux, that may be used in the adaptation to an existing or future transport layer. This structure provides significant flexibility in deploying MPEG-4 in a variety of communications environments.
- The Sync layer adapts ES data for communication across the DAI, providing timing and synchronization information, as well as fragmentation and random access information. The Sync Layer on the receiver side extracts this timing information to enable synchronized decoding and, subsequently, composition of the ES data. Its syntax is configurable and can also be empty.
- The Compression Layer recovers data from its encoded format (ES) and performs the necessary operations to reconstruct the original information. It incorporates the

audiovisual object decoders. The decoded information is then used by the terminal's compositions, rendering and presentation subsystems.

Elementary Stream Management (ESM)

The term ESM is used to refer to the entire set of functionalities needed to describe, express relations between and affect synchronization among such datastreams. Within the extensive set of tools defined by MPEG-4, the ESM tools play a critical role in joining several building blocks together. ESM provides an alternative to the scene description language in that it limits the streaming resources of a presentation to the scene.

The ESM part of Systems also specifies means to identify and name ESs so that they can be referred to in a scene description and can be attached to individual objects. This association is performed in object descriptors that are transmitted in their own ESs. Object descriptors are separate from the scene description itself, thus simplifying editing and remultiplexing of MPEG-4 content. The descriptors associated with audiovisual objects are nodes in the scene related to ES identifiers. An additional mapping is required to resolve these identifiers to actual transport layer channels (for example, port numbers). How this mapping is performed depends on the Delivery Layer instance that is actually used. In accordance with the goal of allowing the use of any Delivery Layer, MPEG-4 does not define this mapping, but rather expects parties that define these delivery layers to define how MPEG-4 content should be mapped to their design in a way that they consider most appropriate.

A hierarchically structured set of descriptors constitutes the major building block of the object description framework. The highest-level descriptor is the OD itself. It serves as a shell that aggregates a number of other descriptors. Most prominent among these are the ES descriptors which describe individual streams. Additional auxiliary information can be attached to an OD, either to describe the content conveyed by these streams in a textual form object content information, or to do IPMP.

Main components of the OD are depicted in Figure 5.31. The linking to the scene is achieved in a two-stage process. First, there is a numeric identifier, called the object descriptor (OD_ID), that labels the object descriptor. This identifier is referenced by the scene description stream and thus links the OD to the scene. The second stage involves the actual binding of ESs identified and described by the ES descriptors included in this OD, using another identifier, which is part of the ES descriptor.

Example 5.10 In the most simple case, an OD contains just one ES descriptor that identifies the audio stream that belongs to the audio source node by which this OD is referenced, as illustrated in Figure 5.32a. The same OD may as well be referenced from two distinct scene description nodes as shown in Figure 5.32b.

Example 5.11 Within a single OD, it is possible to have two or more ES descriptors: one identifying a low bit rate audio stream and another identifying a higher bit rate stream with the same content (Figure 5.33a). In this case, the terminal, or rather the user, has a choice between

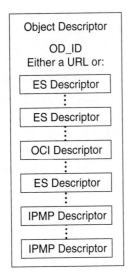

Figure 5.31 Main components of the OD. ©2000 ISO/IEC.

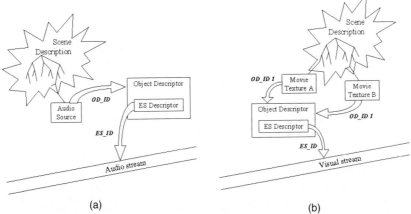

Figure 5.32 ES reference through OD (a) from one scene description node and (b) from two scene description nodes. ©2000 ISO/IEC.

two audio qualities. Specifically, for audio, it is also possible to have multiple audio streams with different languages that can be selected according to user preferences (Figure 5.33b).

It is possible to describe within one OD a set of streams that corresponds to a scalable or hierarchical encoding of the data that represents an audiovisual object. In that case, it is necessary to signal not only the properties of the individual ES, but also their interdependencies. Simple and more complex ES dependence indication is shown in Figure 5.34. In Figure 5.34a, each stream depends on the previous one, but, in Figure 5.34b, the same base layer stream is referenced by two other streams, providing, for example, quality improvements in the temporal domain (temporal scalability) and in the spatial domain (spatial scalability). The precise instruc-

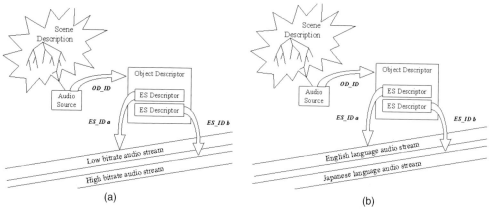

Figure 5.33 Reference to multiple ESs though one OD (a) and audio application for multiple languages (b). ©2000 ISO/IEC.

tions on how the decoder has to handle each individual stream in such a case is incorporated in a decoder-specific info subdescriptor that is contained in each ES descriptor.

ODs are the glue between the scene description and the ESs. An ES descriptor (a part of an OD) is associated with each ES and contains all the information needed to find it, describe it and advise the receiver which resources need to be set up to decode it. ODs are transported in an ES of their own. They are never sent just as is, but are always encapsulated in so-called OD commands.

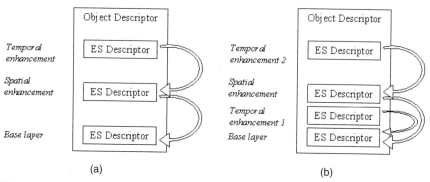

Figure 5.34 Simple (a) and more complex ES dependence indication (b). ©2000 ISO/IEC.

Auxiliary Descriptors and Streams

An OD points to auxiliary streams associated to this set of media streams. There are three types of such streams. First, semantic information about the content of an audiovisual object that is fed by these streams may be covered by means of OCI. Then, IPMP information may be attached. Finally, clock-reference streams may be used to convey time-base-information. Another type of auxiliary information is the QoS descriptors, which may be attached to ES descriptors. The design of the descriptors as self-describing entities constitutes a generic extension mechanism that may be used by ISO or specific applications to attach arbitrary descriptive information to an audiovisual object.

Object content information consists of a set of OCI descriptors that communicate a number of features of the audiovisual object that is constructed by the ESs associated with a given OD. There is a descriptor with keywords, possibly to be used with search engines, a textual description of the content, language and content rating information. These descriptors may be included directly in ODs to indicate static OCI properties. An OD for an audio object may contain a set of different language streams that is activated according to user preferences. In case of OCI changes over the lifetime of the media streams associated with this OD, it is possible to attach an OCI stream to the OD. The OCI stream conveys a set of OCI events that are qualified by their start time and duration. This means that an MPEG-4 presentation may be further decomposed semantically based on the actual content presented, which includes OCI streams as well as scene description streams. This influences their scope. As an example, attaching an OCI stream to a media stream as well as to a whole subscene is presented in Figure 5.35.

The IPMP framework consists of a fully standardized Intellectual Property Identification (IPI) descriptor as well as IPMP descriptors and IPMP streams with a standardized shell and not with a standardized content. The core IPMP framework consists of IPMP descriptors and IPMP streams.

IPMP framework has been engineered in a way that should allow the coexistence of multiple IPMP systems that govern the conditional access to specific content streams or an entire presentation. IPMP descriptors convey proprietary information that may help to decrypt ESs or that contain authorization or entitlement information evaluated by unequal proprietary IPMP subsystems in the receiving terminal. IPMP descriptors may occasionally be changed over time.

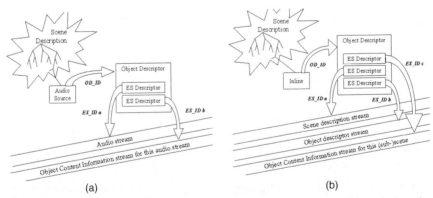

Figure 5.35 Attaching an OCI stream to (a) a media stream or to (b) a whole subscene. ©2000 ISO/IEC.

However, IPMP information that needs frequent updates in a streaming fashion should be conveyed in IPMP streams. IPMP streams also keep IPMP information separate from the original MPEG-4 information.

The QoS descriptor is an additional descriptor that may occur in ES descriptors. It aims to specify the requirements that a specific ES has on the QoS of the transport channel for this stream. The most obvious QoS parameter is the stream priority and a possibility to signal predefined QoS scenarios, most notably guaranteed and best effort channels. QoS descriptors at the ES level have an obvious use in interactive scenarios where the receiving terminal may select individual ESs based on their traffic and other QoS requirements as well as the associated communication cost [5.52].

Because MPEG-4 defines audiovisual objects rather than programs, ISO/IEC decided to rename the "program clock reference" to an object clock reference. An object clock reference is a sample from an object time base. Different objects within the same presentation may have different object time bases, even though this may make tight synchronization impossible [5.53].

Structuring Content by Grouping of Streams

An interesting question is how to use grouping of ESs through one OD to improve the structure of the content. In particular, we want to be able to arrange a large number of elementary streams into groups. Even though initial MPEG-4 applications may only require a small number of streams, the design allows efficient management of content that consists of a large number of streams. As hardware and software capabilities improve, it is only natural that the sophistication of content will also increase.

A number of application scenarios are conceivable and could include the delivery of differentiated quality content to different user groups, delivery of different portions of the content to different user groups and reception of content originating from different sources. Grouping is quite relevant for multicast applications where it should be easy to remove parts of the content somewhere on the way, but in point-to-point applications, it should also be possible to negotiate the desired subset of ESs between client and server.

Managing Content Complexity

Initial ODs convey some indication of profiles and levels of the content referenced by them. These scalar indications allow an immediate decision by the receiving terminal about whether it is able to decode and present the content being received. Due to the potential complexity in terms of the number of scenes, it has also been made possible to indicate such complexity only for the current subscene, that is, excluding parts of the scene that are included through inline nodes [5.52].

In the absence of profile and level indications or at the discretion of the receiving terminal, it is also possible to evaluate the decoder configuration, stream priorities, bit-rate requirements and the dependencies signaled in the ES descriptors. This allows the receiving terminal in an interactive application to request the delivery of a meaningful subset of streams for each media object so that the computational resources of the terminal are not exceeded. Apart from these resource-driven considerations, the terminal or the user needs to evaluate the scene description in order to decide which subset of the media objects advertised in an OD stream are relevant.

Distributed Content-Handling Considerations

MPEG-4 allows the construction of content where different parts may originate from different locations. Furthermore, MPEG-4 makes no assumptions about the type of the underlying communications infrastructure (IP, ATM, broadcast, and so forth). It is then impossible to ensure that use of MPEG-4's distributed content capabilities will result in seamless content presentation in all circumstances.

We expect that content creators, jointly with both content and communication services providers, will create content cognizant of the environment in which it will operate, leveraging the available resources in order to provide a maximum-quality user experience. This is similar to the practice of tuning HTML pages so that they provide both visually rich content as well as reasonable download times across telephone lines. At any rate, the expectation is that the evolution of the Internet and other suitable transport networks will gradually solve these issues [5.52].

System Decoder Model (SDM) for ES Synchronization

An SDM is used to specify the behavior of a receiving MPEG-4 terminal in terms of a timing model and a buffer model. The SDM receives individual ESs from the Delivery Layer through the DAI. MPEG-4 requirements on the end-to-end delivery of data through the DAI, most prominently, a constant end-to-end delay [5.54], are specified. The SDM is shown in Figure 5.36 [5.52]. It consists of the DAI, a number of buffer decoders, composition memories and the compositor. The coder entity for the purposes of the SDM is the Access Unit (AU). Each ES is partitioned in a sequence of such AUs. The AU is the smallest entity to which timing information can be associated. The Sync Layer defines some syntax (the SL packet headers) that carries timing information, that is, time stamps and clock references. This data allows a receiver to determine which portions of different streams are to be composed and, hence, presented at the same time. The syntax of the SL encodes both AU boundaries and the timing information associated with AUs.

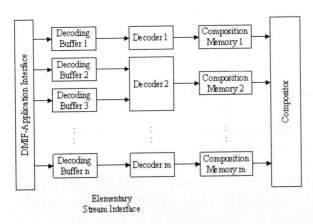

Figure 5.36 SDM.
©2000 ISO/IEC.

The DAI supplies AUs or parts thereof to the decoding buffer that stores the AUs until their decoding time. At that point in time, the SDM assumes instantaneous decoding of the AU, removal of the AU data from the decoding buffer and appearance of the decoded data corresponding to the AU in the associated composition memory. With this model, it is possible for the encoding terminal to know how many decoding buffer resources are available in the receiving terminal for a specific stream at any point in time. The content of decoding buffers is consumed by decoders. In the case of hierarchically encoded audio-video objects, a decoder may be connected to multiple decoding buffers as in the case of Decoder 2 in Figure 5.36. A decoder outputs the decoded data to one composition memory. The decoded data is grouped in composition units. The relation between AUs and composition units is assumed to be known for each specific decoder type. Each composition unit of decoded data is available for composition starting at an indicated composition time, either known implicitly or through an explicit composition time of the subsequent composition unit. Only a minimum of assumptions is made on the compositor for the purpose of defining the SDM. This includes that the compositor instantaneously samples the content of each composition memory of the MPEG-4 terminal.

MPEG-4 Systems BIFS

BIFS provides a complete framework for the presentation engine of an MPEG-4 terminal. It expresses how individual audiovisual objects are to be composed together for presentation on the user's screen and speakers.

As a general paradigm in MPEG-4, all information is conveyed in a streaming manner. The term "ES" refers to data that fully or partially contains the encoded representation of a single audio or visual object, scene description information or control information.

ESs or groups therefore are identified and characterized by ODs. These include the scene description, audiovisual object data and OD streams themselves. The information contained will include the format of the data, for example, a visual face animation stream. Also, the indication of the resources necessary for decoding for example, a profile/level indication, will be included. Alternate representation or scalable encoding using multiple streams for a single audiovisual object can also be signaled by the ODs. For example, an object descriptor may specify a set of elementary streams that jointly contain the compressed representation of an audio signal.

BIFS enables mixing various MPEG-4 media together with 2D and 3D graphics, handling interactivity, and dealing with the local or remote changes of the scene over time.

BIFS is actually composed of four elements:

- The operational elements of the scene, consisting of nodes and routes. These represent the following in particular:
  - Audiovisual objects and their attributes (which define their audiovisual properties)
  - Composition operations
  - Animation of the content
  - Interactive behavior of individual objects by linking event fields to event sink fields between different nodes

- The binary syntax for compressing the node tree as well as the associated routes
- The BIFS-Command protocol, in order to stream scene changes, insert new scenes or objects, delete objects and so forth
- The BIFS-Animation protocol, in order to stream animations of node parameters which is used as a very low overhead mechanism to animate audiovisual objects

A central concept in the MPEG-4 design is transmission and interaction with audiovisual objects of synthetic or natural nature. The Audio and Visual parts of the standard provide the encoding algorithms for individual audiovisual objects. In order to combine these media together into complete presentations, a scene description capability is needed. BIFS provides the input data to the presentation layer of the MPEG-4 terminal. No other scene format covers all the requirements of the MPEG-4 presentation engine. The main concepts driving the design of the BIFS specification are the following:

- *Integration of 2D and 3D synthetic media together in a single format*—By avoiding the burden of mixing multiple media formats together, the content creator has a way to design a complete multimedia content without the hassle of dealing with many different formats, and the end-users benefit from a lighter terminal with state-of-the-art media capabilities.
- *Streaming environment*—All existing scene description formats are designed so that a complete scene has to be downloaded before anything can be viewed on the terminal. In MPEG-4, the terminal is linked to one or several MPEG-4 servers. The scene description, as with any other media, has to be streamed to the client. Allowing the scene to be cut into pieces and streamed to the client, as well as its animation parameters, provides more communication applications. These streaming features are also necessary in order to send new data to the user during the communication.
- *Compression*—Most existing scene representation is in text format, making it editable, but very inefficiently represented in terms of data size. The sizes of scenes are often much smaller than other media. However, complex scenes can be large datasets, sometimes up to several megabytes. Even for smaller scenes, being able to reduce the data size can bring significant improvements in transmission time, especially at low bit rates or in broadcast environments in which the scene has to be transmitted repeatedly. Moreover, animation data can also be streamed in MPEG-4. The efficient compression of such data can significantly reduce the bit rates. A typical example is for a BIFS animation stream that consumes 10 Kb/s, noncompressed data would consume more than 120 Kb/s.

Scene description and stream description are separated strictly in MPEG-4. In particular, the scene description contains no information about the streams that are needed to reconstruct a particular audiovisual object. On the other hand, the stream description contains no information that relates to how an object is to be used within a scene. The scene description is usually consciously authored by the content creator, but the stream description mostly follows from general content creator references, the default setting of an editing tool or even the service provider constraints and rules. The link between the two descriptions is a numeric OD identifier that the scene description uses to point to ODs, which, in turn, provide the necessary information for assembling the ES data required to decode and reconstruct the object at hand.

MPEG-4 specifies a BIFS that is used to describe scene composition information which includes the spatial and temporal locations of objects in scenes. Elements of the scene and relationships between them form the scene graph that must be coded for transmission [5.49]. The scene graph is transmitted at the beginning of the content session and may be dynamically updated as the content plays with a special streams of BIFS update commands. Content developers have wide flexibility to use BIFS in a variety of ways [5.55]. Audio BIFS, like the rest of BIFS, is composed of a number of nodes that can be interlinked to form a scene graph. However, the concept of the audio BIFS scene graph is somewhat different. It is termed an audio subgraph. Although the main (visual) scene graph represents the position and orientation of visual objects in presentation space and their properties such as color, texture and layering, an audio subgraph represents a signal flow graph describing digital signal-processing manipulation.

Designed as a file format for describing 3D models and scenes ("worlds" in VRML terminology), VRML faces some important features that are required for the types of multimedia applications targeted by MPEG-4. In particular, the support for natural video and audio is basic, but the timing model is loosely specified, implying that synchronization in a scene composed of multiple media types cannot be guaranteed. VRML was proposed and developed by the VRML Consortium. Their VRML 2.0 specification became an ISO/IEC international standard in 1998 [5.56].

Both VRML and MPEG-4 BIFS rely on the scene graph to describe the organization of audiovisual material. A scene graph represents a set of hierarchically related nodes. Each node in the visual scene graph represents a visual object (like a cube or image), a property of an object (like the textural appearance of a face or a cube) or a transformation of part of the scene (like a rotation or scaling operation). By connecting multiple nodes together, object-based hierarchies are formed. For example, one node might correspond to the location of a virtual character. The subgraphs, or sets of connected nodes, would represent the head and limbs of a character. By transforming the positions of the limbs, they may be made to move. By transforming the position of the character, all of the subgraphs are automatically transformed as well. Therefore, the character moves, but the limbs stay in the same relative positions. Each node has several fields that detail the properties of the object. For an object node like a cube, the fields give the size and shape of the object. For a property mode, the fields specify particular properties, such as the color of the cube and the image to be texture–mapped to the cube. For a transform node, the fields specify the set of subsidiary nodes that are affected by the transformation, as well as the details of the transform.

BIFS is a binary format, but VRML is a textual format. This is a fundamental difference between the two. Thus, although it is possible to design scenes that are compatible with both

BIFS and VRML, transcoding of the representation formats is required. In what follows, we highlight the functionalities that BIFS adds to the basic VRML set [5.56].

BIFS describes an efficient binary representation of the scene graph information. The coding may be either lossless or lossy. The coding efficiency is derived from a number of classical compression techniques, plus some novel ones. The technique is based on the fact that, given some scene graph data that has been previously received, it is possible to anticipate the type and format of data to be received subsequently.

Example 5.12 The use of context in efficient coding is described in the following listing. Quantization of numerical values is supported, as well as the compression of 2D meshes as specified in MPEG-4 Visual [5.57].

```
Consider a simple coding scheme with the following tags:
          <begin>     - beginning of record
          <end>       - end of record
          <break>     - end of element in record
          <string>    - text string follows
          <number> - number follows
We wish to use this scheme code for a record consisting of first name, last
name and phone number, for example:
          First name: Jim
          Last name: Brown
          Phone:        7771234
With no knowledge context we would need to code this as:
<start><string>Jim<break><string>Brown<break><number>7771234<break><end>
If the context is known, i.e. we know that the structure of the record is
"string, string, number", we do not have to spend bits specifying the type
of each element:
<start>Jim<break>Brown<break>7771234<end>
```

BIFS is designed so that the scene may be transmitted as an initial scene followed by time-stamp modifications to the scene. For dynamic scenes that change over time, this leads to a huge improvement in memory use and reduced latency when compared to equivalent VRML scenes. The BIFS command protocol allows replacement of the entire scene, addition, deletion and replacement of nodes and behavioral elements in the scene graph and modification of scene properties.

A second streaming protocol, BIFS animation, is designed to provide a low-overhead mechanism for the continuous animation of changes to numeric values of the components in the scene. These streamed animations provide an alternative to the interpolator nodes supported in both BIFS and VRML. The main difference is that interpolator nodes contain very large amounts of data that must be loaded in the scene and stored in memory. By streaming these animations, the amount of data that must be held in memory is reduced significantly. Secondly, by removing these data from the scene graph that must be initially loaded, the amount of data that must be processed in order to begin presenting the scene is also reduced.

BIFS includes native support for 2D scenes. This facilitates content creators who want to produce low-complexity scenes, including the traditional television and multimedia industries. Many applications cannot bear the cost of requiring decoders to have full 3D rendering and navigation. This is particularly true where hardware decoders must be of low cost. However, rather than simply partitioning the multimedia world into 2D and 3D, MPEG-4 BIFS allows the combination of 2D and 3D elements in a single scene.

VRML provides simple audio support. This support has been enhanced in MPEG-4. BIFS provides the notion of an audio scene graph, where the audio sources, including streaming ones, can be mixed. It also provides nodes to interlace to the various MPEG-4 audio objects [5.58]. Audio content can even be processed and transformed with special procedural code to produce various sound effects.

BIFS provides support at the scene level for the MPEG-4 facial animation decoder [5.57]. A special set of BIFS nodes exposes the properties of the animated face at the scene level. A full participant of the scene can be integrated with all BIFS functionalities, similarly to any other audio or visual objects.

As MPEG-4 Version 1 augments the virtual reality model of sound in VRML with a versatile abstract-effects model, MPEG-4 Version 2 extends the simple virtual reality model to include two rich and robust techniques for creating virtual audio environments. The first technique is physical, the second one is perceptual. By physical modeling of acoustic environments, we mean processing sound so that the acoustic effects-processing corresponds to the visual scene. Modeling of the acoustic environments is bound to the physical reality defined by the visual scene. Audiovisual interaction is one of the important features of virtual acoustics. The aim is a virtual environment where auditory and visual events are related. Audiovisual objects change both the auditory and visual characteristics, according to their position, orientation, materials and visibility in a scene. The source model in a virtual acoustic environment includes the sound content and directivity properties of the emitter, which can be modeled efficiently using digital filters. The environment model aims at reproducing the effects of the surrounding space (listening room, concert hall, metro station, and so forth). The most efficient approaches are time-domain hybrid methods combining ray tracing and an image source method for direct sound and early reflections with late reverberation modeling based on statistical parameters [5.59]. The listener model is closely related to the method of reproducing the auditory sensation. Different 3D processings are needed for different types of reproduction like headphone, stereophonic and multichannel loudspeaker listening.

In perceptual techniques, creation and modification of environment sound characteristics are based upon perceptual parameterization. Perceptual parameters have recently been introduced into the MPEG-4 Version 2 as another method of creating environmental acoustic effects in the scene, independent of the visual (and physical) reality. The creation of environmental acoustic effects is enabled separately for each sound source, together with characterization of the perceptual quality of the source and the environment in a 3D space. High-level perceptual parameters (source presence and brilliance, room reverberance, heaviness, liveness, and envelopment) are used to derive low-level energy parameters for the control of direct sound and the different parts of the room impulse response, that is, the directional and diffuse early reflections, as well as the late reverberation [5.59, 5.60, 5.61]. The high-level parameters have been derived based on subjective testing of perceived room acoustic quality. Based on these parameters, a real-time spatial sound-processing scheme has been derived [5.61]. This enables computationally efficient, yet perceptually relevant, 3D audio rendering.

The scene description can be dynamically changed at any time. An initial scene description is provided at the beginning of an MPEG-4 stream. It can be as simple as a single node or as complex as one wants (within limits that are established for ensuring conformance). BIFS commands are used to modify a set of properties of the scene at a given time. It is possible to insert, delete and replace nodes, fields and routes as well as to replace the entire scene. For continuous changes of the parameters of the scene, BIFS animation can be used; it specifically addresses the continuous update of the fields of a particular node. BIFS animation is used to integrate different kinds of animation, including the ability to animate face models as well as meshes, 2D and 3D positions, rotations, scale factors and color attributes. The BIFS animation information is conveyed in its own elementary stream.

5.5.3 DMIF

DMIF is a session protocol for the management of multimedia through generic delivery technologies [5.35]. Different from its predecessor, MPEG-4 has been targeted since the beginning to adopt to multiple operating scenarios (local retrieval, remote interaction, broadcast or multicast) and delivery technologies. The design choice was to abstract the functionality that the delivery layer has to provide and to focus the MPEG-4 Systems activity on the common features. The goal is still to produce effective solutions. However, a demarcation has been drawn between the aspects that can be managed uniformly and independently from the aspects that can be managed uniformly and independently from the delivery technology (included in MPEG-4 Systems), and those that instead are impacted by the delivery technology and by the operating scenario (included in MPEG-4 DMIF). This demarcation line is named DAI. The idea is shown in Figure 5.37 along with the generic MPEG-4 layered model, which comprises the Compression Layer, the Sync Layer (that is part of systems) and the Delivery Layer (DMIF) [5.54].

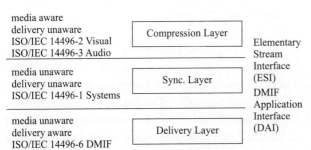

Figure 5.37 MPEG-4 layered model.

The Compression Layer performs media encoding and decoding of ESs [5.62, 5.63]. The Sync Layer manages ESs and their synchronization and hierarchical relations [5.64]. The Delivery Layer ensures transparent access to content irrespective of delivery technology [5.51]. The functionality provided by DMIF is expressed by an interface called DAI and is transmitted into protocol messages. These protocol messages may differ based on the network on which they operate. The QoS is also considered in the DMIF design, and the DAI allows the DMIF user to specify the requirements for the desired stream. The DMIF specification provides hints on how to perform tasks on a few network types, such as the Internet. The DAI is also used for succeeding broadcast material and local files. This means that a single, uniform interface is defined to access multimedia contents on a multitude of delivery technologies. It is appropriate to state that the integration framework of DMIF covers three major technologies: interactive network technology, broadcast technology and storage technology. This is shown in Figure 5.38. Fulfilling the previous requirements represents an improvement to favor the development of truly multimedia applications. The identification of a common interface, from an application's point of view, to today's and tomorrow's networks, encompassing both the QoS-enabled networks and the best-effort model, is the first DMIF requirement. This is particularly useful in relation to QoS issues.

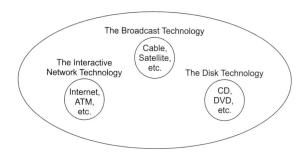

Figure 5.38 The delivery integration of three major technologies [5.36]. ©2001 ISO/IEC.

The second, less obvious requirement of the DMIF reference architecture is to also hide the operational scenario details to the application. It means managing the access to locally or remotely retrieved streams, as well as broadcast and multicast streams, through a common interface to the delivery system. The reasons for these requirements are not that evident, because broadcast content is certainly designed with different criteria than content meant to be retrieved interactively.

The DMIF communication architecture is represented in Figure 5.39. The picture shows how the different operational scenarios are uniformly modeled, through the identification of four basic blocks: originating application, originating DMIF, target DMIF and target application. The elements in the user part are meant to be part of the originating terminal, and the elements in the bottom are part of the target terminal in the case of a remote interactive scenario. The originating

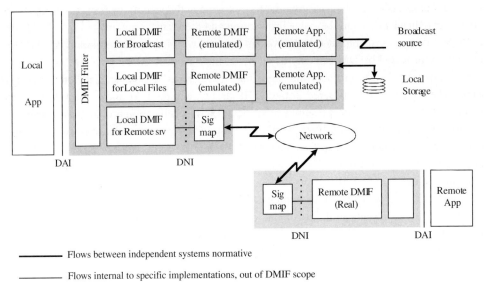

Figure 5.39 DMIF communication architecture [5.36]. ©2001 ISO/IEC.

DMIF module is meant to work in cooperation with the target DMIF module to provide a session-level service. The distinction between originating and target DMIF modules in the local retrieval and broadcast scenario is a bit artificial, but has been left for the uniformity with the remote interactive scenarios

The originating application is the actual application in the terminal, for example an MPEG-4 browser or a multimedia-conferencing application. It is assumed that it has, in all cases, a counterpart, the target application. The originating application interacts with the target application through DMIF. In the case of remote interactive operational scenarios, the two applications reside on separate hosts, and the communication between the two is regulated by some signaling protocol, not known by the applications themselves. The DMIF filter module is identified in the DMIF specification to highlight the potential benefit of the architecture. It represents a sort of container for the various DMIF instances available in a terminal, and its role is to select the appropriate DMIF instance to provide a certain service. Another module is the signaling module (Sig map). This element applies only to the DMIF instances for remote interactive scenarios and is kept separate from the other DMIF elements in order to highlight the role of the DMIF Network Interface (DNI). Particularly for the case of remote interactive systems, DMIF factorizes features that are instead specific for a certain network (such as signaling protocols). The DNI represents the border between the generic and specific tasks of a DMIF instance for remote interactive scenarios.

The DAI is a semantic API that derives directly from the requirement of hiding the delivery technology and operational scenario details from the application. Thus, local and remote retrieval are not different from multicast or broadcast. An MPEG-4 browser would be able to access and present multimedia content uniformly and independently of the operational scenario. In the MPEG-4 context, the DAI formalizes the demarcation line between Systems and DMIF, and it separates the elements and tools of MPEG-4 that are conceptually network unaware from those that instead relate with the delivery technology (covered by DMIF).

DMIF allows the concurrent presence of one or more DMIF instances, each one targeted for a particular delivery technology, in order to provide support in the same terminal multiple delivery technologies and even multiple scenarios (broadcast, local storage, and remote interactive). Multiple delivery technologies may be activated by the same application, which could therefore seamlessly manage data sent by broadcast networks, local file systems and remote interactive peers.

DMIF Computational Model

When an application requests the activation of a service, it uses the service primitives of the DAI and creates a service session. The DMIF implementation then contacts its corresponding peer (that conceptually can be either a remote peer or a local emulated peer) and creates a network session with it. Network sessions have network-wide significance, and service sessions instead have local meaning. The acquisition between them is maintained by the DMIF Layer. In the case of broadcast and local storage scenarios, the way that the network session is created and then managed is out of the scope of this specification. In the case of a remote interactive scenario instead, DMIF uses the native signaling mechanism for that network to create and then manage the network session, for example, ATM signaling. The application peers then use this session to create connections that are used to transport application data, for example, MPEG-4 ESs.

When an application needs a channel, it uses the channel primitives of the DAI. DMIF translates these requests into connection requests that are specific to the particular network implementation. In the case of broadcast and local storage scenarios, the way that the connections are created and then managed is out of the scope of this specification. In the case of a networked scenario instead, DMIF uses the native signaling mechanisms for that network to create those connections. The application then uses these connections to deliver the service.

Figure 5.40 provides a high-level view of a service activation and the beginning of data exchange. The high-level walkthrough consists of the following steps in which DMIF is involved. The originating application requests the activation of a service to its local DMIF instance. A communication path between the originating application and the originating DMIF

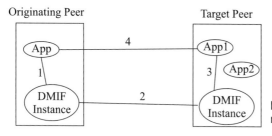

Figure 5.40 DMIF computational model [5.36]. ©2001 ISO/IEC.

instance is established in the control plane (1) and is associated to a locally meaningful service session.

The originating DMIF instance contacts the target DMIF instance. A communication path between these two is established in the control plane (2) and is associated with a unique network session.

The target DMIF instance identifies the target application and forwards the service activation request. A communication path between the target DMIF instance and the target application is established in the control plane (3) and is associated with a locally meaningful service session.

The peer applications create channels (requests flowing through communication paths 1, 2 and 3). The resulting channels in the user plane (4) will carry the actual data exchanged by the applications.

The DMIF specification defines an architecture that is open to future evaluations in the delivery technology and that is able, if actually implemented in the terminals, to protect investments in the development of multimedia applications.

5.5.4 MPEG-4 Video

Digital video is replacing analog video in many existing applications. A prime example is the introduction of Digital Television (DTV). Another example is the progressive replacement of analog video cassettes by DVD as the preferred medium to watch movies. MPEG-2 has been one of the key technologies that enabled the acceptance of these new media. In these applications, digital video will initially provide functionalities similar to analog video, that is the content is represented in digital form instead of analog with direct benefits, such as improved quality and reliability, but the content remains the same to the user. However, after the content is in the digital domain, new functionalities can easily be added. This allows the user to view, access and manipulate the content in completely new ways. The MPEG-4 standard provides key technologies that will enable such functionality.

The MPEG-4 Visual standard consists of a set of tools that enable applications by supporting several classes of functionalities. The most important features covered by the MPEG-4 standard can be summarized in three categories, as shown in Figure 5.41.

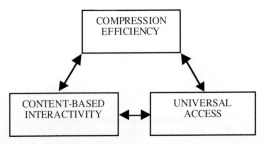

Figure 5.41 Functionalities in MPEG-4 Visual standard.

Compression efficiency has enabled applications such as DTV and DVD. Improved compression efficiency and coding of multiple data streams will increase the acceptance of applications based on MPEG-4 Video. Content-based interactivity is one of the most important novelties offered by MPEG-4. Coding and representing video objects rather than video frames enables content-based applications. Robustness in error-prone environments allows MPEG-4 encoded content to be accessible in a wide range of media, such as mobile networks as well as wired connections. In addition, object-based temporal and spatial scalabilities allow the user to decide where to use sparse resources, which can be the available bandwidth, but also the computing capacity or power consumption. To support some of these functionalities, MPEG-4 should provide the capability to represent arbitrarily shaped video objects. Each object can be encoded with different parameters and at different qualities. The shape of a video object can be represented in MPEG-4 by a binary or a gray-level (alpha) plane. The texture is coded separately from its shape. To increase robustness to errors, special provisions are taken into account at the bit-stream level to allow fast resynchronization and efficient error recovery. The MPEG-4 Video has been explicitly optimized for three bit rate ranges: below 64 Kb/s, 64 to 384 Kb/s and 384 Kb/s to 4 Mb/s. Both CBR and VBR are supported.

An MPEG-4 Video bit stream provides a hierarchical description of a visual scene as shown in Figure 5.42. Each level of the hierarchy can be accessed in the bit stream by special code values called start codes. The hierarchical levels that describe the scene most directly are Visual Object Sequence (VS), Video Object (VO), Video Object Layer (VOL), Group of Video Object Planes (GOV) and Video Object Plane (VOP).

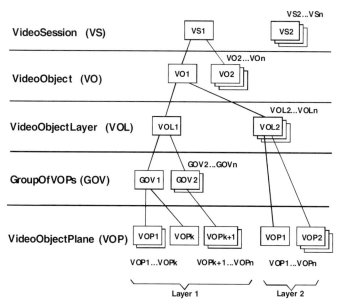

Figure 5.42 MPEG-4 video bit stream logical structure.
©2000 ISO/IEC.

By VS, we mean the complete MPEG-4 scene, which may contain any 2D or 3D natural or synthetic objects and their enhancement layers.

VO corresponds to a particular 2D object in the scene. In the most simple case, this can be a rectangular frame, or it can be an arbitrarily shaped object corresponding to an object or background of the scene.

VOL provides support for scalable coding. Each video object can be encoded in scalable (multilayer) or nonscalable form (single layer), depending on the application, represented by the VOL. Going from the coarse to fine resolution, a VO can be encoded using spatial or temporal scalability. There are two types of VOs: the VOL that provides full MPEG-4 functionality and a reduced functionality VOL, which is the VOL with short headers [5.66]. Each VO is sampled in time, and each time sample of a VO is a VOP. VOPs can be grouped together to form a GOV.

The GOV groups together VOPs. GOVs can provide points in the bit stream where VOPs are encoded independently from each other. Thus, they can provide random access points into the bit stream.

VOP is a time sample of a VO [5.67]. VOPs can be encoded independently of each other or dependent on each other by using motion compensation. In the most common way, the VOP contains the encoded video data of a time sample of a VO. In that case, it contains motion parameters, shape information and texture data. These are encoded using macroblocks [5.68]. It can also be used to code a sprite. A sprite is a VO that is usually larger than the display video presented over time. It is used to represent large, more or less static areas, such as backgrounds. Sprites are encoded using macroblocks. Each macroblock contains four luminance blocks and two chrominance blocks. Each block contains 8x8 pixels and is encoded using the DCT [5.69]. A macroblock carries the shape information, motion information and texture information. Figure 5.43 illustrates the general block diagram of MPEG-4 coding and decoding based on the VOs. Each video object is coded separately. For reasons of efficiency and backward compatibility, video objects are coded through their corresponding VOPs in a hybrid coding scheme. MPEG-4 coding for natural video will also provide tools that enable a number of other functionalities such as object scalability, spatial and temporal scalabilities, sprite overlays, error resilience, and so forth. MPEG-4 Video is capable of coding conventional rectangular video as well as arbitrarily shaped 2D objects in a video scene. It will be able to code video ranging from very low spatial and temporal resolutions in progressive scanning format up to very high spatial and temporal resolutions for professional studio applications, including interlaced video.

In what follows, we will discuss coding tools (shape coding, motion estimation and compensation, texture coding and multifunctional coding), error resilience, sprite coding and scalability.

Shape-Coding Tools for MPEG-4 Natural Video

Besides the shape information available for the VOP in question, the shape-coding scheme also relies on motion estimation to compress the shape information [5.70]. The shape and location of VOPs may vary from one VOP to the next. The shape may be conveyed either implicitly or explicitly. With implicit shape coding, the irregularly shaped object is placed in front of a colored

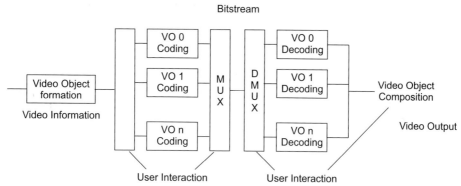

Figure 5.43 General block diagram of MPEG-4 video coding and decoding [5.62]. ©1999 ISO/IEC.

background known to receiver. A rectangular VOP containing both the object and background is coded and transmitted. The decoder retrieves the object by simple chromakeying.

Explicit shape is represented by a rectangular alpha plane that covers the object to be coded. An alpha plane may be binary if only the shape is of interest ("0" for background or "1" for object). It may be also gray level (up to 8 bits per pixel) to indicate various levels of partial transparency for the object. If the alpha plane has a constant gray value inside the object area, that value can be sent separately, and the alpha plane can be coded as a binary alpha plane. One of the most promising algorithms for coding alpha planes is arithmetic coding.

In the MPEG-4 Visual standard, two kinds of shape information are considered as inherent characteristics of a VO. These are referred to as binary and gray shape information. By binary shape information, one means label information that defines which partitions (pixels) of the support of the object belong to the VO at a given time. This kind of information is most commonly represented as a matrix with the same size as that of the bounding box of a VOP. Every element of the matrix can take one of the two possible values, depending on whether the pixel is inside or outside of the VO. Gray-scale shape is generalization of the concept of binary shape providing a possibility to represent transparent objects and to reduce aliasing effects.

Motion Estimation and Compensation

Motion estimation and compensation are commonly used to compress video sequences by exploiting temporal redundancies among frames. The main difference between MPEG-4 and other standards concerning motion compensation is that the block-based techniques used in the other standards have been adopted to the VOP structure in MPEG-4. Three modes for encoding an input VOP are provided by MPEG-4 as shown in Figure 5.44. Namely, Intra VOPs (I-VOPs) are coded without any information from other VOPs. Predicted VOPs and Bidirectional Interpolated VOPs (P- and B-VOPs) are predicted based on I- and/or P-VOPs.

A VOP may be encoded independently of any other VOP. In this case, the encoded VOP is called an I-VOP. A VOP may be predicted using motion compensation based on another previ-

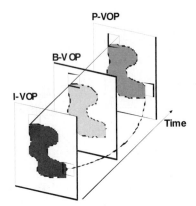

Figure 5.44 The three modes of VOP encoding. ©1999 ISO/IEC.

ously decoded VOP. Such VOPs are called P-VOP. A VOP may be predicted based on past as well as future VOPs. Such VOPs are called B-VOPs.

Motion Estimation (ME) is necessary only for coding P-VOPs and B-VOPs. ME is performed only for macroblocks in the bounding box of the VOP in question. If a macroblock lies entirely within a VOP, ME is performed based on block matching of 16x16 macroblocks as well as 8x8 blocks in advance prediction mode. This results in one motion vector for the entire macroblock and one for each of its blocks.

For P- and B-VOPs, the motion vectors are first differentially coded based on up to three motion vectors of previously transmitted blocks. The exact number depends on the allowed range of the vectors. The maximum range is selected by the encoder and transmitted to the decoder. Variable length coding is then used to encode the motion vectors. Here, for each block of the macroblock, the motion vectors of neighboring blocks are considered. This includes the motion vector of the current block and its four neighbors. Each vector provides an estimate of the pixel value. The actual predicted value is then a weighted average of all these estimates.

Texture-Coding Tools

Coding of texture for arbitrarily shaped regions with a shape described with an alpha map is different than traditional methods. For example, intraframe coding, forward prediction motion compensation and bidirectional motion-compensation are used. This gives rise to the definitions of I-VOPs, P-VOPs and B-VOPs for VOPs that are intracoded, forward predicted or bidirectionally predicted, respectively.

The texture information of a VOP is present in the luminance Y and two chrominance components, C_b, C_r, of the video signal. In the case of an I-VOP, the texture information resides directly in the luminance and chrominance components. In the case of motion-compensated VOPs, the texture information represents the residual error remaining after motion compensation. For encoding the texture information, the standard 8x8 block-based DCT is used. To encode an arbitrarily shaped VOP, an 8x8 grid is superimposed on the VOP. Using this grid, 8x8 blocks that are internal to the VOP are encoded without modifications. Blocks that straddle the VOP are called boundary blocks and are treated differently from internal blocks. The transformed blocks are

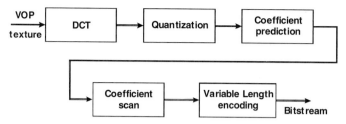

Figure 5.45 Block diagram of VOP texture-coding process. ©1999 ISO/IEC.

quantized, and individual coefficient prediction can be used from neighboring blocks to further reduce the entropy values of the coefficients. This is followed by a scanning of the coefficients to increase the average run length between coded coefficients. Then, the coefficients are encoded by variable-length encoding. This process is illustrated in Figure 5.45.

Internal video texture blocks and padded boundary blocks are encoded using a 2D 8x8 block-based DCT. The DCT coefficients are quantized as a lossy compression step. There are two methods of quantization. The first method uses one of two available quantization matrixes to modify the quantization step size depending on the spatial frequency of the coefficient. The second method uses the same quantization step size for all coefficients. The average energy of the quantized coefficients can be further reduced by using prediction from neighboring blocks. The prediction can be performed from either the block above, the block to the left, or the block above to the left. Candidate blocks for coefficient prediction are illustrated in Figure 5.46. The direction of the prediction is adaptive and is selected based on comparison of horizontal and vertical DC gradients (increase or reduction in its value) of surrounding blocks A, B and C. There are two types of predictions, DC prediction and AC prediction. The DC prediction is performed for the DC coefficient only and is either from the DC coefficient of block A or from the DC coefficient of block C. As for AC prediction, either the coefficients from the first row or the coefficients from the first column of the current block are predicted from the cosited coefficients of the selected candidate blocks.

Before the coefficients are run-length coded, a scanning process is applied to transform the 2D data into a 1D string. Three different scanning methods are possible: zig-zag scan, alternate-horizontal scan and alternate-vertical scan. In zig-zag scan, the coefficients are read out

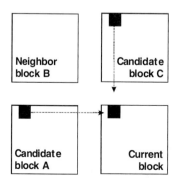

Figure 5.46 Blocks for coefficients prediction. ©1999 ISO/IEC.

diagonally, and, in alternate-horizontal scan, the coefficients are read out with an emphasis on the horizontal direction first. Alternate-vertical scan is similar to the horizontal direction scan, but are applied in the vertical direction. If there is DC prediction in the horizontal direction, the alternate-vertical scan is used, and if DC prediction is performed from the vertical direction, the alternate-horizontal scan is used. Zig-zag scan is used if there is no DC prediction.

The run-length encoding can use two different VLC tables and use the value of the quantizer to determine which VLC table is used [5.65].

Multifunctional Coding

Multifunctional coding refers to features other than coding efficiency. For example, object-based spatial and temporal scalabilities are provided to enable broad-based access across a variety of networks and facilities. This can be useful for Internet and database applications. Spatial scalability with two layers and temporal scalability with two layers are shown in Figure 5.47 and Figure 5.48, respectively.

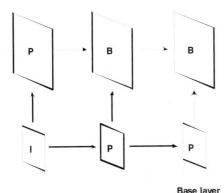

Figure 5.47 Spatial scalability with two layers. ©1999 ISO/IEC.

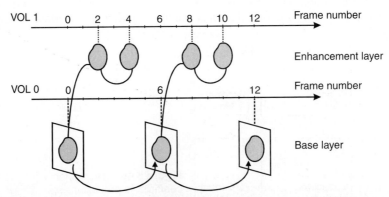

Figure 5.48 Temporal scalability with two layers [5.62]. ©1999 ISO/IEC.

For mobile multimedia applications, spatial and temporal scalabilities are useful for channel bandwidth scaling for robust delivery.

Multifunctional coding also addresses multiview and stereoscopic applications, as well as representations that enable simultaneous coding and tracking of objects for surveillance and other applications. Besides the aforementioned applications, a number of tools are being developed for segmentation of a video scene into objects and for suppressing noise.

Sprite Coding

A sprite consists of those regions of a VO that are present in the scene throughout the video segment. As an example, a background sprite consists of all pixels belonging to the background of a camera-panning sequence. Sprites have been included in MPEG-4 mainly because they provide high efficiency in such cases. For any given instant of time, the background VOP can be extracted.

The texture information for a sprite is represented by one luminance component and two chrominance components. The three components are processed separately. The methods used for processing the chrominance components are the same as those used for the luminance components, after appropriate scaling. Shape and texture information for a sprite are encoded as for an I-VOP. Static sprites are generated before the encoding process begins using the original VOPs. The decoder receives each static sprite before the rest of the video segment. The static sprites are encoded in such a way that the reconstructed VOPs can be generated easily by warping the quantized sprite with the appropriate parameters.

Several possibilities are envisaged for the transmission of sprites. One way is to transmit only a portion of the sprite in the beginning. The transmission portion should be sufficient for reconstructing the first few VOPs. The remainder of the sprite is transmitted, piecewise, as required or as the bandwidth allows. Another method is to transmit the entire sprite in a progressive fashion, starting with a low-quality version and gradually improving its quality by transmitting residual images. A combination of these methods can also be used in practice.

Scalability

Scalability essentially means that the compressed bit stream can be manipulated in a simple manner in order to satisfy constraints on such parameters as bit rates, display resolutions and frame rates or for decoding hardware complexity [5.6]. Generally speaking, this manipulation consists of the extraction of relevant subsets from the compressed bit stream, each of which should represent an efficient compression of the video sequence at the same resolution and distortion [5.71]. The value of scalable compression lies in the fact that the bit stream may be manipulated at any point after the compressed bit stream has been generated. This is significant because, in many important applications, advance knowledge of constraints on resolutions, bit rates or decoding complexities may not be available during compression.

Spatial scalability and temporal scalability are both implemented using multiple VOLs. Consider the case of two VOLs: the base layer and the enhancement layer. For spatial scalability, the enhancement layer improves upon the spatial resolution of a VOP provided by the base layer.

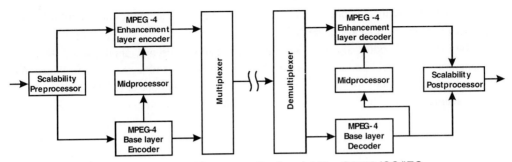

Figure 5.49 Block diagram of MPEG-4 generalized scalability. ©1999 ISO/IEC.

In the case of temporal scalability, the enhancement layer may be decoded if the desired frame rate is higher than that offered by the base layer. Thus, temporal scalability improves the smoothness of motion in the sequence. Object scalability is naturally supported by the VO-based scheme.

MPEG-4 uses a generalized scalability framework to enable spatial and temporal scalabilities. To enable the various scalabilities, this framework allows the inclusion of separate modules. The block diagram of the MPEG-4 generalized scalability framework is presented in Figure 5.49. The specific algorithms implemented in the preprocessor, midprocessor and postprocessor depend upon the type of scalability being enabled. A scalability preprocessor is used to implement the desired scalability. It operates on VOPs. In the case of spatial scalability, the preprocessor down-samples the input VOPs to produce the base layer VOPs that are processed by the VOP encoder. The midprocessor takes the reconstructed base-layer VOPs and up-samples them. The difference between the original VOP and the output of the midprocessor forms the input to the encoder for the enhancement layer. To implement temporal scalability, the preprocessor separates out the frames into two streams. One stream forms the input for the base-layer encoder, and the other is processed by the enhancement layer encoder. The midprocessor is bypassed.

The base-layer VOPs are encoded in the same way as in the nonscalable case discussed in previous sections. VOPs of the enhancement layer are encoded as P-VOPs or B-VOPs.

In scalability coding of a VOP, if the enhancement layer is temporally coincident with an I-VOP in the base layer, it could be treated as a P-VOP. VOPs in the enhancement layer that are coincident with P-VOP in the base layer could be coded as B-VOPs. VOPs in the base layer must be encoded before their corresponding VOPs in the enhancement layer, because the base layer serves as the reference for the enhancement layer.

In temporal scalability, the frame rate of the visual data is enhanced. Enhancement layers carry information to be visualized between the frames of the base layer. The enhancement layer may act in two ways:

- The enhancement layer improves the resolution of only a portion of the base layer.
- The enhancement layer improves the resolution of the entire base layer.

Error Resilience

Error resilience is needed, to some extent, in all transmission media. For an example, due to the rapid growth of mobile communications, wireless networks are typically prone to error and usually operate at relatively low bit rates, for example, less than 64 Kb/s. Both MPEG and ITU-T are working on error-resilience methods, including forward error correction, automatic request for retransmission, scalable coding, slice-based bit-stream partitioning and motion-compensated error correction.

MPEG-4 provides several mechanisms to allow error resilience with different degrees of robustness and complexities [5.39]. These mechanisms are offered by tools providing means for resynchronization, error detection, data recovery and error concealment. There are four error resilience tools in MPEG-4 Visual: resynchronization, data partitioning, header extension code and reversible variable-length codes.

The most frequent way to bring error resilience to a bit stream is resynchronization. It consists of inserting unique markers in the bit stream so that, in the case of an error, the decoder can skip the remaining bits until the next marker and restart decoding from that point on. MPEG-4 allows for insertion of resynchronization markers after an approximately constant number of coded bits or video packets.

The data-partitioning method separates the bits for coding of motion information and those for the texture information. In the event of an error, more efficient error concealment may be applied when, for example, the error occurs on the texture bits only by making use of the decoded motion information.

A header extension code represents the binary codes that allow an optional inclusion of redundant header information, which is vital for correct decoding of video. In this way, the chances of corruption of header information and complete skipping of large portions of the bit stream will be reduced. Reversible Variable Length Coding (RVLC) further reduces the influence of error occurrence on the decoded data. RVLCs are code words that can be decoded in forward as well as backward manners. In the event of an error and skipping of the bit stream until the next resynchronization marker, it is possible to still decode portions of the corrupted bit stream in the reverse order to limit the influence of the error.

Relationship Between Natural and Synthetic Video Coding

The visual objects may have natural or synthetic content, including arbitrary shape VOs, special synthetic objects such as the human face and body and generic 2D/3D objects composed of primitives like rectangles, spheres or indexed face sets, which define an object surface by means of vertexes and surface patches. The synthetic VOs are animated by transforms and special-purpose animation techniques, such as face/body animation and 2D mesh animation. The representation of synthetic VOs in MPEG-4 is based on the prior VRML standard [5.39, 5.56, 5.72, 5.73].

The advent of networked multimedia, for example, across the Internet, has resulted in an increased need for interactive retrieval and composition of natural and synthetic visual content on demand, as well as reuse of archived content at the desktop. This in turn, necessitates new content-based visual representations and standards that allow compression, manipulation, search, browsing and distribution of synthetic and natural VOs, as well as means to synchronize and composite them at the user terminal [5.74]. An important question in object-based video representation is how to extract or segment video objects. Fortunately, a large portion of material used in movie and television productions today is blue screened (chromakeyed), which is when individual VOs are captured as separate camera shots against a blue background. Composition of natural video clips with synthetic objects with known alpha planes, for example, text overlays, channel logos, animation characters and so on, is commonplace. For conventional video sources, an automatic layering method by motion segmentation was proposed [5.75].

The 2D and 3D mesh models were introduced in the image processing and compression literature for motion compensation. In the computer graphics world, meshes have long been used for 3D shape modeling. A 2D mesh is a planar graph that partitions a 2D image region into polygonal patches. The vertexes of the polygonal patches are referred to as node points. Mostly, the patches are triangles or quadtriangles, leading to triangular or quadrilateral meshes, respectively. Mesh-based motion modeling differs from block-based motion modeling in that the patches overlap neither in the reference frame nor in the current frame. Block-based motion modeling and mesh-based motion modeling are represented in Figure 5.50. Arrows show corresponding patches in the reference image (left) and the current image (right). In both cases, backward motion estimation is shown, by which patches are searched in the reference image that best matches a given patch in the current frame. Polygonal patches in the current frame are deformed by the movements of the node points into polygonal patches in the reference frame. The texture inside each patch in the reference frame is warped [5.76].

In content-based video compression, manipulation and indexing, the shape, motion and texture of each arbitrary-shaped VO need to be modeled and encoded independently. In the MPEG-4 Video verification model, the shape of a VOP is represented by a bitmap, called an alpha plane (binary or gray scale), and the text (color) of the VOP is represented by a texture plane [5.75]. The values of pixels in the texture plane are defined in the corresponding alpha-plane pixel as nonzero.

The 2D mesh representation of VOs enables the following functionalities: VO compression, VO manipulation and content-based video indexing.

VO compression. Mesh modeling may improve compression efficiency in two ways. Namely, the mesh model provides better motion compensation, that is, the translation-block model, which may result in less blocking artifacts at lower bit rates. Alternatively, we can choose to transmit texture maps only at selected key frames and to animate these texture maps for the intermediate frames without sending any predicted error image.

VO manipulation. By VO manipulation, we mean augmented reality, synthetic-object transformation/animation and spatiotemporal interpolation.

Augmented reality means merging virtual (computer-generated) images with real moving images (video) to create enhanced display information. The computer-generated images must remain in perfect registration with the moving real images, which is why tracking is needed.

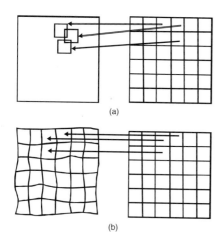

Figure 5.50 (a) 2D block-based motion modeling and (b) 2D quadrilateral mesh-based motion modeling [5.76].
©1998 ISO/IEC.

Synthetic-object transformation/animation replaces a natural video object in a video clip with another VO. The replacement VO may be extracted from another natural video clip or may be transfigured from a still-image object using the motion information of the object to be replaced, which is why temporally continuous motion representation is needed.

As for spatiotemporal interpolation, mesh motion modeling provides more robust motion-compensated temporal interpolation compared to block-based motion modeling.

Content-based video indexing. Mesh representation enables animated key snapshots for a moving visual synopsis of an object. Also, it provides accurate object trajectory information that can be used to retrieve VOs with specific motion. Finally, it provides vertex-based object shape representation, which is more efficient than the bitmap representation for shape-based retrieval.

Synthetic Images

MPEG-4 uses VRML as a starting point for its synthetic image specification. MPEG-4 adds a number of additional capabilities. The first addition is a synthetic Face and Body (FAB) animation capability, which is a model-independent definition of artificial FAB animation parameters. With these parameters, one can represent facial expressions, body positions, mouth shapes, and so forth. Planned capabilities include 3D feature point positions and 3D head and body control meshes for animation and texture mapping of face, body and personal characteristics. Also planned is a text-driven mouth animation to be combined with a text-to-speech capability for a complete text-to-talking-head implementation. Another capability being studied is for texture mapping of real image information onto artificial models, such as the FAB model. For this, wavelet-based texture coding is being considered. An advantage of wavelet-based coding is the relative ease of adjusting the resolution of the rendering.

Associated with FAB coding is a triangular mesh modeling capability to handle any type of 2D or 3D synthetic or natural shape. This also facilitates integration of text and graphics onto synthetic and natural images.

Integration of Face Animation with Natural Video

The face object specified by MPEG-4 is a representation of the human face structured in a way that the visual manifestations of speech are intelligible. The facial expressions allow recognition of the speaker's mood and reproduction of a real speaker as faithfully as possible [5.40]. To fulfill these objectives, MPEG-4 specified three types of facial data: Facial Animation Parameters (FAPs), Facial Definition Parameters (FDPs) and FAP Interpolation Table (FIT).

FAPs allow one to animate a 3D facial model available at the receiver. The way by which this model is made available at the receiver is not relevant. FAPs allow the animation of key feature points in the model, independently or in graphs, as well as the reproduction of visemes and expressions.

FDPs allow one to configure the 3D facial model to be used at the receiver, either by adopting a previously available model or by sending a new model.

FIT allows one to define the interpolation rules for the FAPs that have been interpolated at the decoder. The 3D model is then animated using the FAPs sent and the FAPs interpolated according to the FIT.

Although FAPs continuously provide visual information associated with the behavior of the 3D model, FDPs provide model configuration information, which is typically sent only once. For this reason, FAPs are coded as an individual ES—the facial animation stream—while FDPs are fully coded as BIFS nodes and thus are sent in the BIFS stream.

Figure 5.51 shows a general block diagram of a 3D facial animation system, which may fit many applications. For some applications, FAPs and FDPs are extracted from real video input. For many other applications, the analysis phase does not exist. This means that FAP and FDP data are artificially edited to fulfill a certain goal by synthesizing the necessary data. Assuming that FAPs and FDPs follow the standardized bit stream syntax and semantics, the way by which FAPs and FDPs are generalized is completely irrelevant to the receiving terminal.

MPEG-4 specifies a face model in its neutral state, a number of future points on this neutral face as reference points, and a set of FAPs, each corresponding to a particular facial action deforming a face model in its neutral state. Deforming a neutral face model according to some

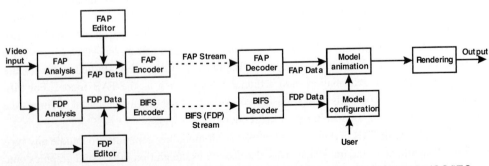

Figure 5.51 General block diagram of a 3D facial animation system [5.39]. ©1998 ISO/IEC.

specified FAP values at each time instant generates a facial animation sequence. The terminal can either use its own animation rules or can download a face model and the associated Facial Animation Tables (FAT) to have a customized animation behavior. Because the FAPs are required to animate faces of different sizes and proportions, the FAP values are defined in Facial Animation Parameters Units (FAPUs). The FAPU is computed from spatial distances between major facial features on the model in its neutral state [5.77]. The FAPU allows interpolation of the FAPs on any facial model in a consistent way, producing results in terms of expression and speech pronunciation. The fractional units used for various FAPs are chosen to allow enough precision for the corresponding FAP.

FAPs

The FAPs are based on the study of minimal perceptible actions and are closely related to muscle actions [5.78]. FAPs were designed to allow the animation of faces, reproducing movements, expressions, emotions and speech pronunciation. The chosen set of FAPs represents a complete set of basic facial movements, allowing terminals to represent most natural facial actions.

The FAP set includes 68 FAPs: 66 low-level parameters associated with the lips, jaw, eyes, mouth, cheek, nose, and so forth, and 2 high-level parameters [5.40]. Although low-level FAPs are associated with movements of key facial zones, typically referenced by a future point, as well as with rotation of the head and eyeballs, visemes and expressions represent more complex actions, typically associated with a set of FAPs. Although the encoder knows the reference feature point for each low-level FAP, it does not precisely know how the decoder will move the model vertexes around that feature point. This is the FAP interpolation model, which describes the specification of the precise changes in the model vertexes corresponding to each FAP.

Table 5.9 shows an excerpt from the FAP list, including number and name, a short description, the specification of the associated FAPU, whether the FAP is unidirectional or bidirectional, the definition of the movement direction for positive values, the group number, the FDP subgroup number and the default quantization step size. MNS is mouth-nose separation [5.39].

With FAP, it is possible to select among six different expressions: joy, sadness, anger, fear, disgust and surprise. Visemes are the visual analog to phonemes and allow the efficient rendering of visemes for better speech pronunciation as an alternative to hearing them represented using a set of low-level FAPs [5.79]. Zero-valued FAPs correspond to a neutral position at the beginning of a session. All FAPs are expressed as displacement from the positions defined for the neutral face. Thus, it is essential to start from neutral faces that are as similar as possible. According to the specification [5.40], the neutral face is characterized by having all muscles relaxed; eyelids tangent to the iris; pupils as one-third of the iris diameter; lips in contact; mouth closed with the upper teeth touching the lower ones and the tongue flat, horizontal and with its tip touching the boundary between upper and lower teeth. Note that the 68 parameters are categorized into 10 groups related to parts of the face. FAP groups are represented in Table 5.10.

The FAP set contains two high-level parameters, visemes and expressions (FAP group 1). A viseme (FAP1) is a visual correlate to a phoneme. Only 14 static phonemes that are clearly distinguished are included in the standard set. Visemes and related phonemes are given in Table 5.11.

Table 5.9 Excerpt of the FAP Specification Table [5.39].

	FAP Name	FAP Description	Units	Uni- or Bidirectional	Motion	Group	FDP Subgroup Number	Default Quantization Step
1	Viseme	Set of values determinining the mixture of two visemes (e.g. pbm, fv, th)	na	na	na	1	na	1
2	Expression	Set of values determinining the mixture of two facial expressions	na	na	na	1	na	1
3	Open_jaw	Vertical jaw displacement (does not affect mouth opening)	MNS	U	down	2	1	4
4	Lower_t_midlip	Vertical top middle inner lip displacement	MNS	B	down	2	2	2
5	Raise_b_midlip	Vertical bottom middle inner lip displacement	MNS	B	up	2	3	2

©1998 ISO/IEC.

Table 5.10 FAP groups [5.39].

Group	Number of FAPs
1: Visemes and expression	2
2: Jaw, chin, inner lowerlip, cornerlips and midlip	16
3: Eyeballs, pupils and eyelids	12
4: Eyebrow	8
5: Cheeks	4
6: Tongue	5
7: Head rotation	3

Table 5.10 FAP groups [5.39]. (Continued)

Group	Number of FAPs
8: Outer lip positions	10
9: Nose	4
10: Ears	4

©1998 ISO/IEC.

Table 5.11 Visemes and related phonemes [5.39].

Viseme No.	Phonemes	Example
0	none	na
1	p, b, m	put, bed, mill
2	f, v	far, voice
3	T, D	think, that
4	t, d	tip, doll
5	k, g	call, gas
6	tS, dZ, S	chair, join, she
7	s, z	sir, zeal
8	n, l	not, lot
9	R	Red
10	A	Car
11	E	Bed
12	I	Tip
13	Q	Top
14	U	Book

©1998 ISO/IEC.

In order to allow for coarticulation of speech and mouth movement, the shape of the mouth of a human is not influenced by the current phoneme, but also the previous and the next phoneme. In MPEG-4, transition from one viseme to the next is defined by blending only two visemes with a weighting factor.

The expression parameter FAP2 defines the six primary facial expressions as shown in Table 5.12. In contrast to visemes, facial expressions are animated by a value defining the excitation of the expression. The facial expression parameter values are defined by textual descriptions. The expression parameter allows an efficient means of animation faces. They are high-level animation parameters. A face model designer creates them for each face model.

One way to achieve redundancy reduction is to send only a subset of active FAPs. This subset is then used to determine the values of other FAPs. Such FAP interpolation exploits the symmetry of a human face or the a priori knowledge of articulation functions. For example, the top-inner lip FAPs can be sent and then used to determine the lip-outer-lip FAPs. The inner-lip FAPs would be mapped to the outer-lip FAPs by interpolation.

Table 5.12 Primary facial expressions for FAP2 [5.79].

No.	Expression Name	Textual Description
1	Joy	The eyebrows are relaxed. The mouth is open, and the mouth corners are pulled back toward the ears.
2	Sadness	The inner eyebrows are bent upward. The eyes are slightly closed. The mouth is relaxed.
3	Anger	The inner eyebrows are pulled downward and together. The eyes are wide open. The lips are pressed against each other or opened to expose the teeth.
4	Fear	The eyebrows are raised and pulled together. The inner eyebrows are bent upward. The eyes are tense and alert.
5	Disgust	The eyebrows and eyelids are relaxed. The upper lip is raised and curled, often asymmetrically.
6	Surprise	The eyebrows are raised. The upper eyelids are wide open, and the lower eyelids are relaxed. The jaw is opened.

©1999 IEEE.

Example 5.13 MPEG-4 defines a generic face model in its neutral state (Figure 5.52), with the following properties:

- Gaze is in the direction of z axis.
- All face muscles are relaxed.
- Eyelids are tangent to the iris.

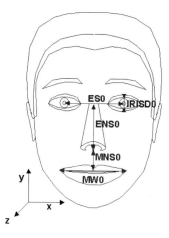

Figure 5.52 A face model in its neutral state and the feature points used to define FAP units [5.39]. ©1998 ISO/IEC.

- The pupil is one-third of the diameter of the iris.
- Lips are in contact, and the line of the lips is horizontal and the same height of lip corners.
- The mouth is closed and the upper teeth touch the lower ones.
- The tongue is flat and horizontal with the lip or tongue touching the boundary between upper and lower teeth.

In order to define face animation parameters for an arbitrary face model, MPEG-4 defines FAP units that serve to scale facial animation parameters for any face model. FAPUs are defined as a fraction of the distances between key facial features. These features, such as eye separation, are defined on a face model that is in the neutral state. The FAPU allows interpolation of the FAPs on any facial model in a consistent way, producing reasonable results in terms of expression and speech pronunciation. The measurement units are shown in Table 5.13.

Table 5.13 FAPUs and their definitions [5.39].

Measurement	Description	Measurement Unit
IRISD0	Iris diameter (by definition it is equal to the distance between upper and lower eyelids) in neutral face	IRISD=IRISD0/1024
ES0	Eye separation	ES=ES0/1024
ENS0	Eye-nose separation	ENS=ENS0/1024
MSN0	Mouth-nose separation	MNS=MNS0/1024

© 1998 ISO/IEC.

Table 5.13 FAPUs and their definitions [5.39]. (Continued)

Measurement	Description	Measurement Unit
MW0	Mouth width	MW=MW0/1024
AU	Angle unit	10E-05 rad

© 1998 ISO/IEC.

Face Model

Every MPEG-4 terminal that is able to decode FAP streams has to provide an MPEG-4 compliant face model that is animate. Usually, this is a model proprietary to the decoder. The encoder does not know about the look of the face model. Using an FDP node, MPEG-4 allows the encoder to specify completely the face model to animate. This involves defining the static geometry of the face model in its neutral state using a scene graph, defining the surface properties and defining the animation rules using FATs that specify how this model gets deformed by the facial animation parameters. Alternatively, the FDP node can be used to calibrate the proprietary face model of the decoder.

A decoder may choose to specify the location of all or some feature points. Then, the decoder is supported to adapt its own proprietary face model such that the model conforms to the feature point positions. Because MPEG-4 does not specify any algorithm for adapting the surface of the proprietary model to the new feature point locations, we cannot specify the subjective quality of a face model after its adaptation. Face model adaptation also allows for the downloading of texture maps for the face. In order to specify the mapping of the texture map onto the proprietary face model, the encoder sends texture coordinates for each feature point. Each texture coordinate defines the location of one feature point on the texture map. This does not allow for precise texture mapping at important features like eyelids or lips. This process of adapting the feature point locations of a proprietary face model according to encoder specifications is referred to as face model calibration. MPEG-4 does not specify any minimum quality of the adapted face model. Therefore, sometimes, this process is called face model adaptation. A method for face model adaptation has been proposed [5.80, 5.81]. An iterative approach based on radial basis functions for scattered data interpolation is used. For each feature point of the proprietary model, a region of interest is defined. When a feature point moves, it deforms the model within this region of interest. The advantage of face model adaptation over downloading a face model from the encoder to the decoder is that the decoder can adapt its potentially very sophisticated model to the desired shape.

In order to download a face model to the decoder, the encoder specifies the static geometry of the head model with a scene graph using MPEG-4 BIFS. VRML and BIFS describe scenes as a collection of nodes arranged in a scene graph. Three types of nodes are of particular interest for the definition of a static head model. A group node is a container for collecting child objects. It allows for building hierarchical models. For objects to move together as a group, they need to

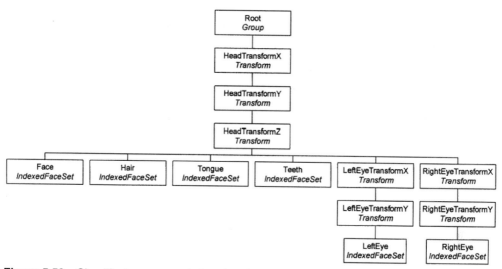

Figure 5.53 Simplified scene graph for a head model. ©1999 ISO/IEC.

be in the same transform group. The transform node defines geometric affine 3D transformations like scaling, rotation and translation that are performed on its children. When transform nodes contain other transforms, their information settings have a cumulative effect. Nested transform nodes can be used to build a transformation hierarchy. An indexed face set node defines the geometry (3D mesh) and surface attributes (color and texture) of a polygonal object. Texture maps are coded with the wavelet coder of the MPEG still-image coder [5.62].

The simplified scene graph for a head model is shown in Figure 5.53. Nested transforms are used to apply rotations about the x,y and z axes, one after another. Embedded into these global head movements are the rotations for the left and right eyes. Separate indexed face sets define the shape and the surface of the face, hair, tongue, teeth, left eye and right eye, thus allowing for separate texture maps. Because the face model is specified with a scene graph, this face model can be easily extended to a head and shoulder model. The surface of the face can be specified using colors or still images to define texture-mapped models. The shape of the face models can be generated using interactive models, scanners or image analysis software [5.82].

Coding of FAPs

MPEG-4 provides two tools for coding FAPs. Coding of quantized and temporally predicted FAPs using an arithmetic coder allows for low-delay FAP coding. Alternatively, DCT coding of a sequence of FAPs introduces a larger delay, but achieves higher coding efficiency. Figure 5.54 shows the block diagram for low-delay encoding of FAPs. The first set of FAP values, FAP_0, in time instant 0 is coded without prediction (intracoding). The value of a FAP at time instant k, FAP_k, is predicted using the previously decoded value FAP_{k-1}. The prediction error e' is quantized using a quantization step size QP multiplied by a quantization parameter FAP_QUANT with 0<FAP_QUANT<31. FAP_QUANT is identical for all FAP values at one time instant k. Using

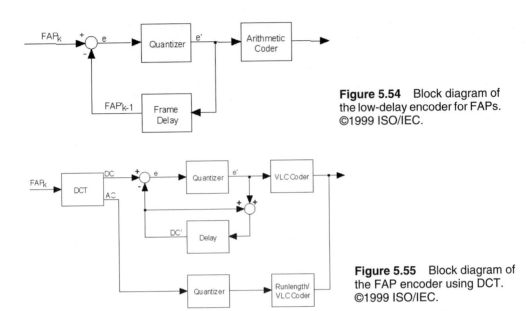

Figure 5.54 Block diagram of the low-delay encoder for FAPs. ©1999 ISO/IEC.

Figure 5.55 Block diagram of the FAP encoder using DCT. ©1999 ISO/IEC.

FAP-dependent quantization step size QP, FAP_QUANT ensures that quantization errors are subjectively evenly distributed between different FAPs. The quantized prediction error e' is arithmetically encoded using a separate adaptive probability model for each FAP. Because the encoding of the current FAP value depends only on one previously coded FAP value, this coding scheme allows for low-delay communications. At the decoder, the received data is arithmetically decoded, dequantized and added to the previously decoded value in order to recover the encoded FAP value. When using FAP_QUANT>15, the subjective quality of the animation deteriorates significantly such that it can be recommended not to increase FAP_QUANT greater than 15 [5.79].

The second tool that is provided for encoding FAPs is the DCT applied to 16 consecutive FAP values. This introduces a significant delay in the coding and decoding processes. Hence, this coding method is mainly useful for applications where animation parameter streams are retrieved from a database. After computing the DCT of 16 consecutive values of one FAP, DC and AC coefficients are coded differently as shown in Figure 5.55.

The DC value is coded predictively using the previous DC coefficient as prediction, and the AC coefficients are directly coded. The AC coefficients and the prediction error of the DC coefficient are uniformly quantized. The quantizer step size can be controlled. The ratio between the quantizer step size of the DC coefficients and the AC coefficients is set to 1:4. The quantized AC coefficients are encoded with one VLC word defining the number of zero coefficients prior to the next nonzero coefficient and one VLC for the amplitude of this nonzero coefficient. The handling of the decoded FAPs with respect to masking and interpolation is not changed. In contrast to the arithmetic coder, the DCT coder is not able to code FAPs with near-lossless quality. At low data rates, the DCT coder requires up to 50% less data rate than the arithmetic coder at the price of an increased coding delay. This advantage in coding efficiency disappears with increasing fidelity of the coded parameters.

FIT

The encoder may allow the decoder to extrapolate the values of some FAPs from the transmitted FAPs [5.83]. Alternatively, the decoder can specify the interpolation rules using FIT. A FIT allows a smaller set of FAPs to be sent for facial animation. This smaller set can then be used to determine the values of other FAPs, using a rational polynomial mapping between parameters. For example, the top inner-lip FAPs can be sent and then used to determine the top outer-lip FAPs. The inner-lip FAPs would be mapped to the outer-lip FAPs using a rational polynomial function that is specified in the FIT.

Integration of Face Animation and Text-to-Speech (TTS) Synthesis

A block diagram showing the integration of TTS synthesizer into an MPEG-4 face animation system is shown in Figure 5.56. Synchronization of a FAP stream with TTS synthesizers using the TTSI is only possible if the encoder sends timing information. This is due to the fact that a conventional TTS system driven by text only behaves as an asynchronous source. Given a TTS stream that contains text in binary form, the MPEG-4 TTSI decoder decodes the text and prosody information according to the interface defined for the TTS synthesizer. The synthesizer creates speech samples that are handed to the compositor. The compositor presents audio and, if required, video to the user. The second output interface of the synthesizer sends the phonemes of the synthesized speech as well as the start time and duration information for each phoneme to the phoneme/bookmark-to-FAP converter [5.77, 5.84]. The converter translates the phonemes and timing information into FAPs that the face rendered uses in order to animate the face model. The precise methods of how the converter derives visemes from phonemes is left to the implementation of the decoder. This allows a coarticulation model at the decoder that uses the current, previous and next phonemes in order to derive the current mouth shape. Bookmarks in the text of TTS are used to animate facial expression and non-speech related parts of the face [5.77, 5.85]. The start time of a bookmark is derived from its position in the text. When the TTS finds a bookmark in the text, it sends this bookmark to the

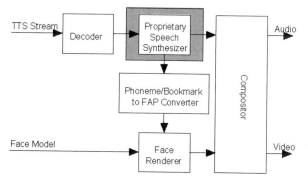

Figure 5.56 Integration of TTS synthesizer into an MPEG-4 face animation system. ©1999 ISO/IEC.

phoneme/bookmark-to-FAP-converter at the same time as it sends the first phoneme of the following word. The bookmark defines the start point and duration of the transition to new FAP amplitude.

BIFS for Facial Animation

In MPEG-4 the scene description information is represented using a parametric methodology [5.39]. The description consists of an efficiently encoded hierarchy (tree) of nodes with attributes and other information, including event sources and targets [5.86]. Leaf nodes in the tree correspond to particular audio or visual objects, and intermediate nodes perform grouping, transformation and other operations. The MPEG-4 scene description framework is partly based on the VRML, which has been significantly extended to address streaming and synchronization issues. To offer complete support for face and body animations, BIFS defines a set of face and body nodes. The most important BIFS nodes for facial animation are shown in Figure 5.57. In order to use face animation in the context of MPEG-4 Systems, a BIFS scene graph has to be transmitted to the decoder. The minimum scene graph contains a face node and an FAP node. The FAP decoder writes the amplitude of the FAP into fields of the FAP node. The FAP node might have the children viseme and expression, which are FAPs requiring a special syntax. This scene graph would enable an encoder to animate the proper face model of the decoder. If a face model is to be controlled from a TTS system, an audio source node needs to be attached to the face node. In order to download a face model to the decoder, the face node requires an FDP node as one of its children. This FDP node contains the position of the feature points in the downloaded model, the scene graph of the model and the face definition table, the face definition mesh and face definition transform nodes required to define the action caused by FAPs. Figure

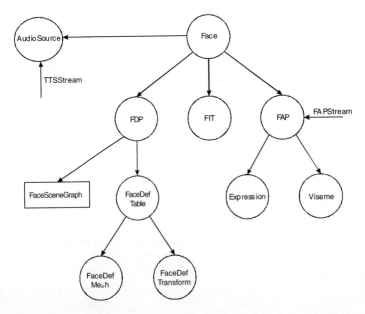

Figure 5.57 Nodes of a BIFS scene graph that are used to describe and animate a face.
©1999 ISO/IEC.

5.57 shows how these nodes relate to each other. The face graph contains the scene graph of the static face. Here, it is assumed that the streams are already decoded. The FIT node, when specified, allows a set of unreceived FAPs to be defined in terms of a set of received FAPs. The standard specifies processes that involve the reading of node values, for example, FAPs and then the writing of output values to nodes in the face hierarchy.

2D Mesh Coding

MPEG-4 version 1 supports 2D uniform or content-based (nonuniform) mesh representation of arbitrary visual objects, which includes an efficient method for animation of such meshes [5.87, 5.88].

In content-based video compression, manipulation, and indexing, the shape, motion and texture of each arbitrary-shaped VO need to be modeled and encoded independently. In the MPEG-4 video verification model, the shape of a VOP is represented by a bitmap, called alpha plane (binary or gray scale), and the text (color) of VOP is represented by a texture plane. The values of pixels in the texture plane are defined in the corresponding alpha plane pixel as non-zero. The motion of the VO is represented by a translational block model. A 2D mesh object is a representation of 2D deformable geometric shape, with which synthetic VOs may be created during a composition process at the decoder by spatially piecewise warping of existing VOPs or still texture objects. The instances of mesh objects at a given time are called Mesh Object Planes (MOPs). The geometry of MOPs is coded lossless [5.77]. Temporally and spatially predictive techniques and variable-length coding are used to compress 2D mesh geometry. The coded representation of a 2D mesh object includes representation of its geometry and motion.

The mesh model offers a versatile alternative, whereby the motion and shape of a VO are modeled in a unified framework, which can also be extended to the 3D object modeling. The 2D mesh modeling corresponds to nonuniform sampling of the motion field at a number of salient feature points (node points) along the contour and interior of the VO. Content-based mesh modeling may require transmission of geometry overhead, unlike block modeling, which requires no such overhead. If the first content-based mesh is designed based on the original VOP, the initial mesh geometry has to be transmitted in addition to all node motion vectors. The mesh geometry needs to be transmitted only once, because subsequent forward motion estimation is based on the most recently updated mesh [5.88].

The 2D mesh representation of a VO enables the following:

- VO compression
- VO manipulation
- Content-based video indexing

Mesh modeling may improve compression efficiency in two ways. Namely, the mesh model provides better motion compensation than the translational block model, which may result in less blocking artifacts at lower bit rates. Alternatively, we can choose to transmit texture maps only at selected key frames and to animate these texture maps without sending any prediction error image for the intermediate frames. This is also known as self-transfiguration of selected key frames using 2D mesh information. VO manipulation deals with augmented reality, synthetic-object transfiguration/animation and spatiotemporal interpolation. On the other hand, in content-based video indexing, mesh representation does the following: enables animated key snapshots for a moving visual synopsis of objects; provides accurate object trajectory information that can be used to retrieve VOs with specific motion and provides vertex-based object shape representation, which is more efficient than the bitmap representation for shape-based object retrieval.

VO Tracking

A VO tracking procedure is shown as a block diagram in Figure 5.58. The 2D mesh design is also presented. Dotted boxes denote optional steps. A feedback loop is designed for the initial VOP. The initial mesh can have a uniform or content-based geometry. A 2D triangular mesh or a MOP is a planar graph that partitions a VOP or its bounding box into triangular patches. The vertexes of each patch are called node points. A 2D mesh object, which consists of a sequence of MOPs, is compactly represented by mesh geometry at some key (intra) MOPs and mesh motion vectors at all other (inter) MOPs. The mesh geometry refers to the location of the node points in the key MOPs. The 2D mesh animation is accomplished by propagating the 2D mesh defined on key MOPs using one motion vector per node point per object plane until the next key MOP. Both mesh geometry and motion (animation) information are predictively coded for an efficient binary representation [5.77]. Mesh-based motion modeling differs from block-based motion modeling (that is used in natural video object coding) in that the triangular patches overlap neither in the reference frame nor in the current frame. Triangular patches in the current frame are mapped onto triangular patches in the reference frame. On the other hand, the texture inside each patch in the reference frame is warped onto the current frame using a parametric mapping, such as affine mapping, as a function of the node point motion vectors. This process is called texture mapping, which is an integral part of mesh animation [5.77]. A uniform mesh is designed over a rectangular region, which is generally the bounding box of the VOP [5.78]. It is specified in terms of five parameters: the number of nodes in the horizontal and vertical directions, the horizontal and vertical dimensions of each rectangular cell in half pixel units and the triangle split code that specifies how each cell is divided into two triangles. A content-based mesh may be designed to fit exactly on the corresponding VOP. The procedure consists of three steps: (1)

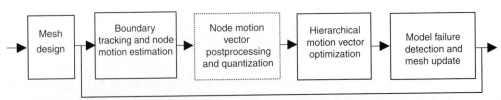

Figure 5.58 The 2D mesh design and tracking procedure. ©1999 ISO/IEC.

approximation of the VOP contour by a polygon through selection of N_b boundary node points, (2) selection of N_i interior node points and (3) Delaunay triangulation to define the mesh topology [5.89]. There are various methods for approximation of arbitrary-shaped contours by polygons [5.74].

Motion data of the 2D mesh may represent the motion of a real VO for natural VO compression and manipulation applications or may be synthetic for animation of a still texture map. In the former case, the motion of a natural VO may be estimated by forward mesh tracking. The latter requires special-purpose tools and/or artistic skills. In forward mesh tracking, we search the current VOP for the best matching locations of the node points of the previous (intra or inter) mesh, thus tracking image factures until the next intra MOP. This procedure applies for both uniform and content-based meshes. Various techniques have been proposed for node motion estimation for forward mesh tracking. The simplest method is to form blocks that are centered around the node points and then employ a closed-form solution or block matching to find motion vectors at the node points independently [5.87, 5.90]. Alternatively, hexagonal matching [5.90] and closed-form matching [5.91] techniques find the optimal motion vector at each node under the parametric warping of all constraints at the expense of more computational complexity [5.92]. Another method is iterative gradient-based optimization of node point locations, taking into account that image features and mesh provide significantly improved performance and robustness in enforcing constraints to avoid foldovers [5.93, 5.94].

2D-Mesh Object Encoder/Decoder

A simplified architecture of an encoder/decoder supporting a 2D-mesh object is depicted in Figure 5.59. A mesh analysis module extracts the 2D mesh data, which is then encoded by the mesh encoder. The coded mesh representation is embedded in a BIFS ES. At the receiver, the 2D mesh decoder is invoked automatically by the BIFS animation code [5.77]. A 2D mesh object can be used together with a VO or a still-texture object encoder/decoder.

Mesh data consists of a list of node locations (x_n, y_n) where n is the node index ($n=0,\ldots,N-1$) and a list of triangles t_m where m is the triangle index ($m=0,\ldots,M-1$). Each triangle t_m is specified by a triplet $<i,j,k>$ of the indexes of the node points that are vertexes of that triangle. The syntax of the compressed binary representation of intra and inter MOPs and the semantics of the

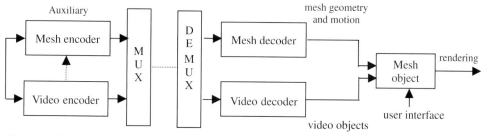

Figure 5.59 Scalable architecture with a video and mesh encoder. ©1999 ISO/IEC.

sented. The composition process may be very simple, as in direct linear mixing, or very complex, with arbitrary effects-processing code downloaded and multiple sound objects presented spatially using 3D audio.

MPEG-4 Natural Audio Coding

The tools defined by MPEG-4 natural audio coding can be combined to implement different audio-coding algorithms. A set of different algorithms has been defined to establish optimum coding efficiency for the broad range of anticipated applications [5.108]. Figure 5.65 shows assignment of codecs to bit-rate ranges. The following lists the major algorithms of MPEG-4 natural audio [5.109]:

- General audio coding for medium and high qualities
- Twin VQ additional coding tools to increase the coding efficiency at very low bit rates
- HVXC low-rate clean speech coder
- CELP telephone speech/wideband speech coder

In addition to the coding tools used for the basic coding functionality, MPEG-4 provides techniques for additional features like bit-rate stream scalability.

General Audio Coding (Advanced Audio Coding Based)

This key component of MPEG-4 Audio covers the bit rate range of 16 Kb/s per channel up to bit rates higher than 64 Kb/s per channel. Figure 5.66 shows the arrangement of the building blocks of an MPEG-4 GA encoder in the processing chain. The same building blocks are present in a decoder implementation, performing the inverse processing steps.

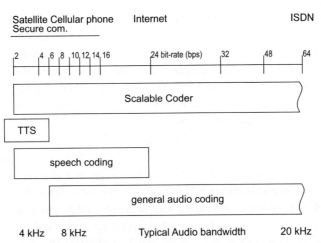

Figure 5.65 Assignment of codecs to bit-rate ranges. ©2000 ISO/IEC.

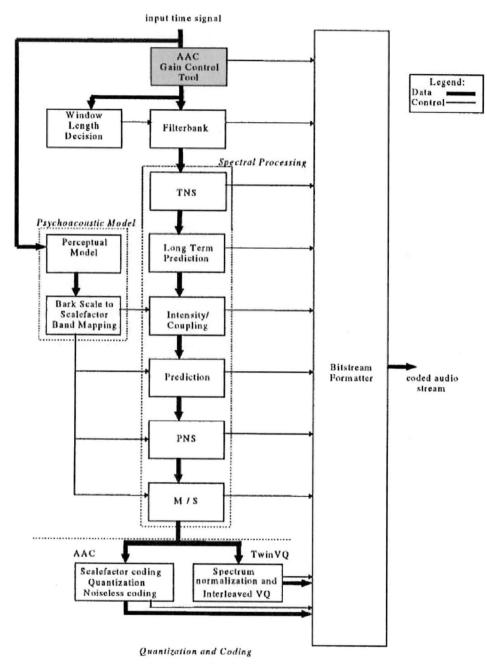

Figure 5.66 Building blocks of the MPEG-4 GA coder. ©2000 ISO/IEC.

The filter bank in MPEG-4 GA is derived from MPEG-2 AAC, that is, it is an MDCT supporting block lengths of 2,048 points and 256 points, which can be switched dynamically. Compared to previously known transform-coding schemes, the length of the long block transform is rather high, offering improved coding efficiency for stationary signals. MPEG-4 GA supports an additional mode with a block length of 1,920/240 points to facilitate scalability with the speech-coding algorithms in MPEG-4 Audio. All blocks are overlapped by 50% with the preceding and the following blocks. For improved frequency selectivity, the incoming audio samples are windowed before applying the transform. MPEG-4 AAC supports two different window shapes that can be switched dynamically. The two different window shapes are a sine-shaped window and a Kaiser-Bessel-Derived (KBD) window offering improved far-off rejection compared to the sine-shaped window. An important feature of the time-to-frequency transform is the signal adaptive selection of the transform length. This is controlled by analyzing the short time variance of the incoming time signal. To ensure block synchronization between two audio channels with different block-length sequences, eight short transforms are performed in a row using 50% overlap each and using specially designed transition windows at the beginning and the end of a short sequence. This keeps the spacing between consecutive blocks at a constant level of 2,048 input samples. For further processing, the spectral data in the quantization and coding parts of the spectrum are arranged in the so-called scale factor bands roughly reflecting the bark scale of the human auditory system. The frequency-domain prediction improves redundancy reduction of stationary signal elements. Because stationary signals can nearly always be found in long transform blocks, it is not supported in short blocks. The actual implementation of the prediction is a second-order backward adaptive lattice structure, independently calculated for every frequency line. The use of the predicted values instead of the original ones can be controlled on a scale factor bound-basis and is decided based on the achieved prediction gain in that band. To improve stability of the predictors, a cyclic-reset mechanism is applied that is synchronized between encoder and decoder using a dedicated bit-stream element. The required processing power of the frequency-domain prediction and the sensitivity to numerical imperfections make this tool hard to use on fixed-point platforms. Additionally, the backward adaptive structure of the predictor makes such bit streams quite sensitive to transmission errors.

Long-Term Prediction (LTP), newly introduced in MPEG-4, is an efficient tool for reducing the redundancy of a signal between successive coding frames. This tool is especially effective for the parts of the signal that have a clear pitch property. Because the long-term predictor is a forward adaptive predictor (prediction coefficients are sent as side information), it is inherently less sensitive to round off numerical errors in the decoder or bit errors in the transmitted spectral coefficients. The adaptive quantization of the spectral values is the main source of the bit-rate reduction in all transform coders. It assigns a bit allocation to the spectral values according to the accuracy demands determined by the perceptual model, realizing the irrelevancy reduction. The key components of the quantization process are quantization function and the noise shaping that is achieved through the scale factors. The quantizer used in MPEG-4 GA has been designed similar to the one used in MPEG 1/2 Layer 3. It is a nonuniform quantizer. The main advantage over a conventional uniform quantizer is the implicit noise shaping that this quantization creates. The absolute quantizer step size is determined by a specific bit-stream element. It can be adjusted in 1.5 dB steps. To improve the subjective quality of the coded signal, the noise is further shaped by scale factors. They are used to amplify the signal in certain spectral regions (the scale factor bands) to increase the SNR in these bands. Thus the spectral values usually need more bits to be coded afterward. Like the global quantizer, the step size of the scale factors is 1.5 dB. To reconstruct the original spectral values in the decoder properly, the scale factors have to be transmitted within the bit stream. MPEG-4 GA uses an advanced technique to code the scale factors as efficiently as possible. It exploits the fact that scale factors usually do not change too much from one scale-factor band to another. Thus, differential encoding provides some advantage. It also uses a Huffman code to reduce the redundancy further within the scale-factor data.

The noiseless coding kernel within an MPEG-4 GA encoder tries to optimize the redundancy reduction within the spectral data coding. The spectral data is encoded using a Huffman code that is selected from a set of available codebooks according to the maximum quantized value. The set of available codebooks includes one to signal that all spectral coefficients in the respective scale factor band are 0, implying that neither spectral coefficients nor a scale factor are transmitted for that band. To find the optimum trade-off between selecting the optimum table for each scale factor band and minimizing the number of data elements to be transmitted, an efficient grouping algorithm is applied to the spectral data.

The basic idea of Temporal Noise Shaping (TNS) relies on the duality of time and frequency domains. TNS uses a prediction approach in the frequency domain to shape the quantization noise over time. It applies a filter to the original spectrum and quantizes this filtered signal. Additionally, quantized filter coefficients are transmitted in the bit stream. These are used in the decoder to undo the filtering performed in the encoder, leading to a temporally shaped distribution of quantization noise in the decoded audio signal. TNS can be viewed as a postprocessing step of the transform, creating a continuous signal adaptive filterbank instead of the conventional two-step switched-filterbank approach. The actual implementation of the TNS approach within MPEG-4 GA allows for up to three distinct filters applied to different spectral regions of the input signal, further improving the flexibility. A feature newly introduced into MPEG-4 GA is the Perceptual Noise Substitution (PNS) [5.110]. The technique of PNS is based on the observation that one noise sounds like the other. This means that the actual fine structure of a noise signal is of minor importance for the subjective perception of such a signal. Instead of transmitting the actual spectral components of a noisy signal, the bit stream would just signal that this frequency region is noiselike and gives some additional information on the total power in that band. PNS can be switched on a scale-factor band basis. Even if there are just some spectral regions with a noisy structure, PNS can be used to save bits. In the decoder, a randomly generated noise will be inserted into the appropriate spectral region according to the power level signaled with the bit stream.

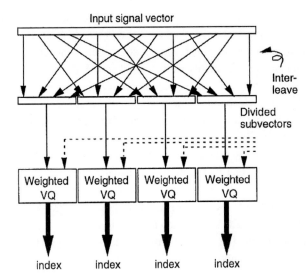

Figure 5.67 Twin VQ [5.40]. ©1998 ISO/IEC.

Twin VQ

To increase coding efficiency for coding of musical signals at very low bit rates, twin VQ-based coding tools are part of the GA coding systems in MPEG-4 Audio. The basic idea is to replace the conventional encoding of scale factors and spectral data used in MPEG-4 AAC by an interleaved VQ applied to a normalized spectrum [5.111, 5.112]. The basic idea of the weighted interleaved vector quantization (Twin VQ) scheme is represented in Figure 5.67. The input signal vector (spectral coefficients) is interleaved into subvectors. These subvectors are quantized using vector quantizers. Twin VQ can achieve a higher coding efficiency at the cost of always creating a minimum amount of loss in terms of audio coding.

Speech Coding in MPEG-4 Audio

Most of the recent speech-coding algorithms can be categorized as spectrum coding or hybrid coding. Spectrum coding models the input speech signal based on a vocal tract model, which consists of a signal source and a filter as shown in Figure 5.68. A set of parameters obtained by analyzing the input signal is transmitted to the receiver. Hybrid coding synthesizes an approximate speech signal based on a vocal tract model. A set of parameters used for this synthesis is modified to minimize the error between the original and the synthesized speech signals. A best parameter set can be searched for by repeating this analysis-by-synthesis procedure. The obtained set of parameters is transmitted to the receiver as the compressed data after quantization. In the decoder, a set of parameters for source and LP synthesis filtering is recovered by inverse quantization. These parameter values are used to operate the same vocal tract model as in the encoder. Figure 5.69 gives a block diagram of hybrid speech coding. Source and LP synthesis filters correspond to those in Figure 5.68. The error between the input signal and the synthesized signal is weighted by a Perceptual Weighted (PW) filter. The filter has a frequency

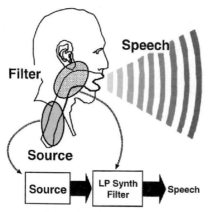

Figure 5.68 Vocal tract model. ©2000 ISO/IEC.

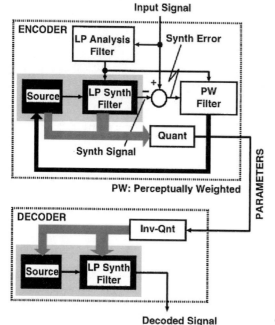

Figure 5.69 Hybrid speech coding. ©2000 ISO/IEC.

response that takes the human auditory system into consideration. Thus, a perceptually best parameter selection can be achieved.

The MPEG-4 natural speech-coding toolset provides a generic coding framework for a wide range of applications with speech signals. Two different bandwidths are covered [5.58]: 4 KHz and 7 KHz. The MPEG-4 natural speech-coding toolset contains two algorithms: HVXC

Table 5.14 Specifications of MPEG-4 natural speech-coding tools [5.113].

HVXC	Parameter	
Sampling frequency	8 KHz	
Bandwidth	300-3400 Hz	
Bit rate [bit/s]	2,000 and 4,000	
Frame size	20 ms	
Delay	33.5-56 ms	
Features	Multi-bit-rate coding Bit-rate scalability	
CELP	**Parameter**	**Parameter**
Sampling frequency	8 KHz	16 KHz
Bandwidth	300-3,400 Hz	50-7,000 Hz
Bit rate [bit/s]	3,850-12,200 (28 bit rates)	10,900-23,800 (30 bit rates)
Frame size	10-40 ms	10-20 ms
Delay	15-45 ms	15-26.75 ms
Features	Multi-bit-rate coding Bit-rate scalability Bandwidth scalability	

©1998 ISO/IEC.

and CELP. The specifications of the MPEG-4 natural speech-coding toolset are summarized in Table 5.14, and Figure 5.70 represents the corresponding toolset.

MPEG-4 is based on tools that can be combined according to the user needs. HVXC consists of the Line Spectral Pair (LSP), VQ and harmonic VQ tool. The Regular Pulse Excitation (RPE) tool, Multipulse Excitation (MPE) tool and LSP VQ tool form CELP. The RPE tool is allowed only for the wideband mode because of its simplicity at the expense of the quality. The LSP VQ tool is common both in HVXC and CELP.

HVXC. A basic block diagram of HVXC is presented in Figure 5.71. LP analysis to find the LP coefficients is first performed. Quantized LP coefficients are supplied to the inverse LP filter to find the prediction error. The prediction error is transformed into a frequency domain,

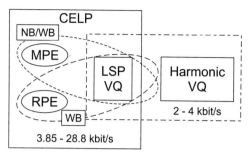

NB: Narrow Band   RPE: Regular Pulso Excitation
WB: Wide Band   MPE: Multipulse Excitation
VQ: Vector Quantization

Figure 5.70 MPEG-4 natural-speech coding toolset. ©2000 ISO/IEC.

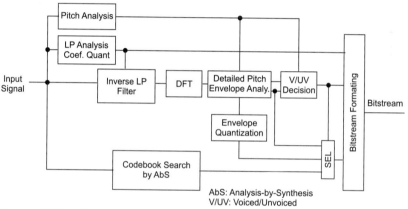

AbS: Analysis-by-Synthesis
V/UV: Voiced/Unvoiced

Figure 5.71 HVXC block diagram. ©2000 ISO/IEC.

and the pitch and the envelope of the spectrum are analyzed. The envelope is quantized by weighted VQ in voiced sections. In unvoiced sections, a closed-loop search of an excitation vector is arrived at.

CELP. Figure 5.72 shows a block diagram of CELP. The LP coefficients of the input signal are first analyzed and then quantized to be used in an LP synthesis filter driven by the output of the excitation codebooks. Encoding is performed in two steps. LTP coefficients are calculated in the first step. In the second step, a perceptually weighted error between the input signal and the LP synthesis filter is minimized. This minimization is achieved by searching for an appropriate code vector for the excitation codebooks. Quantized coefficients, as well as indexes to the code vectors of the excitation codebooks and the LTP coefficients, form the bit stream. The LP coefficients are quantized by vector quantization and the excitation can be either MPE or regular pulse excitation RPE [5.114]. MPE and RPE both model the excitation signal by multiple pulses. However, a distance exists in the degrees of freedom for pulse positions. MPE allows more freedom of the interpulse distance than RPE, which has a fixed-interpulse distance. Thanks to such a flexible interpulse distance, MPE achieves better coding quality than RPE [5.113]. On

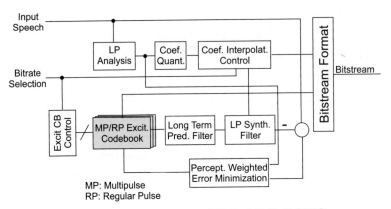

Figure 5.72 Block diagram of coding CELP. ©2000 ISO/IEC.

Table 5.15 The excitation signal types of MPEG-4/CELP.

Excitation	Bandwidth	Features
MPE	Narrow, wide	Quality, scalability
RPE	Wide	Complexity

the other hand, RPE requires fewer computations than MPE by trading off its coding quality. Such a low computational requirement is useful in the wideband coding where the total computation should naturally be higher than in the narrowband coding. The excitation signal types of MPEG-4/CELP are summarized in Table 5.15.

Scalability in MPEG-4 Natural Audio

Bit-stream scalability is the ability of an audio codec to support an ordered set of bit streams that can produce a reconstructed sequence. Moreover, the codec can output useful audio when certain subsets of the bit stream are decoded. The minimum subset that can be decoded is called the base layer. The remaining bit streams in the set are called enhancement or extension layers. Depending on the size of the extension layers, there exists large-step or small-step (granularity) scalability. Small-step scalability denotes enhancement layers of around 1 Kb/s or smaller. Typical data rates for the extension layers in a large-step scalable system are 16 Kb/s or more. A scalability in MPEG-4 Natural Audio largely relies on differences, either in time domain, or, as in the case of AAC layers of the spectral lines, it is in frequency domain.

Synthetic Audio in MPEG-4

Natural and synthetic audio are not unrelated methods for transmitting sound. As sound models in perceptual coding grow more sophisticated, the boundary between decompression and synthesis becomes somewhat blurred [5.115]. In [5.106, 5.116], the relationships among parametric models of sound, digital sound creation and transmission, perceptual coding, parametric compression and various techniques for algorithmic synthesis have been discussed.

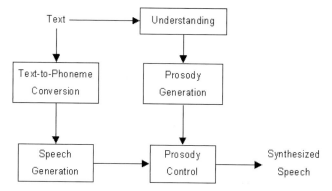

Figure 5.73 The interaction between text-to-phoneme conversion text understanding and prosody generation and application. ©2000 ISO/IEC.

TTS systems generate speech sound according to given text. This technology enables the translation of text information into speech so that the text can be transferred through speech channels such as telephone lines. TTS systems consist of the multiple processing modules shown in Figure 5.73 [5.117]. Such a system accepts text as input and generates a corresponding phonemes sequence. Phonemes are the smallest units of human language. Each phoneme corresponds to one sound used in speech. About 120 phonemes are sufficient to describe all human languages. The phoneme sequence is used to generate a basic speech sequence without prosody, that is, without pitch, duration and amplitude variations. In parallel, a text-understanding module analyzes the input for phrase structure and inflections. Using the result of this processing, a prosody-generation module creates the proper prosody for the text [5.118]. Finally, a prosody-control module changes the prosody parameters of the basic speech sequence according to the results of the text-understanding module, yielding synthesized speech. Today, TTS systems are used for many applications, including automatic voice-response systems (the "telephone men" systems), email reading and information services for the visually handicapped [5.118]. The applications of TTS are expanding in telecommunications, personal computing and the Internet. Current research in TTS includes voice conversion (synthesizing the sound of a particular speaker's voice), multilanguage TTS and enhancement of the naturalness of speech through modern sophisticated voice models and prosody generators.

Text, that is, a sequence of words written in some human language, is a widely used representation for speech data in standalone applications. However, it is difficult with existing technology to use text as a speech representation in a multimedia bit stream for transmission. The MPEG-4 TTSI is defined so that speech can be transmitted as a bit stream containing text. It also provides interoperability among TTS synthesizers by standardizing a single bit stream format for this purpose.

Synthetic speech is becoming a rather common media type. It plays an important role in various multimedia application areas. For instance, by using TTS functionality, multimedia content with narration can be easily created without recording natural speech. In MPEG-4, a single common interface for TTS systems is standardized. This interface allows speech information to be transmitted in the International Phonetic Alphabet (IPA) or in a textural (written) form of any language. The MPEG-4 TTSI is a hybrid/multilevel scalable TTSI that can be considered a superset of the conventional TTS framework. This extended TTSI can use prosodic information taken from natural speech in addition to input text and can thus generate much higher quality synthetic speech. As well as an interface to TTS synthesis systems, MPEG-4 specifies a joint coding method for phonemic information and FAPs. The MPEG-4 TTSI has important functionalities both as an individual code and in synchronization with the facial animation techniques described in Tekalp and Ostermane [5.77]. The basic TTSI format is extremely low bit rate. The synthesized speech with predefined prosody will deliver emotional content to the listener. One of the important features of the MPEG-4 TTSI is the ability to synchronize synthetic speech with the lip movements of a computer generated talking head. In this technique, the TTS synthesizer generates phoneme sequences and their duration and communicates them to the facial animation visual object decoder so that it can control the lip movement. Through the MPEG-4 elementary synchronization capabilities, the MPEG-4 TTSI can perform synthetic motion picture dubbing [5.52]. The MPEG-4 TTSI decoder can use the system clock to select an adequate speech location in a sentence and communicates this to the TTS synthesizer, which assigns appropriate duration for each phoneme. Using this method, synthetic speech can be synchronized with the lip shape of the moving image. An overview of the MPEG-4 TTSI decoding process showing the interaction between the syntax parser, the TTS synthesizer, and the face animation decoder is given in Figure 5.74. The TTS synthesizer and face-decoder blocks are not normatively described and operate in a terminal-dependent manner [5.117].

The architecture of the decoder can be described as a collection of interfaces. Upon receiving a multiplexed MPEG-4 bit stream, the demux passes coded MPEG-4 TTSI ESs to the syntactic decoder. Other ESs are passed to other decoders. Receiving a coded MPEG-4 TTSI bit stream, the synthetic decoder passes a number of different pieces of data to the TTS synthesizer. The input type specifies whether TTS is being used as a standalone function or in the synthetizer with facial animation or motion-picture dubbing. The control commands sequence specifies the language, gender, age and speech rate of the speaking voice. The input text specifies the character string for the text to be synthesized. Auxiliary information, such as IPA phoneme symbols

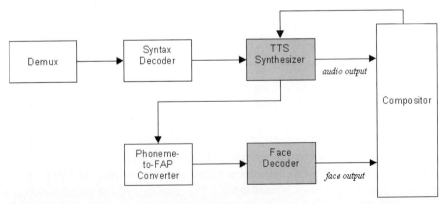

Figure 5.74 Block diagram of the MPEG-4 TTSI decoder. ©2000 ISO/IEC.

(which allow text in a language foreign to the decoder to be synthesized), lip shape patterns and trick-mode commands, is also passed along the interface between the synthetic decoder and the TTS synthesizer. The synthesizer constructs a speech sound and delivers it to the audio composition system. The interface from the compositor to the TTS synthesizer allows the local control of the synthesized speech by users. Using this interface and an appropriate interactive scene, users can start, stop, rewind and fast-forward the TTS system. Controls can also allow changes to the speech rate, pitch range, gender, and age of the synthesized speech by the user. The TTS synthesizer and the face animation can be driven synchronously by the same input control stream, which is the text input to the MPEG-4 TTSI. From this input stream, the TTS synthesizer generates synthetic speech, and, at the same time, phoneme symbols, phoneme durations and word boundaries generate the phoneme/bookmark-to-FAP converter, which generates relevant facial animation. In that way, the synthesized speech and facial animation are synchronized when they enter the scene composition framework.

The tool that provides audio synthesis capability in MPEG-4 is termed the structured audio coder [5.106, 5.119]. MPEG-4 structured audio is a codec like the other audio tools in MPEG-4. A sound transmission is decomposed into two pieces: a set of synthesis algorithms that describe how to create sound and a sequence of synthesis controls that specifies which sounds to create. The synthesis model is not fixed in the MPEG-4 terminal. The standard specifies a framework for reconfigurable software synthesis. Like the other MPEG-4 media types, a structured audio bit stream consists of a decoder configuration header that tells the decoder how to begin the decoding process and then a stream of bit-stream access units that contain the compressed data. In structured audio, the decoder configuration header contains the synthesis algorithms and auxiliary data, and the bit-stream access units contain the synthesis control instructions.

Audio BIFS

The part of BIFS controlling the composition of a sound scene is called audio BIFS. It provides a unified framework for sound scenes that use streaming audio, interactive presentation, 3D spatialization and dynamic download of custom signal-processing effects [5.55, 5.120].

Audio BIFS contains significant advances in quality and flexibility over VRML audio. There are two main modes of operation that audio BIFS is intended to support: virtual-reality and abstract-effects compositing. In virtual-reality compositing, the goal is to re-create a particular acoustic environment as accurately as possible. Sound should be presented spatially according to its location relative to the listener in a realistic manner. Moving sounds should have a Doppler shift. Distant sounds should be attenuated and low-pass filtered to simulate the absorptive properties of air. Sound sources should radiate sound unevenly, with a specific frequency-dependent directivity pattern. This type of scene composition is most suitable for virtual world applications and video games where the application goal is to immerse the user in a synthetic environment. In abstract-effects compositing, the goal is to provide content authors with a rich suite of tools from which artistic considerations can be used to choose the right effect for a given situation.

A schematic diagram for the overall audio system in MPEG-4 is shown in Figure 5.75. Sound is conveyed in the MPEG-4 bit stream as several ESs that contain coded audio. There are four ESs in the sound scene. Each of these ESs contains a primitive media object, which, in the case of audio, is a single-channel or multichannel sound that will be composited into the overall scene. The MPEG-4 audio system shows the interaction between decoding, scene description and audiovisual synchronization. The conceptual flow is from the bottom of the figure to the top. At the bottom, two multiplexed MPEG-4 bit streams, each from a different server, convey several ESs containing compressed data. Each bit stream is demultiplexed. A total of four ESs is produced. The ESs are decoded using various MPEG-4 decoders into four primitive media objects containing uncompressed PCM scene graphs and are presented to the listener as though they emanate from the sound nodes. The BIFS part and the audio BIFS part of the scene graph are separated, but there is no technical difference between them. Namely, audio BIFS is just a subset of BIFS. Audio BIFS consists of a number of nodes that are interlinked in a scene graph. An audio BIFS scene graph is termed an audio subgraph. The audio BIFS nodes are summarized in Table 5.16, which describes their function in an audio scene. Each node has several fields that specify the parameters of operation of the node. In MPEG-4 BIFS, these fields and their operation are carefully quantized and transmitted in a binary data format for maximum compression of the scene graph.

5.5.6 Profiles and Levels in MPEG-4

Profiles and levels in MPEG-4 serve two main purposes:

- Ensuring interoperability between MPEG-4 implementations
- Allowing conformance to the standard to be tested

Profiles exist not only for the audio and visual parts of the standard (audio profiles and visual profiles), but also for the systems part of the standard, in the form of graphics profiles, scene graph profiles and an object descriptor profile. Different profiles are created for different application environments. The policy for defining profiles is that they should enable as many applications as possible while keeping the number of different profiles low [5.7].

Media profiles describe the object types that can be used to create the scene and tools that can be used to create those object types.

Visual Object Types

Five different object types represent video information:

- The Simple object type is an error-resilient rectangular natural VO of arbitrary height/width ratio, developed for low bit rates. It uses relatively simple and inexpensive coding tools, based on I-VOPs and P-VOPs.
- The Simple Scalable object type is a scalable extension of Simple, which gives temporal and spatial scalability using Simple as the base layer. The enhancement layer is still rectangular.

MPEG-4—Coding of Audiovisual Objects

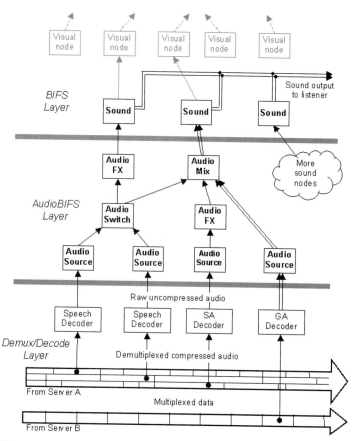

Figure 5.75 The MPEG-4 audio system.

Table 5.16 The audio BIFS nodes.

Name	Function
AudioSource	Attach sound decoder to scene graph
AudioMix	Mix M channels of sound into N channels
AudioSwitch	Select subset of M input channels of sound
AudioDelay	Delay sounds for synchronization
AudioFX	Apply algorithmic signal-processing effects

© 1998 ISO/IEC.

Table 5.16 The audio BIFS nodes. (Continued)

Name	Function
AudioBuffer	Cache sound for use in interactive playback
Sound	Position sound in 3D virtual environment
Sound2D	Position sound in 2D scene
Group	Group multiple nodes together for hierarchical transformation
ListeningPoint	Specify location of virtual listener in scene
TermCap	Query terminal for available playback resources

© 1998 ISO/IEC.

- The Core object type uses a tool superset of Simple, giving better quality through the use of bidirectional interpolation, and it has binary shape. It supports scalability based on sending extra P-VOPs.
- The Main object type is the VO that gives the highest quality. Compared to Core, it also supports gray-scale shape, sprites and interlaced content in addition to progressive material.
- The N-bit object type is equal to the Core object type, but it can vary the pixel depth from 4 to 12 bits for the luminance as well as the chrominance planes.

One special object type represents still natural visual information. This is the Still Scalable Texture object type. It gives an arbitrary shape still image that uses wavelet coding for scalability and incremental download and build up.

The following object types use synthetic tools, some of which are in combination with natural video texture:

- The Animated 2D Mesh object type combines the synthetic mesh with natural video. The natural video coding uses the same tools as the Core object type. This video can be mapped onto the mesh and deformed by moving the points in the mesh.
- The Basic Animated Texture object type allows mesh animation with arbitrary shape still images.
- The Simple Face object type has the tools for facial animation. This object type does not define what the face looks like. The animation can be applied to any local model of choice.

Visual Profiles

The visual profiles determine which visual object types can be present in the scene:

- The Simple Profile accepts only objects of type Simple and was created with low-complexity applications in mind. The first use is mobile use of (audio) visual services, and the second is putting very low-complexity video on the Internet. Also, small camera devices recording moving video to, for example, disk or memory chips, can make good use of this profile. The Simple Profile has three levels with bit rates from 64 to 384 Kb/s. The levels also define the maximum total surface for the objects and the amount of macroblocks per second that the decoder needs to be able to decode. Further, they define the size of various (hypothetical) buffers needed for decoding.
- The Simple Scalable Profile can supply scalable coding in the same operational environments as foreseen for Simple and has two levels defined.
- The Core Profile accepts Core and Simple objects types. It is useful for higher quality interactive services, combining good quality with limited complexity and supporting arbitrary shape objects. Also, mobile broadcast services should be supported by this profile. The maximum bit rate is 384 Kb/s in Level 1 and 2 Mb/s in Level 2. The number of macroblocks is chosen such that a scene using this typical session size can have overlapping objects and still be filled.
- The Main Profile was created with broadcast services in mind, addressing progressive as well as interlaced material. It combines the highest quality with the versatility of arbitrarily shaped objects using gray-scale coding. The highest level accepts up to 32 objects (of Simple, Core or Main type) for a maximum total bit rate of 38 Mb/s.
- The N-bit profile is useful for areas that use terminal imagers, such as surveillance applications. Medical applications may also want to use enhanced pixel depth giving a larger dynamic range in color and luminance. It accepts objects of types Simple, Core and N-bit profiles.
- The Simple Face Profile accepts only objects of type Simple Face. Depending on the level, either one or a maximum of four faces can appear in the scene, for example, for a virtual meeting. Bit rates remain very low. Even for the second level, 32 Kb/s is more than adequate for driving the four faces.
- The Hybrid Profile allows combining both natural and synthetic objects in the same scene while keeping complexity reasonable. On the natural side, it compares to the Core Profile, and on the synthetic side, it adds animated meshes, scalable textures and animated faces. This is a rich set of tools for creating attractive hybrid natural and synthetic content. This profile can be used to place real objects into a synthetic world and to add synthetic objects to a natural environment.
- The Basic Animated Texture Profile allows animation of still pictures and facial animation. Attractive content can be created at very low bit rates.

Audio Object Types

For coding natural sound, MPEG-4 includes AAC and twin VQ algorithms. The following object types exist:

- The AAC Main object type has multichannel capability to give five full channels plus a separate low-frequency channel in one object. It is very similar and compatible with the AAC Main profile that is defined in MPEG-2 (ISO/IEC 13818-7). MPEG-4 AAC adds the perceptual noise-shaping tool.
- The MPEG-4 AAC Low Complexity object type is a low complexity version of the AAC Main Object type.
- The MPEG-4 AAC Scalable Sampling Rate object type is the counterpart to the MPEG-2 AAC Scalable Sampling Rate profile.
- The MPEG-4 AAC LTP object type is similar to the AAC Main object type with the LTP replacing the MPEG-2 AAC predictor. This gives the same efficiency with significantly lower implementation costs.
- The AAC Scalable object type allows a large number of scalable combinations, including combinations with twin VQ and CELP coder tools as the core coders. It supports only mono or two-channel stereo sound.
- The Twin VQ object type is based on fixed-rate VQ instead of the Huffman coding used in AAC. It operates at lower bit rates than AAC and supports mono and stereo sound.
- The CELP object type uses CELP. It supports 8 KHz and 16 KHz sampling rates at bit rates from 4 to 24 Kb/s. CELP bit streams can be coded in a scalable way using bit-rate scalability and bandwidth scalability.
- The HVXC object type gives a parametric representation of 8 KHz mono speech at fixed bit rates between 2 and 4 Kb/s and below 2 Kb/s using a variable bit-rate mode, and supports pitch and speed changes.
- The TTSI object type gives an extremely low bit-rate phonemic representation of speech. Bit rates range from 0.2 to 1.2 Kb/s. The synthesized speech can be synchronized with a facial animation object.

A number of different object types exists for synthetic sound:

- The Main Synthetic object type collects all MPEG-4 structured audio tools. Structured audio is a way to describe methods of synthesis [5.117]. Sound can be described at 0 Kb/s to 3 to 4 Kb/s for extremely expressive sounds in MPEG-4 Structured Audio format.
- The Wavetable Synthesis object type is a subset of the Main Synthetic object type. It provides relatively simple sampling synthesis.
- The General Musical Instrument Digital Interface (MIDI) object type gives interoperability with existing content. Unlike the Main Synthetic or Wavetable

Synthesis object types, it does not give completely predictable sound quality and decoder behavior.

- The Null object type provides the possibility to feed raw PCM data delivery to the MPEG-4 audio compositor in order to allow mixing in of local sound at the decoder. The support for this object type is in the compositor.

Audio Profiles

There are only four different audio profiles. The application area for the Speech profile can be deduced from its name. Two levels are defined, determining whether either 1 or a maximum 20 objects can be present in the audio scene. A prime reason for defining the Scalable profile was to allow good quality, reasonable complexity, low-bit-rate audio on the Internet and an environment in which bit rate varies from user to user and from one minute to the next. Scalability allows making optimal use of available, and even dynamically changing, bandwidth while having to encode and store the material only once. The Scalable profile has four levels that restrict the amount of objects in the scene, the total amount of channels and the sampling frequency. The highest level employs the novel concept of complexity units. The Synthetic profile groups all the synthetic object types. The main application areas are found where good quality sounds are needed at very low data rates while the sound source usually uses microphones. Three levels define the amount of memory for the data, the sampling rates, amount of TTSI objects and some further processing restrictions. The Main profile includes all object types. It is useful in environments where processing power is available to create very rich, highest quality audio scenes that may combine microphone-recorded sources with synthetic ones. Example applications are the DVD and multimedia broadcast. This profile has four levels, defined in terms of complexity units. There are two different types of complexity units: Processor Complexity Units (PCU) specified in millions of operations/s, and RAM Complexity Units (RCU) specified in terms of number of kwords.

Graphics

Graphics profiles regulate which of the graphics and textual elements can be used to build a scene. They are expressed in terms of BIFS nodes. Three hierarchical graphics profiles are defined in MPEG-4: Simple 2D, Complete 2D and Complete. Simple 2D provides the basic functionalities needed to create a visual scene with visual objects without giving additional graphical elements. Complete 2D allows elements like bitmaps, backgrounds, circles, boxes and lines, all in a flat space. These elements have characteristics like line width and color. Complete contains the full set of BIFS graphics nodes with which complete and elaborate 3D graphics can be created. It adds to Complete 2D, for example, the sphere, the cone, 3D boxes, directional lighting, and so forth.

Systems Profiles

Here, we present the Scene Graph profiles and Object Descriptor profiles. The Scene Graph profile defines what types of transformational capabilities need to be supported to the terminal. This is defined in terms of the scene graph elements (BIFS nodes) that the decoding terminal needs to be able to understand. Examples are translations, 3D rotations and elements like input sensors with which interactive behavior can be created. The Scene Graph profiles follow a structure similar to the graphics profiles, with one profile added for audio-only scenes. Thus, we have Audio, Simple 2D, Complete 2D, and Complete. The target applications are very much the same as those for the Graphics profiles with Simple 2D providing placement capabilities, Complete 2D adding for instance rotation and Complete giving the capability to do arbitrary transformations in a 3D space.

The Object Descriptor profile specifies allowed configurations of the object descriptor and Sync Layer tools [5.121]. The object descriptor contains all descriptive information, and the Sync Layer tool provides the syntax to convey, among others, tuning information for ESs. The main reason for wanting to subject the Object Descriptor to profiling lies in reducing the amount of asynchronous operations and the necessary permanent storage.

5.6 MPEG-4 Visual Texture Coding (VTC) and JPEG 2000 Image Compression Standards

With the increasing use of multimedia communication systems, image compression requires higher performance and new features. JPEG 2000 is an emerging standard for still-image compression. It is not only intended to provide rate distortion and subject image quality performance superior to existing standards, but also to provide functionality that current standards can either not address efficiency or not address at all [5.122]. The compression advantages of JPEG 2000 are a direct result of the inclusion into the standard of a number of advanced and attractive features, including progressive recovery, lossy/lossless compression and region of interest capabilities. These features lay the foundation for JPEG 2000 to provide tremendous benefits to a range of industries. Some of the applications that will benefit directly from JPEG 2000 are image archiving, Internet, Web browsing, document imaging, digital photography, medical imaging and remote sensing.

Functionally, JPEG 2000 includes many advanced features:

- Component precision of 1 to 127 bits/sample (signed or unsigned)
- Components that may each have a different precision and subsampling factor
- Use of image data that may be stored compressed or uncompressed
- Lossy and lossless compression
- Progressive recovery by fidelity or resolution
- Tiling
- Error resilience
- Region-of-interest coding
- Random access to an image in a spatial domain
- Security

Image compression must not only reduce the necessary storage and bandwidth requirements, but also must allow extraction for editing, processing and targeting particular devices and applications. JPEG 2000 allows extraction of different resolutions, pixel fidelities, regions of interest, components and more, all from a single compressed bit stream. This allows an application to manipulate or transmit only the essential information for any target device from any JPEG 2000 compressed source image.

Some of the technology highlights for JPEG 2000 are the following:

- Wavelet subband coding
- Reversible integer-to-integer and nonreversible real-to-real wavelet transforms
- Reversible integer-to-integer and nonreversible real-to-real multicomponent transforms
- Bit-plane coding
- Arithmetic coding
- Use of Embedded Block Coding with Optimized Truncation (EBCOT) coding scheme
- Code-stream syntax similar to JPEG
- File format syntax

In what follows, the structure of the JPEG 2000 standard is presented together with performance and the complexity comparisons with existing standards.

5.6.1 JPEG 2000 Development Process

The JPEG 2000 project was motivated by submission of the Compression with Reversible Embedded Wavelet (CREW) algorithms to an earlier standardization effort for lossless and near-lossless compression (known as JPEG-LS) [5.123, 5.124].

With the continual expansion of multimedia and Internet applications, the needs and requirements of the technologies used grew and become quite complex. A call for technical contributions was issued in March 1997, requesting compression technologies be submitted to an evaluation during the November 1997 WG1 meeting in Sydney, Australia [5.125]. The Wavelet/Trellis Coded Quantization (WTCQ) algorithm ranked first overall in both the subjective and objective evaluations. It was further decided that a series of core experiments would be conducted to evaluate WTCQ and other technologies in terms of JPEG 2000 desired features and in terms of algorithm complexity. Results from the first round of core experiments were presented at the March 1998 WG1 meeting in Geneva. Based on these experiments, it was decided to create a JPEG 2000 VM, which would lead to a reference implementation of JPEG 2000. The VM was modified in each meeting based on experiments performed between meetings. Results from Round 1 core experiments were selected to modify WTCQ in the first release of the VM (VM0).

The basic ingredients of the WTCQ algorithm are the discrete wavelet transform, Trellis Coded Quantization (TCQ) using step sizes chosen with a Lagrangian rate-allocation procedure [5.126, 5.127] and binary arithmetic coding. The embedding principle asserts the encoded bitstream should be ordered in a way that maximally reduces Mean Square Error (MSE) per bit transmitted [5.123, 5.128, 5.129, 5.130, 5.131]. In WTCQ, embedding is provided by the bit-plane coding. The bit-plane coding operates on TCQ indexes (trellis quantized wavelet coefficients) in a way that enables successive refinement. This is accomplished by sending bit planes in decreasing order from most to least significant. To exploit spatial correlation within bit planes, spatial context models are used. In general, the context can be chosen within a subband and across subbands. The WTCQ bit-plane coders avoid the use of intersubband contexts to maximize flexibility in scalable decoding and to facilitate parallel implementation. WTCQ also includes a binary mode, a classification of coefficients, multiple decompositions (dyadic, packet and others) and difference images to provide lossless compression [5.132].

Additions and modifications to VM0 continued for several meetings. VM2 supported user-specified floating point and integer transforms, as well as user-specified decompositions (dyadic, uniform and so forth). As a simpler alternative to the Lagrangian rate allocation, a fixed quantization table (Q-table) was included. This is analogous to the current JPEG standard [5.133, 5.134, 5.135]. When a Q-table is used, precise rate control can still be obtained by truncating the (embedded) bit stream. In addition to TCQ, scalar quantization was included in VM2. For integer wavelets, scalar quantization with the step size 1 was employed (no quantization), which allowed progression to lossless in the manner of CREW. Rate control for integer wavelets was accomplished by embedding, and a lossless compression scheme was available from the fully decoded embedded bit stream. Other features, such as tiling, region of interest coding and decoding, error resilience and approximate wavelet transforms with limited spatial support were added to the VM. Several refinements were made to the bit-plane coder. The major changes were the deinterleaving of bit planes and improvements to the context modeling. Within a given bit plane of each subband, the bits were deinterleaved into three subplanes of the following types: bits predicted to be newly significant, refinement bits and bits predicted to remain insignificant. The idea of subplanes was first reported in 1998 and was motivated by rate-distortion concerns [5.131]. Also, it is desirable to have the bits with the steepest rate-distortion slopes appear first in an embedded bit stream. The VM2 bit-plane coder has no intersubband dependencies such as those used in zero tree-based schemes [5.128, 5.130]. In VM2, all coding was carried out using context-dependent binary arithmetic coding [5.136]. It should be noted that, when encoding a particular bit, neither significance prediction nor context-modeling stages can use any information that would not be available at the decoder when that bit needs to be decoded. Thus, for wavelet coefficients that are noncausal with respect to the scan pattern, only information from more significant bit planes is used.

EBCOT included the idea of dividing each subband into rectangular blocks of coefficients and performing the bit-plane coding independently on these codeblocks. This partitioning reduces memory requirements in both hardware and software implementations, and it also provides a certain degree of (spatial) random access to the bit stream. EBCOT also included an efficient syntax for forming the sub-bit plane of multiple code blocks into packets. EBCOT was adopted for inclusion in VM3 [5.137].

During the March 1999 WG1 meeting in Korea, the MQ coder (submitted by Mitsubishi) was adopted as the arithmetic coder for JPEG 2000. This coder is functionally similar to the QM coder available as an option of the original JPEG standard. The MQ coder has some useful bit-stream creation properties, is used in the JBIG-2 standard, and should be available on a royalty and fee-free basis for ISO standards.

In fact, one goal of WG1 has been the creation of a Part 1, which could be used entirely on a royalty and fee-free basis. This is essential for the standard to gain wide acceptance as an inter-change format. At the same time, as changes were being made to the internal coding algorithms, syntax wrapping of the compressed data was developed. This syntax is made up of a sequence of markers compatible with those of the original JPEG [5.135] and with features added to allow the identification of relevant portions of the compressed data. One Annex of the JPEG 2000 standard contains an optional minimal file format to include information such as the color space of the pixels and intellectual property (copyright) information for images. The inclusion of this Annex will prevent the proliferation of the property file format that happened with the original JPEG. This optional file format is extensible, and Part 2 defines storage of many additional types of metadata. The standardization process has already produced the WD and the CD documents [5.138, 5.139]. Final Draft International Standard (FDIS) was produced in August 2000, and finally JPEG 2000 was produced in December 2000.

Part 2 became an International Standard in October 2001. Division of the standard between Part 1 and Part 2 is shown in Table 5.17. It lists the various components of the compression system and the extensions likely for Part 2. For example, Part 1 will require one floating-point wavelet (9,7) and one integer wavelet (5,3), and Part 2 will allow multiple wavelets, including user-defined ones [5.140].

Part 3 of JPEG 2000 is known as MotionJPEG 2000. MotionJPEG has been a commonly used method of editing high-quality video without the existence of an ISO standard. This technology became an International Standard in November 2001 and should have important application in the next generation of digital cameras and elsewhere. In addition, MotionJPEG 2000 will allow support for both lossless and lossy compression in a single codec.

Part 4 of the JPEG 2000 Standard addresses conformance-testing issues, which is a key function to assure the interoperability of various implementations of the standard by the widest community of developers.

Part 5 of the JPEG 2000 standard contains reference software that implements the basic features of Part 1 and is due to be disseminated for both noncommercial and commercial users. A CD of Part 5 of the standard was also produced, together with two releases of reference code: one Image Power' JasPer codec written in the C language and the other in the Java programming language. Due to increased interest in JPEG 2000 and its growing importance in various applications and business models, two more parts were added to the standard. Part 6 became an IS in May 2002. Part 7 is scheduled for a later date.

Part 6, Mixed Raster Content (MRC), will deal with file format issues for compound images based on a mixed-raster approach, allowing JPEG 2000, JBIG 2 and other coding schemes to be mixed in a common environment.

Part 7 will contain a Technical Report (TR) outlining guidelines for minimum support of Part 1 of the standard. It addresses issues facing hardware implementations of the Part 1 standard in Application-Specific Integrated Circuits (ASIC) and Field Programmable Gate Array (FPGA) applications.

Table 5.17 Division of the JPEG-2000 standard between Part 1 and Part 2 [5.140].

Technology	Part 1	Part 2
Bit stream	Fixed and variable length markers.	New markers can be skipped by a Part 1 decoder.
File format	Optional. Provide intellectual property (for example, copyright) information, color or tone-space for image and general method of including metadata.	Allow metadata to be interleaved with coded data. Define types of metadata.
Arithmetic coder	MQ coder.	MQ coder.
Coefficient modeling	Independent coding of fixed-size blocks within subbands. Division of coefficients into three sub-bit planes. Grouping of sub-bit planes into layers.	Special models for binary or graphic data.
Quantization	Scalar quantizer with dead-zone and truncation of code blocks.	Trellis coded quantization.
Transformation	Low complexity (5,3) and high performance Daubechies (9,7). Mallat decomposition.	Many more filters, perhaps user-defined filters. Packet and other decompositions.
Component decorrelation	Reversible Component Transformation (RCT), YCrCb transform.	Arbitrary point transform or reversible wavelet transform across components.
Error resilience	Resynchronization markers.	Fixed-length entropy coder, repeated headers.
Bit-stream ordering	Progressive by tile part, then SNR, or resolution or component.	Out of order tile parts.

©2000 IEEE.

5.6.2 Overview of Still-Image Coding Standards

In order to present an analytical study of JPEG 2000 functionalities, the following standards will be overviewed: MPEG-4 VTC [5.141], JPEG [5.136], JPEG-LS [5.142], and Portable Network Graphics (PNG) [5.143]. JPEG is one of the most popular coding techniques in imaging applications ranging from Internet to digital photography. Both MPEG-4 VTC and JPEG-LS are very recent standards that have started appearing in various applications. Although PNG is not formally a standard and is not based on the state-of-the-art techniques, it is becoming increasingly popular for Internet-based applications. Although JPEG 2000 supports coding of bilevel and paletted color images, we restrict ourselves to continuous tone because it is one of the most popular image types. Other image coding standards are JBIG [5.144] and JBIG 2 [5.145]. These are known for providing good performance for bilevel images, but they do not support an efficient coding of continuous tone images with a large enough number of levels [5.146].

MPEG-4 VTC

MPEG-4 VTC is the algorithm used in the MPEG-4 standard in order to compress the texture information in photo-realistic 3D models. Because the texture in a 3D model is similar to a still picture, this algorithm can also be used for compression of still images [5.141]. It is based on the DWT, scalar quantization, zero-tree coding and arithmetic coding. MPEG-4 VTC supports SNR scalability through the use of different quantization strategies: Single Quantization (SQ), Multiple Quantization (MQ) and Bilevel Quantization (BQ). SQ provides no SNR scalability. MQ provides limited SNR scalability, and BQ provides generic SNR scalability. Resolution scalability is supported by the use of band-by-band (BB) scanning instead of traditional zero-tree scanning, or Tree-Depth (TD), which is also supported. MPEG-4 VTC also supports coding of arbitrary-shaped objects by means of a shape-adaptive DWT, but does not support lossless coding. Several objects can be encoded separately, possibly at different qualities, and then composited at the decoder to obtain the final decoded image.

JPEG

This is a very well known ISO/ITU-T standard created in the late 1980s. There are several modes defined for JPEG, including baseline, lossless, progressive and hierarchical. Baseline mode is the most popular and supports lossy coding only. It is based on the 8x8 block DCT, zig-zag scanning, HVS weighting uniform scalar quantization and Huffman coding. The lossless mode is based on a predictive scheme and Huffman coding [5.135]. The progressive and hierarchical modes of JPEG are both lossy and differ only in the way that the DCT coefficients are coded or computed, respectively, when compared to the baseline mode. They allow a reconstruction of a lower quality or lower resolution version of the image by partial decoding of the compressed bit stream. Progressive mode encodes the quantized coefficients by a mixture of spectral selection and successive approximation, and hierarchical mode uses a pyramidal approach to computing the DCT coefficients in a multiresolution way.

JPEG-LS. JPEG-LS is the latest ISO/ITU-T standard for lossless coding of still images. It also provides for near-lossless compression. It is based on adaptive prediction, context modeling and Golomb coding. In addition, it features a flat-region detector to encode these in run-lengths. Near-lossless compression is achieved by allowing a fixed maximum sample error. This algorithm was designed for low complexity while providing high lossless compression ratios. However, it does not provide support for scalability, error resilience or any such functionality.

PNG

PNG is a World Wide Web Consortium (W3C) recommendation for coding of still images. It is based on a predictive scheme and entropy coding. The entropy coding uses the Deflate algorithm of the popular Zip file compression utility, which is based on LZ77 coupled with Huffman coding. PNG is capable of lossless compression only and supports gray scale, paletted color and true color, an optional alpha plane, interlacing and other features.

5.6.3 Significant Features of JPEG 2000

The JPEG 2000 standard provides a set of features that are of vital importance to many high-end and emerging applications by taking advantage of new technologies. It addresses areas where current standards fail to produce the best quality of performance and provides capabilities to markets that currently do not use compression. The most significant features are the possibility to define regions of interest in an image, the spatial and SNR scalabilities, the error resilience and the possibility of intellectual property rights' protection. All these features are incorporated within a unified algorithm [5.146].

Region of Interest (ROI) Coding

One of the features included in JPEG 2000 is the ROI coding. In accordance with this, certain ROIs of the image can be coded with better quality than the rest of the image (background). The ROI scaling-based method scales up the coefficients so that the bits associated with the ROI are placed in higher bit planes. During the embedded coding process, those bits are placed in the bit-stream before the non-ROI parts of the image. Thus, the ROI will be decoded, or refined, before the rest of the image. Regardless of the scaling, a full decoding of the bit stream results in a reconstruction of the whole image with the highest fidelity available. If the bit stream is truncated, or the encoding process is terminated before the whole image is fully encoded, the ROI will have a higher fidelity than the rest of the image [5.147]. The ROI approach defined in JPEG 2000 Part 1 allows ROI encoding of arbitrary-shaped regions without the need of shape information and shape decoding.

Scalability

In general, scalable coding of still images means the ability to achieve coding of more than one resolution and/or quality simultaneously. Scalable image coding involves generating a coded bit stream in a manner that facilitates the derivation of images of more than one resolution and/or quality by scalable decoding. Reasoning that many applications require images to be simultaneously available for decoding at a variety of resolutions or qualities, the architecture supports scalability. If a bit stream is truly scalable, decoders of different complexities, from low-performance decoders to high-performance decoders, can coexist. Although low-performance decoders may decode only small partitions of the bit stream producing basic quality, high-performance decoders may decode much more and produce significantly higher quality. The most important types of scalability are SNR scalability and spatial scalability [5.138, 5.139].

SNR scalability is intended for use in systems with the primary common feature that a minimum of two layers of image quality are necessary. It involves generating at least two image layers of the same spatial resolution, but of different qualities, from a single image source. The lower layer is coded by itself to provide the basic image quality. The enhancement layer, when added back to the lower layer, generates a higher quality reproduction of the input image.

Spatial scalability involves generating at least two spatial resolution layers from a single source so that the lower layer is coded by itself to provide the basic spatial resolution but the enhancement layer employs the spatially interpolated lower layer and carries the full spatial resolution of the input image source. Both types of scalability are very important for Internet and database access applications and bandwidth scaling for robust delivery. The SNR and spatial scalability types include the progressive and hierarchical coding modes already defined in the current JPEG. An additional advantage of spatial and SNR scalability types is their ability to provide resilience to transmission errors because the most important data of the lower layer can be sent across a channel with better error performance, while the less critical enhancement layer data can be sent across a channel with poor error performance.

Error Resilience

To improve the performance of transmitting compressed images across the error-prone channels, error-resilient bit-stream syntax and tools are included in this standard. The error-resilience tools deal with channel errors using data partitioning and resynchronization, error detection and concealment and QoS transmission based on priority [5.138, 5.139]. Many applications require the delivery of image data across different types of communication channels. Typical wireless communication channels give rise to random and burst bit errors. Internet communications are prone to loss due to traffic congestion.

IPRs

An optional file format (JP2) for the JPEG 2000 compressed image data has been defined by the standard. This format has provisions for both image and metadata and a mechanism to indicate the tone scale or color space of the image. This is a mechanism by which readers may recognize the existence of IPR information in the file. Also, it is a mechanism by which metadata, including vendor-specific information, can be included in the file.

5.6.4 Architecture of JPEG 2000

A block diagram of the JPEG 2000 encoder and decoder is illustrated in Figure 5.76.

At first, the discrete transform is applied on the source image data. The transform coefficients are then quantized and entropy coded before forming the output bit stream. The decoder is the reverse of the encoder. The code stream is first entropy decoded, dequantized and inverse discrete transformed, resulting in the reconstructed image data. Depending on the wavelet transform and the applied quantization, JPEG 2000 can be both lossy and lossless. This standard works on image tiles. The term "tiling" refers to the partition of the original (source) image into rectangular nonoverlapping blocks called tiles. They are compressed independently as though they were entirely distinct images. The process of tiling, DC level shifting and DWT of each image tile component is shown in Figure 5.77. This is the strongest form of spatial partitioning because all operations, including component mixing, wavelet transform, quantization and entropy coding, are performed independently on the different tiles of the image.

All tiles have exactly the same dimensions, except maybe those that abut the right and lower boundaries of the image. The nominal tile dimensions are exact powers of two. Tiling reduces memory requirements and constitutes one of the methods for the efficient extraction of a region of the image. Prior to computation of the forward DWT on each tile, all samples of the image tile component are DC level shifted by subtracting the same quantity (that is, the component depth) from each sample. The tile components are decomposed into different decomposition levels using a wavelet transform. These decomposition levels contain a number of subbands populated with coefficients that describe the horizontal and vertical spatial frequency character-

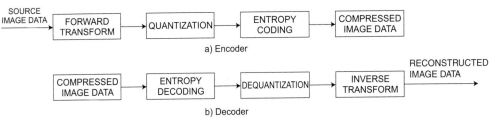

Figure 5.76 Block diagram of JPEG 2000 a) encoder and b) decoder architecture [5.35].

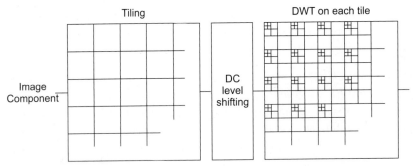

Figure 5.77 Tiling, DC level shifting and DWT on each image tile component [5.125]. ©2001 IEEE.

istics of the original tile component planes as shown in Figure 5.77. The coefficients provide local frequency information. A decomposition level is related to the next decomposition level by spatial powers of two. To perform the forward DWT, the standard uses a 1D subband decomposition of a 1D set of samples into low-pass samples and high-pass samples, representing a down-sampled residual version of the original set needed for the perfect reconstruction of the original set from the low-pass set. Any user supplied wavelet filter banks may be used [5.147, 5.148]. The DWT can be irreversible or reversible. The default irreversible transformation is implemented by means of the Daubechies 9/7-tap filter. The analysis and the corresponding synthesis filter coefficients are given in Table 5.18. The default reversible transformation is implemented by means of the 5-tap/3-tap filter, the coefficients of which are given in Table 5.19.

Table 5.18 Daubechies 9/7 analysis and synthesis filter coefficients [5.154].

Analysis filter coefficients		
I	Low-pass filter $h_L(i)$	High-pass filter $h_H(i)$
0	0.6029490182363579	1.115087052456994
+/-1	0.2668641184428723	-0.5912717631142470
+/-2	-0.07822326652898785	-0.05754352622849957
+/-3	-0.01686411844287495	0.09127176311424948
+/-4	0.02674875741080976	N/A
Synthesis filter coefficients		
I	Low-pass filter $h_L(i)$	High-pass filter $h_H(i)$
0	1.115087052456994	0.6029490182363579
+/-1	0.5912717631142470	-0.2668641184428723
+/-2	-0.05754352622849957	-0.07822326652898785
+/-3	-0.09127176311424948	0.01686411844287495
+/-4	N/A	0.02674875741080976

©2000 IEEE.

Figure 5.78 Periodic symmetric extension of the signal ABCDEFG [5.154]. ©2000 IEEE.

The standard supports two filtering models: convolution based and lifting based. For both modes the signal should be first extended periodically as shown in Figure 5.78. The periodic symmetric extension is used to ensure that, for filtering operations that take place at both boundaries of the signal, one signal sample exists and spatially corresponds to each coefficient of the filter mask. For illustration purposes, it has been assumed that, for the signal ABCDEFG, we have A>B>...>G. The number of additional samples required at the boundaries of the signal is filter-length dependent [5.149].

Convolution-based filtering consists of creating a series of dot products between the two filter masks and the extended 1D signal. Lifting-based filtering consists of a sequence of very simple filtering operations for which alternatively odd sample values of the signal are updated with a weighted sum of even sample values. On the other hand, even sample values are updated with a weighted sum of odd sample values [5.139, 5.150]. For the reversible (lossless) case, the results are rounded to integer values.

Example 5.15 The lifting-based filtering for the 5/3 analysis filter is achieved as follows:

$$y(2n+1) = x_{est}(2n+1) - \left\lceil \frac{x_{est}(2n) + x_{est}(2n+2) - 1}{2} \right\rceil \qquad (5.3)$$

Table 5.19 5/3 analysis and synthesis filter coefficients [5.154].

	Analysis filter coefficients	
I	Low-pass filter $h_L(i)$	High-pass filter $h_H(i)$
0	6/8	1
+/-1	2/8	-1/2
+/-2	-1/8	N/A
	Synthesis filter coefficients	
I	Low-pass filter $h_L(i)$	High-pass filter $h_H(i)$
0	1	6/8
+/-1	1/2	-2/8
+/-2	N/A	-1/8

©2000 IEEE.

$$y(2n) = x_{est}(2n) + \left\lfloor \frac{y(2n-1) + y(2n+1) + 2}{4} \right\rfloor \qquad (5.4)$$

where x_{est} is the extended input signal, y is the output signal, and $\lfloor a \rfloor$ and $\lceil a \rceil$ indicate the largest integer not exceeding a and the smallest integer not exceeded by a, respectively.

Quantization is the process by which the coefficients are reduced in precision. This operation is lossy unless the quantization step is 1 and the coefficients are integers as produced by the reversible integer 5/3 wavelet. Each of the transform coefficients $a_b(u,v)$ of the subband b is quantized to the value $q_b(u,v)$ according to the formula

$$q_b(u,v) = sign(a_b(u,v)) \left\lfloor \frac{|a_b(u,v)|}{\Delta_b} \right\rfloor \qquad (5.5)$$

where Δ_b is the quantization step [5.122]. The dynamic range depends on the number of bits used to represent the original image tile component and on the choice of the wavelet transform. All quantized transform coefficients are signed values even when the original components are unsigned. These coefficients are expressed in a sign-magnitude representation prior to coding. Although Part 1 of the JPEG 2000 standard uses only simple scalar dead-zone quantization, significant data size reduction can also be obtained by throwing away portions of the data. Part 2 of the standard contains a TCQ [5.151, 5.152]. This technology has a fairly high encoding cost, but adds a minimal amount of complexity for a decoder and produces higher quality images, and sometimes does a better job visually. However, the improvement may not be noticeable in terms of SNR. Each subband of the wavelet decomposition is divided into rectangular blocks, called code blocks, which are coded independently using arithmetic coding. A binary arithmetic entropy coder called the MQ coder is used to provide compression of symbols' output by the context model. The complexity and compression are much higher than the typically used Huffman coder in JPEG [5.139]. The code blocks are coded one bit plane at a time, starting with the most significant bit plane with a nonzero element to the least significant bit plane. For each bit-plane in a code block, a special code-block scan pattern is used for each of three passes. Each coefficient bit in a bit plane is coded in only one of the three passes. A rate distortion optimization method is used to allocate a certain number of bits to each block.

JPEG 2000 supports multiple-component images. Different components need not have the same bit depths nor need they have all been signed or unsigned. The standard supports two different component transformations: one Irreversible Component Transformation (ICT) and one RCT. It is usual that the input image has three components: Red, Green and Blue (RGB). The block diagram of the JPEG 2000 multiple-component encoder is shown in Figure 5.79. Here, C_1, C_2 and C_3 represent, in general, the color-transformed output components. If needed, prior to applying the forward-color transformation, the image component samples are DC level shifted. The ICT may be used only for lossy coding. It can be seen as an approximation of a YCbCr transformation of the RGB components.

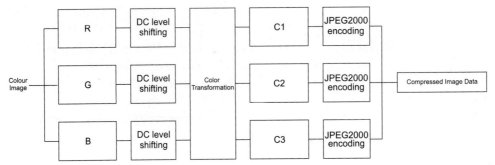

Figure 5.79 JPEG 2000 multiple-component encoder.

The forward and the inverse ICT transformations are achieved from the following equations:

$$\begin{pmatrix} Y \\ Cb \\ Cr \end{pmatrix} = \begin{pmatrix} 0.299 & 0.587 & 0.114 \\ -0.1687 & -0.33126 & 0.5 \\ 0.5 & -0.41869 & -0.08131 \end{pmatrix} \begin{pmatrix} R \\ G \\ B \end{pmatrix} \quad (5.6)$$

$$\begin{pmatrix} R \\ G \\ B \end{pmatrix} = \begin{pmatrix} 1.0 & 0 & 1.402 \\ 1.0 & -0.34413 & -0.71414 \\ 1.0 & 1.772 & 0 \end{pmatrix} \begin{pmatrix} Y \\ Cb \\ Cr \end{pmatrix} \quad (5.7)$$

The RCT may be used for lossy or lossless coding. It is a decorrelating transformation, which is applied to the three first components of an image. Three goals are achieved by this transformation:

- Color decorrelation for efficient compression
- Reasonable color space with respect to the human visual system for quantization
- Ability of having lossless compression, that is, exact reconstruction with finite integer precision

For the RGB components, the RCT can be seen as an approximation of a YUV transformation. The forward and inverse RCT is performed by means of

$$\begin{pmatrix} Yr \\ Ur \\ Vr \end{pmatrix} = \begin{pmatrix} \left\lfloor \dfrac{R+2G+B}{4} \right\rfloor \\ R-G \\ B-G \end{pmatrix} \quad (5.8)$$

that is,

$$\begin{pmatrix} G \\ R \\ B \end{pmatrix} = \begin{pmatrix} Yr - \left\lfloor \dfrac{Ur+Vr}{4} \right\rfloor \\ Ur+G \\ Vr+G \end{pmatrix} \quad (5.9)$$

Part 1 of the JPEG 2000 standard contains the YCrCb transform used in the original JPEG standard. It also includes an RCT useful for lossless compression of three-component color imagery. Part 2 contains the ability to do an arbitrary joint transform to decorrelate components. This is essential for good compression on multi- and hyperspectral imagery. A standard block diagram of a JPEG 2000 coder and the correspondence with the Annexes of the WD of the standard are shown in Figure 5.80. An encoder starts at the left of the figure with an image and produces a code stream at the right. A decoder works in the opposite direction.

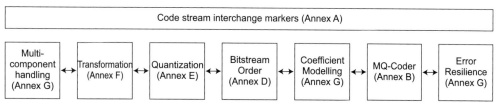

Figure 5.80 Standard block diagram of a JPEG 2000 coder and the correspondence with the Annexes of the working draft [5.159]. ©1998 ISO/IEC.

5.6.5 JPEG 2000 Bit Stream

JPEG 2000 provides better rate-distortion performance than the original JPEG standard for any given rate. The improvements in the near visually lossless realm are more modest, approximately 20% [5.140].

There are four basic dimensions of progression in the JPEG 2000 bit stream: resolution, quality, spatial location and component. Different types of progression are achieved by the ordering of packets within the bit stream. Although this provides an important mechanism for spatial progression, we assume for simplicity that the image consists of a single tile. Each packet is then associated with one component (say i), one layer (j), one resolution level (k) and one packet partition location (m).

Example 5.16 Let us say that an image is divided into tiles, and each tile is transformed. The subbands of a tile are divided into packet partition locations. Finally, each packet partition location is divided into code blocks. This is illustrated in Figure 5.81 where 12 code blocks of one packet partition location at resolution level 2 of a 3-level dyadic wavelet transform are given. The packet partition location is emphasized by heavy lines. The division of one packet partition location into 12 code blocks is also shown.

Figure 5.82 depicts one packet for the packet partition location illustrated in the previous figure. Each of the 12 code blocks can contribute a different number of sub-bit planes (possibly zero) to the packet. Empty packet bodies are allowed.

A packet can be interpreted as one quality increment for one resolution level at one spatial location. Packet partition locations correspond roughly to spatial locations. A layer is then a collection of packets: one from each packet partition location of each resolution level. A layer then can be interpreted as one quality increment for the entire image at full resolution. Each layer

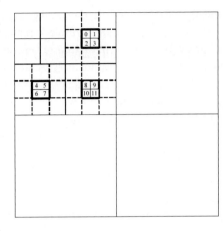

Figure 5.81 Twelve code blocks of one packet partition location at resolution level 2 of a three-level dyadic wavelet transform. The packet partition location is presented by heavy lines.

Packet Header	n_0 sub-bitplanes from code-block 0	n_1 sub-bitplanes from code-block 1	....	n_{11} sub-bitplanes from code-block 11

Figure 5.82 The composition of one packet partition location with 12 code-blocks.

provides more bits of some of the wavelet coefficients. The role of layers in providing progression by SNR has been described [5.140]. The layers need not be designed specifically for optimal SNR progression. JPEG 2000 does not explicitly define a method of subsampling color components as JPEG does. A JPEG 2000 encoder could place all the high frequency bands of the color components in the last layer. A decoder that did not receive high frequency subbands could use a simplified transform to save computational complexity. For images with significant color edges, some bits of the color coefficients might be saved in earlier layers.

The JPEG 2000 bit stream contains markers that identify the progression type of the bit stream. Other markers may be written that store the length of every packet in the bit stream. To change a bit stream from progressive by resolution to progressive by SNR, a parser can read all the markers, change the type of progression in the markers, write the lengths of the packets out in the new order, and write the packets themselves out in the new order. There is no need to run the MQ coder or the context model or even to decode the block-inclusion information. The complexity is only slightly higher than a pure copy operation.

If the ROIs are known in advance, that is, at encode time, JPEG 2000 provides additional methods of providing greater image quality in the foreground verses the background. First, all the code blocks that contain coefficients affecting the ROI can be identified, and the bit planes of these coefficients can be stored in higher layers relative to other coefficients. Thus, a layer-progressive bit stream can naturally send the ROI with higher quality than the background. In addition, an explicit ROI can be defined, and those coefficients that affect the ROI can be shifted and coded as if they were in their own set of bit planes. For an encoder, this allows individual coefficients to be enhanced rather than entire code blocks, If the ROIs are not known at encoding time, there are still several methods for a smart server to provide exactly the right data to a client requesting a specific region. The simplest method to provide access to spatial regions of the image (which are not known at encoding time) is for the encoder to tile the image. Because tiling divides the image spatially, any region desired by the client will lie within one or more tiles. Tiles as small as 64 by 64 are useable, but tiles this small increase the bit rate noticeably. Tiles greater than 256 by 256 samples have almost no compression performance impact, but offer less flexible access for small regions. All of the parsing operations on the whole image can selectively be applied to specific tiles. Other tiles could be discarded or transmitted at a much lower quality. The bitstream contains the length of each tile, so it is always possible to locate the desired tiles with minimal complexity. Similarly, packet partitions can be extracted from the bit stream for spatial access. The length information is still stored in the tile header, and the data corresponding to a packet partition location is easily extracted. Finer grain access is possible by parsing individual code blocks. As in the case of packet partition locations, it is necessary to determine which code blocks affect which pixel locations. The correct packets containing these code blocks can be determined from the progression-order information. Finally, the location of the compressed data for the code blocks can be determined by decoding the packet headers.

All uncompressed tiled image formats allow regions of an image to be edited, and only those tiles affected need to be rewritten to disk. With compression, the compressed size of an edited tile can change. Because of the flexibility in quantization in JPEG 2000, it is possible to truncate an edited tile to fit in the previous size. Alternatively, Part 2 of the standard allows out-of-order tiles within the bit stream, so an edited tile could be rewritten at the end of the bit stream. The main header of a JPEG 2000 bit stream contains the width and height of the image. It also contains a horizontal and vertical offset for the start of the image. This allows the image to be cropped to a subrectangle of the original without requiring a forward and inverse wavelet transform for recompression. All tiles inside the newly cropped image need not be changed at all, and tiles on the edge of the new image need only to have the code blocks on the edges recorded. New tile headers and packet headers are written to the bit stream (no wavelet transform). Finally, the integer nature (5,3) of the wavelet allows an image or partition of an image to be compressed multiple times with the same quantization with no additional loss. This is the only time that the decompressed sample values are not clipped when they fall outside the full dynamic range (for example, 0 to 255 for 8-bit images).

5.6.6 Compression Efficiency Comparisons

Compression efficiency is one of the top priorities in the design of image products [5.153]. In Santa-Cruz and Ebrahimi [5.146], lossless and lossy progressive compression and efficiency results to evaluate how well the algorithms code different types of imagery are presented. The support of progressive coding is included, too. The algorithms have been evaluated with seven images from the JPEG 2000 test set, covering various types of imagery. The images "Bike" (2048x2560) and "Café" (2048x2560) are natural, "Cmpnd" (512x768) and "Chart" (1688x2347) are compound documents consisting of the text, photographs and computer graphics, "Aerial2" (2048x2048) is an aerial photography, "Target" is a computer-generated image and "US" (512x448) is an ultra scan. All these images have a depth of 8 bits per pixel.

Table 5.20 summarizes the lossless compression efficiency of lossless JPEG (L-JPEG), JPEG-LS, PNG and JPEG 2000 for all the test images. For JPEG 2000, the reversible DWT filter, refered to as $J2K_R$, has been used. In the case of L-JPEG, optimized Huffman tables and the predictor yielding the best compression performance have been used for each image. For PNG, the maximum compression setting has been used, and for JPEG-LS, the default options were chosen. MPEG-4 VTC is not considered because it does not provide a lossless functionality. It can be seen that, in almost all cases, the best performance is obtained by JPEG-LS. JPEG 2000 provides, in most cases, competitive compression ratios with the added benefit of scalability. PNG performance is similar to JPEG 2000. PNG provides the best results for the "Target" image. JPEG-LS and PNG achieve much larger compression ratios for the "Cmpnd" image. This image contains, for the most part, block text on a white background. PNG performs the best although this is solely due to the very large compression ratio that it achieves on target. However, JPEG-LS provides the best compression ratio for most images. To conclude, as far as lossless compression is concerned, JPEG 2000 seems to perform reasonably well in terms of its ability to deal efficiently with various types of images. However, in specific types of images such as "Cmpnd," JPEG 2000 is outperformed by far by JPEG-LS. This result is even more striking, noting that JPEG-LS is a significantly less complex algorithm.

Table 5.20 Lossless compression ratios for seven test images [5.146].

Image	$J2K_R$	JPEG-LS	L-JPEG	PNG
Bike	1.77	1.84	1.61	1.66
Café	1.49	1.57	1.36	1.44
Cmpnd1	3.77	6.44	3.23	6.02
Chart	2.60	2.82	2.00	2.41
Aerial2	1.47	1.51	1.43	1.48
Target	3.76	3.66	2.59	8.70
US	2.63	3.04	2.41	2.94
Average	2.50	2.98	2.09	3.52

©2000 IEEE.

The rate distortion behavior of lossy (nonreversible) JPEG 2000 and progressive JPEG is depicted in Figure 5.83 for a natural image. It is seen that JPEG 2000 significantly outperforms the JPEG scheme [5.122]. It can be calculated that, for similar Peak SNR (PSNR) quality, JPEG 2000 compresses almost twice as much as JPEG.

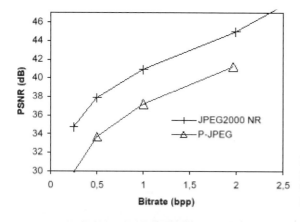

Figure 5.83 Rate distortion results for progressive JPEG 2000 versus progressive JPEG for a natural image [5.122]. ©2000 IEEE.

(a)

(b)

Figure 5.84 Reconstructed images compressed at 0.125 bpp by means of (a) JPEG and (b) JPEG 2000 [5.122]. ©2000 IEEE.

The superiority of JPEG 2000 can be subjectively judged with the help of Figure 5.84 where the reconstructed image "Hotel" (720x576) is shown. Both images were compressed at a rate of 0.125 bpp using JPEG and JPEG 2000. The degradation of the image in Figure 5.84(a) is evident.

In order to evaluate the error-resilience features offered by the different standards, a transmission channel with random errors has been simulated [5.146] together with the evaluation of the average reconstructed image quality after decompression. Table 5.21 shows the results for JPEG 2000 with irreversible wavelet transform and JPEG baseline. Only the results of "Café" image are shown. The behavior is very similar for the other images. As it can be seen, the recon-

Table 5.21 PSNR, in dB, corresponding to average Root MSE (RMSE) of 200 runs of the decoded "Café" image when transmitted across a noisy channel with various bit error rates (BER) and compression bit rates for JPEG baseline and JPEG 2000 (J2K) [5.146].

bpp	IS	BER 0	BER 1E-06	BER 1E-05	BER 1E-04
0.25	J2K	23.06	23.00	21.62	16.59
	JPEG	21.94	21.79	20.77	16.43
0.5	J2K	26.71	26.42	23.96	17.09
	JPEG	25.40	25.12	22.95	15.73
1.0	J2K	31.90	30.75	27.08	16.92
	JPEG	30.34	29.24	23.65	14.80
2.0	J2K	38.91	36.38	27.23	17.33
	JPEG	37.22	30.68	20.78	12.09

©2000 IEEE.

structed image quality under transmission errors is higher for JPEG 2000 than JPEG across all encoding bit rates and error rates. However, at low bit rates (0.25 and 0.5 bpp), the quality of JPEG 2000 decreases more rapidly than JPEG as the errors increase, although the absolute quality is always higher. Concerning the visual quality at moderately low error rates (that is, 1E-06), for JPEG 2000 it is much higher when compared to JPEG. It should also be noted that at higher error rates (that is, 1E-04), the reconstructed image quality in JPEG 2000 is almost constant across all bit rates. This is due to the fact that, in JPEG 2000, each subband block is coded by bit planes. When the error rate is high, almost all blocks are effected in the most significant bit planes, which are transmitted first. When particular bit planes are affected in a block, lower bit planes cannot be decoded and are therefore useless. In the case of JPEG, the problem is even worse. The higher the encoding bit rate means the lower the decoded quality. This can be explained by the fact that, when an 8x8 block is affected by a transmission error, the entire block is basically lost. The higher the encoding bit rate means the more bits it takes to code a block. Therefore, the probability of a block being hit by an error and lost is higher for the same bit error rate. In other words, in JPEG, the density of error protection decreases with an increase in bit rate.

Many applications require features in a coding algorithm other than simple compression efficiency. This is often referred to as functionalities. Table 5.22 summarizes the comparison from a functionality point of view. A functionality matrix is provided in Santa-Cruz and Ebrahimi [5.146]. It indicates the set of supported features in each standard.

This table clearly shows that JPEG 2000 is the standard offering the richest set of features in an efficient manner and within an integrated algorithmic approach. MPEG-4 VTC (also JPEG 2000) is able to produce progressive bit streams without any noticeable overhead. However, the

Table 5.22 Functionality matrix. A + indicates that it is supported. The more +s means the more efficiently or better it is supported. A - indicates that it is not supported [5.146].

Functionality	JPEG 2000	JPEG-LS	JPEG	MPEG-4 VTC	PNG
Lossless compression performance	+++	++++	+[1]	-	+++
Lossy compression performance	+++++	+	+++	++++	-
Progressive bit streams	+++++	-	++[2]	+++	+
ROI coding	+++	-	-	+[3]	-
Arbitrary-shaped objects	-	-	-	++	-
Random access	++	-	-	-	-
Low complexity	++	+++++	+++++	+	+++
Error resilience	+++	++	++	+++	+
Noniterative rate control	+++	-	-	+	-
Genericity[4]	+++	+++	++	++	+++

[1] Only using the lossless mode of JPEG.
[2] Only in the progressive mode of JPEG.
[3] Tile-based only.
[4] Ability to compress different types of imagery efficiently across a wide range of bit rates.

©2000 IEEE.

latter provides more progressive options and produces bit streams that are parseable and that can be rather easily reorganized by a transcoder. Along the same lines, JPEG 2000 also provides random access to the block level in each subband, thus making it possible to decode a region of the image without having to decode it as a whole. These features could be very advantageous in applications such as digital libraries.

Error Resilience

Concerning error resilience, JPEG 2000 offers higher protection than JPEG, as shown in the previous section. MPEG-4 VTC also offers error-resilience features and, although it could not be evaluated, the support should be in-between JPEG and JPEG 2000. JPEG-LS does not offer any particular support for error resilience besides restart markers and has not been designed with it in mind. As for PNG, it offers error detection, but no concealment possibilities.

Overall, one can say that JPEG 2000 offers the richest set of features and provides superior rate-distortion performance. However, this comes at the price of additional complexity when compared to JPEG and JPEG-LS, which might be currently perceived as a disadvantage for some applications, as was the case for JPEG when it was first introduced.

5.7 MPEG-7 Standardization Process of Multimedia Content Description

MPEG-7, formally named Multimedia Content Description Interface, is the standard that describes multimedia content so that users can search, browse and retrieve the content more efficiently and effectively than they could by using existing mainly text-based search engines [5.155]. It is a standard for describing the features of multimedia content. The word "features" or "descriptions" represents a rich concept that can be related to several levels of abstraction. Descriptions vary according to the types of data. Furthermore, different types of descriptions are necessary for different purposes of categorization. MPEG-7 will specify a standard set of descriptors that can be used to describe various types of multimedia information. Also, MEPG-7 will standardize ways to define other descriptors as well as structures for the descriptors and their relationships. This description will be associated with the content to allow fast and efficient searching for material of the user's interest. A language to specify description schemes, that is, a DDL, will be standardized, too. Audiovisual material that has MPEG-7 data associated with it can be indexed and searched for. This material includes still pictures, graphics, 3D models, audio, speech, video and information about how these elements are combined in a multimedia presentation. Special cases of these general data types may include facial expressions and personal characters [5.155, 5.156].

Different people want to use the audiovisual information for various purposes. However, before the information can be used, it must be located. At the same time, the increasing availability of potentially interesting material makes this search more difficult. This challenging situation led to the need for a solution to the problem of quickly and efficiently searching for various types of multimedia material of interest to the user. The MPEG-7 standard wants to answer to this need and to provide the solution [5.157].

MPEG-7 is rather different from the other MPEG standards because it does not define a way to represent data with the objective to reconstruct the data as faithfully as possible like MPEG-1, MPEG-2 and MPEG-4 did. The increasingly pervasive role that audiovisual sources are destined to play in our lives and the growing need to have these sources further processed make it necessary to develop forms of audiovisual information representation that go beyond the simple waveform or pixel-based, frame-based (such as MPEG-1 and MPEG-2) or even object-based (such as MPEG-4) representations. This necessitates forms of representation that allow some degree of interpretation of the information's meaning, which can be passed on to, or accessed by, a device or a computer code. The people active in defining the MPEG-7 standard represent broadcasters, equipment and chip manufacturers, digital content creators and managers, telecommunication service providers, publishers, IPR managers and researchers.

5.7.1 Objective of the MPEG-7 Standard

The objective of MPEG-7 is to set a standard for the description of multimedia material. This includes speech, audio, video, still pictures and 3D models. It also includes information about how these elements are combined in a multimedia scene, presentation or document. MPEG-7 will define a number of elements: descriptors, description schemes, a DDL, system tools and coding schemes for the descriptions. These are the normative elements of MPEG-7. These parts need to be specified to ensure the interoperability between MPEG-7-enabled systems. Before defining these elements, we will first address the definitions of the key concepts of data and feature [5.158]. Data refers to Audio Visual information that will be described using MPEG-7, regardless of storage, coding, display, transmission medium or technology. Examples are an MPEG-4 stream; a video tape; a CD containing music, sound or speech; a picture printed on paper, and an interactive multimedia presentation on the Web.

A feature is a distinctive characteristic of the data that signifies something to somebody. Some examples are color of an image, pitch of a speech segment, rhythm of an audio segment, camera motion in a video, style of a video, the title of a movie, the actors in a movie, and so forth.

A Descriptor (D) is a representation of a feature. It defines the syntax and the semantics of the feature representation. Possible descriptors are the color histogram, the average of the frequency components, the motion field, the text of the title, and so forth. A D value is an instantiation of a D for a given dataset or subset.

A Description Scheme (DS) specifies the structure and semantics of the relationships between its components, which may be both Ds and DSs schemes. Examples are a movie, temporally structured as scenes and shots, including some textural descriptors at the scene level, and color, motion and audio Ds at the shot level. A description consists of a DS structure and the set of descriptor values that describe the data. A coded description is a description that has been encoded to fulfill relevant requirements, such as compression efficiency, error resilience, random access, and so forth. DDL is a language that allows the creation of new description schemes and possibly, descriptors. It also allows the extension and modification of existing description schemes. Figure 5.85 gives a graphical view of the relation between the different MPEG-7 elements and their relations [5.157].

MPEG-7 addresses many applications and many types of usage. The standard will address real-time and non-real-time applications, interactive and unidirectional (broadcast) and online as well as offline usage. In this context, a real-time environment means that descriptions are being associated with the content while it is being captured. MPEG-7 descriptions will support query modalities such as text-based only, subject navigation, interactive browsing, visual navigation and summarization, search by example, and using features and sketches.

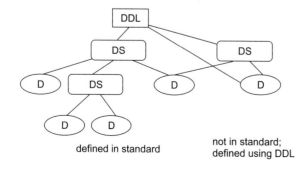

Figure 5.85 Elements of the MPEG-7 standard [5.156]. ©2000 ISO/IEC.

Which modality to use depends on the task at hand and the application environment. The description definition of a piece of multimedia material does not exist. First, more than one descriptor may exist to represent the same feature, fulfilling different requirements. Second, and more important in this context, the exact description depends very much on the application and the user. MPEG-7 will not define what description is right for a certain body of content, but only gives the tools to represent such a description. In this sense, MPEG-7 follows the policy adopted for previous MPEG standards: The analysis engine and the encoder will not be standardized. An analogy exists with MPEG-4 that can represent arbitrary-shaped VOs. MPEG-4 only specifies the syntax and semantics to represent a shape and defines how that representation should be decoded, regardless of how it was obtained. The same applies to the systems that make use of the MPEG-7 descriptions, such as search engines and filters, which reside on the other end of a possible MPEG-7 processing chain. The scope of the MPEG-7 standard using a simplified processing chain is shown in Figure 5.86. MPEG-7 will have the possibility to denote spatiotemporal entities in non-MPEG-4 content.

Figure 5.86 A possible processing chain and scope of the MPEG-7 standard [5.156]. ©2000 ISO/IEC.

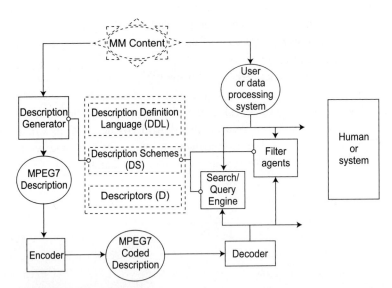

Figure 5.87 Representation of possible applications using MPEG-7 [5.156]. ©2000 ISO/IEC.

In comparison with other available or emerging solutions for multimedia description, MPEG-7 can be characterized by the following:

- *Its generality*—The capability to describe content from many application environments
- *Its object-based data model*—The capability of independently describing individual objects within a scene
- *Its integration of low- and high-level features/descriptors into a single architecture*—The capability to combine the power of both types of descriptors
- *Its extensibility provided by the DDL*—The capability to keep growing, to be extended to new application areas, to answer newly emerging needs and to integrate novel description tools

Figure 5.87 explains a hypothetical MPEG-7 chain in practice. The circular boxes depict tools that are implementing functions, such as encoding or decoding, and the square boxes represent static elements, such as a description. The dotted boxes in the figure encompass the normative elements of the MPEG-7 standard. Besides the descriptors themselves, the database structure plays a crucial role in the final retrieval's performance. To allow the desired fast judgment about whether the material is of interest, the indexing information will have to be structured, for example, in a hierarchical or associative way.

5.7.2 Status of the MPEG-7 Standard

Currently, MPEG-7 concentrates on the specification of description tools together with the development of the MPEG-7 reference software, known as XM (eXperimentation Model).

The MPEG-7 Audio group develops a range of description tools from generic audio descriptors to more sophisticated description tools like spoken content. Generic audio description tools will allow the search for similar voices, by searching similar envelopes and fundamental frequencies of a voice sample against a database of voices. The spoken content description scheme is designed to represent the output of a great number of state-of-the-art automatic speech-recognition systems, containing both words and phoneme representations and transition likelihoods. This alleviates the problem of out-of-vocabulary words, allowing retrieval even when the original words were wrongly decoded.

The MPEG-7 Visual group is developing four groups of description tools: color, texture, shape and motion. Color and texture description tools will allow the searching and filtering of visual content (images, graphics and video) by dominant color or textures in the same (arbitrarily shaped) regions or the whole image. Shape description tools will facilitate query by sketch or by contour similarity in an image database or, for example, searching trademarks in registration databases. Motion description tools will allow searching of videos with similar motion patterns that can be applicable to news or to surveillance applications.

The Multimedia Description Schemes group is developing the description tools dealing with generic and audiovisual and archival features. Its central tools deal with content management and content description. Content management description tools cover the viewpoints of media creation and production and usage. Media description tools allow searching for preferred storage formats, compression qualities and aspect ratios, among others. Creation and prediction description tools cover the typical archival and credit information (for example, title, creators and classification). Usage description tools deal with description related to the use of the described content (for example, rights, broadcasting, dates and places, availability, audience and financial data). The content description covers both structural and conceptual viewpoints. Structural description tools provide segmentation, both spatial and temporal, of the content. Among other functionalities, this allows assigning descriptions to different regions and segments and providing importance rating of temporal segments and regions. Conceptual description tools allow providing of a semantic-based description besides the content description and content management description tools. Other tools target content organization, navigation, access and user preferences.

5.7.3 Major Functionalities in MPEG-7

The MPEG-7 standard consists of the following parts [5.157, 5.158]:

- *Part 1: Systems*—The tools that are needed to prepare MPEG-7 descriptions for efficient transport and storage and to allow synchronization between content descriptions. There are also tools related to managing and protecting intellectual property.
- *Part 2: DDL*—The language for defining new description schemes and perhaps also new descriptors.
- *Part 3: Audio*—The descriptors and description schemes dealing with only audio descriptors.
- *Part 4: Visual*—The descriptors and description schemes dealing with visual descriptors.
- *Part 5: Multimedia Description Schemes*—The descriptors and description schemes dealing with generic features and multimedia descriptions.
- *Part 6: Reference Software*—A software implementation of relevant parts of the MPEG-7 standard.
- *Part 7: Conformance*—Guidelines and procedures for testing conformance of MPEG-7 implementations.

In what follows, the major functionalities that the different parts of the MPEG-7 standard offer are described.

MPEG-7 Systems

This part includes the tools that are needed to prepare MPEG-7 descriptions for efficient transport and storage and to allow synchronization between content and tools related to managing and protecting intellectual property. It defines the terminal architecture and the normative interfaces. The information representation specified in the MPEG-7 standard provides the means to represent coded multimedia content description information. The entity that makes use of such coded representation of the multimedia content is referred to as a terminal. The architecture of a terminal making use of MPEG-7 representations is depicted in Figure 5.88. A transmission/storage medium refers to the lower layers of the delivery infrastructure. These layers deliver multiplexed streams to the delivery layer. The transport of the MPEG-7 data can occur on a variety of delivery systems. For example, this includes MPEG-2 transport streams, IP or MPEG-4 (MP4) files or streams. The Delivery Layer encompasses mechanisms allowing synchronization, framing and multiplexing of MPEG-7 content. The MPEG-7 architecture allows conveying data such as queries or request back from the terminal to the transmitter or server. The Delivery Layer provides MPEG-7 ES to the Compression Layers. They consist of consecutive individually accessible portions of data named access units. An access unit is the smallest data entity to which timing information can be attributed. Description information is either a complete description of the multimedia content or a fragment of the description. MPEG-7 data can be represented either in textual format, in binary format or in a mixture of the two formats, depending on application usage. The syntax of the textual format is defined in Part 2 (DDL) of the standard [5.157]. The syntax of the binary format for MPEG-7 data (BiM) is defined in Part 1 (Systems) of the standard. At the Compression Layer, the flow of access units (either textual or binary encoded) is parsed, and the content description is reconstructed. The MPEG-7 binary stream can be either parsed by the BiM parser and transformed in textual format and then transmitted in textual format to further reconstruction processing, or the binary stream can be parsed by the BiM parser and then transmitted in proprietary format to further processing. MPEG-7 access units are further structured as commands encapsulating the scheme or the description information. Commands provide the dynamic aspects of the MPEG-7 content. They allow a description to be delivered in a single chunk or to be fragmented in small pieces. They allow basic operations on the MPEG-7 content, such as updating a descriptor, deleting part of the description or adding new DDL structure.

MPEG-7 normative interfaces are presented in Figure 5.89. Content refers either to essence or to content description. An MPEG-7 binary/textual encoder transforms the content into a compliant format. A textual format interface describes the format of the textual units. The MPEG-7 textual decoder consumes a flow of such access units and reconstructs the content description in a normative way. A binary format interface describes the format of the binary access units. The MPEG-7 binary decoder consumes a flow of such access units and reconstructs the content description in a normative way. An MPEG-7 binary/textual decoder transforms data into a content description.

The question often arises as to how proof can be established that the binary representation and textual representation provide dual representations of the content. The process is described in Figure 5.90. In addition to the elements described in MPEG-7 normative interfaces, the validation process involves the definition of a canonical representation of a content description. The validation process works as follows. A content description is encoded in a lossless way in textual and in binary format, generating two different representations of the same entity. The two

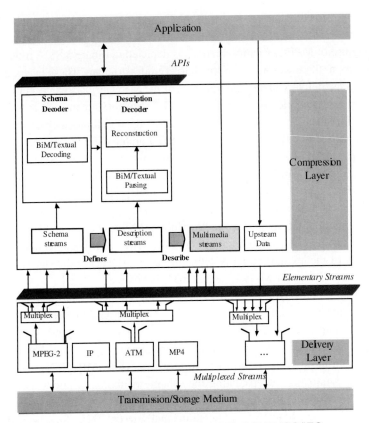

Figure 5.88 MPEG-7 architecture [5.156]. ©2000 ISO/IEC.

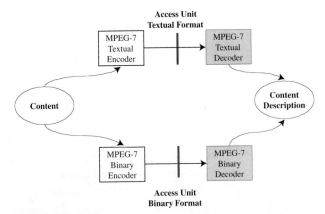

Figure 5.89 MPEG-7 normative interfaces [5.156]. ©2000 ISO/IEC.

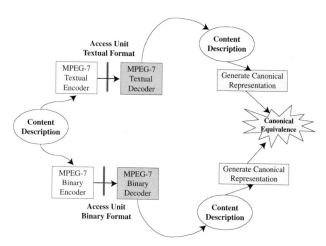

Figure 5.90 Validation process [5.156].©2000 ISO/IEC.

encoded descriptions are decoded with their respective binary and textual decoders. The two canonical descriptions will be equivalent.

MPEG-7 DDL

The main tools used to implement MPEG-7 descriptions are the DDL, DSs and Ds. Ds bind a feature to a set of values. DDs are models of the multimedia objects and of the data model of the description. They specify the types of the Ds that can be used in a given description and the relationships between these Ds or between other DSs. The DDL forms a core part of the MPEG-7 standard. It provides the solid descriptive foundation by which users can create their own DSs and D. The DDL defines the syntactic rules to express and combine DSs and Ds. The DDL is a schema language to represent the results of modeling audiovisual data, that is, DSs and Ds. The DDL must satisfy the MPEG-7 DDL requirements. It has to be able to express spatial, temporal, structural and conceptual relationships between the elements of Ds and DSs. It must provide a rich model for link and references between one or more Ds and the data that they describe. In addition, it must be platform and application independent as well as human and machine readable. The general consensus within MPEG-7 is that it should be based on Extensible Markup Language (XML) syntax. The XML schema language has been selected to provide the basis for the DDL [5.157]. As a consequence of this decision, the DDL can be broken down into the following logical normative components: the XML scheme structural language components and XML scheme data type language components.

MPEG-7 Audio

MPEG-7 Audio CD comprises the following technologies: the audio description framework, sound effect description tools, instrument timbre description tools, spoken content description, uniform silence segment and melodic Ds to facilitate query-by-humming. Four sets of audio description tools that roughly represent application areas are integrated in the CD sound effects, musical instrument timbre, spoken content and melodic contour [5.157].

The sound effects Ds and DSs are a collection of tools for indexing and categorization of general sound effects. Support for automatic sound-effect identification and indexing is included, as well as tools for specifying taxonomy of sound classes and tools for specifying an ontology of sound recognizers. Such recognizers may be used to index and segment sound tracks automatically.

Timbre Ds aims to describe perceptual features of instrument sounds. Timbre is currently defined in the literature as the perceptual features that make two sounds having the same pitch and loudness sound different. The aim of the timbre DS is to describe these perceptual features with a reduced set of Ds. The Ds relate to notions such as *attack, brightness* or *richness* of sound.

The Spoken Content DS consists of combined word and phone lattices for each speaker in an audio stream. By combining the lattices, the problem of out-of-vocabulary words is greatly alleviated, and retrieval may still be carried out when the original decoding was in error. The DS can be used for two broad classes of retrieval scenarios: indexing into and retrieval of an audio stream and indexing of multimedia objects annotated with speech.

The Melody Contour DS is a compact representation for melodic information that allows for efficient and robust melodic similarity matching, for example, in query by humming. The Melody Contour DS uses a five-step contour (representing the interval difference between adjacent notes) in which intervals are quantized. The Melody Contour DS also represents basic rhythmic information by storing the number of the nearest whole beats of each note, which can dramatically increase the accuracy of matches to a query.

Several technologies combine to form the low-level audio D framework. One of the foundations is the scale tree, which allows (generally temporal) series of Ds to be represented in a scalable way. The low-level Ds to fit into this foundation include temporal envelope, spectral envelope, harmonicity, spectral centroid and fundamental frequency.

MPEG-7 Visual

MPEG-7 Visual description tools included in CD/XM consist of basic structures and Ds that cover color, texture, shape, motion, localization and other basic visual features. Each category consists of elementary and sophisticated Ds [5.157]. There are five visual-related basic structures: grid layout, time series, multiview, spatial 2D coordinates and temporal interpolation. The grid layout is a splitting of the image into a set of equally sized rectangular regions so that each region can be described separately. Each region of the grid can be described in terms of other Ds such as color texture. Furthermore, the D allows assignment of the subdescriptors to all rectangular areas, as well as to an arbitrary subset of rectangular regions.

The 2D/3D D specifies a structure that combines 2D Ds representing a visual feature of a 3D object seen from different view angles. The D forms a complete 3D view-based representation of the object. Any 2D visual D, such as, for example, contour shape, region shape, color or texture can be used. The 2D/3D D allows the matching of 3D objects by comparing their views, as well as comparing pure 2D views with 3D objects.

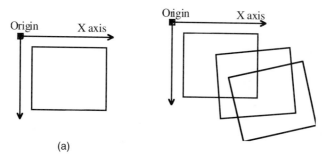

Figure 5.91 Local (a) and integrated coordinate (b) system [5.156]. ©2000 ISO/IEC.

A time series D defines a temporal series of Ds in a video segment and provides image-to-video frame matching and video-frames-to-video-frames matching functionalities. Two types of time series are available: regular time series and irregular time series. In the former, Ds locate regularly (with constant intervals) within a given time span. This enables a simple representation for the application that requires low complexity. On the other hand, Ds locate irregularly (with various intervals) within a given time span in the latter. This enables an efficient representation for the application that has the requirement of narrow transmission bandwidth or low storage capability. These are useful in particular to build Ds that contain the time series of Ds. The spatial 2D coordinates description defines a 2D spatial coordinate system to be used in other Ds/DSs when relevant. It supports two kinds of coordinate systems: local and integrated, as shown in Figure 5.91. In a local coordinate system, all images are mapped to the same position. In an integrated coordinate system, each image (frame) may be mapped to different areas. The integrated coordinate system can be used to represent coordinates as a mosaic of a video shot.

The temporal interoperation D describes a temporal interpolation using connected polynomials. This can be used to approximate multidimensional variable values that change with time, such as an object position in a video. The description size of the temporal interpolation is usually much smaller than describing all values. As an example, 25 real values are represented by 5 linear interpolation functions and 2 quadratic interpolation functions. Real data and interpolation functions are shown in Figure 5.92. The beginning of the temporal interpolation is always aligned to time 0.

The color Ds are color space, color quantization, dominant colors, scalable color, color-structure, color layout and group of frames/group of pictures color descriptor.

The feature is the color space that is to be used in other color-based Ds. In the current descriptions, the following color spaces are supported: RGB, YCrCb, Hue Saturation Value (HSV) and linear transformation matrix with reference to RGB and monochrome.

The color quantization D defines the quantization of a color space, and it supports uniform and nonuniform quantizers as well as lookup tables. Great flexibility is provided for a wide range of applications. For a meaningful application in the context of MPEG-7, this descriptor has to be combined with others to express the meaning of the values of a color histogram.

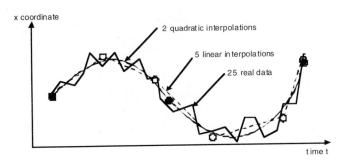

Figure 5.92 Real data and interpolation functions [5.157]. ©2000 ISO/IEC.

The dominant color D is most suitable for representing local (object or image region) features where a small number of colors are enough to characterize the color information in the region of interest. Whole images are also applicable (for example, flag images or color trademark images). Color quantization is used to extract a small number of representing colors in each region/image.

The scalable color D is a color histogram in HSV color space, which is encoded by a Haar transform. Its binary representation is scalable in terms of bit-representation accuracy across a broad range of data rates. The scalable color description is useful for image-to-image matching and retrieval based on a color feature. Retrieval accuracy increases with the number of bits used in the representation.

The color structure D is a color feature D that captures both color content (similar to a color histogram) and information about the structure of this content. Its main functionality is image-to-image matching, and its extended use is for still-image retrieval. The extraction methods embed color structure information into the descriptor by taking into account the colors in a local neighborhood of pixels instead of considering each pixel separately. The color structure D provides additional functionality and improved similarity-based image-retrieval performance for natural images compared to the ordinary color histogram.

The color layout D specifies the spatial distribution of colors for high-speed retrieval and browsing. It targets not only image-to-image matching, and video-clip-to-video-clip matching, but also layout-based retrieval for color, such as sketch-to-image matching which is not supported by other color Ds. This D can be applied either to a whole image or to any part of an image. This D can also be applied to arbitrarily shaped regions [5.160].

The group of frames/group of pictures color D extends the scalable color D that is defined for a still image to the color description of a video segment or a collection of still images.

There are three texture Ds: homogeneous texture, texture browsing and edge histogram. Homogeneous texture has emerged as an important visual primitive for searching and browsing through large collections of similar-looking patterns. An image can be considered as a mosaic of homogeneous textures so that these texture features are associated with the looking patterns. An image can be considered as a mosaic of homogeneous textures so that these texture features associated with the regions can be used to index the image data. The homogeneous texture D provides a precise quantitative description of a texture that can be used for accurate search and retrieval in this respect.

The texture browsing D is useful for representing homogeneous texture for browsing type applications, and requires only 12 bits (maximum). It provides a perceptual characterization of texture, similar to a human characterization, in terms of regularity, coarseness and directionality. The computation of this D proceeds similarly to the homogeneous texture D. First, the image is filtered with a bank of orientation and scale-tuned filters (modeled using Gabor functions); from the filtered outputs, two dominant texture orientations are identified. Three bits are used to represent each of the dominant orientations. This is followed by analyzing the filtered image projections along the dominant orientations to determine the regularity (quantized to 2 bits) and coarseness (92 bits x 2). The second dominant orientation and second scale feature are optional. This D, combined with the homogeneous texture Descriptor, provides a scalable solution to representing homogeneous texture regions in images.

The edge histogram D represents the spatial distribution of five types of edges, namely, four directional edges and one nondirectional edge. Because edges play an important role for image perception, they can retrieve images with similar semantic meaning. Thus, it primarily targets image-to-image matching (by example or by sketch), especially for natural images with nonuniform edge distribution. In this context, the image-retrieval performance can be significantly improved if the edge histogram D is combined with other Ds such as the color histogram D. Besides having the best retrieval, performances considering this D alone are obtained by using the semiglobal and the global histograms generated directly from the edge histogram D as well as the local ones for the matching process.

There are four shape Ds: object region-based shape, contour-based shape, 3D shape and 2D-3D multiple view. The shape of an object may consist of either a single region or a set of regions as well as some holes in the object, as illustrated in Figure 5.93. Because the region-based shape D makes use of all pixels constituting the shape within a frame, it can describe any shape. This includes not only a simple shape with a single connected region as in Figure 5.93 (a) and (b), but also a complex shape that consists of holes in the object or several disjoint regions as illustrated in Figure 5.93 (c), (d) and (e), respectively. The region-based shape D not only can describe such diverse shapes efficiently in a single D, but can also describe robust to minor deformations along the boundary of the object. Figure 5.93 (g), (h) and (i) are very similar shaped images for a cup. The differences are at the handle. Shape (g) has a crack at the lower handle, and the handle in (i) is filled. The region-based shape D considers (g) and (h) similar, but different from (i) because the handle is filled. Figures 5.93 (j through l) show the part of video sequence where two disks are being separated. With the region-based D, they are considered similar. The descriptor is also characterized by its small size, fast extraction time and matching.

The contour-based shape D captures characteristic shape features of an object or region based on its contour. It uses so-called curvature scale-space representation, which captures perceptually meaningful features of the shape.

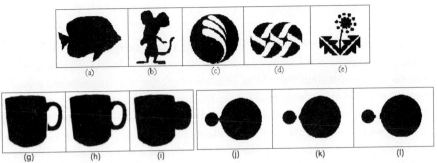

Figure 5.93 Examples of various shapes. The black pixel within the object corresponds to 1 in an image, and the white background corresponds to 0 [5.156]. ©2000 ISO/IEC.

The object-based shape D is based on the curvature scale-space representation of the contour. This representation has a number of important properties:

- It captures characteristic features of the shape very well, enabling similarity-based retrieval.
- It reflects properties of the perception of the human visual system and offers good generalization.
- It is robust to nonrigid motion.
- It is robust to partial occlusion of the shape.
- It is robust to perspective transformations, which result from the changes of the camera parameters and are common in images and video.

Considering the continuous development of multimedia technologies, virtual worlds and augmented reality, 3D contents have become a common feature of today's information systems. Most of the time, 3D information is represented as polygonal meshes. MPEG-4, within the SNHC subgroup, considered this issue and developed technologies for efficient 3D mesh-model coding. Within the framework of the MPEG-7 standard, tools for intelligent content-based access to 3D information are needed. The main MPEG-7 applications targeted here are search and retrieval and browsing of 3D model databases.

The proposed 3D shape D described in detail aims at providing an intrinsic shape description of 3D mesh models. It exploits some local attributes of the 3D surface.

There are four motion Ds: camera motion, object motion trajectory, parametric object motion and motion activity [5.161].

The camera motion D characterizes 3D camera motion parameters. It is based on 3D camera motion parameter information, which can be automatically extracted or generated by capture devices. The camera motion D supports the following well-known basic camera operations: fixed, panning (horizontal rotation), tracking (horizontal transverse movement, also called traveling in the film industry), tilting (vertical rotation), booming (vertical transverse movements), zooming (change of the focal length), dollying (translation along the optical axis) and rolling (rotation around the optical axis).

The motion trajectory of an object is a simple, high-level feature defined as the localization, in time and space, of one representative point of this object. This D shows usefulness for content-based retrieval in object-oriented visual databases. It is also of help in more specific applications. In a given context with a prior knowledge, trajectory can enable many functionalities. In surveillance, alarms can be triggered if some object has a trajectory identified as dangerous (for example, passing through a forbidden area, being unusually quick, and so forth). In sports, specific actions (for example, tennis rallies taking place at the net) can be recognized. Such a description also allows enhancing data interactions/manipulations. Semiautomatic multimedia editing can be performed, or a trajectory can be stretched or shifted to adopt the object motion to any given sequence global context.

The camera motion D is essentially a list of keypoints (x,y,z,t) along with a set of optional interpolating functions that describe the path of the object between keypoints in terms of acceleration. The speed is implicitly known by the keypoints' specification. The keypoints are specified by their time instant and either their 2D or 3D Cartesian coordinates, depending on the intended application. The interpolating functions are defined for each component $x(t)$, $y(t)$ and $z(t)$ independently.

Parametric motion models have been extensively used within various related image-processing and analysis areas, including motion-based segmentation and estimation, global motion estimation, and mosaic and object tracking. Parametric motion models have already been used in MPEG-4 for global motion estimation and compensation and sprite generation. Within the MPEG-7 framework, motion is a highly relevant feature related to the spatiotemporal structure of a video and concerning several MPEG-7 specific applications, such as storage and retrieval of video databases and hyperlinking purposes. Motion is also a crucial feature for some domain-specific applications that have already been considered within the MPEG-7 framework, such as language indexing. The basic underlying principle consists of describing the motion of objects in video sequences as a 2D parametric model. Specifically, affine models include translations, rotations, scaling and combinations of them, planar perspective models make it possible to take into account global deformations associated with perspective projections and quadratic models make it possible to describe more complex movements [5.162].

The parametric model is associated with arbitrary (foreground or background) objects, defined as regions (group of pixels) in the image over a specified time interval. In this way, the object motion is captured in a compact manner as a set of a few parameters. Such an approach leads to a very efficient description of several types of motions, including simple translations, rotation and zooming, or more complex motions, such as combinations of these elementary motions [5.163, 5.164].

Defining appropriate similarity measures between motion models is mandatory for effective motion-based retrieval. It is also necessary for supporting both level queries, useful in query-by-example scenarios, and high-level queries such as "search for object approaching the camera," for "search for object describing a rotational motion," "search for object translating left," and so forth.

A human watching a video or animation sequence perceives it as being a slow sequence, fast-paced sequence, action sequence, and so forth. The activity D captures this intuitive notion of intensity of action or pace of action in a video segment. Examples of high activity include scenes such as goal scoring in a soccer match, scoring in basketball games, a high-speed car chase, and so forth. On the other hand, scenes such as news reader shot, an interview scene, a still shot, and so forth, are perceived as low action shots. Video content in general spans the gamut from high to low activity, so we need a D that enables us to express accurately the activity of a given video sequence/shot and to cover comprehensively the aforementioned gamut. The activity D is useful for applications such as video repurposing, surveillance, fast browsing, dynamic video summarization, content-based query, and so forth. For example, we could slow down the presentation frame rate if the activity D indicates high activity to make the high activity viewable. Another example of an application is finding all the high-action shots in a news video program example, which can be viewed both as browsing and abstraction.

As for localization, there exists the region locator and spatiotemporal locator. The region locator D enables localization of regions within images or frames by specifying them with a brief and scalable representation of a box or a polygon. On the other hand, the spatiotemporal locator describes spatiotemporal regions in a video sequence, such as moving-object regions, and provides localization functionality. An example of spatiotemporal regions is shown in Figure 5.94.

The main application for these locaters is hypermedia, which displays the related information when the designed point is inside the object. Another main application is object retrieval by checking whether the object has passed particular points. This can be used for surveillance. The spatiotemporal locator can describe both spatially connected and nonconnected regions.

Among others Ds, we will mention here the face recognition D [5.165]. It can be used to retrieve face images that match a query face image. The D represents the projection of a face vector onto a set of basis vectors that span the space of possible face vectors. The face recognition feature set is extracted from a normalized face image. This normalized face image contains 56 lines with 46 intensity values in each line. The center of two eyes in each face image are located on the 24th row and the 16th and 31st column for the right and left eye, respectively. This normalized image is then used to extract the 1D face vector that consists of the luminance pixel values from the normalized face image arranged into a 1D vector, using a raster scan starting at the top-left corner of the image and finishing at the bottom-right corner of the image. The face

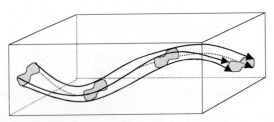

Figure 5.94 An example of a spatiotemporal region.

recognition feature set is then calculated by projecting the 1D face vector onto the space defined by a set of basis vectors.

MPEG-7 MMDSs

The main focus of the MMDS is to standardize a set of description tools (Ds and DSs) dealing with generic and multimedia entities. Generic entities are features that are used in audio, visual and text descriptions and are therefore generic to all media. These are, for example, vector, histogram, time and so forth. Apart from this set of generic description tools, more complex description tools are standardized. They are used whenever more than one medium needs to be described (audio and video). These description tools can be grouped into five different classes according to their functionalities:

- *Content description*—Representation of perceivable information
- *Content management*—Information about media features and the creation of the use of the audiovisual content
- *Content organization*—Representation of the analysis and classification of several audiovisual contents
- *Navigation and access*—Specification of summaries and variations of the audiovisual content
- *User interaction*—Description of user preferences pertaining to the consumption of the multimedia material

MPEG-7 provides DSs for content descriptions. These elements describe the structure (regions, video frames and audio segments) and semantics (objects, events and abstract notions). Structural aspects describe the audiovisual content from the viewpoint of its structure. The structure DSs are organized around segment DSs that represent the spatial, temporal or spatiotemporal structure of the audiovisual content. The segment DS can be organized into a hierarchical structure to produce a table of contents for accessing or an index for searching the audiovisual content. The segments can be further described on the basis of perceptual features using MPEG-7 Ds for color, texture, shape, motion, audio features and semantic information using textual annotations. Conceptual aspects describe the audiovisual content from the viewpoint of real-world semantics and conceptual notions. The semantic DSs involve entities, such as objects, events, abstract concepts and relationships. The structure DSs and semantic DSs are related by a set of links, which allows the audiovisual content to be described on the basis of both content structure and semantics together [5.157].

MPEG-7 also provides DSs for content management. Together, these elements describe different aspects of creation and production, media coding, storage and file formats and content usage. The functionality of each of these classes of DSs is given as creation information, usage information and media description. Creation information describes the creation and production of audiovisual content. The creation information describes the creation and classification of the audiovisual content and other material that is related to the audiovisual content. The creation information provides a title, textual annotation and creation information, such as creators, creation locations and dates. The classification information describes how the audiovisual material is classified into categories, such as gear, subject, purpose, language and so forth. It also provides review and guidance information, such as age classification, subjective review, parental guidance and so forth. Finally, the related material information describes whether other audiovisual material exists that is related to the content being described. Usage information describes the usage information related to the audiovisual content, such as user rights, availability, usage record and financial information. The usage information is typically dynamic in that it is subject to change during the lifetime of the audiovisual content. A media description describes the storage media, such as the compression, coding and storage format of the audiovisual data. The media information DSs identify the master media, which is the original source from which different instances of the audiovisual content are produced. The instances of the audiovisual content are referred to as media profiles. Each media profile is described individually in terms of the encoding parameters, storage media information and location [5.157].

MPEG-7 provides encryption schemes for organizing and modeling collections of audiovisual content, segments, events and/or objects and for describing their common properties. The collections can be further described using different models and statistics in order to characterize the attributes of the collection members [5.157]. The collection structure DS describes collections of audiovisual content or pieces of audiovisual material such as temporal segments of video. The collection structure DS groups the audiovisual content, segments, events or objects into collection clusters and specifies properties that are common to the elements.

MPEG-7 provides DSs that facilitate navigation and access of audiovisual content by specifying summaries, views, partitions and variations of multimedia data. The MPEG-7 summary DSs provide summaries and abstracts of audiovisual content to enable efficient browsing and navigating of audiovisual data. The MPEG-7 space and frequency views provide views of the audiovisual data in the space or frequency domain, which allows multiresolution and progressive access. The MPEG-7 variation DSs specify the relation between different variations of audiovisual material, which allow adaptive selection of the different variations of the content under different terminal and delivery conditions.

Finally, the best set of MPEG-7 DSs deals with user interaction. The user preference information describes user frequencies pertaining to the consumption of the multimedia material. This allows, for example, matching between user preferences and MPEG-7 content descriptions in order to facilitate personalization of audiovisual content access, presentation and consumption. The user preference DS allows the specification of preferences for different types of content and modes of browsing, including context dependencies in terms of time and place. The user preference DS also allows weighting of the relative importance of different preferences. The user preference DS allows the specification of the privacy characteristics of the preferences and whether preferences are subject to update, such as by an agent who automatically learns through interaction with the user.

MPEG-7 Reference Software (XM)

The XM software is the simulation platform for the MPEG-7 Ds, DSs, Coding Schemes (CSs), and DDL. Besides the normative components, the simulation platform also needs some nonnormative components, essentially to execute some procedural code to be executed on the data structure. The data structures and the procedural code together form the applications. The XM *applications* are divided into two types: the server applications and the client applications. The server applications are used to extract the D data from the media data. The extracted D data is coded and written to an MPEG-7 bit stream. To create a server application, the media data has to be specified. This is done by a database file containing all the names of media files for which the DS and DSs should be extracted and stored. The processing of the database file, containing the input file information is performed by the general XM-component/class Image IO. The Image IO class also includes the loading of media data. The loaded media data is stored in an object of a media class. From the media data, the D data is extracted using an extraction class. First, the application has to give the references of the media and D data (addresses of the VOs) to the extraction object. Then, the extraction is performed. The next step is the coding of the D data into its binary representation. Analogous to the extraction, a coding scheme class is used. The output of the coding is put in a bit stream, which is implemented with the class Encoder File IO. The Encoder File IO class gets the bits from the coding scheme and writes them to the MPEG-7 database file [5.157].

The client application performs the search in the MPEG-7 coded database by computing the distance between the query D and all reference Ds of the database. Therefore, one D, the query D, is extracted in the same way as in the server application except that the coding is not performed. The reference Ds are all extracted from the MPEG-7 bit stream. The Decoder FileIO class is used to read the data from the bit-stream file. Each decoded D is stored in a D class object of the array containing all Ds of the database. Before decoding, the references of the Decoder FileIO object and of the D object have to be given to the coding scheme (which includes the encoder and the decoder). Now the query D and the reference Ds are available. The matching is performed by the search object of the search class with one reference descriptor after the other. Prior to the matching, the addresses of both Ds have to be given to the search object. The matching returns the distance between both Ds. For easier processing of the results, the number of the reference D and its distance to the query D are given to an object of the MatchList class. This class stores the numbers and the distances of the best matches. At the end of the client application, the *n* best matches are printed at the output. The filename of a specific reference D can be computed by its number and the database file containing all the media filenames, which was used for the extraction of the database in the server application.

MPEG-7 Conformance

MPEG-7 conformance includes the guidelines and procedures for testing conformance of MPEG-7 implementations.

5.7.4 Applications Enabled by MPEG-7

Many applications, services and domains will benefit from the MPEG-7 standard. One of the first efforts in the development of the standard was collecting requirements from an application point of view. MPEG-7 has helped the requirements process to make the distinction between push and pull applications. Pull roughly refers to interactive queries, and push denotes the situation in which information is available in streamed form, such as in broadcast [5.166].

Currently, the MPEG-7 Applications document distinguishes between these push and pull applications and another category, the so-called specialized professional and control applications, such as biomedical and remote sensing. Some examples of applications are as follows [5.166, 5.167]:

- *Digital libraries*—Many examples fall under this category. They include video libraries, image catalogs, musical dictionaries, future home multimedia databases, and so forth.
- *Multimedia directory services*—An example is The Yellow Pages.
- *Broadcast media selection*—This includes radio channel, TV channel and Internet broadcast search and selection.
- *Multimedia editing*—Personalized electronic news services and media authoring are examples.
- *Universal access to multimedia content*—This is allowing content to scale to access conditions and devices in an intelligent way.
- *Automated processing of multimedia information*—This is an automated analysis of the output from a surveillance camera, where this output has already been segmented and where MPEG-7 descriptions of the objects have been generated.
- The potential applications are spread across the following application domains:
- Film, video and radio archives
- Professional editing and journalism (for example, searching speeches of a certain politician using his name, his voice or his face; real-time markup of incoming programs)
- Education and cultural services (for example, history museums, art galleries, and so forth)
- Tourist information
- Entertainment (for example, searching for a game or karaoke)
- Investigation services (for example, human characteristics recognition and forensics)
- Computer vision and information systems
- Geographical information systems, remote sensing (for example, cartography, ecology, natural resource management, and so forth) and surveillance (for example, traffic control, surface transportation, nondestructive testing in hostile environments, and so forth)
- Biomedical applications

- Shopping (for example, searching for clothing and fashions)
- Architecture, real estate and interior design
- Social (for example, dating services)

These lists are not exhaustive, but they are considered to provide a set of representative requirements for the standard so that other applications are also enabled. New applications are being added to the list as the work continues. An example of an application that has recently been given much attention is that of universal access to multimedia content.

5.8 MPEG-21 Multimedia Framework

The aims of starting MPEG-21 are the following:

- To understand if and how various components fit together
- To discuss which new standards may be required, if gaps in the infrastructure exist and when the above two points have been reached.
- To accomplish the integration of different standards

The digital marketplace, which is founded upon ubiquitous international communication networks such as the Internet, rewrites existing business models for trading physical goods with new models for distributing and trading digital content electronically. In this new marketplace, it is becoming increasingly difficult to separate the different IPRs that are associated with multimedia content.

The latest MPEG project, MPEG-21 Multimedia Framework, has been started with the goal to enable transparent and augmented use of multimedia resources across a wide range of networks and devices.

The basic elements of the framework are the following:

- Digital Items, which are structured digital objects with a standard representation, identification and metadata within the MPEG-21 framework
- Users of all entities that interact in the MPEG-21 environment or make use of MPEG-21 digital items

A digital item is a very broad concept. Let us see how this applies to music compilation. This digital item may be composed of music files or streams, associated photos, videos, animation graphics, lyrics, scores and MIDI files, but could also contain interviews with singers, news related to the song, statements by opinion makers, ratings of agencies, positions in the hit list and so forth. Most important, it could contain navigational information driven by user preferences and, possibly, bargains related to each of these elements.

The meaning of a user in MPEG-21 is very broad and is by no means restricted to the end-user. Therefore, an MPEG-21 user can be anybody who creates content, provides content, archives content, rates content, enhances or delivers content, aggregates content, syndicates content, sells content to end-users, consumes content, subscribes to content, regulates content or facilitates or regulates transactions that occur from any of the previous examples.

The work carried out so far has identified seven technologies that are needed to achieve the MPEG-21 goals. They include the following [5.168]:

- *Digital item declaration*—A uniform and flexible abstraction and interoperable schema for declaring digital items
- *Content representation*—How the data is represented as different media
- *Digital item identification and description*—A framework for identification and description of any entity regardless of its nature, type or granularity
- *Content management and usage*—The provision of interfaces and protocols that enable creation, manipulation, search, access, storage, delivery and (re)use of content across the content distribution and consumption value chain
- *Intellectual property management and protection*—The means to enable content to be persistently and reliably managed and protected across a wide range of networks and devices
- *Terminals and networks*—The ability to provide interoperable and transparent access to content across networks and terminal installations
- *Event reporting*—The metrics and interfaces that enable users to understand precisely the performance of all reportable events within the framework

The relationship of the seven technologies in the framework is depicted in Figure 5.95.

To carry out the necessary tasks, MPEG has identified the following method of work. First, it is necessary to define a framework supporting the vision. This is being done by drafting a TR that describes the complete scope of the multimedia framework and that identifies the critical technologies of the framework. The TR explains how the components of the framework are related and identifies the goals that are not currently filled by existing standardized technologies. The next step is the involvement of other relevant bodies in this effort. This is necessary because some technologies needed for MPEG-21 are not MPEG specific and are better dealt with by other bodies. For each of the technologies that is not yet available, MPEG will do one of the following:

- Develop them if MPEG has the necessary expertise.
- Otherwise, engage other bodies to achieve their development.

Last, the actual integration of the technologies has been performed. The TR was approved in December 2001.

MPEG-21 Multimedia Framework

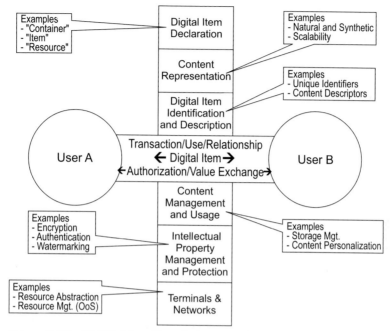

Figure 5.95 MPEG-21 multimedia framework [5.168]. ©2001 ISO/IEC.

5.8.1 Audiovisual Content Representation Issues

MPEG-21 intends to address the traditional audiovisual content representation issues, but with one major difference. Content can no longer be seen as essence (what the user has traditionally consumed) or as metadata (the description of the essence), but as an integrated whole.

A second area of the MPEG-21 standard is explained by the consideration that the way the user interacts with this complex world is difficult to separate from the way the user acquires the right to access the content. MPEG-21 will therefore also identify the interfaces with content access in such a way that content protection—a necessity for the holders of the rights in order to retain control of their assets—is transparent to the end-user.

A third area of the MPEG-21 standard is caused by the pace of advances in content digitization that makes more urgent than ever the need to define a more comprehensive way to identify and describe content as compared to the simple solutions provided by MPEG-2 and MPEG-4. The problem is twofold. There is a need to define content identification and description at the semantic level (what the identification and description code means), but also the way this information is carried by the content itself needs to be defined. The obvious requirement is that content identification and description should be carried out in such a way that it cannot be tampered with. Here we are talking about such technologies as watermarking or fingerprinting techniques.

MPEG-21 presents itself as the enabler of the world of multimedia that has been in the making for some years and to which several bodies, MPEG not being the last, have provided some basic technology elements.

This richness of solutions is part of the problem. In fact, there is no big picture to describe how these elements, either in existence or under development, relate to each other. The first goal of MPEG-21 is then to understand if and how these various components fit together and to see which new standards may be required. This material is in a TR.

MPEG-21 will contain other parts that will be normative. These will address the technologies that are needed in an electronic-trading environment founded upon ubiquitous networks that are encouraging new business models for trading digital content. The digital nature of content will make the distinction between content types less clear as their integration in new products and services makes the traditional boundaries between the different industries less distinct. What will remain unchanged is the value, both commercial and intrinsic, of the digital asset resources, and what will be enhanced are the new possibilities presented by the tools that enable players on the delivery chain to create, collect, package, distribute, store, access, consume, use and reuse content. The underpinning vision is one of a multimedia framework that supports transactions that are interoperable and highly automated to support these new types of commerce. A slogan for this could be "six billion creators, value addresses, publishers, retailers, consumers and resellers all seamlessly connected by networks and all operating through interoperable standards."

5.8.2 Description of a Multimedia Framework Architecture

As previously stated, the functionalities of a multimedia framework architecture have been grouped into architectural elements, even though some overlap exists between these and the purposes of standardization. At first, user requirements are formulated that are relevant for all seven architectural elements:

- To satisfy the experience of all types of users in the multimedia framework through the extension of all existing members of the value chain (creators, rights holders, distributors and consumers of digital items).
- To ensure that the increasing sophistication of technological solutions does not undermine the user experience.
- To achieve interoperability of systems through the integration of the components of the multimedia framework. A component is the binding of a resource to all of its relevant descriptors.
- To provide the means to protect the intellectual properties of all categories of users.
- To ensure that the privacy of users will be respected.

MPEG-21 Digital Item Declaration

The goal is to establish a uniform and flexible abstraction and interoperable schema for defining digital items. A digital item is a structured digital object with a standard representation, identification and description within the MPEG-21 framework. This entity is also the fundamental unit of distribution and transaction within the MPEG-21 framework.

Content Representation
MPEG-21 provides content representation technology to represent efficiently any content of all the relevant data types of natural and synthetic origin, or any combination thereof, in a scalable and error-resilient way. The various elements in a multimedia scene will be independently accessible, synchronizable and multiplexed and will allow various types of interaction.

Digital Item Identification and Description
By digital item identification, we mean a token that is uniquely designated and enables the recognition of a digital item, its organization and attributes. The information used to describe a digital item is called a digital item description. The framework for the identification and description is interoperable and integrated to provide the following:

- Accuracy, reliability and uniqueness of identification
- Seamless identification of any entity regardless of its nature or type of granularity
- Persistent and efficient methods for the association of identifiers with digital items
- Security and integrity of identification and description that survive all kinds of manipulations and alterations
- Automated processing of rights transactions and content location, retrieval and acquisition

Content Management and Usage
The MPEG-21 multimedia framework should provide interfaces and protocols that enable creation, manipulation, search, access, storage, delivery and (re)use of content (which can be any media data and descriptive data) across the content distribution and consumption value chain, with emphasis on improving the interaction model for users with personalization and content management. The previous should be supported both when the user is performing these functions and when the functions are delegated to nonhuman entities (such as agents). In this context, content management should not be understood as managing the rights of the content.

Intellectual Property Management and Protection
The MPEG-21 multimedia framework should provide a multimedia digital rights management framework that does the following:

- Enables all users to express their rights and interests in, and agreements related to, digital items and to have assurance that those rights, interests and agreements will be persistently and reliably managed and protected across a wide range of networks and devices.
- Enables, to the extent possible, the capture, codification, dissemination and reflection of updates of relevant legislation, regulations, agreements and cultural norms that

together create the setting and generally accepted societal platform for commerce involving digital items.
- Provides, to the extent possible, a uniform technical and organizational foundation for domain governance organizations that govern (on behalf of all users of digital items) the behavior of devices, systems and applications involved in interacting with digital items and services that provide transactional support within the MPEG-21 framework.

Terminals and Networks

With terminals and networks, we achieve interoperable transport access to distributed advanced multimedia content by scheduling users from network and terminal installation, management and implementation issues. This will enable the provision of network and terminal resources on demand to form user communities where multimedia content can be created and shared. This is always with the agreed/contracted quality, reliability and flexibility, allowing the multimedia applications to connect arbitrary sets of users such that the quality of the user experience will be guaranteed.

This implies the following at a minimum:

- Networks should provide content-transport functions according to a QoS contract established between the user and the network.
- Terminals and networks should provide scalable execution functions as requested by content.
- Access to network and terminal resources will happen through interfaces.

Event Reporting

MPEG-21 should provide metrics and interfaces that enable users to understand precisely the performance of all reportable events within the framework. Such event reporting then provides users a means of acting on specific interactions, as well as enabling a vast set of out-of-scope processes, frameworks and models to interoperate with MPEG-21. Event reporting creates a standardized set of metrics and interfaces with which to describe the temporally unique events and interactions within MPEG-21.

5.8.3 Requirements for Digital Item Declaration

Within any system (such as MPEG-21) that proposes to facilitate a wide range of actions involving digital items, there is a strong need for a concrete representation of any individual digital item. Clearly, there are many kinds of content, and probably just as many possible ways of representing it. This presents a strong challenge to design a powerful and flexible model for digital items that accommodate the many types of content, as well as any and all new forms that the content may assume in the future. Such a model is truly useful only if it yields a schema that can be used to represent unambiguously and to communicate interoperably about any digital items defined within the model.

The framework must support the following global requirements for a digital item declaration [5.169]:

- Hierarchies of containers and digital items must be capable of being efficiently searched and traversed. A container represents a potentially hierarchical structure that allows items to be grouped.
- Media resources and the digital item declaration are fully separable, meaning that the representation of an item can be communicated and processed in the absence of local copies of the associated media resources.
- The framework must enable the robust processing, validation and manipulation of digital items and containers.
- Identification and revision management of elements in a digital item or container must be supportable in an open and extensible manner.
- Digital items and containers may contain individual elements that are associated with multiple locations within the definition/hierarchy.

The framework must support a definition of containers that does the following:

- Support the unrestricted construction and description of hierarchical groupings of digital items.
- Support the metaphor of shelves for the organization and management of collections of digital items.
- Support the metaphor of packages for the transfer and delivery of digital items.

The framework must support a definition of digital items that supports the construction of compilations of items that fully preserves the structure and properties of the subitems.

The framework must support definitions of components that do the following:

- Associate a media resource of any type or format with an item.
- Provide an externally identifiable target for links from a media resource.

The framework must support a definition of descriptors that does the following:
- Associate a component or statement of any description scheme with any element. A statement is a literal textual value that contains information, but not an asset.
- Is capable of being described by other descriptors.

Also, the framework must support the configuration and atomization of digital items that enable a flexible mechanism for defining decision trees.

5.9 ITU-T Standardization of Audiovisual Communication Systems

The ITU is involved in the preparation of standards for the Global Information Infrastructure (GII). It is a focal point of the converging technologies and industries and will provide various multimedia applications. When considering the transition from national to GII, it is natural that the attention of those involved with telecommunications, such as entrepreneurs, service providers and regulators, should be focused on questions of technology, market opportunity, industry structure, investment risks and potential returns. Global interconnectivity and interoperability are some of the key issues of GII and require global standards. There have been many activities on GII-related standardization since the GII was first advocated in 1994 [5.170]. The standardization activity on GII in ITU-T was initiated by ITU-T SG13 in July 1995. Focusing on standardization, the ITU can be seen as a GII facilitator by providing global interconnectivity and interoperability through its standards. It is generally accepted that the essential global standards must address market needs, must not impair or restrict the creativity of equipment manufacturers, information providers and service providers and must provide a realistic and stable base for the envisioned information infrastructure. Global specifications are universally seen as necessary for a timely, successful GII. Such standards can achieve applicability and interoperability and can meet the market requirements for cost effectiveness, QoS, and support for cultural diversity. For the development of GII, the ITU is ideally placed to be an integrator, linking nonindustrial and developed countries.

5.9.1 ITU-T Standardization Process

Without a common language that both the transmitter and the receiver understand, communication is impossible. For multimedia communication that involves transmission of video data, standards play an even more important role. A video coding standard not only has to specify a common language, formally known as the bit-stream syntax, but the language also has to be efficient. Efficiency has two aspects. One is that the standard has to support a good compression algorithm that brings down the bandwidth requirement for transmitting the video data. The other is that the standard has to allow efficient implementation of the encoder and the decoder, that is, the complexity of the compression algorithm has to be as low as possible. Suppose some multimedia content needs to be transmitted from a source to a destination. The success of the communication is mainly determined by whether the source and the destination understand the same language. Adoption of standards by equipment manufacturers and service providers results in higher volume and hence lowers the cost. In addition, it offers consumers more freedom of choice among manufacturers and therefore is highly welcomed by the consumers.

There are two major types of standards. Industrial or commercial standards are mainly defined by mutual agreement among a number of companies. Sometimes these standards can become very popular in the market, become the de facto standards and are widely accepted by the other companies. The other type is called a voluntary standard that is defined by volunteers in open committees. The agreement on these standards has to be based on the consensus of all committee members. These standards are usually driven by the market needs. At the same time, they need to stay ahead of the development of technologies.

For multimedia communication, there are two major standard organizations: ITU-T, and the ISO. For example, recent video-coding standards defined by these two organizations are summarized in Table 5.23. These standards differ mainly in the operating bit rates due to the applications that they were originally designed for, but all standards can essentially be used for all applications at a wide range of bit rates. In terms of coding algorithms, all standards follow a similar framework.

Table 5.23 Video-coding standards.

Standards organization	Video-coding standard	Typical range of bit rates	Typical applications
ITU-T	H.261	px64 Kb/s, p=1,2,...,30	ISDN video phone
ISO	IS 11172-2 MPEG-1 Video	1.2 Mb/s	CD-ROM
ISO	IS 13818-2 MPEG-2 Video	4-80 Mb/s	SDTV, HDTV
ITU-T	H.263	64 Kb/s or below	PSTN video phone
ISO	CD 14496-2 MPEG-4 Video	24-1024 Kb/s	Interactive audio/video
ITU-T	H.263 Version 2	< 64 Kb/s	PSTN video phone
ITU-T	H.26L	< 64 Kb/s	Network-friendly packet-based video

In the past, most video compression and coding standards were developed with a specific application and networking infrastructure in mind. For example, ITU-T recommendation H.261 was optimized for use with interactive audiovisual communication equipment, for example, a videophone [5.171], and in conjunction with the H.320 series of recommendations as multiplex and control protocols on top of ISDN [5.173]. Consequently, the H.261 designers made various design choices that limit the applicability of H.261 to this particular environment. The original H.263 was developed for video-compression rates below 64 Kb/s. This was the first international standard for video compression that would permit video communications at such low rates [5.174, 5.175]. After H.263 was completed, it become apparent that incremental changes could be made to H.263 that would visibly improve its compression performance. It was thus decided in 1996 that a revision to H.263 would be created that incorporated these incremental improvements. ITU-T recommendation H.263 Version 2 (abbreviated as H.263+) is the very first international standard in the area of video coding that is specifically designed to support the full range of both circuit-switched and packet-switched networks [5.176, 5.177, 5.178]. H.263+ contains functionalities that improve the quality of video transmission in error-prone environments and nonguaranteed QoS networks. H.26L is an ongoing standard activity that is searching for advanced coding techniques that can be fundamentally different from H.263.

A number of trends and directions have been identified that need to be taken into account by both the ITU-T and other standards development organizations when developing their programs, program priorities and liaison or partnership arrangements. Telecommunication networks are currently providing voice and data services worldwide with a high level of reliability and defined QoS and are based on different network technologies with interworking among them. Extension of the networks to include broadband capabilities is based on ATM technology. ATM is also being enhanced to provide not only for connection-oriented services, but also to meet requirements of connectionless network capabilities and services supported by these capabilities. Networks based on IPs provide a platform that allows users connected to different network infrastructures to have a common set of applications and to exchange data with an undefined QoS. The IP suite is evolving to include voice, data and video applications with defined QoS. Additionally, terrestrial radio, cable and satellite networks are providing local broadcast entertainment services and are also evolving to provide interactive voice, data and video services.

5.9.2 Audiovisual Systems (H.310, H.320, H.321, H.322, H.323, and H.324)

Audiovisual services provide real-time communication of speech together with visual information between two or more end-users. The visual information is typically moving pictures, but may be still pictures, graphics or any other form. The ITU-T Study Group 15 has been standardizing the audiovisual communication systems in various network environments. The first set of such standards, called *Recommendations* in the ITU-T, was formally established in December 1990 for narrow-band ISDN (N-ISDN), which provides digital channels of 64 Kb/s (B channel), 384 Kb/s (H0 channel), and 1,536/1,920 Kb/s (H11/H12 channel) [5.171]. Recommendation H.320 describes a total system stipulating several other recommendations to which respective constituent elements, such as audio coding, video coding, multimedia multiplexing and system control should conform. Its major target applications are video conferencing and videotelephony, but other applications are not excluded [5.173]. Since then, multipoint and security enhancements of N-ISDN systems have been developed [5.179]. In parallel with this, a new standardization activity was initiated in July 1990 towards broadband and high-quality audiovisual systems by forming an experts group. The first videoconferencing standard to emerge was H.320, which was defined for a circuit-switched narrow-band ISDN environment at bandwidths ranging from 64 Kb/s to more than 2 Mb/s. The ITU-T standards are also addressing videoconferencing in different network environments, such as H.324 for POTS [5.180], the H.323 LANs [5.181] and H.310 suite for ATM [5.182].

Table 5.24 shows the network environments for which audiovisual communication systems have been defined and gives the numbers of the ITU-T recommendations that specify these

ITU-T Standardization of Audiovisual Communication Systems

Table 5.24 Audiovisual communication systems in various network environments [5.171].

Network	GSTN	N-ISDN	Guaranteed QoS LANs	Nonguaranteed QoS LANs	ATM (B-ISDN, ATM LANs)
Channel capacity	Up to 28.8 Kb/s	Up to 1,536 or 1,920 Kb/s	Up to 6/16 Mb/s	Up to 10/100 Mb/s	Up to 600 Mb/s
Characteristics	Ubiquitous	Circuit based Existing	Similar to N-ISDN	Packet-loss prone	Future basic network
Total system (date of first approval)	H.324 (03/96)	H.320 (12/90)	H.322 (03/96)	H.323 (11/96)	H.310 (11/96) H.321 (03/96)
Audio coding	G.723.1	G.711 G.722 G.728	G.711 G.722 G.728	G.711 G.722 G.723.1 G.728	G.711 G.722 G.728 ISO/IEC 11172-3
Video coding	H.261 H.263	H.261	H.261	H.261 H.263	H.261 H.262
Data	T.120, etc.	T.120, etc.	T.120, etc.	T.120, etc.	T.120, etc.
System control	H.245	H.242	H.242	H.245	H.242 (for H.321) H.245 (for native H.310)
Multimedia multiplex and synchronization	H.223	H.221	H.221	H.225.0 TCP/IP etc.	H.222.0 H.222.1
Call setup signaling	National standards	Q.931	Q.931	Q.931 H.225.0	Q.931

©1997 IEEE.

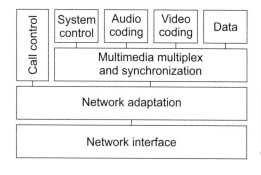

Figure 5.96 General protocol stack of H-series audiovisual communication terminal [5.171]. ©1997 IEEE.

systems, as well as their constituent elements. A general protocol stack model is shown in Figure 5.96. Note that system control and data may be directly on the top of network adaptation.

The General Switched Telephone Network (GSTN) system has been studied by a separate experts group for low bit-rate coding. It has produced not only the total system recommendation H.324 [5.180], but also the improved video-coding recommendation H.263 [5.175], the improved audio coding recommendation G.723.1 [5.183] and the multimedia multiplexing scheme defined in recommendation H.223 [5.184].

H.320 Standard

For all of the ITU standards, interoperability with the H.320 standard is mandatory. However, this interoperability is achieved through a gateway that, in some cases, must perform translations between different signaling protocols, different compression standards and different multiplexing schemes. The variations of signaling, compression and multiplexing for the various standards are due to the differing characteristics of the underlying networks to which each standard applies. Various aspects of the H.320 standard are clearly reusable in these network environments, but other aspects are not. For example, the audio- and video-compression algorithms or variations of them are being adopted by most of the new standards. However, different multiplexing schemes are being developed to better suit each network. One aspect of the H.320 standard is the centralized approach to multiparty conferencing. The H.320 standard defines a central conference server called a Multipoint Control Unit (MCU) to enable multiparty calls [5.185]. Each participant in the call contacts the MCU directly, which then controls the conference. This paradigm is clearly suited to the point-to-point connectivity nature of ISDN networks. Although not available everywhere, the ISDN is today's most widely used circuit-switched network for interactive multimedia communication. Video-telephone and videoconferencing systems based on the H.320 family of ITU-T recommendations are still the only affordable, medium- to high-quality videoconferencing solutions for most business users.

Standards for Audiovisual Services across ATM H.310 and H.321

The next generation public network is envisaged as, BISDN which is based on ATM [5.186]. Customer premises networks are also moving to ATM so that seamless network services can be provided [5.187]. ATM networks provide many opportunities for new and improved services, but they also pose new problems that must be solved before these services can be offered [5.6]. ATM network characteristics are summarized in Table 5.25.

The objective of standardizing audiovisual communication systems in ATM environments was to allow interoperability among different systems and interoperability with terminals connected to other networks while taking advantage of the opportunities and alleviating the limitations of ATM. In particular, it was an essential requirement that the new generation systems should interwork with the existing ones, that is, the ATM audiovisual systems should be able to interwork with H.320 systems situated in the N-ISDN. This could be achieved in many ways, including switching the elementary media coding to a common coding (H.261, G.711, and so forth) and using an intermediate gateway.

Table 5.25 ATM network characteristics.

Properties	Characteristics
Opportunities	Availability of high bandwidths Flexibility in bandwidth usage Variable bit-rate capability Service integration Use of Cell Loss Priority (CLP) Multipoint distribution in the network Flexibility in multimedia multiplexing or multiple connections
Limitations	Cell loss Cell delay variation (jitter) Packetization delay Usage parameter control (peak and/or average rates)

One of the greatest opportunities of BISDN is the high bandwidth available, which may be up to several hundred Mb/s compared to only 1.5 or 2 Mb/s for N-ISDN. Generally, higher bandwidth brings higher quality.

A video-coding standard, ITU-T recommendation H.262 has been established that gives pictures of broadcast television quality at around 5 to 10 Mb/s. High-quality stereo sound with subjective quality equal or close to that of a CD is obained at 384 Kb/s or lower bit rates by using MPEG-1 Audio [5.13]. Its extension to multichannel and lower sampling frequencies is also available as MPEG-2 Audio [5.16]. Hence, the ATM audiovisual systems should be able to realize high quality.

Another outstanding feature of the ATM network is its capability to achieve service integration. Cells and virtual channels can transport any type of information media after they are digitized and packetized. Different types of services can share the same network. This is seen by the user as an opportunity to access a number of different services through a single terminal. Hence, the ATM audiovisual communication systems should be able to cover as many applications as possible in a harmonized way. Possible applications are as follows:

- Conversational services (videoconferencing, videotelephony and distance learning)
- Distributive services (TV broadcasting and intracompany TV)
- Retrieval services (video on demand and network database)
- Messaging services (video mail)
- Video transmission (point-to-point transport of video programs)
- Video surveillance (road traffic monitoring)

The high bandwidth of the ATM network also provides a capability of low delay for conversational services. N-ISDN audiovisual systems use H.261 video coding, which incurs a buffering delay of at least four times the frame period (133 ms) plus any display delay due to picture skipping. The ATM audiovisual systems should significantly improve the end-to-end delay, so a target of less than about 150 ms has been set. This value corresponds to the acceptable-for-most user-applications level of specification in ITU-T recommendation G.114 for one-way transmission time [5.188].

To meet the previous requirements, ITU-T SG15 has developed the following two recommendations for audiovisual communication systems in ATM environments:

- H.321—Adaptation of H.320 visual telephone terminals to BISDN environments [5.189]
- H.310—Broadband audiovisual communication systems and terminals [5.182]

Recommendation H.321 specifies the adaptation of H.320 visual telephone terminals to B-ISDN environments, thus satisfying the requirement that ATM terminals should interwork with those connected to N-ISDN. Recommendation H.310 includes the H.320/H.321 interoperation mode, but also defines a native mode, which takes advantage of the opportunities provided by ATM, to provide higher quality audiovisual communication systems. Although N-ISDN allows only a small number of transfer rates, quantized to being multiples of 64, 384, 1,536 and 1,920 Kb/s, BISDN allows a wide, almost infinite, range of transfer rates. This provides an obvious benefit of flexibility, but also causes a potential interoperability problem. It may happen that one terminal supports a group of transfer rates and another supports a different group of transfer rates with no value in common. Recommendation H.310 solves this problem by first defining the transfer rate to be a multiple of 64 Kb/s and then by mandating the two rates: 96x64=6,114 Kb/s and 144x64=9,216 Kb/s. Other optional transfer rates can be negotiated through the H.245 capability exchange procedures [5.190].

The two mandatory rates correspond to the H.262 Main profile at Main level medium-quality services and high-quality services, respectively [5.171].

Audiovisual communication requires the following phases to be completed before audiovisual communication can take place:

- Initial VC setup
- Capability exchange to identify the available common modes of operation
- Additional VC setup or bandwidth modification of the initial VC
- Logical channel establishment for audiovisual and data communication

Due to the wide flexibility of the ATM networks, the following two solutions have been considered for the basic model of H.310 (Figure 5.97):

- *Solution One*—A VC is initially used for only H.245 signaling, and audio is subsequently adjusted to accommodate audiovisual signals as well.
- *Solution Two*—The initial VC is used for H.245 signaling, after which another VC is set up for audiovisual signals.

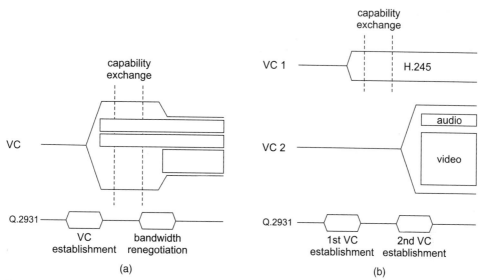

Figure 5.97 Two alternatives for H.310 start-up procedures (a) Solution One and (b) Solution Two [5.171]. ©1997 IEEE.

Solution One has two advantages. It provides audio immediately when the connection is made and may offer cheaper connection charges. However, it has many disadvantages. Two compatible terminals may not always connect at the first attempt if the calling terminal selects the wrong AAL, and the bandwidth may have to be renegotiated if the initial bandwidth is not equal to that actually required, which is a likely problem in asymmetric audiovisual channel configurations.

Solution Two, which avoids the disadvantages of Solution One by negotiating the capabilities of the second VC by running H.245 on the first VC, was adopted as the basic mode of operation of H.310. The initial VC is symmetrical with 64 Kb/s bandwidth and AAL5. Solution One may be added in the future if it is considered beneficial.

All H.310 terminals, when operating in the natural ATM mode, use the H.245 control protocol. Other audiovisual devices that use ATM transport may not use H.245.

Standard H.322–Guaranteed QoS LAN Systems

The proposal that the ITU should have a recommendation covering the provision for LANs and for videotelephony and video conferencing facilities, equivalent to those specified by recommendation H.320 for N-ISDN, was made at the September 1993 meeting of the SG15 Experts Group for ATM Video Coding. At a subsequent meeting the Working Party mandated the Experts Group to begin studies and to produce a draft recommendation under the working number H.322. Even before the ITU-T had commenced its work on H.327, the Institute for Electrical and Electronic Engineers (IEEE) had been developing this new LAN standard. Originally, it was known as Iso-

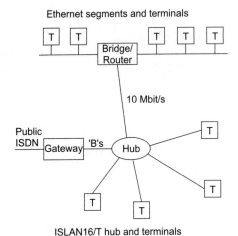

Figure 5.98 Typical configuration of H.322 using ISLAN 16-T [5.171]. ©1997 IEEE.

chronous Ethernet, frequently abbreviated to ISO-Enet, but was later renamed ISLAN 16-T [5.191]. Typical configuration of H.322 using ISLAN 16-T is shown in Figure 5.98.

ISLAN 16-T can be considered as an upgrade to the conventional 10 Mb/s Ethernet. It does require that terminals are star-wired to a central hub. However, the majority of recent Ethernet installations has this physical configuration even though the logical one is the traditional linear segment. As the 16 in its name infers, the bit rate is 16 Mb/s and, because each user has an individual link to the hub, all of this 16 Mb/s is available to that user and is not shared with others. It can be configured in two modes.

In the first mode, the 16 Mb/s is divided into a 10 Mb/s portion and a 6 Mb/s portion. The former is used with conventional Ethernet protocols and therefore gives complete compatibility with existing Ethernet software. The user can continue to access the file servers and to communicate with other Ethernet terminals as before. The remaining 6 Mb/s appears as 96 B channels, each using 64 Kb/s, plus one signaling channel.

In the second mode, the entire 16 Mb/s appears as B channels plus a signaling channel. Initially, this mode is unlikely to be used often because it offers no compatibility with Ethernet, and few applications need more than 6 Mb/s. However, this mode is well matched to H.262, which has a maximum rate of 15 Mb/s, and this may be important in the future.

A connection between the ISLAN 16-T hub and the existing Ethernet carries the 10 Mb/s traffic. To access a wider population of H.320 terminals, some B channels are connected between the new hub and a gateway to the ISDN.

The H.322 gateway unit need not carry out any function that is specific to the H.320 signals passing through it. Consequently, it is not restricted to serving only H.322 terminals on its LAN side, but can equally handle generic ISDN terminals. It is expected that the first H.322 gateways will have N-ISDN interfaces, but, as BISDN becomes widespread, interfaces to that will become more prevalent. Such a configuration will then permit good- to high-quality video at 2 Mb/s to be sent to and received from remote locations across the public BISDN. However,

H.322 does not mandate any minimum number of simultaneous calls that it can support or any minimum number of simultaneous calls to or from the ISDN.

To provide multipoint calls, an MCU can be added. Placing this at the gateway gives much flexibility over the balance between the number of participants on-LAN and off-LAN. Such an MCU can be of the switched type as exemplified by recommendation H.231 or the continuous presence type [5.192].

It is also possible to provide unidirectional audiovisual services to multiple recipients as in recommendation H.331 [5.193]. Such MCU and broadcast facilities may be provided by additional units separate from the gateway, or they may be combined into the gateway itself. H.322 mandates neither their presence nor the method of implementation.

ITU-T H.323 Standard

This recommendation specifies equipment and systems for visual telephony on nonguaranteed QoS LANs. It covers those situations where the transmission path includes one or more LANs, which may not provide a guaranteed QoS equivalent to that of N-ISDN. Examples of this type of LAN include Ethernet, Fast Ethernet, Token Ring [5.194] and Fiber Distributed Data Service (FDDS) [5.195].

The primary design considerations in the development of H.323 were the following:

- Interoperability, especially with N-ISDN and H.320
- Control of access to the LAN to avoid congestion
- Multipoint call models
- Scalability from small- to medium-sized networks.

H.323 terminals may be used in multipoint configurations and may interwork with H.310 terminals on BISDN, H.320 terminals on N-ISDN, H.321 terminals on BISDN, H.322 terminals on guaranteed QoS LANs, H.324 terminals on GSTN and wireless networks and V.70 terminals on GSTN [5.196].

Provisions have been made to provide a gatekeeper that performs admission control for H.323 terminals within its zone that are attempting to gain access to the LAN. The criteria used by the gatekeeper to allow such access are nonstandardized. In addition, the gatekeeper can limit the bandwidth that a terminal uses and can control the call model used, which also affects bandwidth usage.

The zone of the H.323 system is presented in Figure 5.99. The scope of H.323 does not include the LAN itself or the transport layer that may be used to connect various LANs

Only elements needed for interaction with the Switched Circuit Network (SCN) are within the scope of H.323. The combination of the H.323 Gateway, the H.323 terminal and the out-of-scope LAN appears on the SCN as an H.320, H.310 or H.324 terminal. Recommendation H.323 describes the total system and its components, including terminals, gateways, gatekeepers, multipoint controllers, multipoint processors and MCUs. Recommendation H.225.0 describes the underlying protocols used for media packetization and control in the H.323 system [5.197].

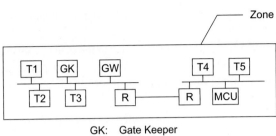

GK: Gate Keeper
GW: Gate Way
MCU: Multipoint Control Unit
R: Router
T: Terminal

Figure 5.99 Zone of the H.323 system [5.171]. ©1997 IEEE.

A zone is the collection of all terminals, gateways, and multipoint control units (MCUs) managed by a single gatekeeper. A zone includes at least one terminal and may or may not include gateways or MCUs. A zone has one and only one gatekeeper. Multiple LAN segments that are connected using routers or other devices may be in the same zone.

Example 5.17 Figure 5.100 is an example of the H.323 terminal. It shows the user equipment interfaces, video codec, audio codec, telematic equipment, H.255.0 layer, system control functions and the interface to the LAN. All H.323 terminals have a system control unit, H.225.0 layer, network interface and an audio codec unit. The video codec unit and user data applications are optional.

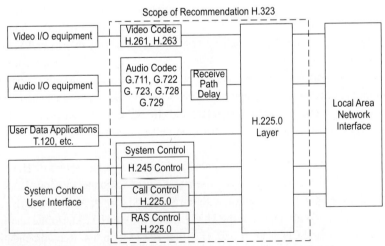

Figure 5.100 H.323 terminal equipment [5.171]. ©1997 IEEE.

H.324 Standard

If the telephony network is to be used for multimedia conferencing, most commercial products rely on the H.324 protocol hierarchy to ensure interoperability. The ITU-T recommendation H.324 entitled "Terminals for low bit rate multimedia communications" [5.180] provides an overview of PSTN multimedia terminals and references and all other ITU-T recommendations that are necessary to build such a terminal in a standard conformant way. For this reason, the standardization community refers to an H.324 family of recommendations.

H.324-based systems are free to negotiate protocol parameter values, such as packet size and allows round-trip delay. Such values, and especially the observed round-trip delay, substantially impact the performance of the error-resilience modes. To minimize packetization delay, packets with small sizes are usually negotiated. A typical payload size for Adaptation Layer (AL3) video packets in H.324 systems is 128 bytes. This is a number often used in various interoperability tests of commercial H.324 systems [5.198].

Unfortunately, many of today's H.324 systems have a significant end-to-end delay, due mainly to the integration of H.324 protocol mechanisms within a PC operating system environment, which is usually not optimized for real-time applications. The typical end-to-end delay for video in such a system can be well above 0.5 s. Therefore, AL3 retransmission should be avoided if interactive use, and thus a reasonable delay, is to be achieved. Because the special modem protocols used for H.324 do not perform any of their own error control or correction, significantly high bit-error rates can occur. Modems allow us, however, to optimize the trade-off between error rates and bit rates. System designers use this mechanism to gain optimal performance based on their design considerations.

5.9.3 Video-Coding Standards (H.261, H.263 and H.26L)

There are two approaches to understanding video-coding standards. One approach is to focus on the bit-stream syntax and to try to understand what each layer of the syntax represents and what each bit in the bit stream indicates. This approach is very important for manufacturers who need to build equipment that is compliant to the standard. The other approach is to focus on coding algorithms that can be used to generate standard-compliant bit streams and to try to understand that each component does not specify any encoding algorithms. The latter approach provides a better understanding of video-coding techniques as a whole, not just the standard bit-stream syntax.

H.261 Standard

This standard is defined by the ITU-T Study Group (SG) 15 for video telephony and video-conferencing applications [5.199]. After a reorganization within ITU-T in early 1997, SG 16 is the new group for video-coding standards. H.261 emphasizes low bit rates and the low coding delay. It was originated in 1984 and intended to be used for audiovisual services at bit rates around p×384 Kb/s, where p is between 1 and 5. In 1988, the focus shifted, and it was decided to aim at bit rates around p×64 Kb/s, where p is from 1 to 30. Therefore, H.261 also has an informal name called px64. H.261 was approved in December 1990. The coding algorithm used in H.261 is basically a hybrid of motion compensation to remove temporal redundancy and transform coding to reduce spatial redundancy. Such a framework forms the basis of all video-coding standards that were developed later. Therefore, H.261 has very significant influence on many other existing and evolving video-encoding standards.

Digital video is composed of a sequence of pictures, or frames, that occur at a certain rate. For H.261, the frame rate is specified to be 30,000/1,001 (approximately 29.97) pictures per second. Each picture is composed of a number of samples. These samples are often referred to as pixels (picture elements) or pels. For a video-coding standard, it is important to understand the picture sizes that the standard applies to and the position of the samples. H.261 is designed to deal with two picture formats: the CIF and QCIF. In the still-image mode as defined in H.261, four times the currently transmitted video is used. For example, if the video format is CIF, the corresponding still-image format is 4 CIF. Table 5.26 summarizes a variety of picture formats supported by H.261 and H.263.

Table 5.26 A variety of picture formats supported by H.261 and H.263 [5.176].

Parameter	Sub-QCIF	QCIF	CIF	4CIF	16CIF
Number of pixels per line	128	176	352	704	1,408
Number of lines	96	144	288	576	1,152
Uncompressed bit rates	4.4 Mb/s	9.1 Mb/s	37 Mb/s	146 Mb/s	584 Mb/s

©1997 ITU-T.

H.261 is designed for videotelephony and videoconferencing, in which typical source material is composed of scenes of talking persons, so-called head and shoulder sequences, rather than general TV programs that contain a lot of motion and scene changes.

In H.261, each sample contains a luminance component called Y and two chrominance components called Cb and Cr. In particular, black is represented by $Y=16$, white is represented by $Y=235$, and the range of Cb and Cr is between 16 and 240, with 128 representing color difference (that is, gray). A particular format, as shown in Table 5.26, defines the size of the image, hence the resolution of the Y pels. The chrominance pels, however, typically have a lower resolution than the luminance pels in order to take advantage of the fact that human eyes are less sensitive to chrominance than luminance. In H.261, the Cb and Cr pels are specified to have half the resolution, both horizontally and vertically, of that of the Y pels. This is commonly referred to as 4:2:0 format. Each Cb or Cr pel lies in the center of four neighboring Y pels, as shown in Figure 5.101. Note that block edges lie in-between rows or columns of Y pels [5.199].

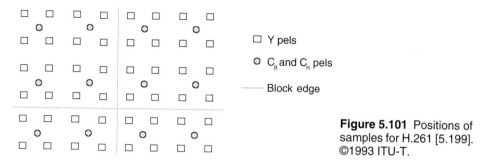

Figure 5.101 Positions of samples for H.261 [5.199]. ©1993 ITU-T.

Typically, we do not code an entire picture all at once. Instead, it is divided into blocks that are processed one by one both by the encoder and the decoder in a scan order as shown in Figure 5.102. This approach is often referred to as block-based coding.

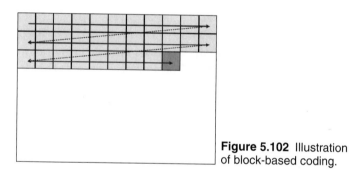

Figure 5.102 Illustration of block-based coding.

In H.261, a block is defined as a group of 8x8 pels. Because of the downsampling in the chrominance components as mentioned earlier, one block of Cb pels and one block of Cr pels correspond to four blocks of Y pels. The collection of these six blocks is called a Macroblock (MB), as shown in Figure 5.103 with the order of blocks marked as 1 to 6. An MB is treated as one unit in the coding process.

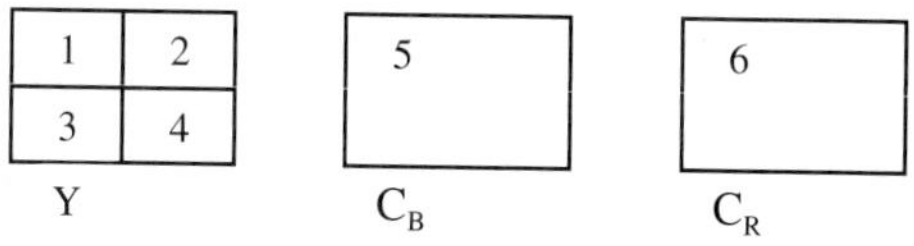

Figure 5.103 A macroblock [5.199]. ©1993 ITU-T.

A number of MBs are grouped together and called a Group of Blocks (GOB). For H.261, a GOB contains 33 MBs, as shown in Figure 5.104. The resulting structures for a picture, in the CIF case and the QCIF case, are shown in Figure 5.105 [5.199].

1	2	3	4	5	6	7	8	9	10	11
12	13	14	15	16	17	18	19	20	21	22
23	24	25	26	27	28	29	30	31	32	33

Figure 5.104 GOB for H.261 [5.199]. ©1993 ITU-T.

GOB 1	GOB 2
GOB 3	GOB 4
GOB 5	GOB 6
GOB 7	GOB 8
GOB 9	GOB 10
GOB 11	GOB 12

CIF

GOB 1
GOB 3
GOB 5

QCIF

Figure 5.105 GOB structures in CIF and QCIF case for H.261 [5.199]. ©1993 ITU-T.

Compression of video data typically is based on two principles: the reduction of spatial redundancy and the reduction of temporal redundancy. H.261 uses the DCT to remove spatial redundancy [5.200] and motion compensation to remove temporal redundancy [5.201].

The coding algorithm used in H.261 can be summarized into block diagrams in Figure 5.106 and Figure 5.107 [5.199]. At the encoder, the input picture is compared with the previously decoded frame with motion compensation. The difference signal is DCT transformed and quantized and then entropy-coded and transmitted. At the decoder, the decoded DCT coefficients

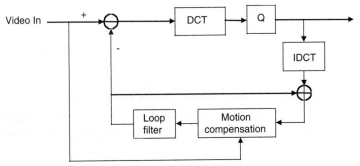

Figure 5.106 Block diagram of a video encoder used in H.261 [5.199]. ©1993 ITU-T.

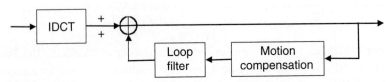

Figure 5.107 Block diagram of a video decoder used in H.261 [5.199]. ©1993 ITU-T.

are IDCT-transformed and then added to the previously decoded picture with motion compensation. Because the prediction of the current frame is composed of blocks at various locations in the reference frame, the prediction itself (or simply called the predicted frame) may contain coding noise and blocking artifacts. These artifacts may cause higher prediction errors. It is possible to reduce the prediction errors by passing the predicted frame through a low-pass filter before it is used as the prediction for the current frame. This filter is referred to as a loop filter (optional) because it operates inside the motion compensation loop.

As in all video-coding standards, H.261 specifies only the bit-stream syntax and how a decoder should interpret the bit stream to decode the image. Therefore, it specifies only the design of the decoder, not how the encoding should be done. For example, an encoder can simply decode to use only zero motion vectors and let transform coding take all of the burden of coding the residual. This may not be an efficient encoding algorithm, but it does generate a standard compliant bit stream.

H.261 has been included in several ITU-T H-series terminal standards for various network environments. One example is H.320 that is mainly designed for narrow-band ISDN terminals [5.173]. H.320 defines the systems and terminal equipment that use H.261 for video coding; H.221 for frame multiplexing; H.242 for signaling protocol [5.202] and G.711, G.722 and G.728 for audio coding. H.261 can also be used in other terminal standards including H.321, H.322 and H.324.

H.263 Standard

The activities of H.263 started around November 1993, and the standard was adopted in March 1996. The main goal of this endeavor was to design a video-coding standard suitable for applications with bit rates lower than 64 Kb/s (the so-called very-low bit-rate applications). For example, when sending video data across the PSTN and the mobile network, the video bit rates typically range from 10 to 24 Kb/s. During the development of H.263, it was identified that the near-term goal would be to enhance H.261 using the same general framework, and the long-term goal would be to design a video-coding standard that may be fundamentally different from H.261 in order to achieve further improvement in coding efficiency. As the standardization activities moved along, the near-term effort became H.263 and H.263 Version 2, and the long-term effort is now referred to as H.26L [5.174], previously called H.263L.

In essence, H.263 combines the features of H.261 together with MPEG and is optimized for very low bit rates. In terms of SNR, H.263 can provide a 3 to 4 dB gain over H.261 at bit rates below 64 Kb/s. In fact, H.263 provides superior coding efficiency to that of H.261 at all bit rates. When compared with MPEG-1, H.263 can give a 30% bit-rate saving.

Because H.263 was built on top of H.261, the main structures of the two standards are essentially the same. Therefore, we focus only on the differences between the two standards. The major differences include the following:

- H.263 supports more picture formats and uses a different GOB structure.

- H.263 uses half-pel motion compensation, but does not use loop filtering (optional) as in H.261.
- H.263 uses 3D VLC for coding of DCT coefficients.
- In addition to the basic coding algorithm, four options in H.263 that are negotiable between the encoder and the decoder provide improved performance.
- H.263 allows the quantization step size to change at each MB with less overhead.

In addition to CIF and QCIF as supported by H.261, H.263 also supports sub-QCIF, 4CIF and 16CIF (Table 5.26). Chrominance subsampling and the relative positions of chrominance pels are the same as those defined in H.261. However, H.263 uses different GOB structures (Figure 5.108). Unlike H.261, a GOB in H.263 always contains at least one full row of MBs.

A major difference between H.261 and H.263 is the half-pel prediction in the motion compensation. This concept is also used in MPEG. The motion vectors in H.261 can have only integer values, but H.263 allows the precision of motion vectors to be at a half of a pel. For example, it is possible to have a motion vector with values (4.5, -2.5). When a motion vector has noninteger values, bilinear interpolation is used to find the corresponding pel values for prediction.

The coding of motion vectors in H.263 is more sophisticated than that in H.261. The motion vectors of three neighboring MBs (the left, the above and the above right) as shown in

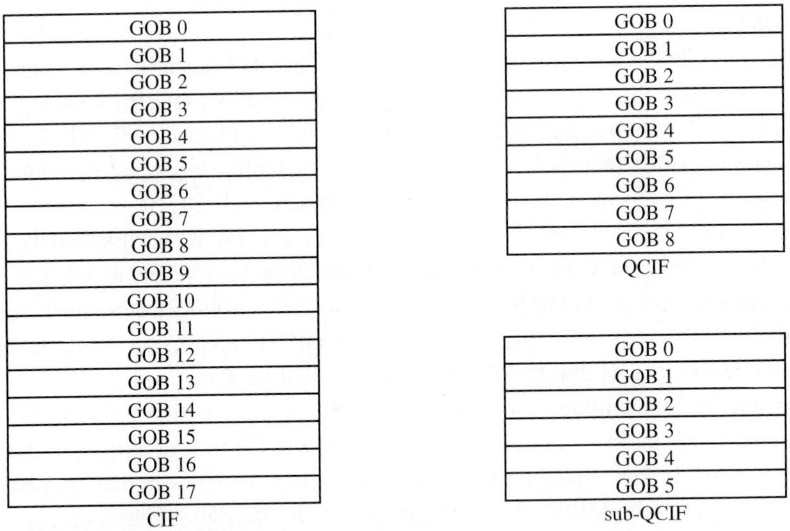

| GOB 0 |
| GOB 1 |
| GOB 2 |
| GOB 3 |
| GOB 4 |
| GOB 5 |
| GOB 6 |
| GOB 7 |
| GOB 8 |
| GOB 9 |
| GOB 10 |
| GOB 11 |
| GOB 12 |
| GOB 13 |
| GOB 14 |
| GOB 15 |
| GOB 16 |
| GOB 17 |

CIF

| GOB 0 |
| GOB 1 |
| GOB 2 |
| GOB 3 |
| GOB 4 |
| GOB 5 |
| GOB 6 |
| GOB 7 |
| GOB 8 |

QCIF

| GOB 0 |
| GOB 1 |
| GOB 2 |
| GOB 3 |
| GOB 4 |
| GOB 5 |

sub-QCIF

Figure 5.108 GOB structures for H.263 [5.174]. ©1996 ITU-T.

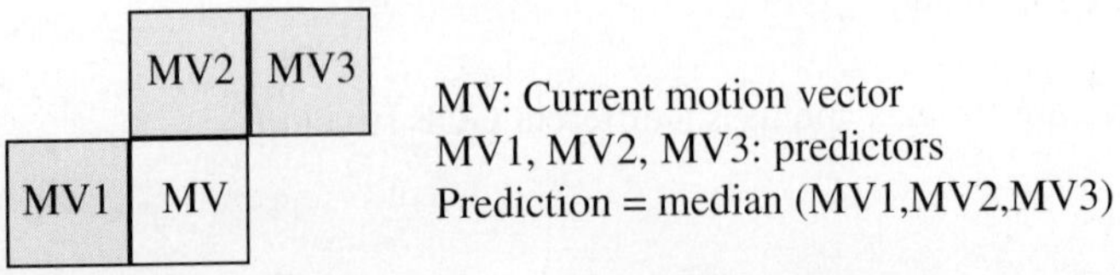

MV: Current motion vector
MV1, MV2, MV3: predictors
Prediction = median (MV1,MV2,MV3)

Figure 5.109 Prediction of motion vectors in H.263 [5.174]. ©1996 ITU-T.

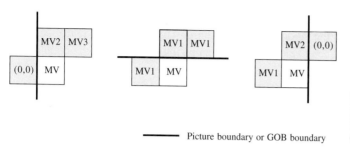

Figure 5.110 Motion vector prediction at picture/GOB boundaries in H.263 [5.174]. ©1996 ITU-T.

Figure 5.109 are used as predictors. The median of the three predictors is used as the prediction for the motion vector of the current block, and the prediction error is coded and transmitted. However, around a picture boundary or GOB boundary, special cases are needed (Figure 5.110). When only one neighboring MB is outside the picture boundary or GOB boundary, a zero motion vector is used to replace the motion vector of that MB as the predictor. When two neighboring MBs are outside, the motion vector of the only neighboring MB that is inside is used as the prediction.

H.263 specifies four options that are negotiable between the encoder and the decoder. At the beginning of each communication session, the decoder signals the encoder as to which of these options the decoder has the capability to decode. If the encoder also supports some of these options, it may enable those options. However, the encoder does not have to enable all the options that are supported by both the encoder and decoder. The four options in H.263 are the unrestricted motion vector mode, the syntax-based arithmetic-coding mode, the advanced prediction mode and the PB-frame mode.

In the unrestricted motion vector mode option, motion vectors are allowed to point outside of the picture boundary. In this case, edge pels are repeated to extend to the pels outside so that prediction can be done. A significant coding gain can be achieved with unrestricted motion vectors if there is movement around picture edges, especially for smaller picture formats like QCIF and sub-QCIF. In addition, this mode allows a wider range of motion vectors than H.261. Large motion vectors can be very effective when the motion in the scene is caused by heavy motion, for example, motion due to camera movement.

In syntax-based arithmetic coding, arithmetic coding is used, instead of VLC tables, for entropy coding. Under the same coding condition, using arithmetic coding will result in a bit stream different from the bit stream generated by using a VLC table, but the reconstructed frames and the SNR will be the same. Experiments show that the average bit-rate savings is about 3 to 4% for interframes and about 10% for intrablocks and frames [5.203].

In the advanced prediction mode, Overlapped Block Motion Compensation (OBMC) [5.204] is used to code the luminance of P-pictures, which typically results in less blocking artifacts. This mode also allows the encoder to assign four independent motion vectors to each MB, that is, each block in one MB can have an independent motion vector. In general, using four motion vectors gives better prediction since one motion vector is used to represent the movement of an 8x8 block instead of a 16x16 MB. Of course, this implies more motion vectors and hence requires more bits to code the motion vectors. Therefore, the encoder has to decide when to use four motion vectors and when to use only one. Finally, in the advanced prediction mode, motion vectors are allowed to cross picture boundaries as is the case in the unrestricted motion vector mode.

In the PB-frame mode, a PB-frame consists of two pictures coded as one unit, as shown in Figure 5.111. The first picture, called the P-picture, is a picture predicted from the last decoded picture. The last decoded picture can be either an I-picture, a P-picture or the P-picture of a PB-frame. The second picture, called the B-picture (B for bidirectional), is a picture predicted from both the last decoded picture and the P-picture that is currently being decoded. As opposed to the B-frames used in MPEG, PB frames do not need separate bidirectional motion vectors. Instead, forward vectors for the P-picture are scaled and added to a small delta vector to obtain vectors for the B-picture. This results in less bit-rate overhead for the B-picture. For relatively simple sequences at low bit rates, the picture rate can be doubled with this mode with minimal increase in the bit rate. However, for sequences with heavy motion, PB-frames do not work as well as B-pictures. Also, note that the use of PB-frame mode increases the end-to-end delay, so it may not be suitable for two-way interactive communication.

As in H.261, H.263 can be used in several terminal standards for different network environments. One example is H.324 [5.180] that defines audiovisual terminals for the traditional PSTN. In H.324, a telephone terminal uses H.263 as the video codec, H.223 as the multiplexing protocol, H.245 as the control protocol [5.190], G.723.1 for speech coding at 5.3/6.3 Kb/s and V.34 for the modem interface. H.324 is sometimes used to refer to the whole set of standards. H.263 can also be used in other terminal standards, such as H.323, which is designed for LANs without guaranteed QoS.

The ITU-T H.261/H.263 video-compression standards were designed for real-time coding and decoding for videoconferencing across constant bit-rate connections. In addition to low-coding delay, these standards extend the notion of I and P frames to I and P blocks within a frame. A block is typically an 8x8 pixel region. These algorithms achieve smoother processing

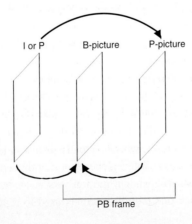

Figure 5.111
The PB-frame mode.

requirements and bandwidth profiles by essentially distributing an I-frame across several frames. I-block insertion is required to recover from the accumulation of DCT mismatch errors and for more efficient encoding when acute scene changes eliminate the advantage of difference encoding. For example, H.263 defines a maximum forced update period of 132 transmitted MB, within which each block must be updated at least once.

H.261/H.263 error resilience is achieved by inserting an I-block asynchronously and incrementally to refresh the image. This scheme, however, does not have regular synchronization points. I-block insertions are scattered across different frames. A block refreshed by an I-block insertion in the current frame might get corrupted again in the next frame if it is decoded using a corrupted motion vector reference. In fact, certain motion patterns could cause an error to propagate indefinitely [5.205].

As for the distribution interval, it can be reduced by shortening the forced updating period at the expense of a higher data rate or lower quality. This problem could be solved by retransmitting the lost data at the expense of increased delays.

H.263+ (H.263 Version 2) Standard

H.263+ is a revision of the original 1999 version of the H.263 ITU-T recommendation [5.176]. The original H.263 was developed for video compression at rates lower than 64 Kb/s and more specifically at rates lower than 33.4 Kb/s (V.34 modem). This was the first international standard for video compression that would permit video communications at such a low rate [5.174, 5.175]. H.263+ (often called H.263 Version 2) contains approximately 12 new features that do not exist in H.263. These include new coding modes that improve compression efficiency, support for scalable bit streams, several new features to support packet networks and error-prone environments, added functionality and support for a variety of video formats. Among the new features of H.263+, one of several that correct design inefficiencies of the original H.263 recommendation, is modified quantization mode. This mode has four key elements:

- Indication for larger quantizer changes from macroblock to macroblock to better react to rate-control requirements
- The ability to use a finer chrominance quantizer to better preserve chrominance fidelity
- The capability to support the entire range of quantized coefficient values rather than having to clip values greater than 128
- Explicitly restricting the representation of quantized transform coefficients to those that can reasonably occur

The second modification of the original H.263 is motion vector range. When H.263+ mode is invoked, the range is generally larger and depends on the frame size. Motion vector ranges in H.263+ are shown in Table 5.27.

Another modification to the original H.263 recommendation is the addition of a rounding term to the equation for half-pel interpolation. The rounding term toggles from frame to frame, thus eliminating this rounding bias and thereby reducing the artifact noticeably. Finally, H.263+

Table 5.27 Motion vector ranges in H.263+ [5.176].

Frame sizes up to	Motion vector range
352 x 288	[-32, 31.5]
704 x 576	[-64, 63.5]
1,408 x 1,152	[-128, 127.5]
Widths up to 2,048	Horizontal range [-256, 255.5]

©1997 ITU-T.

supports a wider variety of input video formats than H.263. In addition to five standard sizes, arbitrary frame sizes, in multiples of four from (32x32) to (2,048x1,152) can be supported.

One feature of H.263 Version 2 is that it extends the possible source formats specified in H.263. These extensions include the following:

Higher Picture Clock Frequency (PCF)—This allows picture clock rates higher than 30 frames per second. This feature helps to support additional camera and display technologies.

Custom picture formats—It is possible for the encoder and the decoder to negotiate a custom picture format, which is not limited by a number of fixed formats anymore. The number of lines can be from 4 to 1,152 as long as it is divisible by 4, and the number of pels per line can be from 4 to 2,048 as long as it is divisible by 4.

Custom Pixel Aspect Ratios (PARs)—This allows the use of additional PARs other than those used in CIF (11:12), SIF (10:11) and the square (1:1) aspect ratio. All custom PARs are shown in Table 5.28.

Table 5.28 Custom PARs [5.176].

PAR	Pixel width: pixel height
Square	1:1
CIF	12:11
525 type for 4:3 picture	10:11
CIF for 16:9 picture	16:11
525 type for 16:9 picture	40:33
Extended PAR	m:n, m and n are relatively prime

©1997 ITU-T.

Among the new negotiable coding options specified by H.263 Version 2, five of them are intended to improve coding efficiency:

Advanced intracoding mode—This is an optional mode for intracoding. In this mode, intrablocks are coded using a predictive method. A block is predicted from the block to the left or the block above, as shown in Figure 5.112. For isolated intrablocks for which no prediction can be found, the prediction is simply turned off.

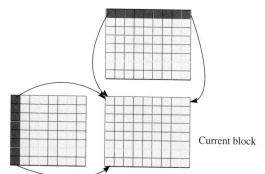

Current block

Figure 5.112
Advanced intracoding mode.

Alternate inter-VLC mode—This mode provides the ability to apply a VLC table originally designed for intra coding to intercoding, where there are often many large coefficients, by simply using a different interpretation of the level and the run.

Modified quantization mode—This mode improves the flexibility of controlling the quantizer step size. It also reduces the quantizer step size for chrominance quantization in order to reduce the chrominance artifacts. An extension of the range of values of the DCT coefficient is also provided. In addition, by prohibiting certain unreasonable coefficient representations, this mode increases error detection performance and reduces decoding complexity.

Deblocking filter mode—In this mode, an adaptive filter is applied across the 8x8 block edge boundaries of decoded I- and P-pictures to reduce blocking artifacts. The filter affects the picture that is used for the prediction of subsequent pictures and thus lies within the motion-prediction loop, similar to the loop filtering in H.261.

Improved PB-frame mode—This mode deals with the problem that the PB-frame mode in H.263 cannot represent large motion very well. It provides a mode with more robust performance under complex motion conditions. Instead of constraining a forward motion vector and a backward motion vector to come from a single motion vector as in H.263, the improved PB-frame mode allows them to be totally independent as in the B-frames of MPEG.

The following optional modes are designed to address the needs of mobile video and other unreliable transport environments:

Slice structured mode—In this mode, a slice structure replaces the GOB structure. Slices have more flexible shapes and may appear in any order within the bit stream for a picture. Each slice has a specified width. The use of slices allows a flexible partitioning of the picture in contrast to the fixed partitioning and fixed transmission order required by the GOB structure. This can provide enhanced error resilience and can minimize the video delay.

Reference picture selection mode—In this mode, the reference picture does not have to be the most recently encoded picture. Instead, any temporally previous picture can be referenced. This mode can provide better error resilience in unreliable channels, such as mobile and packet networks, because the codec can avoid using an erroneous picture for future reference.

Independent segment decoding mode—This mode improves error resilience by ensuring that any error in a certain region of the picture does not propagate to other regions.

The temporal, SNR, and spatial scalability modes support layered-bit-stream scalability in three forms, similar to MPEG-2. Bidirectionally predicted frames, which are the same as those used in MPEG, are used for temporal scalability by adding enhancement frames between other coded frames. This is shown in Figure 5.113. A similar syntactical structure is used to provide an enhancement layer of video data to support spatial scalability by adding enhancement information for construction of a higher-resolution picture, as shown in Figure 5.114. Finally, SNR scalability is provided by adding enhancement information for reconstruction of a higher-fidelity picture with the same picture resolution, as in Figure 5.115. Furthermore, different scalabilities can be combined together in a very flexible way (Figure 5.116).

Two other enhancement modes are described in H.263 Version 2:

Reference picture resampling mode—This allows a prior-coded picture to be resampled, or warped, before it is used as a reference picture (Figure 5.117). The warping is defined by four

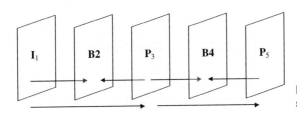

Figure 5.113 Temporal scalability [5.176]. ©1997 ITU-T.

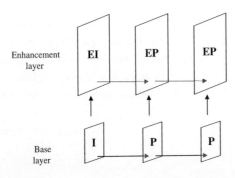

Figure 5.114 Spatial scalability [5.176]. ©1997 ITU-T.

ITU-T Standardization of Audiovisual Communication Systems

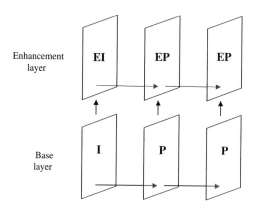

Figure 5.115 SNR scalability [5.176]. ©1997 ITU-T.

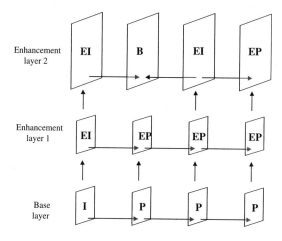

Figure 5.116 Multilayer scalability [5.176]. ©1997 ITU-T.

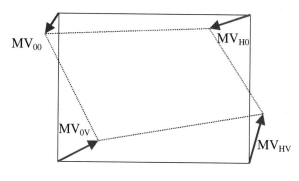

Figure 5.117 Reference picture resampling [5.176]. ©1997 ITU-T.

motion vectors that specify the amounts of offset of the four corners of the reference picture. This mode allows an encoder to switch smoothly between different encoded picture sizes, shapes and resolutions. It also supports a form of global motion compensation and special-effect image warping.

Reduced-resolution update mode—This mode allows the encoding of interframe difference information at a lower spatial resolution than the reference frame. It gives the encoder the flexibility to maintain an adequate frame rate by encoding foreground information at a reduced spatial resolution while holding onto a higher-resolution representation of more stationary areas of a scene.

One important feature of H.263 Version 2 is the usage of supplemental information, which may be included in the bit stream to signal enhanced display capabilities or to provide tagging information for external usage. For example, it can be used to signal a full-picture or partial-picture freeze or a freeze-release request with or without resizing. It can be used to label a snapshot, the start and end of a video segment and the start and end of a progressively refined video. The supplemental information may be present in the bit stream even though the decoder may not be capable of providing the enhanced capability to use it, or even to interpret it properly. In other words, unless a requirement to provide the requested capability has been negotiated by external means in advance, the decoder can simply discard anything in the supplemental information. Another use of the supplemental enhancement information is to specify a chroma key for representing transparent and semitransparent pixels [5.206].

Another round of H.263 extensions has created a third generation of the H.263 syntax, informally called H.263++. Four key technical areas were identified for further investigation toward possible later standardization: variable transform type, adaptive arithmetic coding, error-resilient VLC tables and deringing filtering.

H.263++ Standard Development

The H.263++ development effort is intended for near-term standardization of enhancements to produce a third version of the H.263 video codec for real-time communications and related non-conversional services [5.207]. Key technical areas showing potential for performance gain of H.263++ are the following [5.208, 5.209]:

Error-resilient data partitioning—Creation of a data partitioned and layered protection structure for the coded data and a longer resynchronization codeword to improve the detectability and to reduce the probability of false detection.

4x4 block-size motion compensation—Long-term picture memories; rate-distortion optimization alterations; motion optimization alterations; a new type of deblocking filter; a new type of intraspatial prediction and some VLC alterations for transform coefficients, motion vectors and coded block pattern.

- *Adaptive quantization*—Rate-distortion optimized quantization and truly optimal rate-distortion trellis encoding for an additive distortion measure

- *Enhanced reference picture selection*—Multiframe motion-compensated prediction and modified interframe prediction method
- *Enhanced scalability*—New P-picture types in enhancement layers
- *IDCT mismatch reduction*—Integer inverse transform
- *Deblocking and deringing filters*—Directional classifications and identifications of outlying values of block corner pixels for spatial treatment
- *Error concealment*—Provides error tolerance

H.26L Standard

The long-term recommendation H.26L (previously called H.263L) is scheduled for standardization in the year 2002 and may adopt a completely new compression algorithm. H.26L is an effort to seek efficient video-coding algorithms that can be fundamentally different from the MC-DCT framework used in H.261 and H.263. When completed, it will be a video-coding standard that provides better quality and more functionalities than existing methods. The first call for proposals for H.26L was issued in January 1998. According to the call for proposals, H.26L is aimed at very low bit-rate, real-time, low end-to-end delay coding for a variety of source materials [5.210]. It is expected to have low complexity permitting software implementation, enhanced error robustness (especially for mobile networks) and adaptive rate-control mechanisms. The schedule for H.26L activities is shown in Table 5.29.

Table 5.29 Schedule for H.26L [5.210].

Year	Schedule
Jan 1998	Call for proposals
Nov 1998	Evaluation of the proposals
April 1999	First test model of H.26L (TML1)
1999–2001	Collaboration phase
Oct 2001	Determination
July 2002	Decision

©1998 ITU-T.

The following technical proposals were evaluated in response to the call for proposals for H.26L [5.211]:

- Modified prediction/transform-based method
- Vector quantization with block approximation either by reference to a codebook or by motion compensation from a previous frame
- Loop-filtering method for reducing the blocking artifacts, corner outliner and ringing noise
- Adaptive scalar quantizer scheme using nonzero-level codebooks

- DCT-based embedded video coder using rearrangement of DCT coefficients
- Rough segmentation affine motion compensation scheme, vector quantization and multishape DCT
- Data partitioning using a data-reordering algorithm
- Video coding using long-term memory for multiple reference frames and affine motion-compensated prediction

Multihypothesis (MH) motion pictures are an extension of P-pictures proposed for H.26L. Each block of an MB can be compensated by a linear combination of two motion-compensated blocks. Conventional B-pictures also employ two linearly combined motion-compensated blocks, but one motion-compensated signal (hypothesis) originates from a future reference frame. In contrast to B-pictures, MH pictures use temporally previous pictures for prediction and cause no extra coding delay. In addition, decoded MH pictures are also used for reference to predict future MH pictures.

MH pictures are shown in Figure 5.118. Two blocks of temporally previous pictures are used to predict the current MH picture. MH pictures are also used for reference to predict future MH pictures. MB modes for MH pictures are presented in Figure 5.119. For each block, a Multihypothesis Block Pattern (MHBP) indicates one or two hypotheses. Seven MH MB types are added to the standardized MB types for intercoding. The additional seven types allow MH motion-compensated prediction for seven different block sizes.

MH pictures as well as P-pictures use temporally previous pictures for prediction. Each block can be compensated by one hypothesis (conventional motion compensation) or two hypotheses. An MHBP indicates one hypothesis or two hypotheses for each block. The MHBP is dependent on the MB type. When the MHBP indicates one hypothesis for a block in the MB, motion vector data and a reference frame parameter are specified. When the MHBP indicates two hypotheses for a block, two motion vectors and two reference frame parameters are indicated.

The additional MH MB types use the code numbers of the universal VLC as specified in Table 5.30.

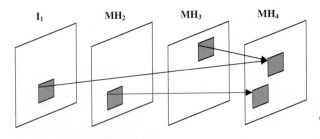

Figure 5.118 M++ pictures [5.212]. ©2001 ITU-T.

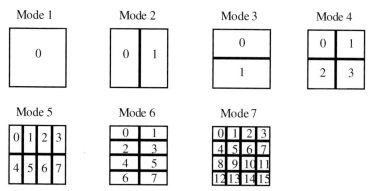

Figure 5.119 MB modes for MH motion pictures [5.212]. ©2001 ITU-T.

Table 5.30 MB types for the MH pictures [5.212].

Code number	MB type
0	Skip
1	16x16
2	MH 16x16
3	MH 16x8
4	MH 8x16
5	MH 8x8
6	MH 8x4
7	MH 4x8
8	MH 4x4
9	8x8
10	16x8
11	8x16
12	8x4
13	4x8
14	4x4
15	Intra4x4

©2001 ITU-T.

Table 5.31 MB types for the MH pictures [5.212].

Code number	MHBP
0	One hypothesis
1	Two hypotheses

©2001 ITU-T.

The MHBP uses the universal VLC as shown in Table 5.31.

One-hypothesis blocks, as well as two-hypothesis blocks, are independent of the block size individual reference frame parameters and reference frame parameter pairs. The universal VLC code numbers are used for signaling the reference frames.

The encoder has to determine the MH MB type and the number of hypotheses for each block. For a one-hypothesis block, the motion vector data and reference frame parameter are determined by rate-constrained long-term memory motion estimation. An integer-pel accurate estimate for all reference frames is redefined to half-pel and quarter-pel accuracy.

For a two-hypothesis block, the motion vectors and reference frame parameters are determined by rate-constrained MH motion estimation.

Rate-constrained MH motion estimation is performed by an iterative algorithm. The solution for the one-hypothesis block is used for initialization. The algorithm works as follows:

1. One hypothesis is fixed and long-term-memory motion estimation is applied to the complementary hypothesis so that the MH rate-distortion costs are minimized.
2. The complementary hypothesis is fixed, and the first hypothesis is optimized.

These two steps are repeated until convergence. Usually, the algorithm converges very fast after one to two iterations.

For the conditional motion estimation, an integer-pel accurate estimate for all reference frames is refined to half-pel and quarter-pel accuracies.

The computational complexity of a two-hypothesis block is just 2 to 4 times the complexity of a one-hypothesis block as the iterative algorithm is initialized with the one-hypothesis solution. However, this complexity can be further reduced by efficient search strategies.

For decoding MH pictures, the decoder has to add two motion-compensated signals. When the MH block pattern indicates a two-hypothesis block, the pixel values of the two motion-compensated blocks are added and divided by 2 (integer division). No additional memory is required.

5.9.4 ITU-T Speech-Coding Standards

The ITU has standardized three speech coders, which are applicable to low-bit-rate multimedia communications. ITU recommendation G.729 8 Kb/s CS-ACELP has a 15 ms algorithmic codec delay and provides network-quality speech [5.213]. It was originally designed for wireless applications, but is applicable to multimedia communication as well. Annex A of recommendation G.729 is a reduced-complexity version of the CS-ACELP coder. It was designed explicitly for simultaneous voice and data applications that are prevalent in low-bit-rate multimedia communications. These two coders use the same bit stream format and can interpolate. The ITU recommendation G.723.1 6.3 and 5.3 Kb/s speech coder for multimedia communications was designed originally for low-bit-rate videophones. Its frame size of 30 ms and one-way algorithmic codec delay of 37.5 ms allow for a further reduction in bit rate compared to the G.729 coder. In applications where low delay is important, the delay of G.723.1 may be too large. However, if the delay is acceptable, G.723.1 provides a lower-complexity alternative to G.729 at the expense of a slight degradation in quality.

An enormous number of new speech coders have been standardized. For example, in the 1995-96 time period, three new international standards (ITU G.729, G.729A and G.723.1) and three new regional standards (enhanced full-rate coders for European and North American mobile systems) have emerged.

Speech quality as produced by a speech coder is a function of bit rate, complexity, delay and bandwidth. Hence, when considering speech coders, it is important to review all these attributes. It is important to realize that there is a strong interaction between all these attributes and that they can be traded off against each other. Additional factors that influence the selection of a given speech coder are availability and licensing conditions the way that the standard is specified. Most speech coders operate at a fixed bit rate regardless of the input signal characteristics. Because multimedia speech coders share the channel with other forms of data, it is better to make the coder variable rate. For simultaneous voice and data applications, a good compromise is to create a silence compression scheme as part of the coding standard. A common solution is to use a fixed rate for active speech and a low rate for background noise. Silence compression consists of two main algorithms. The first is a Voice Activity Detector (VAD), which determines if the input signal is speech or some sort of background noise. If the signal is declared speech, it is coded at the full fixed bit rate. Sometimes no bits are transmitted at all. The second algorithm, Comfort Noise Generation (CNG), is invoked at the receiver to reconstruct the main characteristics of the background noise. Obviously, the performance of the VAD is critical to the overall speech quality. The CNG scheme must be designed in such a way that the encoder and decoder stay synchronized, even if there are no bits transmitted during some interval.

The delay of a speech-coding system usually consists of three major components. Most low-bit-rate speech coders process a frame of speech data at a time. The speech parameters are updated and transmitted for every frame. In addition, to analyze the data properly, it is sometimes necessary to analyze data beyond the frame boundary. This is referred to as look-ahead. Hence, before the speech can be analyzed, it is necessary to buffer a frame's (plus look-ahead) worth of data. The resulting delay is referred to as algorithmic delay. This is the only delay component that cannot be reduced by changing the implementation. All other delay components depend on the implementation. The second major contribution comes from the time that it takes the encoder to analyze the speech and the decoder to reconstruct the speech. This is referred to as processing delay. It depends on the speed of the hardware used to implement the coder. The sum of the algorithmic and processing delays is called the one-way codec delay. The third component is the communication delay, which is the time it takes for an entire frame of data to be transmitted from the encoder to the decoder. The total of these three delays is the one-way system delay. Maximum values of 400 ms for the one-way system delay can be tolerated if there are no echoes. However, new testing methodologies revealed that, for the case of communication, it is preferable if the one-way delay is below 200 ms. In many applications, such as teleconferencing, it is necessary to bridge several callers so that each person can hear all the others. For speech coders, this means decoding each bit stream, summarizing the decoded signals and then re-encoding the sum signals. This process not only doubles the delay, but it also reduces the speech quality due to the multiple encodings.

Speech coders are often implemented on (or share) special purpose hardware, such as DSP chips. Their attributes can be described as computing speed in MIPS, RAM and ROM. Currently, speech coders requiring less than 15 MIPS are thought of as low complexity. Those requiring 30 MIPS or more are considered high complexity. From the system designer's point of view, more complexity results in higher costs and greater power usage. For portable applications, greater power usage means reduced time between battery recharges or using larger batteries, which means more expense and weight. Thus, complexity is an important factor.

In what follows, we give an overview of the standardization process for the three ITU coders. We start with a description of how the requirements are set. This is illustrated by the specifics for each of the speech-coder attributes: bit rate, delay, complexity and quality. In most standardization procedures, it is common to specify the requirements using the Terms of Reference (ToR). This document not only contains a schedule, but also specifies the performance requirements and objectives.

Bit Rate

For G.729, one of the ToR requirements was that the speech coder should operate at 8 Kb/s. This rate was selected in part because it fits the range of first-generation digital cellular standards, from 6.7 Kb/s in Japan to 7.95 Kb/s in the United States to 13 Kb/s in Europe. For G.723.1, the ToR requirement was that the speech coder should operate at lower than 9.6 Kb/s. As it turned out, all the coders tested ranged between 5.0 and 6.8 Kb/s. For the Digital Simultaneous Voice and Data (DSVD) coder, the ToR requirements for bit rate were derived from the amount of speech data that could be carried across a 14.4 Kb/s modem. The bit rates of the five candidate coders submitted for this standard were all near 8 Kb/s. None of the three coders had a silence compression scheme as part of the main body of the recommendation. Subsequent work created silence compression schemes for both G.723.1 and G.729, which are included as Annexes to each recommendation.

Delay

This is a major difference between G.723.1 and G.729. The ToR requirement for delay for G.729 was discussed for more than a year. Initially, it was a maximum one-way codec delay of 10 ms.

Later, the frame size was allowed to grow to 16 ms. G.729 has a 5 ms look-ahead. Assuming a 10 ms processing delay and a 10 ms transmission delay, the one-way system delay for G.729 is 35 ms. The principal application of G.723.1 is low-bit-rate videophones, which typically operate at 5 frames/s or fewer. This rate equates to a video frame period of 200 ms. The final version of G.723.1 has a look-ahead of 7.5 ms, making a one-way system delay of 97.5 ms. In deliberating on the delay requirements for a DSVD coder, SG14 was cognizant of the delay inherent in V.34 modems. These modems often have one-way delays greater than 35 ms.

Complexity

In formulating the requirements for G.729, the trade-off discussed involved delay and complexity. The ITU-R was concerned about creating a coder that would be too complex with too high a delay. Ultimately, they accepted a delay target that allowed a significant reduction in complexity compared with the G.728 coder. The MIPS are reduced to around 17. The amount of RAM required is 3,000 words, 50 % more than G.728. Much of this extra memory usage is due to the use of larger frames. G.723.1 is of lower complexity than G.729 (14.6 MIPS at 5.3 Kb/s and 16 MIPS at 6.3 Kb/s), and uses 2,200 words of RAM. The requirements for use of the DSVD coder were 10 MIPS, 2,000 words of RAM and 10,000 words of ROM.

Quality

Table 5.32 gives speech-quality performance requirements and objectives for G.729. It does not include requirements unrelated to speech quality, such as bit rate, delay and complexity, which are also discussed in the ToR. The first requirement is that, for error-free conditions, a single encoding should be rated not worse than 32 Kb/s (G.726). In separate testing of G.729, G.723.1 and the DSVD version of G.729, all three coders met this requirement. In Degradation Category Rating (DCR) tests, subjects seem to equate different with worse. As a result, G.729 received lower scores than G.726 for DCR testing. However, if Absolute Category Rating (ACR) tests were performed, the MOS of G.729 was never significantly worse than G.726 and was sometimes better. The testing of G.723.1 and G.726A was less extensive. The second requirement concerned speech quality with noisy channels. For 10^{-3} random bit-error rate, the speech quality should again be no worse than G.726 under similar conditions.

All three coders encode music signals, but the quality of the music is poor. The reason for this is that Linear Prediction Analysis by Synthesis (LPAS) coders rely on pitch prediction to achieve high coding efficiency. Most music signals lack a pitch structure, and all the coding burden has to be carried by the excitation and low-order LP.

The overall performance of the three coders was similar. It seemed that the G.723.1 and G.729A (DVD version of G.729) coders are slightly less robust for background noises and tandem conditions. Their performance for clean speech and general robustness is sufficient enough that the ITU sees fit to recommend them for use in simultaneous voice and data applications, such as low-bit-rate multimedia communications.

Table 5.32 Speech quality performance requirements and objectives for G.729 [5.213].

Parameter	Requirements	Objectives
Quality (without bit errors)	Not worse than 32 Kb/s G.726	N/A
Random bit errors BER<10^{-3}}	Not worse than G.726	Equivalent to 32 Kb/s G.726
Detected frame erasures	No more than 0.5 MOS	N/A
Random and bursty 3%	Degradation from 32 Kb/s ADPCM without errors	
Undetected burst errors	N/A	For further study
Level dependency	Not worse than 32 Kb/s G.726	As low as possible
Talker dependency	Not worse than 32 Kb/s G.726	N/A
Capability to transmit music	N/A	No annoying effects generated
Tandeming capability for speech	Two asynchronous codings with a total distortion of <4 asynchronous 32 Kb/s G.726	3 asynchronous codings with a total distortion <4 Asynchronous 32 Kb/s G.726
Tandeming with other ITU standards	<4 asynchronous 32 Kb/s G.726	Synchronous tandeming property
Tandeming with regional DMR standards	For further study	N/A

Idle channel noise
  Weighted
  Single frequency
	
	For further study
	Not worse than 32 Kb/s G.726
		
		Not worse than 32 Kb/s G.726
		N/A

Capability to transmit signaling/ information tones	DTMF, CCITT Nos.5, 6, 7, CCITT R2, Q.35, Q.23, V.25	Distortion as low as possible

©1996 IEEE.

5.9.5 Multimedia Multiplex and Synchronization Standards

ESs, such as audio, video, data, video frame synchronous control and indications signals, each of which may be internationally standardized or private, are multiplexed into a serial packet stream according to H.222.0. H.222.1/H.222.0 functions include multiplexing timebase recovery, media synchronization, jitter removal, buffer management, security and access control, subchannel signaling and trick modes which are mechanisms to support video recorder-like control functionality, for example, fast forward, rewind and so forth. Recommendation H.222.1 specifies elements and procedures from the generic H.222.0 for their use in ATM environments and also specifies codepoints and procedures for ITU-T-defined ESs [5.214]. H.222.1 allows the use of both the H.222.0 program stream and the H.222.0 transport stream. Only single-program transport streams are allowed [5.171]. A particular call may consist of multiple program streams or transport streams, each carried in separate ATM virtual channels and all referring to a common system time clock. Subchannel signaling is the process by which a subchannel for audio, video and other ESs is established and released between send and receive H.222.1 entities. Although H.222.1 specifies an unacknowledged signaling procedure using the Program Stream Map for H.222.0, Program Streams and Program Specific Information for H.222.0 Transport Streams, as well as an acknowledged signaling procedure, it offers improved call-phase synchronization and reliability.

ITU-T Recommendation H.221

This recommendation is entitled "Frame structure for a 64-1920 Kb/s channel in audiovisual teleservices" [5.215]. H.221 is the multiplex and bending protocol for H.320 terminals. Up to 30 ISDN B channels can be bundled together to form a superchannel with a bit rate of $n*64$ Kb/s. The media channels for audio and video and the data information are multiplexed onto the supperchannel. For audio and video information, H.221 does not perform any error control, but relies completely on the error resilience of the media coding, which is possible because of ISDN's isochronous nature and its low error rates. The protocol offers only a bit-oriented, unprotected, point-to-point transport service.

ITU-T Recommendation H.223

This recommendation is entitled "Multiplexing protocol for low bit rate multimedia communication" [5.216]. Three different types of ALs are available that have different characteristics in terms of error probability and delay. Low-delay channels allow higher error rates, and reliable channels might have indefinitely long delays. AL1 and AL2 serve different duties; AL3 is designed for the use with coded video. Video data is encapsulated in small, variable-length packets (typically around 100 bytes, although larger packet sizes can be negotiated). A 16-bit Cyclic Redundancy Check (CRC) for each packet allows error detection. The packetization overhead for each packet is 1 to 3 bytes, plus error control information of AL3. AL3 includes an optional retransmission protocol, which sometimes allows the retransmission of a lost or corrupted packet. The retransmission of AL3 relies on the fast arrival of the confirmation messages, which give indications about correctly transmitted packets. These confirmation messages arrive at the sender of the original message with twice the one-way delay because a complete roundtrip of data and confirmation are necessary. The retransmission of the damaged packet after notification will incur a third one-way delay, resulting in a one-way delay three times .

ITU-T Recommendation H.225

Recommendation H.225.0 describes the means by which audio, video, data and control are associated, coded and packetized for transport between H.323 terminals on a nonguaranteed QoS LAN, or between H.323 terminals and H.323 gateways, which in turn may be connected to H.320, H.324 or H.310/H.321 terminals on NISDN, GSTN or BISDN, respectively. This gateway, terminal configuration and procedures are described in H.323, and H.225.0 covers protocols and message formats. The scope of H.225.0 communication is between H.323 terminals and H.323 gateways on the same LAN, using the same transport protocol, as shown in Figure 5.120. This LAN may be a single segment or ring or, if logical, could be an enterprise data network compromising multiple LANs bridged or routed to create one interconnected network. H.225 makes use of RTP/RTCP for media-stream packetization and synchronization for all underlying

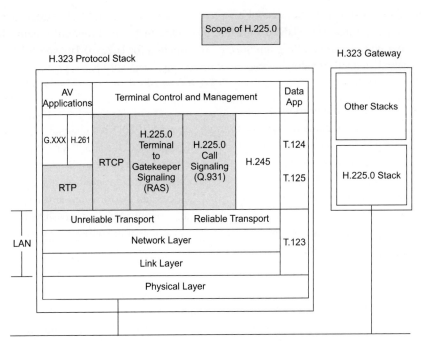

Figure 5.120 Scope of H.225.0 [5.171]. ©1997 IEEE.

LANs. H.225.0 assumes a call model where initial signaling on a non-RTP transport address is used for a call establishment and capability negotiation followed by the establishment of one or more RTP/RTCP connections. Thus, H.323/H.225.0 constitutes a profile for RTP/RTCP using the IETF terminology [5.171]. The general approach of H.225.0 is to provide a means of synchronizing packets that makes use of the underlying LAN/transport facilities. H.225.0 does not require all media and control to be mixed into a single stream, which is then packetized. The framing mechanisms of H.221 are not used for the following reasons:

- Not using H.221 allows each media to receive different error treatments as appropriate.
- H.221 is relatively sensitive to the loss of random groups of bits packetization allowing greater robustness in the LAN environment.

H.225 terminals send audio and video using RTP through unreliable channels to minimize delay. Error concealment or another recovery action must be applied to overcome lost packets; in general, audio/video packets are not transmitted because this would result in excessive delay. It is assumed that bit errors are detected in the lower layers, and errored packets are not sent up to H.225.0. Note that audio/video and call signaling/H.245 control are never sent on the same channel and do not share a common message structure. H.225.0 terminals are capable of sending and receiving audio and video on separate transport addresses using separate instances of RTP to allow for media-specific frame sequence numbers and separate QoS treatment for each medium.

However, an optional mode, where audio and video packets are mixed in a single frame and are sent to a single transport address, is for further study.

Common Control Protocol H.245

This recommendation defines messages and procedures for the exchange of control information between multimedia terminals. H.245 specifies terminal-to-terminal signaling to determine the coding and decoding capabilities of the remote terminal, following the establishment of the network connection, and to coordinate the assignment and release of terminal resources throughout the call. H.245 has been defined as a generic recommendation that is suitable for use in a range of multimedia terminal applications. Recommendations H.310 and H.323 described previously, it has also been adopted in H.324, the terminal for the GSTN and in V.70, which is for the multimedia terminal using modems. H.245 will also be used for the mobile terminal recommendation with working number H.324/M.

H.245 was structured by defining three main sections: syntax, semantics and procedures. In the first of these, ASN.1 notation was used to define generic syntax for messages [5.190]. The semantics section defines the meaning of syntax elements and provides syntactic constraints. Interaction between the different protocol entities is only through communication with the H.245 user. H.245 provides a number of different services to the H.245 user. Services may be applicable to a specific terminal recommendation. Some of these services are as follows:

- *Capability exchange*—Before multimedia communication begins, a terminal must be aware of the remote terminal's capabilities to receive and decode the multimedia signals. Terminal video capabilities may be one of, or a range of, H.261, H.262 or H.263 video. Audio capabilities may be one of the ITU-T G-series recommendations, such as G.711, G.722 or MPEG-1 or MPEG-2 Audio. Data capabilities include the T.120-series of recommendations. Audiovisual multiplexing capabilities are also recognized. Multiplexing capabilities include specification of the H.221 Program Stream or Transport Stream and ATM AL type and parameters.
- *Logical channels signaling procedures*—Following the capability exchange, but before the actual transmission of multimedia signals, the terminal coding and decoding resources are assigned using logical channel signaling. A logical channel number simply represents a specific channel in the system multiplex. Logical channel numbers are unique in each direction of transmission for a particular call.
- *Control and indication signals*—Messages are defined to carry the control and indication signals defined in H.221 and H.320 [5.215]. These are intended for various purposes, including maintenance loops, video/audio active/inactive signals, fast update request, source switching in multipoint applications and so forth.

It was expected from early in its development that H.245 would be a living document and would have additional features added to it from time to time, either to make it suitable for use in new terminal recommendations or simply to provide additional functionality to existing terminal recommendations. Thus, H.245 syntax has been designed to be extensible. This has been achieved by the use of extension markers in the syntax, as well as a protocol identifier field that indicates the version of H.245 that is being used. Extension markers allow syntax to be added so that earlier H.245 decoder implementations can skip the additional syntax, without understanding it, and continue to decode normally [5.218]. The protocol identifier field will be used to indicate more substantive changes to the recommendation, such as the definition of new procedures or the addition of new syntax that must be understood by the remote terminal. When the message containing the protocol identifier field is received and an earlier version of the H.245 is indicated, a terminal must restrict its use of messages and procedures to those of the earlier version. The set of procedures for each protocol entity in H.245 is specified using the Specification and Description Language (SDL) diagrams. The SDL diagrams define not only normal operations, but also actions to be taken in the event of exception conditions, such as the reception of unexpected messages.

5.10 IETF and Internet Standards

The IETF is an open international standardization body of network designers, operators, vendors and researchers focused on the development of Internet standard protocols for use on the Internet and intranets. The IETF is focused on the development of protocols used on IP-based networks. It consists of many working groups and is managed by the Internet Engineering Steering Group (IESG), Internet Architecture Board (IAB) and Internet Society (ISOC). The IETF is different from most standardization bodies in that it is a totally open community with no formal membership [5.219]. One of the strengths of the Internet is its global connectivity [5.220]. For this connectivity, it is essential that all the hosts on the Internet interoperate with one another and understand the common protocol at various layers. The Internet standardization process of IETF under the ISOC is the key to the success of GII over IP-based networks such as the Internet.

5.10.1 IETF Standardization Process

The Internet by definition is a complex mingle of networks based in the TCP/IP protocols. The whole structure of Internet management makes the prediction of its evolution complex. The Internet is largely self-developed by its own users, and it is too complex to predict changes about the relationships between the Internet and other technologies.

The ISOC was officially formed in January 1992. It was formed by a number of people with long-term involvement in the IETF, in order to provide an institutional home and financial support for the Internet standardization process [5.221]. Today, the ISOC is a nonprofit, nongovernmental, international, professional membership society with more than 100 organizations and 6,000 individual members in more than 100 countries. It provides leadership in addressing issues that confront the future of the Internet and is the organization home for the groups responsible for Internet infrastructure standards, including the IETF and the IAB.

ISOC aims to ensure the beneficial, open evolution of the global Internet and its related interworking technologies through leadership in standards, issues and education. The Society's individual and organizational members are bound by a common stake in maintaining the viability and global scaling of the Internet. The Society is governed by a board of trustees elected by its membership around the world, and the Board is responsible for approving appointments to the IAB from among the nominees submitted by an IETF nominating committee.

The IAB is the technical advisory group of the ISOC. It is chartered to provide oversight of the architecture of the Internet and its protocols and to serve in the context of the Internet standards process as a final appealing body. The IAB is responsible for approving appointments to the IESG from among the nominees submitted by the IETF nominating committee.

The IESG is responsible for technical management of IETF activities and the Internet standards' process. As part of the ISOC, it administers the Internet standards' process according to the established rules and procedures. The IESG is directly responsible for the actions associated with entry and movement along the standards track, including final approval of specifications as Internet standards. The IESG is composed of the IETF Area Directors (ADs) and the chairperson of the IETF, who also serves as the chairperson of the IESG. Representative of the increasingly larger span of the Internet is the fact that the IESG has established formal liaison with the ATM Forum and the ITU-T.

The IETF is a loosely self-organized group of people who make technical and other contributions to the engineering and evolution of the Internet and its technologies. It is open to any interested individual. The actual technical work of the Internet is mostly done inside the IETF. It is the principal body engaged in the development of new Internet standard specifications, although it is not itself a part of the ISOC. Much of the work is handled through mailing lists, because the IETF holds meetings only three times per year.

The IETF is composed of individual Working Groups (WGs), which are organized by topics into several areas, each of which is coordinated by one or more ADs. These are the members of the IESG. Nominations to the IAB and the IESG are made by nominating committee members.

At present, the IETF is organized into the following areas:

- *Applications area*—Issues related to applications, other than security and networks
- *General area*—Internal IETF organization issues
- *Internet area*—Improvements on the TCP/IP protocols for increased usage and versatility
- *Operations and management area*—Concerned with management and operation control of the Internet
- *Routing area*—Internet routing protocol issues
- *Security area*—Support for security across all areas
- *Transport area*—Transport of different payloads across IP and the IP transport by other media
- *User services area*—A forum for people interested in all levels of user services and the quality of information available to users of the Internet

Each area is further divided into WGs, ranging from a couple to several dozen.

The Internet Assigned Number Authority (IANA) is the central coordinator for the assignment of unique parameter values for IPs. The IANA is chartered by the ISOC to act as the clearinghouse to assign and coordinate the use of numerous IP parameters. IANA functions as the top of the pyramid for Domain Name System (DNS) and Internet address assignment, establishing policies for these functions [5.221].

Request for Comments (RFC) started in 1969 as a series of notes about the Internet and then the Advanced Research Agency Network (ARPANET). The specification documents of the IP suite, as defined by the IETF and its steering group (IESG), are now formally published as RFCs [5.222]. There are several categories of RFCs, from informational to standard. Furthermore, there are different standardization degrees [5.223]. For a given RFC to reach the full Internet standard statue, stable implementations in multiple, independent, and interoperable versions are required.

The Internet standardization process is managed by IESG [5.224]. The existence of interoperable running implementations is the key requirement for advancement of the process. A document may take two paths in order to become an RFC. The first path is through the IETF. The IETF WGs develop documents that may be approved for publication as RFCs by ADs. Another path for an RFC to go through is for it to be individually submitted to the RFC editor.

The very first step, however, is for a document to become an Internet draft so that it may be distributed, read and commented on. These drafts, as well as all IETF documents, should be focused, handling few points of doubt. If required, a subject can be separated into different components and each treated separately in a different WG. If it is required, a WG can be created in a very fast way, after an initial session, in order to assess its interest. When created, a WG has a very well-defined charter and publishes its goals and milestones. There is no formal voting process inside the WG, and the results are achieved by consensus, often after discussing results of different demonstrations.

WGs are loosely co-ordinated through their ADs, besides mutual interests that their participants may share. Most of the work is being done by volunteers, and the IETF policy of accepting only working implementations for final standards makes the final approval of a particular WG extremely dependent on its real utility to the overall Internet community. Thus, Internet standards are always de facto standards although their widespread usage in the Internet may vary strongly. The whole structure is based on the active participation and interest of its volunteers, regardless of their individual motivations; it is an extremely fluid process when confronted with a more traditional telecommunications standards fora.

Although existing groups have published goals, these are usually short lived (about 1 to 2 years) and very well defined. The technical implementation to reach these goals is entirely dependent on the individual WG participants and on proven implementations of their proposals. Furthermore, WGs can be created and terminated according to their participants' ideas. Thus, it is complex to predict Internet evolution and future activities.

The whole structure of Internet management makes the prediction of its standards' evolution complex. Nevertheless, some of these areas are of more interest to AD and to issues relevant to cohesion and integration of architectures, and trends can be identified inside these areas. Clearly, the Transport area and the Operations and Management area have more impact on future network integration, although some of the work being progressed in the Internet and Routing and Security areas may be of some influence also.

5.10.2 Internet Network Architecture

Telecommunication networks and computer networks have been developed from different perspectives. Telecommunication networks have relied on circuit switching. The circuits have provided either a constant bandwidth or a constant data rate. When telephone circuit switches became so complicated that new services were required three or four years of switch software modification, the telecommunications industry developed a new architecture called Intelligent Network (IN) to facilitate the introduction of new services. IN defines interface-to-switch call-processing software so that central computers at a service control point can instruct the switch on how to handle a call.

Computer networks have adopted packet switching, which facilitates statistical multiplexing of burst data transmissions of different sources. Furthermore, computer networks have relied on the processing power of customer premises' equipment to control the network.

Although the global telephone network was originally designed to support one service (voice), the Internet architecture was designed to support a broad range of data communications services. In addition, IP was designed to operate across a wide range of network technologies. Like other network architectures, the Internet has a layered set of protocols. IP is simple, and it defines an addressing plan and a packet delivery service. An effort is made to deliver each packet, but there are no guarantees concerning the transmit time or even the reliability. Many protocols can run on top of IP [5.225]. The most common one is TCP, which provides a guaranteed delivery service. IP does not guarantee the delivery of packets, and TCP/IP is subject to unpredictable delays. As the Internet expands, new protocols such as RTP and RSVP are being developed. Another group of IP-based protocols supports multicasting, which increases the efficiency of network utilization for applications such as Internet radio and videoconferencing. Multimedia applications are being transferred from server to client, and people are experimenting with voice and video real-time connections across the Internet [5.6].

The Internet reference network architecture is composed of end nodes (hosts) linked by subnetworks as shown in Figure 5.121. All the hosts belonging to the same subnetwork exchange data directly. The crossing of subnetwork boundaries is enabled by means of intermediate nodes. Hosts and routers exchange data by means of the IP, which is the universal protocol used by the heterogeneous network components to offer a unified abstraction of the network service.

The IP is capable of offering a network service in which the information is packaged in data units named packets or datagrams. The network offers no assurance on the delivery of the packets to the intended recipient (best-effort service). Intermediate nodes decide where to route a packet addressed to a given destination on the basis of routing tables built by exchanging information with other intermediate nodes by means of custom protocols, such as Routing Informa-

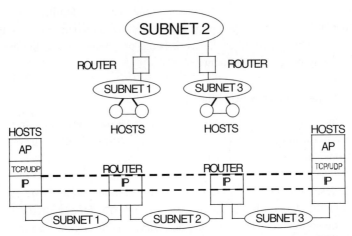

Figure 5.121 Internet network architecture [5.226]. ©1997 IEEE.

tion Protocol (RIP), OSPF and Border Gateway Protocol (BGP). The Internet Control Message Protocol (ICMP) supports IP by offering some basic control capabilities, such as sending reachable packets and asking an upstream packet source to slow down the packet transmission rate in the event of congestion. The end-to-end protocols add the capability to multiplex or demultiplex multiple flows of packets at the end nodes, and they may add reliability. In particular, UDP offers an unreliable service and adds a differentiation of packet flows within a host by means of port numbers [5.223]. The TCP offers a reliable sequenced delivery of byte streams on top of the datagram service offered by the IP. TCP is a connection-oriented protocol that enables connections with significance only at the end nodes. A windowing scheme is applied to enforce flow control, that is, to avoid the overrunning of slow receivers, as well as to allow the traffic source to adapt to network overload. In particular, the number of outstanding packets, that is, the number of packets that a source is entitled to transmit while waiting for an acknowledgement, is timed according to a probing of available bandwidth.

5.10.3 Internet Protocols

The IP has been the foundation of the Internet and virtually all multivendor private internetworks. A protocol, known as IPv6 (IP version 6) has been defined to ultimately replace IP [5.225]. The driving motivation for the adaptation of a new version of IP was the limitation imposed by the 32-bit address field in IPv4. Previous versions of IP (1 through 3) were successively defined and replaced to reach IPv4. Version 5 was assigned to the Stream Protocol, which runs parallel to IPv4 in some routers, which explains the use of the label "version 6."

Until recently, the Internet and most other TCP/IP networks have primarily provided support for rather simple distributed applications, such as file transfer, electronic mail, and remote access using Telnet. However, today, the Internet is increasingly becoming a multimedia, application-rich environment led by the huge popularity of the Web. At the same time, corporate networks have branched out from simple email and file transfer applications to complex client/server environments and intranets that mimic the applications available on the Internet. All of these developments have outstripped the capability of IP-based networks to supply needed functions and services. An interworked environment needs to support real-time traffic, flexible congestion control schemes and security features.

IP provides the functionality for interconnecting end systems across multiple networks. For this purpose, IP is implemented in each end system and in routers, which are devices that provide connection between networks. Higher-level data at a source and system are encapsulated in an IP Protocol Data Unit (PDU) for transmission. This PDU is then passed through one or more networks and connecting routers to reach the destination end system. The router must be able to cope with a variety of differences among networks, including addressing schemes, maximum packet sizes, interfaces and reliability.

The networks may use different schemes for assigning addresses to devices. For example, an IEEE802 LAN uses either 16-bit or 48-bit binary addresses for each attached device. An X.25 public packet-switching network uses 12-digit decimal addresses (encoded as 4 byte/digit for a 48-bit address). Some form of global network addressing must be provided, as well as a directory service.

Packets from one network may have to be broken into smaller pieces to be transmitted on another network, a process known as fragmentation. For example, Ethernet imposes a maximum packet size of 1,500 bytes. A maximum packet size of 1,000 bytes is common on X.25 networks. A packet that is transmitted on an Ethernet system and picked up by a router for retransmission on an X.25 network may have to segment the incoming packet into two smaller ones.

The hardware and software interfaces to previous networks differ. The concept of a router must be independent of these differences.

Various network services may provide anything from a reliable end-to-end virtual circuit to an unreliable service. The operation of the routers should not be defined on the assumption of network reliability. The operation of the router depends on an IP. IP must be implemented in all stations on all networks as well as on the routers.

Example 5.18 Consider the transfer of a block of data from station X to station Y as shown in Figure 5.122. The IP layer at C receives blocks of data to be sent to Y from TCP in X. The IP layer attaches a header that specifies the global Internet address of Y. That address is in two parts: network identifier and station identifier. Let us refer to this block as the IP datagram. IP recognizes that the destination (Y) is on another subnetwork. Therefore, the first step is to send the datagram to a router 1. To accomplish this, IP hands its data unit down to a Logical Link Control (LLC) with the appropriate addressing information. LLC creates an LLC PDU that is sent down to the Media Access Control (MAC) layer. The MAC layer constructs a MAC packet with a header that contains the address of router 1. Next, the packet travels through the LAN to router 1. The router removes the packet and LLC headers and trailers and analyzes the IP header to determine the ultimate destination of the data, in this case Y. The router must now make a routing decision. There are two possibilities:

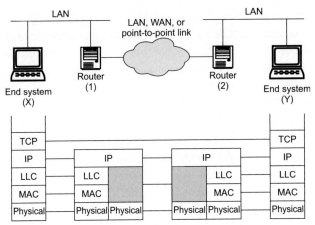

Figure 5.122 Protocol architecture including IP [5.225].
©1996 IEEE.

- The destination station Y is connected directly to one of the subnetworks to which the router is attached.
- To reach the destination, one or more additional routers must be traversed. In this example, the datagram must be routed through router 2 before reaching the destination.

The IP datagram to router 2 passes to router 1 through the intermediate network. For this purpose, the protocols of that network are used. If the intermediate network is an X.25 network, the IP data unit is wrapped in an X.25 packet with appropriate addressing information to reach router 2. When this packet arrives at router 2, the packet header is stripped off. The router determines that this IP datagram is destined for Y, which is connected directly to a subnetwork to which the router is attached. The router therefore creates a packet with a destination address of Y and sends it out onto the LAN. Finally, the data arrives at Y, where the packet, LLC and Internet headers and trailers can be stripped off. This IP service is unreliable, that is, IP does not guarantee that all data will be delivered or that the delivered data will arrive in the proper order. It is the responsibility of the next higher layer, TCP in this case, to recover from any errors that occur. Because delivery is not guaranteed, there is no particular reliability requirement on any of the subnetwork types. Because the sequence of delivery is not guaranteed, successive datagrams can follow different paths through the Internet. This allows the protocol to react to congestion and failure on the Internet by changing routes.

Classical IP Stack

Figure 5.123 illustrates the classical IP stack, which includes end-user applications such as SMTP for the exchange of electronic mail messages and network-specific applications such as the DNS for the node-naming service [5.6]. Other user application-specific protocols are the remote terminal (Telnet), FTP and Network News Transfer Protocol (NNTP) for exchange of

IETF and Internet Standards

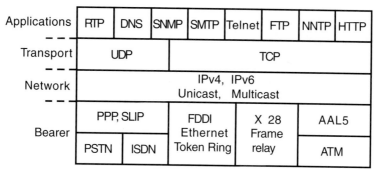

Figure 5.123 IPs [5.226]. ©1997 IEEE.

newsgroup information. The popular Web client/server applications based on the HTTP have been introduced, too. Among the key features of these applications is the capability to access a huge amount of multimedia information distributed worldwide in a transparent and user-friendly manner. The browsing of information is enabled by the user of the hypertext structure, where a document formatted using HTML contains links to other documents. The management of network resources on the Internet is carried out in the frame of the Simple Network Management Protocol (SNMP). That is an application layer protocol running over UDP for resilience designed to exchange management information among network nodes.

IP Version 6

With the shortcomings of the existing IP becoming increasingly evident, a new protocol known as IPv6 (IP version 6) has been defined. An IPv6 protocol data unit (known as a packet) has the general form shown in Figure 5.124. The only header that is required is referred to simply as the IPv6 header. This is of fixed size with a length of 40 octets, compared to 20 octets in the mandatory portion of the IPv4 header.

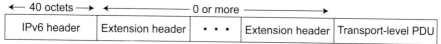

Figure 5.124 IPv6 PDU general form [5.225]. ©1996 IEEE.

The IPv6 header has a fixed length of 40 octets, consisting of the following fields as shown in Figure 5.125.

- *Version (4 bits)*—Indentifies IP version number; the value is 6.
- *Priority (4 bits)*—Indentifies the priority value.
- *Flow label (24 bits)*—May be used by a host to label those packets for which it is requesting special handling by routers within a network.

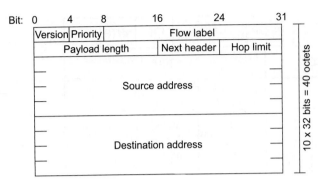

Figure 5.125 IPv6 header [5.225]. ©1996 IEEE.

- *Payload length (16 bits)*—Indentifies the length of the remainder of the IPv6 packet following the header, in octets. In other words, this is the total length of all the extension headers plus the transport-level PDU.
- *Next header (8 bits)*—Identifies the type of header immediately following the IPv6 header.
- *Hop limit (8 bits)*—Identifies the remaining number of allowable hops for this packet. The hop limit is set to some desired maximum value by the source and is decremented by 1 if the hop limit is decremented to zero.
- *Source address (128 bits)*—Identifies the address of the originator of the packet.
- *Destination address (128 bits)*—Identifies the address of the intended recipient of the packet. This may not in fact be intended as the ultimate destination if a routing header is present.

Priority Field

The 4-bit priority field enables a source to identify the desired transmit and delivery priority of each packet relative to other packets from the same source. The field enables the source to identify two separate priority-related characteristics of each packet. First, packets are classified as being part of the traffic for which the source is either providing congestion control or not. Second, packets are assigned one of eight levels or relative priority within each classification.

Congestion-controlled traffic refers to traffic for which the source backs off in response to congestion. An example is TCP. Let us consider what this means. If there is congestion on the network, TCP segments will take longer to arrive at the destination, and, hence, acknowledgments from the destination back to the source will take longer. As congestion increases, it becomes necessary for segments to be discarded en route. The discarding could be done by a router when that router experiences buffer overflow, or it could be done by an individual network allowing the route when a switching node within the network becomes congested. The nature of congestion-controlled traffic is that a variable amount of delay in the delivery of packets, and even for packets to arrive out of order, is acceptable.

Noncongestion-controlled traffic is traffic for which a constant data rate and a constant delivery delay, or at least a relatively smooth data rate and delivery delay, are desirable. Examples are real-time video and audio, for which it makes no sense to retransmit discarded packets. Further, it is important to maintain smooth delivery flow. Eight levels of priority are allocated for this type of traffic from the lowest priority, 8 (most willing to discard), to the highest priority, 15 (least willing to discard). In general, the criterion is how much the quality of the received traffic will deteriorate in the face of some dropped packets. For example, low-fidelity audio, such as a telephone voice conversation, would typically be assigned a high priority. The reason is that the loss of a few packets of audio is readily apparent as clicks and buzzes on the line. On the other hand, a high-fidelity video signal contains a fair amount of redundancy, and the loss of a few packets will probably not be noticeable. Therefore, this traffic is assigned a relatively low priority.

Flow Label

The IPv6 standard defines a flow as a sequence of packets sent from a particular source to a particular (unicast or multicast) destination for which the source desires special handling by the intervening routers. A flow is uniquely identified by the combination of the source address and a nonzero 24-bit flow label. Thus, all packets that are to be part of the same flow are assigned the same flow label by the source. From the source point of view, a flow typically will be a sequence of packets that are generated from a single application instance at the source and that have the same transfer service requirements. A flow may comprise a single TCP connection or even multiple TCP connections. An example of the use of multiple TCP connections is file transfer application, which could have one control connection and multiple data connections. A single application may generate a single flow or multiple flows. An example of the use of multiple flows is multimedia conferencing, which might have one flow for audio and one for graphics windows, each with different transfer requirements in terms of data rate, delay and delay variation.

From the router's point of view, a flow is a sequence of packets that shares attributes that affect how they are handled by the router. These include path, resource allocation, discard requirements, accounting and security attributes. The router may treat packets from different flows in a number of ways, including allocating different buffer sizes, giving different precedence terms of forwarding and requesting different qualities of service from subnetworks.

There is no special significance to any particular flow label. Instead, the special handling to be provided for a packet flow must be declared in some other way. For example, a source might negotiate or request special handling ahead of time from routers by means of a control protocol or at transmission time by information in one of the extension headers in the packet, such as the hop-by-hop options header.

IPv6 Addresses

IPv6 addresses are 128 bits in length. Addresses are assigned to individual interfaces on nodes, not to the nodes themselves. A single interface may have multiple unique unicast addresses. Any of the unicast addresses associated with a node's interface may be used to identify that node uniquely. IPv6 allows three types of addresses:

- *Unicast*—An identifier for a single interface. A packet sent to a unicast address is delivered to the interface identified by that address.
- *Anycast*—An identifier for a set of interfaces (typically belonging to different nodes). A packet sent to an anycast address is delivered to one of the interfaces identified by that address.
- *Multicast*—An identifier for a set of interfaces (typically belonging to different nodes). A packet sent to a multicast address is delivered to all interfaces identified by that address.

Unicast addresses. These addresses may be structured in a number of ways. The following have been identified: provider-based global, link-local, site-local, IPv4-compatible, IPv6 and loopback.

A provider-based global unicast address provides for global addressing across the entire universe of connected hosts. The address has five fields after the packet prefix:

- *Registry identifier*—Identifies the registration authoring that assigns the provider portion of the address.
- *Provider identifier*—Identifies a specific service provider that assigns the subscriber portion of the address.
- *Subscriber identifier*—Distinguishes among multiple subscribers attached to the provider portion of the address.
- *Subnet identifier*—Identifies a topologically connected group of nodes within the subscriber network.
- *Interface identifier*—Identifies a single node interface among the group of interfaces identified by the subnet prefix.

Link-local addresses are to be used for addressing on a single link or subnetwork. They cannot be integrated into the global addressing scheme.

Site-local addresses are designed for local use, but are formatted in such a way that they can be later integrated into the global address scheme. The advantage of such addresses is that they can be used immediately by an organization that expects to transition to the use of global addresses.

A key issue in deploying IPv6 is the transition from IPv4 to IPv6. It is not practical to replace all IPv4 routers on the Internet or a private Internet with IPv6 routers and to replace all IPv4 addresses with IPv6 addresses. Instead, there is a lengthy transition period when IPv6 and IPv4 must coexist. IPv4-compatible IPv6 addresses accommodate this coexistence period. Full coexistence can be maintained as long as all IPv6 nodes employ an Ipv4-compatible address. As general IPv6 addresses come into use, coexistence will be more difficult to maintain.

The unicast address 0: 0: 0: 0: 0: 0: 0: 1 is called the loopback address. It may be used by a node to send IPv6 packet to itself. Such packets are not sent outside a single node.

Anycast addresses. An anycast address enables a source to specify that it wants to contact any one node from a group of nodes using a single address. A packet with such an address will be routed to the nearest interface in the group, according to the router's measure of distance. Anycast addresses are allocated from the same address space as unicast addresses. Thus, members of an anycast group must be configured to recognize that address, and routers must be configured to be able to map an anycast address to a group of multicast interface addresses.

An example of the use of an anycast address is within a routing header to specify an intermediate address along a route. The anycast address could refer to the group of routers associated with a particular provider or particular subnet, thus dictating that the packet be routed through the Internet in the most efficient manner.

Multicast addresses. IPv6 includes the capability to address a predefined group of interfaces with a single multicast address. A packet with a multicast address is to be delivered to all members of the group. A multicast address consists of an 8-bit format prefix of all ones, a 4-bit flags field, a 4-bit scope field and a 112-bit group identifier.

Multicasting is a useful capability in a number of contexts. For example, it allows hosts and routers to send neighbor discovery messages only to those machines that are registered to receive them, removing the necessity for all other machines to examine and discard irrelevant packets. As another example, most LANs provide a natural broadcast capability. A multicast address can be assigned that has a scope of link-local with a group ID configured on all nodes on the LAN to be a subnet broadcast address.

Hop-by-Hop Options Header

This header carries optional information that, if present, must be examined by every router along the path. It consists of next header, header extension length and/or more options as shown in Figure 5.126.

Next header (8 bits) identifies the type of header immediately following this header. The length of header extension is 8 bits. A variable-length field consists of one or more option definitions. Each definition is in the form of three subfields: option type (8 bits), which identifies the option; length (8 bits), which specifies the length of the option data field in octets and option data, which is a variable-length specification of the option. It is actually the lowest-order five bits of the option type field that are used to specify a particular option. The higher-order two bits indicate the action to be taken by a node that does not recognize this option type. The conven-

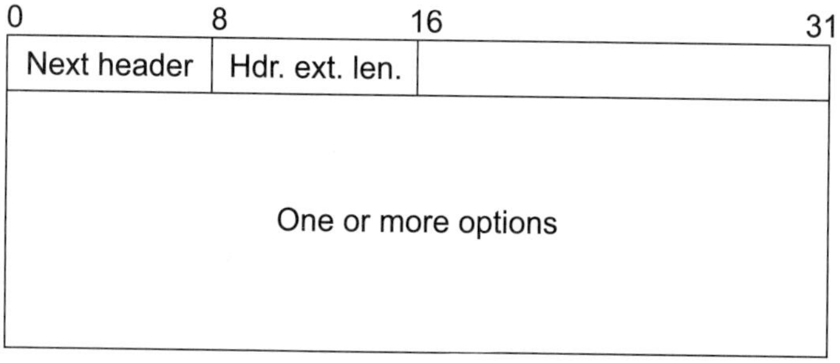

0	8	16	31
Next header	Hdr. ext. len.		
One or more options			

Figure 5.126 Hop-by-hop options header [5.225]. ©1996 IEEE.

tions for the option type field also apply to the destination options header. This header carries optional information that, if present, is examined only by the packet's destination mode.

Fragment Header

IPv6 fragmentation may be performed only by service nodes, not by routers along a packet delivery path. The path-discovery algorithm enables a node to learn the Maximum Transmission Unit (MTU) of the bottleneck subnetwork on the path. With this knowledge, the source node will fragment, as required, for each given destination address. Otherwise, the source must limit all packets to 576 octets, which is the minimum MTU that must be supported by each subnetwork. The fragment header (Figure 5.127) consists of the following:

- *Next header (8 bits)*—Identifies the type of header immediately following this header.
- *Reserved (8 bits)*—Remains for future use.
- *Fragment offset (13 bits)*—Indicates where in the original packet the payload of this fragment belongs. This implies that fragments must contain a data field that is a multiple of 64 bits long.
- *Res (2 bits)*—Remains reserved for future use.
- *M flag (1 bit)*—Identifies that 1=more fragments and 0=last fragment.
- *Identification (32 bits)*—Identifies the original packet uniquely. The identifier must be unique for the packets' source address and destination address for the time during which the packet will remain on the Internet.

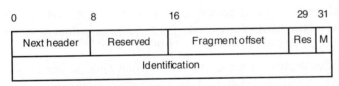

Figure 5.127 Fragment header [5.225]. ©1996 IEEE.

Routing Header

The routing header contains a list of one or more intermediate nodes to be visited on the way to a packet's destination. All routing headers starts with a 32-bit block consisting of four 8-bit fields, followed by routing data specific to a given routing type. A generic routing header is presented in Figure 5.128. The four 8-bit fields are the following:

- *Next header*—Identifies the type of header immediately following this header.
- *Header extension length*—Indentifies the length of this header in 64-bit units, not including the first 64 bits.
- *Routing type*—Identifies a particular routing header variant. If a router does not recognize the routing type value, it must discard the packet.
- *Segments left*—Identifies the number of explicitly listed intermediate nodes still to be visited before reaching the final destination.

IETF and Internet Standards

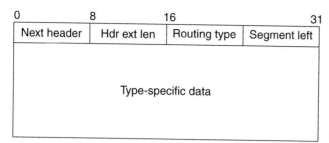

Figure 5.128 Generic routing header [5.225]. ©1996 IEEE.

IPv6 requires an IPv6 node to reverse routes in a packet that it receives containing a routing header in order to return a packet to the sender. Figure 5.129 shows a configuration in which two hosts are connected by two providers, and the two providers are in turn connected by a wireless network. IPv6 has the ability to select a particular provider to maintain connections while mobile and to route packets to new addresses dynamically.

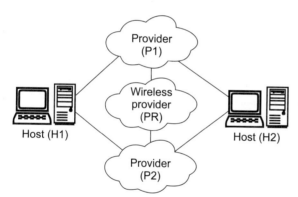

Figure 5.129 Examples of a routing configuration.

IPv6 Security

The Internet community has developed application-specific security mechanisms in a number application areas, including elecetronic mail, network management, Web access and others. However, users have some security concerns that cut across protocol layers. IP-level security encompasses two functional areas: authentication and privacy. The authentication mechanisms ensure that a received packet was in fact transmitted by the party identified as the source in the packet header. In addition, this mechanism ensures that the packet has not been altered in transit. The privacy facility enables communicating nodes to encrypt messages to prevent eavesdropping by third parties. In August 1995, the IETF published five security-related proposed standards that define a security capability at the Internet level. The documents provide the following [5.225]:

- *RFC1825*—Overview of security architecture
- *RFC1826*—Description of a packet authentication extension to IP

- *RFC1827*—Description of a packet encryption extension to IP
- *RFC1828*—Specific authentication mechanism
- *RFC1829*—Specific encryption mechanism

Support for these features is mandatory for IPv6 and optional for IPv4. In both cases, the security features are implemented as extension headers that follow the main IP header. The extension header for authentication is known as the authentication header, and the header for privacy is known as the Encapsulating Security Payload (ESP) header.

The authentication header provides support for data integrity and authentication of IP packets. The authentication header is given in Figure 5.130. It consists of the following fields [5.225]:

- *Next header (8 bits)*—Identifies the type of header immediately following this header.
- *Length (8 bits)*—Identifies the length of authentication data field in 32-bit words.
- *Reserved (16 bits)*—Remains for future use.
- *Security parameter index (32 bits)*—Identifies a security association.
- *Authentication data (variable)*—Identifies an integral number of 32-bit words.

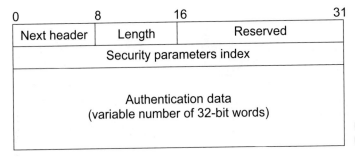

Figure 5.130 Authentication header [5.225]. ©1996 IEEE.

What the authentication data field contains will depend on the authentication algorithm specified. In any case, the authentication data is calculated across the entire IP packet, excluding any fields that may change in transit. Such fields are set to zero for purposes of calculation at both source and destination. The authentication calculation is performed prior to fragmentation at the source and after reassembly at the destination. Hence, fragmentation-related fields can be included in the calculation. The two IP security mechanisms can be combined in order to transmit an IP packet that has both privacy and authentication. Two approaches can be used based on the order in which the two services are applied: encryption before authentication and authentication before encryption.

Figure 5.131a illustrates the case of encryption before authentication. In this case, the entire transmitted IP packet is authenticated, including both encrypted and unencrypted parts. In this approach, the user first applies ESP to the data to be protected. Then, he presents the authen-

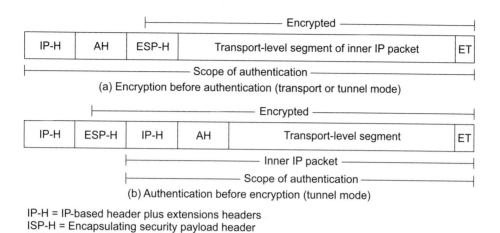

IP-H = IP-based header plus extensions headers
ISP-H = Encapsulating security payload header
E-T = Encapsulating security payload trailing fields
AH = Authentication header

Figure 5.131 Privacy and authentication combination [5.225]. ©1996 IEEE.

tication header and the plain-text IP header(s). There are two subcases: transport-mode ESP and tunnel-mode ESP.

In transport-mode ESP, authentication applies to the entire IP packet delivered to the ultimate destination, but only the transport-layer segment is protected by the privacy mechanisms (encrypted).

In tunnel-mode ESP, authentication applies to the entire IP packet delivered to the other IP destination address. Authentication is performed at that destination. The entire inner IP packet is protected by the privacy mechanisms for delivery to the inner IP destination.

Figure 5.131b illustrates the case of authentication applied before encryption. This approach is appropriate only for tunnel-mode ESP. In this case, the authentication header is placed inside the inner IP packet. This inner packet is both authenticated and protected by the privacy mechanisms. The use of authentication prior to encryption might be preferable for several reasons. First, because the Authentication Header (AH) is protected by ESP, it is impossible for anyone to intercept the message and the AH without detection. Second, it may be desirable to store the authentication information with the message and the destination for later reference. It is more convenient to do this if the authentication information applies to the unencrypted message. Otherwise, the message would have to be re-encrypted to verify the authentication information.

5.10.4 Real-Time Multimedia Transmission across the Internet

The growth of the Internet and intranets has attracted a great deal of attention to the implementation and performance of networked multimedia services that involve the transport of real-time multimedia data streams over nonguaranteed QoS networks based on the IP. Continuing advances in computing technology, together with developments in signal coding and network protocols, have made transmission of real-time multimedia data across the Internet and intranets a viable and important application. An understanding of the Internet multimedia data transmission architecture is beneficial for developing signal-processing applications suitable for this fast growth area. Furthermore, effective design and use of the intermediate protocol layers of this architecture requires in-depth knowledge of both signal processing and networking. Based on their functionalities, the protocols directly related to real-time multimedia data transmission across the Internet can be classified into four categories:

- Signaling
- Session control
- Transport
- Network infrastructure

Signaling

Several protocols can be used for the higher-layer functions of signaling and session control. Signaling includes sending announcements about a multimedia session to prospective participants or inviting selected participants to join a session. In both cases, the details of the session, including the types of compression techniques used for audio and video signals, the number of audio channels, and so forth, may be a part of the signaling message. Generating and handling the responses of the receiver to a signaling message, for example, accept, join, reject, busy, forward and so forth, are handled by signaling protocols, too.

The ability of a receiver to decode the selected payload types and possible negotiations of the capabilities may be covered by signaling. Current protocols supporting signaling for multimedia sessions on the Internet include Session Description Protocol (SDP) for describing multimedia sessions [5.227], Session Announcement Protocol (SAP) [5.228] for announcing the described sessions and Session Initiation Protocol (SIP) for inviting users (human or machine) to participate in multimedia sessions [5.229]. HTTP and URL can be used to announce and describe sessions in a bulletin board format, which may also be considered as a part of a specific type of signaling.

Session Control

This defines the messages and procedures to control the delivery of the multimedia data during an established session. The Real-Time Streaming Protocol (RTSP) addresses tasks such as providing a means for choosing delivery channels and mechanisms, selecting a multimedia data segment for playback and controlling playback or recording properties using controls similar to the familiar ones on VCRs [5.230]. The H.323 standard defined by ITU-T standardizes both signaling and session control for tightly coupled multimedia communications sessions. A discussion of the relation between H.323 and other IPs can be found in [5.231].

Transport

The transport protocol has very tight relationships with the way that the multimedia payload types are organized and used. We discuss here the details of the transport layer based on the RTP [5.232, 5.233].

RTP is designed to deliver various kinds of real-time data across packet networks. It addresses the needs of real-time data transmission only and relies on other well-established network protocols for other communication services, such as routing, multiplexing and timing. RTP is based on the Application Level Framing (ALF) and Integrated Layer Processing (ILP) principles. They dictate using the properties of the payload in designing a data transmission system as much as possible [5.234]. For example, if we know that the payload is MPEG, we should design our packetization scheme based on slices because they are the smallest independently decodable data units for MPEG Video. This approach provides a much more suitable framework for MPEG transmission across networks with high packet loss rates. Also, we can identify and protect the critical information by repeating it frequently or sending it across a reliable channel. The services provided by the RTP are discussed in the following sections.

Payload type identification. The type of the payload contained in an RTP packet is indicated by an integer in a special field at the packet header. The receiver interprets the content of the packet based on this number. Certain common payload types have assigned payload type numbers. For other payloads, this association can be defined externally, for example, through signaling during the starting of a session or with session control protocols. The payload type identification service of the RTP together with the multiplexing services supported by the underlying transport protocol, such as UDP, provides the necessary infrastructure to multiplex a large variety of information effectively. Multicast transmission of several multimedia streams multiplexed together with any other type of information can easily be handled using these services.

RTP allows additional information to be added to its generic headers for each payload type. This information may be used to increase the packet-loss resiliency of the transmission. For example, each RTP packet carrying MPEG Video contains information about the picture type (intra, predictive and bidirectional), motion vector ranges, and so forth, copied from the latest picture header, increasing the decodability of individual packets.

Packet sequence numbering. Each RTP packet that belongs to a stream contains a 16-bit sequence number field that is incremented by one for each packet sent. The sequence numbers make packet loss detection possible because the lower protocol layers need not provide this information. Also, packets received out of order can be reordered using the sequence numbers. The initial sequence number is selected as a random number so that RTP packets do not cause known-plaintext attacks on the encryption that may be used at some later stage of their transmission.

Because the packets may be delivered out of order, receipt of a packet with an out-of-order sequence number does not necessarily imply packet loss. In most applications, a certain number of packets are buffered before starting the playback so that late or out-of-order packets can be used when they arrive. The buffer size depends on the network jitter, and, for interactive real-time applications, the buffer size is limited by the allowed delay.

Time stamping. Each RTP packet carries a 32-bit time stamp that reflects the sampling instant of the first byte in the payload portion of the packet. The interpretation and use of the time stamp are payload dependent. For example, for MPEG ES payloads, the time stamp represents the presentation time of the MPEG picture or audio frame, a portion of which is carried by the packet, based on a 90-KHz clock. It is the same for all packets that make up a picture or audio frame and, in a video stream with B-frames, it is not monotonically increasing. On the other hand, for fixed-rate audio (for example, PCM), the time stamp may reflect the sampling period. If blocks covering n audio samples are read from an input device, the time stamp would be increased by n for each such block, regardless of whether the block is transmitted in a packet or dropped as silent.

The time stamp, together with the information provided by the associated RTCP packets, is to be used for the following:

- Encoder/decoder clock matching
- Synchronization of several sources
- Measurement of packet-arrival jitter

Similar to the initial value of the sequence numbers, the initial value of the time stamp is random in order to make known-plaintext attacks on encryption difficult.

Source identification. The source of each RTP packet is identified by an integer called Synchronization SouRCe (SSRC) identifier included in the packet header. Each sender initially picks a random number for its SSRC. It is the sender's responsibility to detect and resolve collisions when more than one source picks the same number in the same session. The relation between several sources participating in a session, as well as their characterizing names, is established through RTCP.

The delivery-monitoring function of the RTP is carried out using the associated protocol, RTCP. RTCP is based on periodic transmission of control packets from all participants of a session to all other participants using the same distribution mechanism as the RTP data packets. RTCP's main functions are described in the following sections.

Feedback on the quality of distribution and timing. In an RTP session, each sender and each receiver send periodic reports to each session participant. Part of this report contains information on the quality of reception characterized as the following:

- Fraction of the lost RTP packets since the last report
- Cumulative number of packets lost since the beginning of reception
- Packet interarrival jitter
- Delay since receiving the last sender's report.

Sender and receiver reports contain enough information to determine these quantities at each participant's location. This feedback in reception quality is an integral part of RTP, and it is intended to be used for congestion and flow-control purposes as well as network performance input for the adaptive coding applications. Because RTP does not define an explicit flow control mechanism, an RTP application is capable of generating high traffic rates, causing network congestion. It is important to prevent this by analyzing the RTCP packets coming from the receivers so that other network applications are not disturbed.

Sending the feedback reports to all participants makes it possible to determine the extent of network problems. Additionally, a network management entity may monitor the network performance by observing these reports without actively participating in each session.

As for the timing, each sender's periodic RTCP packets contain 64-bit Network Time Protocol (NTP) time stamps, indicating the wall-clock (absolute) time when the RTCP packet was sent. This information can be used in combination with the timing information returned in reception reports from other receivers to measure roundtrip propagation to those receivers. Additionally, the sender's RTCP packet contains an RTP time stamp that corresponds to the same time as the NTP time stamp, but in the same units and with the same random offset as the RTP time stamps of the RTP data packets. This correspondence is to be used for intra- and intermedia synchronization for sources with synchronized NTP time stamps. A detailed discussion of the clock synchronization procedures can be found in Mills [5.235].

Participant identification. Special RTCP messages are used to establish a connection between the real identification of an RTP source, called by its canonical name (CName), and the current SSRC numbers used by it. CNames are very similar to email addresses following the **user name@host** syntax. Also, identification messages carry additional information about the participants, such as their names, email addresses, phone numbers, and so forth.

Control packet transmission scaled to the number of participants. As the number of the session participants increases, unregulated RTCP message traffic may consume significant bandwidth. In order to prevent this, RTCP scales itself by changing its message transmission interval based on the number of session participants. The suggested RTCP bandwidth is less than 5% of the bandwidth allocated for a session. Algorithms to achieve this are discussed in Schulzinne [5.232].

Minimal session control information. This optional functionality can be used for conveying simple session information, such as names of the participants, to everyone.

Network Infrastructure

All real-time multimedia data transmission applications across the Internet depend on both of the fundamental Internet transport protocols, UDP and TCP, for several functions, such as multiplexing, error control, flow control, and so forth. In turn, TCP and UDP depend on the basic IP for the network services support including network addressing [5.236].

The Point-to-Point Protocol (PPP), which defines a standardized method for sending datagrams across communications links such as telephone and ISDN lines, is an integral part of several real-time data transmission applications, such as Internet telephony. Several other protocols addressing specific requirements of real-time delivery are on their path to becoming standards. RSVP, which defines and implements QoS requirements, is important for multimedia data delivery across the Internet.

The lower layer (network infrastructure) protocols have a fundamental impact on the performance and usability of signal-coding techniques in networked applications. For example, if the network offers some service guarantees, such as delay bounds or guaranteed packet deliveries (no loss), signal-coding techniques with no error resilience can be used. If appropriate data flow control is done at the lower layers, application designers need not worry about network-buffer overflows due to short-term high-output data rates as in the case for I-frames in MPEG Video. In many cases, such additional services offered by the lower layers are not free, and a price-performance compromise may be obtained by using layered coding techniques. In this case, specialized services are needed only for transmitting a portion of the encoded data streams.

Multimedia Data for Network Use

It is well known that delivering real-time data encoded in any form across the Internet is possible. Nevertheless, real-time multimedia streams with the properties described in the following sections are more convenient for networked applications.

Natural breakpoints for packetization. Packetizing a stream that has natural breakpoints can be easier and more efficient. As an example, if a picture is JPEG-coded and is presented to a packetizer, the resulting packets contain arbitrary sections of the encoded data. If one of these packets is lost, it will be practically impossible to decode the remaining packets even if they are received, However, if the same JPEG coded picture contains special restart markers indicating starting of independently decodable blocks, a lost packet will not cause such a problem.

Adjustable packet sizes. Different technologies used as parts of the Internet have different frame (largest data unit) sizes. In order to carry a packet that is larger than the smallest frame size allowed on its path, called an MTU, the packet needs to be fragmented and reassembled. If the size of a packet can be changed based on the MTU, fragmentation can be avoided.

Well-defined high-priority information. If certain parts of a data stream are vital for decoding the rest of it, it is preferable to have these parts in easily identifiable and separable sections so that they can be transmitted more reliably.

Flexible rate control. An encoding scheme that has a rate that can easily be changed is useful in adapting its output to network conditions.

Ease of transcoding. The heterogeneity of bandwidths used for Internet access requires using different rates for the same multimedia material. Data streams that can easily be transcoded to change their bandwidth are definitely preferable.

Layered coding. Layered coding is beneficial for two purposes. The first one is to remove the need for transcoding by providing representations of the same multimedia source at different bit rates without noticeably increasing the overall bandwidth. The second benefit is to obtain a price/performance compromise by sending only a portion of a stream through channels.

Resilience to error propagation. Assuming that the packet losses will be unavoidable in the foreseeable future, techniques that prevent or reduce the propagation of data loss effects are preferable.

5.10.5 MPEG-4 Video Transport across the Internet

MPEG-4 is a standard designed for representation and delivery of multimedia information across a variety of transport protocols. It includes interactive scene management, visual and audio representations and systems functionality like multiplexing, synchronization and object-descriptor framework. The MPEG-4 Systems specification defines an architecture and tools to create audiovisual scenes from individual objects. The scene description and synchronization tools are the core of the Systems specification [5.64].

The scene description is encoded separately and is treated as another elementary bit stream. This separation allows for providing different QoSs for a scene description that has very low or no loss tolerance and media streams in the scene that are usually loss tolerant. The MPEG-4 scene description, also referred to as BIFS, is based on VRML and specifies the spatiotemporal composition of scenes. BIFS update commands can be used to create scenes that evolve over time. This architecture allows creation of complex scenes with potentially hundreds of objects. This calls for a high rate of establishment and release of numerous short-term transport channels with the appropriate QoS. DMIF is a general applications and transport delivery framework specified by MPEG-4 [5.43]. DMIF's main purpose is to hide the details of the transport network from the user as well as to ensure signaling and transport interoperability between end systems. In order to keep the user unaware of underlying transport details, MPEG-4 defined an interface between user-level applications and DMIF called DAI. The DAI provides the required functionality for realizing multimedia applications with QoS support. Although DMIF makes transport-independent application development possible, there is also a need to develop transport dependent mappings for MPEG-4 content to be able to use existing infrastructure and support applications that do not use DMIF. The IETF group is specifying payload formats and synchronization schemes for delivering MPEG-4 presentations using RTP [5.237].

Use of RTP

A number of Internet drafts describe RTP packetization schemes for MPEG-4 data [5.238]. Media-aware packetization (for example, video frames split at recoverable subframe boundaries) is a principle in RTP, so it is likely that several RTP schemes will be needed to suit the different kinds of media, audio, video, and so forth. No matter what packetization scheme is used, they must have a number of common characteristics. However, such characteristics depend on the fact that the RTP session contains a single ES or a FlexMux stream. An RTP session contains a single ES with the following characteristics:

- The RTP time stamp corresponds to the presentation time if the earliest access unit is within the packet.
- RTP packets have sequence numbers in transmission order. The payloads logically or physically have SL sequence numbers, which are in decoding order, for each ES.
- The MPEG-4 time scale (clock ticks per second) is the time-stamp resolution in the case of MPEG-4 systems and must be used as the RTP time scale.

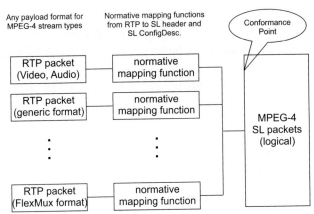

Figure 5.132 RTP-packet-to-SL-packet mapping [5.121]. ©2001 ISO/IEC.

- To achieve a base level of interoperability and to ensure that any MPEG-4 stream may be carried, all senders and receivers must implement a generic payload format.
- Streams should be synchronized using RTP techniques (notable RTCP sender reports). When the MPEG-4 Object Clock Reference (OCR) is used, it is logically mapped to the NTP time axis used in RTCP.
- The RTP packetization schemes may be used for MPEG-4 ES standing alone (for example, without MPEG-4 systems, including BIFS). Each RTP stream is passed through a mapping function, which is specific to the payload format used. This mapping function yields an SL packetization stream. RTP-packet-to-SL-packet mapping is shown in Figure 5.132.

There may be a choice of RTP payload formats for a given stream, for example, as an ES, an SL-packetized stream using FlexMux, and so on. The following is recommended:

- Terminals implementing a given subsystem (for example, video) accept at least an ES and the default SL packing of that stream, if they exist.
- Terminals implementing a given payload format accept any stream over that format for which they have a decoder, even if that packing is not normally the best packing.

Future versions of this specification will identify the single standard RTP packing format for each MPEG-4 stream type.

System Architecture

The MPEG-4 system developed is an end-to-end system consisting of an MEPG-4 server, the DMIF component for signaling and session management on an IP network and an MPEG-4 client for media playback and rendering. Figure 5.133 shows the components of the system.

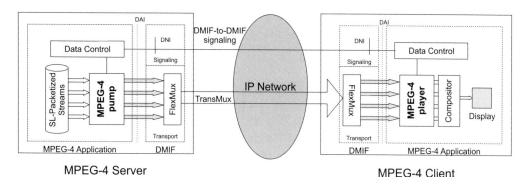

Figure 5.133 MPEG-4 System architecture [5.121]. ©2001 ISO/IEC.

The media data and the media composition data are transmitted to a client as separate streams, typically with different QoS requirements, in the same session. Furthermore, because the number of objects in a presentation can be quite large, the overhead required to manage the session is large. Interactivity makes this problem more complex because the resource required for a session will now depend on the user behavior, essentially when user interaction with objects changes the number of objects in the scene either by adding or deleting objects. The MPEG-4 server consists of an MPEG-4 pump, an object scheduler and a DMIF instance for IP networks. The server delivers SL packets to the DMIF layer, which multiplexes them in a single FlexMux stream and transmits them to the client [5.239]. The complexity of the player (that is, client) has grown as a result of the new features and functionality offered by MPEG-4. The player is responsible for compositing a scene from individual objects in addition to decoding and displaying the object.

A player consists of three logical components: a DMIF instance, ES decoders, and compositor. The DMIF instance is responsible for managing data access from a network or a file. A player typically contains several decoders, each handling a specific ES. ESs are audiovisual streams as well as streams that describe the composition, rendering and behavior of a presentation. Each object in a presentation is carried in a separate ES. Because MPEG-4 presentations can include media objects from several sources, potentially with different clock frequencies, there is an additional burden on the client to track multiple clocks. This is typically done using soft Phase-Locked Loops (PLLs). Because of this additional complexity, a player's performance depends on the complexity of the content. Intelligent resource management and usage are necessary to use the resources, such as memory, efficiently.

The capability to add and remove objects during presentations and to interact with objects differentiates object-based audiovisual presentation. DMIF also supports network independence by providing a DAI. DMIF is also responsible for negotiating the requested QoS for the applications. Typically, streams with the same QoS requirements are multiplexed into a single channel using FlexMux. These FlexMux packets are transported to the other end on the underlying transport network where they are demultiplexed by the peer DMIF entity and passed on to the application. DMIF is also responsible for communicating user interaction commands from Command Ds by means of DAI user-command primitives. These commands are transmitted across the DMIF-to-DMIF signaling channel.

A session is established before a server starts transmitting objects to a client. Session establishment is done by DMIF upon a request from the client. The MPEG-4 presentation to be delivered is selected by the client and communicated to the server as a part of session-establishment messages. MPEG-4 does not specify how a client selects a presentation. As the session establishment continues, the server sends the initial OD to the client. This initial OD contains pointers to the ESs that are part of the session. The client uses this information to request additional channels for the ESs. Each presentation contains at least two ESs, a scene-description stream and an OD stream. In addition to these two ESs, a presentation may include a clock-reference stream, command D stream, IPMP D stream or an OCI stream, in addition to the media stream. Intermedia synchronization is achieved using decoding and composition time stamps contained in SL packets. All ESs received by the client are time stamped. The time stamp indicates the time an access unit is processed by the client.

MPEG-4 Server

Figure 5.134 shows the components of the MPEG-4 server. The server delivers objects in a presentation as scheduled by the scheduler. The ESs, carrying media and media data for the presentation, are stored in the form of SL-packetized streams (SPS). The SL packet header contains the information, such as decoding and composition time stamps, clock references and packet repetition flags.

Scheduling and multiplexing of audiovisual objects in a presentation is a complex problem. Because of the different application domains, no solutions can be directly applied to the problem of scheduling audiovisual objects and also of scheduling in the presence of user interaction that might alter the presentation. The resource constraints that affect the transmission of access units are the channel capacity and buffer capacity at the receiving terminal [5.240].

The scheduler is also useful during the content creation process to determine if the presentation being designed can be scheduled for specific channel rates and client buffer capacity. It may not be possible to find a solution for a given set of resources, that is, the presentation cannot be scheduled with the given resources. In order to create a schedulable presentation, some constraints may be relaxed. In the case of scheduling objects, relaxing a constraint may involve increasing the buffer capacity, increasing the channel capacity, not scheduling some object instances, or removing some objects from a presentation.

It is also necessary to solve the problem online or, in some cases, to compute incremental schedules. Computing a schedule in real time is necessary to support interactive applications. When a user event adds a new object to the presentation, the resulting schedule has to be computed in real time to determine if the event can be supported. A scheduler may also be used offline to determine if an MPEG-4 presentation is suitable for a given channel and buffer capacities.

The MPEG-4 pump talks to a client using DMIF during session setup and delivers access units during the session. The server maintains a list of sessions established with clients and a list

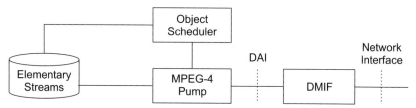

Figure 5.134 MPEG-4 Server components [5.121]. ©2001 ISO/IEC.

of channels for each session. Session identifiers and channel identifiers are used to identify channels in signaling between clients and a server. Network delays and data loss occur in the system if the content is not designed properly. A presentation created without the knowledge of target networks and clients could create long start-up delays, buffer overflows or underflows. This could cause distortion, gaps in media playback and problems with the synchronization of different media streams.

MPEG-4 Client

In order to achieve media synchronization, MPEG-4 defines a system decoder model, which abstracts the behavior of the receiving terminal in terms of synchronization, buffer management and timing for temporally accurate presentations of MPEG-4 scenes. These scenes, which are hierarchical groupings of media objects carrying synthetic or natural content, are composed using the scene description information coded using BIFS. BIFS is augmented by the OD framework. While the scene description declares the spatiotemporal relationship of the media objects, the ODs in the OD framework identify the resources for the ESs that carry these media objects and that associate the streams with the systems decoder model. BIFS and the OD framework form part of the basic building blocks for the architecture of the MPEG-4 client.

Figure 5.135 illustrates the architecture for the implementation of an MPEG-4 client used in the streaming application. The controller manages the flow of control and data information, the creation of buffers and decoders and attaches the transport channels established by the DMIF layer to the decoding buffers. The amount of buffer size to allocate for the decoding buffers is specified in decoder configuration Ds. The SL Manager manages a set of transport channels for receiving BIFS, OD and media streams by binding them to their respective decoding buffers. The SL Manager also provides functionality for forwarding client requests through the DAI to the DMIF process and receiving both control and data information from the server. The compositor encompasses a set of other components for rendering MPEG-4 presentations/scenes and for handling user events from the applications.

First, the client application requests a session establishment with the server and specifies the MPEG-4 presentation to be played. The SL Manager forwards this request to the DMIF layer, which handles the establishment of the session. In case of a successful session establishment, the server provides the initial OD information. This information is passed from the client DMIF through a specific DAI primitive to the SL manager. The initial OD is used for allocating buffers for the scene description (BIFS), OD and command descriptor streams. Then, the presen-

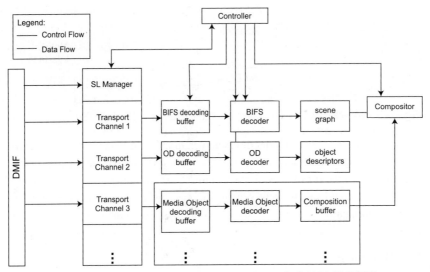

Figure 5.135 Architecture of an MPEG-4 client [5.121]. ©2001 ISO/IEC.

tation controller, which is a part of the compositor module requests, from the SL Manager the establishment of transport channels for the reception of the associated media streams. The SL Manager invokes the respective DAI commands and forwards this request to the DMIF layer. The DMIF establishes the transport channels taking into account the QoS characteristics of the media streams.

After the creation of the transport channels, the client may start receiving the MPEG-4 data. The BIFS data units are decoded to form a scene graph, which is a hierarchical ordering of nodes that describes the media object in a scene. The node attributes define the behavior and appearance of the object as well as their placement in space and time. The OD data units are decoded into a list of ODs used for associating the media streams (that is, ESs) with the media objects and for configuring the system to receive media streams.

The decoded objects in an MPEG-4 presentation are composed and presented by the compositor. The composition process involves positioning the media objects in a scene using scene description and handling the dynamic behavior of scenes.

5.11 Concluding Remarks

Standards play a major role in the multimedia communications because they provide interoperability between hardware and software provided by multiple vendors. However, production of standards is beset by the problem that the many industries having a stake in it have radically different approaches to standardization. The success of the MPEG, ITU-T and IETF standardization approaches is based on a number of concurrent elements.

MPEG is the leading standardization body in audiovisual representation technology. After the great success of the MPEG-1 and MPEG-2 standards, which opened the digital frontiers to audiovisual information and allowed the deployment of high performance services, MPEG-4 standard supports new ways of communication access and interaction with digital audiovisual data, and offers a common technical solution to various services. It also extends to layered coding (scalabilities), multiview (stereoscopic video), shape/texture/motion coding of objects, and animation. Its role extends to the Internet, Web TV, large databases (storage, retrieval and transmission), and mobile networks. The new standard MPEG-7 specifies a standardized description of various types of multimedia information. This description is associated with the content itself, to allow fast and efficient searching for multimedia that is of interest to users. The description can be attached to any kind of multimedia material, no matter what the format of the description is. Stored material that has this information attached to it can be indexed, searched and retrieved. The latest MPEG project MPEG-21 Multimedia Frameworks has been started with the goal to enable transparent and augmented use of multimedia resources across a wide range of networks and devices.

The ITU-T standardization process in multimedia communications deals with video and speech coding as well as multimedia multiplex and synchronization. In the past, most video compression and coding coding standards were developed with a specific application and networking infrastructure in mind. ITU-T Recommendation H.261 was optimized for use with interactive audiovisual communication equipment, e.g., a videophone, and in conjunction with the H.320 series of recommendations as multiplex and control protocols on top of ISDN. Consequently, the H.261 designers made various design choices that limit the applicability of H.261 to this particular environment. The original H.263 was developed for video compression rates below 64 Kbs per second. This was the first international standard for video compression which would permit video communications at such low rates. After H.263 was completed, it become apparent that there were incremental changes that could be made to H.263 that visibly improved its compression performance. It was thus decided in 1996 that a revision to H.263 would be created which incorporated these incremental improvements. ITU-T Recommendation H.263 Version 2 (H.263+) is the very first international standard in the area of video coding which is specifically designed to support the full range of both circuit-switched and packet-switched networks. H.263+ contains functionalities that improve the quality of video transmission in error-prone environments and nonguaranteed quality of service (QoS) networks. H.26L is an ongoing standard activity that is searching for advanced coding techniques that can be fundamentally different from H.263.

The IETF is focused on the development of protocols used on IP-based networks. The IETF is different from most standardization bodies in that it is a totally open community with no formal membership. One of the strengths of the Internet is its global connectivity. For this connectivity, it is essential that all the hosts in the Internet interoperate with one another, understanding the common protocol at various layers. The Internet standardization process of IETF under the Internet Society (ISOC) is the key to the success of multimedia communications over IP-based networks such as the Internet.

CHAPTER 6

Multimedia Communications Across Networks

Chapter Overview

This chapter concentrates on multimedia communications across networks. After an introductory discussion concerning packet audio-video in the network environment, we invoke the concept of video transport across generic networks. We then describe Multimedia transport across ATM networks. This is followed by multimedia across IP networks, including video transmission, traffic specifications for MPEG video transmission on the Internet and bandwidth allocation mechanism. We outline the issues concerning multimedia Digital Subscriber Lines (DSLs). The concepts of Internet access networks are presented and illustrated. We finally discuss special issues relating to multimedia across wireless networks, such as wireless broadband communication for multimedia audiovisual solutions, mobile and broadcasting networks as well as digital TV infrastructure for interactive multimedia services.

6.1 Packet Audio/Video in the Network Environment

Packet-switched networks were invented for carrying computer data because the burst-type nature of such information makes it uneconomical to use continuously connected circuits. Audio and video signals, in contrast, have for many years been carried across fixed-bit-rate circuit-switched connections. However, developments in ATM networks have generated discussions between network and coding specialists concerning the potential advantages of variable bit-rate transmissions across such networks. In recent years, considerable interest has been shown in the general statistical multiplexing of digitally encoded audio and video signals, and particular attention has been given to packet-based systems. A large number of papers and books has appeared in the literature, addressing topics such as delays involved, the associated queuing problems, the effects of packet loss and the regeneration of lost packets.

The increase in communication of multimedia information over the past decades has resulted in many new multimedia processing and communication systems being put into service. The growing availability of optical fiber links and rapid progress in Very Large-Scale Integration (VLSI) circuits and systems have fostered a tremendous interest in developing sophisticated multimedia services with an acceptable cost. Today's fiber technology offers a transmission capacity that can easily handle high bit rates. This leads to the development of networks that integrate all types of information services [6.1]. By basing such a network on packet switching, the services (video, voice and data) can be dealt with in a common format. Packet switching is more flexible than circuit switching in that it can emulate the latter while vastly different bit rates can be multiplexed together. In addition, the network's statistical multiplexing of variable rate sources may yield a higher fixed-capacity allocation [6.2, 6.3, 6.4].

6.1.1 Packet Voice

In comparison to circuit-switched networks, packet switching offers several potential advantages in terms of performance. One advantage is efficient use of channel capacity, particularly for bursty traffic. Although not as bursty as interactive data, speech exhibits some burstiness in the form of talksparts [6.5]. Average talkspart duration depends on the sensitivity of the speech detector, but it is well known that individual speakers are active only about 35 to 45% in typical telephone conversations. By sending voice packets only during talksparts, packet switching offers a natural way to multiplex voice calls as well as voice with data. Another advantage is that call blocking can be a function of the required average bandwidth rather than the required peak bandwidth. In addition, packet switching is flexible. For example, packet voice is capable of supporting point-to-multipoint connections and priority traffic. Furthermore, because packets are processed in the network, network capabilities in traffic control, accounting and security are enhanced. However, packet voice is not without difficulties. Continuous speech of acceptable quality must be reconstructed from a voice packet that experiences variable delays through the network. The reconstruction process involves compensating for the variable delay component by imposing an additional delay. Hence, the packet should be delivered with low-average delay and delay variability.

Speech can tolerate a certain amount of distortion (for example, compression and clipping) but is sensitive to end-to-end delay. The exact amount of maximum tolerable delay is subject to debate. It is generally accepted to be in the range of 100 to 600 ms. For example, the

Packet Audio/Video in the Network Environment

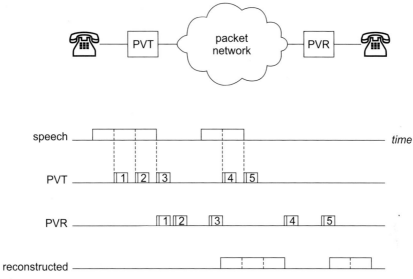

Figure 6.1   Packet voice [6.10]. ©1989 IEEE.

public telephone network has a maximum specification of 600 ms. In order to minimize packetization and storage delays, it has been proposed that voice packets should be relatively short, on the order of 200 to 700 bits, and generally should contain less than 10 to 50 ms of speech [6.6, 6.7, 6.8]. Network protocols should be simplified to shorten voice packet headers (for example, on the order of 4 to 8 bytes) although time stamps and sequence numbers are likely needed. Because a certain amount of distortion is tolerable, error detection, acknowledgements and retransmissions are unnecessary in networks with low error rates. Flow control can be exercised end-to-end by blocking calls. In addition, network switches can possibly discard packets under heavy traffic conditions. In this case, embedded coding has been proposed whereby speech quality degrades gracefully with the loss of information [6.9]. Packet voice is shown in Figure 6.1 [6.10]. It can be seen that the packets are generated at regular intervals during talksparts at the Packet Voice Transmitter (PVT). The reconstruction process at the Packet Voice Receiver (PVR) must compensate for the variable delay component by adding a controlled delay before playing out each packet. This is constrained by some value, D_{max}, which is the specified maximum percentage of packets that can be lost or miss playout. In addition to buffering voice packets, it might be desirable for the PVR to attempt to detect lost packets and to recover their information.

There are two basic approaches to the reconstruction process [6.11, 6.12, 6.13]. In the Null Timing Information (NTI) scheme, reconstruction does not use timing information (that is, time stamps) to determine packet delays through the network. The PVR adds a fixed delay D to the first packet of each talksparts as shown in Figure 6.2.

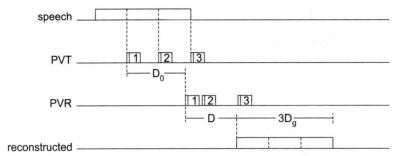

Figure 6.2 NTI reconstruction scheme [6.10]. ©1989 IEEE.

If D_0 is the transit delay of a first packet through the network and D_g is a packet-generation time (assumed to be constant), the total delay of the first packet from entry into the network to playout is

$$D_t = D_0 + D_g \tag{6.1}$$

Subsequent packets in the talkspart are played out at intervals of D_g after the first packet. Therefore, sequence numbers are required to indicate the relative positions of packets in the talkspart. If a packet is not present at the PVR at its playout time, it is considered lost. The choice of D involves a trade-off. Increasing D reduces the percentage of lost packets, but increases total end-to-end delays and the size of the queue at the PVR. D cannot be too large due to the constraint from D_{max} or too small due to P_{loss}. Because D_0 is random, the silence intervals between talksparts are not reconstructed accurately.

Example 6.1 Reconstruction of silences in an NTI scheme is shown in Figure 6.3. Let d and d' denote the values of D_0 for the talksparts preceding and following a silence interval(s). Suppose that d and d' are identically distributed with variance σ^2 and have some positive correlation r. Then, the error in the length of the reconstructed silence is

$$\varepsilon = d - d' \tag{6.2}$$

and has the variance

$$\text{var}(\varepsilon) = 2\sigma^2(1-r) \tag{6.3}$$

which is directly proportional to the variance of packet delays. Evidently, the NTI scheme would be adequate only if a small delay variance could be guaranteed.

Because the scheme depends on the first packet of each talkspart, the loss of a first packet might cause confusion at the PVR.

If delay variability can be significant, a more elaborate reconstruction is necessary. In the Complete Timing Information (CTI) approach, the reconstruction process uses full timing information in the form of time stamps to determine each packet's delay accurately through the net-

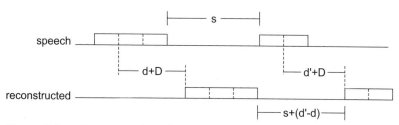

Figure 6.3   Reconstruction of silence scheme [6.10]. ©1989 IEEE.

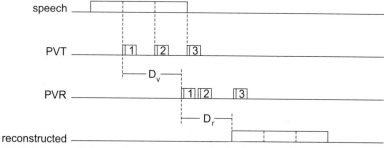

Figure 6.4   CTI reconstruction scheme [6.10]. ©1989 IEEE.

work, denoted D_v. As it can be seen from the Figure 6.4, the PVR adds a controlled delay D so that the total entry-to-playout delay D_t

$$D_t = D_v + D_r \tag{6.4}$$

is as uniform as possible for all packets. In addition to time stamps, sequence numbers are also desirable for detecting lost packets.

There are various choices for the format of the time-stamp fields. The most obvious choice is a global time stamp, but this requires precise synchronization of both PVT and PVR to a global clock. A second choice is to encode the relative time between consecutive packets. This means there is an unknown constant end-to-end-delay. A large time-stamp field is also required because the time between packets could be long. Finally, the time stamp can indicate the delay that a packet has accumulated in transit so far [6.11]. In this case, the time stamp might be more appropriately called a delay stamp. A packet is generated with a delay stamp initialized to zero. Each node increments the delay stamp by the amount of time that the packet has spent in that node, possibly including propagation delays along links as well.

6.1.2   Integrated Packet Networks

The economies and flexibility of integrated networks make them very attractive, and packet network architectures have the potential for realizing these advantages. However, the effective integration of speech and other signals, such as graphics, image and video into an Integrated Packet Network (IPN) can rearrange network design properties. Although processing speeds will continue to increase, it will also be necessary to minimize the nodal per-packet processing requirements imposed by the network design. Data signals must generally be received error free in order to be useful. The inherent structure of speech and image signals and the way in which they are perceived allows for some loss of information without significant quality improvement. This presents the possibility of purposely discarding limited information to achieve some other goal, such as the control of temporary congestion. One of the goals in IPNs is to construct a model that considers the entire IPN (transmitters, packet multiplexers and receivers) as a system to be optimized for higher speeds and capabilities [6.14]. In order to simplify the processing at network nodes, more complex processing at network edges can be allowed. The transmitter forms a packet switch, varying in its importance to the reconstruction of high-quality speech at the receiver. Packet multiplexers discard speech packets according to this delivery priority in order to control overload. The receiver then attempts to regenerate the information contained in any discard packets. Although this model is concerned specifically with speech, the approach can be extended to other structural signals, such as graphics, image and video signals.

A transmitter subsystem is shown in Figure 6.5. The transmitter first classifies speech segments according to models of the speech production process (voiced sounds, fricatives and plosives).

This model-based classification is used to remove redundancy during coding, to assign delivery properties and to regenerate discarded speech packets. After classification, the transmitter removes redundancy from the speech using a coding algorithm based on the determined model. For example, voiced sounds (vowels) could be coded with a block-oriented pitch prediction coder. After coding, the transmitter assigns a delivery priority to each packet based on the quality of regeneration possible at the receiver. In forming packets from speech segments, the delivery priority would be included in the network portion of the packet header. The classification and any coding parameters would be included in the end-to-end portion of the header. Packet multiplexers exist at each outgoing link of each network node as well as at each multiplexed network access point. A packet multiplexer subsystem with the arriving packet discarded is shown in Figure 6.6. Here, λ is the effective arrival rate, and μ represents the effective service

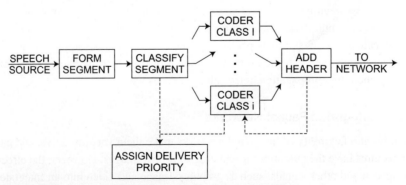

Figure 6.5 Transmitter subsystem [6.14]. ©1989 IEEE.

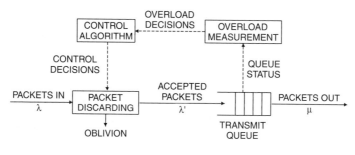

Figure 6.6 Packet multiplexer subsystem with arriving packet discarded [6.14]. ©1989 IEEE.

rate. Each packet multiplexer monitors local overload and discards packets, according to packet delivery priority (read from the network portion of the packet header) and is locally determined by the measure of overload level. It is assumed that arriving packets are discarded. It is also possible to discard already-queued packets. In addition, if error checking is performed by the nodes, any packet (data or speech) found to have an error is discarded.

The receiver decodes the samples in speech packets delivered to it based on the classification and coding parameters contained in the end-to-end header. It also determines the appropriate time to play them out. A receiver subsystem is shown in Figure 6.7. The receiver synchronization problem requires only packet sequence numbers. Global synchronization is administratively difficult, and time stamps must be modified at each packet multiplexer, requiring additional per-packet processing [6.11]. Potential speech detector impairments, such as clipping, are eliminated whenever the network is not overloaded. Even deriving periods of considerable overload, the received quality may be better if at least a few background noise packets are delivered and then used to regenerate noise that is similar in character to the actual noise. If a packet is lost for any reason (for example, discarded by the network because of overload or errors, excessively delayed in the network, and so forth) the receiver must first detect the loss by inspecting sequence numbers of those packets that have been received. It must further make a determination of the class of each lost packet so that the appropriate regeneration model can be applied using the previous header and sample history. A correct class determination will be critical to regenerating the lost information accurately. It can be easily done as follows. In a string of packets with the same class, we can virtually ensure that the first packet will be received by assigning it a high delivery priority. Assuming perfect delivery of these first packets, the class of any lost packet will match the class of the last received packet. Thus, the receiver's class decision can be virtually error free.

In summary, the advantages gained by taking a total system approach to an integrated packet network are as follows:

- A powerful overload control mechanism is provided.
- The structure of speech is effectively exploited.

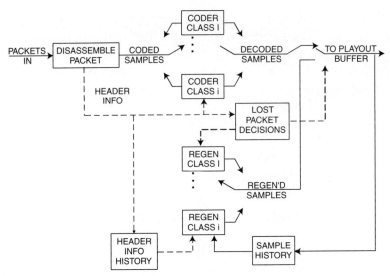

Figure 6.7 Receiver subsystem [6.14]. ©1989 IEEE.

- Extremely simple per-packet processing for overload control is allowed.
- Only one packet per speech segment is required.
- Receiver speech synchronization is simplified.
- Reduced per-packet error processing at packet multiplexers is possible.

6.1.3 Packet Video

Asynchronous transfer of video, which often is referred to as packet video, can be defined as the transfer of video signals across asynchronous Time Division Multiplex (ATDM) networks, such as IP and ATM. The video may be transferred for instantaneous viewing or for subsequent storage for replay at a later time. The former case has requirements on pacing so that the received video data can be displayed in a perceptually continuous sequence. The latter case can be seen as a large data transfer with no inherent time constraints. In addition to the requirement on pacing, the maximal transfer delay may also have bounds from camera to monitor if the video is a part of an interactive conversation or conference. These limits are set by human perception and determine when the delay starts at the information exchange. Parts of the signal may be lost or corrupted by errors during the transfer. This will reduce the quality of the reconstructed video, and, if the degradation is serious enough, it may cause the viewer to reject the service. Thus, the general topics of packet video are to code and to transfer video signals asynchronously under quality constraints.

The synchronous transfer mode combines the circuit-switched routing of telephony networks with the asynchronous multiplexing of packet switching. This is accomplished by establishing a connection (fixed route) through the network before accepting any traffic. The information is then sent in 53-octet long cells. The switches route cells according to address information contained in each cell's five-octet header. Traffic on a particular link consists of randomly interleaved cells belonging to different calls. The network guarantees that all cells of a call follow the same route and, hence, get delivered in the same order as they were sent. The intention is that ATM networks should be able to guarantee the QoS in terms of cell loss and maximum delay, as well as maximum delay variations [6.15].

The IP differs in two major respects from ATM. There is no pre-established route and the packets are of variable length (up to 65,535 octets). IP does not give any guarantees on the delivery of the packets, and they may even arrive out of order if the routing decision is changed during the session. These issues will be addressed by the introduction of IPng in conjunction with RSVP. In IPng, often called IP (version 6), packets contain a 24-bit flow identifier in addition to the source and destination addresses and can be used in routers for operations like scheduling and buffer management to provide service guarantees. Delay and some loss is inevitable during transfers across both ATM and IP networks. The delay is chiefly caused by propagation and queuing. The queuing delay depends on the dynamic load variations on the links and must be equalized before video can be reconstructed. Bit errors can occur in the optics and electronics of the physical layer through thermal and impulsive noise. Loss of information is mainly caused by a multiplexing overload of such magnitude and duration that buffers in the nodes overflow. Video in digital form is a 3D signal. It is a time sequence of equidistantly spaced 2D pictures or frames. Frames can be samples of a real scene captured by a camera or a sensor. They may also be generated by computer graphics. The digitized frames of a video sequence can either be scanned sequentially row by row or be interlaced, where first the odd-numbered rows are scanned from top to bottom followed by the even-numbered rows. If the source produces a signal with RGB components, it is transformed into a YIQ format with one luminance component (Y) and two chrominance, or color, components (I and Q). The structure of a video stream is illustrated in Figure 6.8. The stream consists of frames that may be composed of fields if interlaced scanning is used. The fields are composed of lines of pixels where each pixel consists of color components (each has a fixed number of bits).

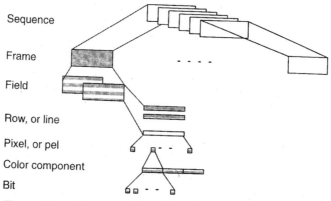

Figure 6.8 Structure of the video signal.

A video communication system is shown in Figure 6.9. The camera continuously captures a scene. The digitized video is passed to an encoder. A function that is often part of the encoder is the bit-rate control, which is used to regulate compression to adapt the bit rate to the channel in the network. Typically, a common reconstruction is that of the access capacity to the network. However, the constraint need not be a single upper limit on the rate, but could be a more general function. It can also be effected by flow-control messages from the network as well as from the receiver. Often the bit-rate control is the information segmentation and framing. A frame is a segment of data with added control information. Segments that are formed at the application level typically constitute the loss unit. Errors and loss in the network load to the loss of one or more application segments. Further segmentation occurs at the network level, where the data is segmented into multiplexing (IP packets or ATM cells), which is the loss unit for the network. The application layer segmentation and framing should simplify the handling of information loss that may occur during the transfer. The network framing is needed to detect and possibly correct bit and burst errors as well as packet or cell losses. The framing thus contains control information that may even include error-control coding. The receiver side performs functions that are reciprocal to the sending functions and may compensate for errors during the transfer. These functions include decoding, error handling, delay equalization, clock synchronization and digital-to-analog conversion.

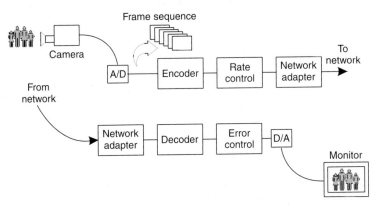

Figure 6.9 An example of a video transmission system.

6.2 Video Transport across Generic Networks

The emergence of digital storage and transmission of video has been driven by the availability of fast hardware at affordable prices, largely thanks to the economies obtained through standardization of compression algorithms. Digital video is being used in many applications ranging from videoconferencing (where two or more parties can carry out an interactive communication) to video on demand (where several users can access the video information stored at a central location). Each application has different requirements in terms of bit rate, end-to-end delay, delay jitter, and so forth. ATM technology is targeted to be used with BISDN and allows flexible and efficient delivery of multimedia data, accommodating many different delays and bit-rate requirements [6.16]. Our goal is not to present an exhaustive survey of issues concerning video transport across generic networks. We choose to concentrate on the general ideas. Details can be found in several reports [6.17, 6.18, 6.19].

Video applications involve real-time display of the decoded sequence. Transmission across a CBR channel requires that there is a constant end-to-end delay between the time that the encoder processes a frame and the time at which that same frame is available to the decoder [6.17]. Because the channel rate is constant, it will be necessary to buffer the variable rate information generated by the video encoder. The size of the buffer memory will depend on the acceptable end-to-end delay. It is necessary to provide a rate-control mechanism that will ensure that the buffer does not overflow and that all the information arrives at the decoder. The basic idea is to lower the video quality for scenes of higher complexity to avoid overflow. The simplest approach to rate control relies on deterministic mapping of each buffer occupancy level to a fixed coder mode of operation [6.18]. Some models of the coder have been proposed to set up coding-rate predictions that are used to drive the buffer control [6.19]. In other cases, ideas from control theory are used to devise the buffer controller [6.20]. In general, the buffer control is designed for a particular encoding scheme, and the scheme-dependent heuristics tend to be introduced [6.21, 6.22]. A more detailed review of the problem in connection with a survey of the algorithms proposed for rate control can be found in Ortega, Ramchandran and Vetterli [6.23]. The goal of all these algorithms is to maximize the received quality given the available resources.

Whereas ISDN offers both circuit-switched and packet-switched channels, video transmission uses a circuit-switched channel. This is the case for most videoconferencing products. In this scenario, the transmission capacity available to the end-user is constant throughout the duration of the call. The main advantage of this approach is its reliability because the channel capacity is guaranteed. On the other hand, in computer networks, video is manipulated just as any other type of data. Video data is packetized and routed through the network sharing the transmission resources with other available services, such as remote login, file transfer, and so forth, which are also built on the top of the same transport protocols. Such systems are being implemented on LANs [6.24] and WANs [6.25, 6.26] with both point-to-point and multipoint connections. The systems are often referred to as best effort because they provide no guarantees on the end-to-end transmission delay and other parameters. In a best effort environment, the received video quality may change significantly over time.

ATM networks seek to provide the best of both worlds by allowing the flexibility and efficiency of computer networks while providing sufficient guarantees to permit reliable transmission of real-time services. Using ATM techniques allows flexible use of capacity, permitting dynamic routing and reuse of bandwidth. Because video compression algorithms produce a variable number of bits, the periods of low activity of one source can be reused by other sources. For example, for N sources (each requires a CBR channel at a rate R bits/s), it might be possible to transmit them together across a single channel with a rate less than RN. This reduction in capacity is the so-called Statistical Multiplexing Gain (SMG). If all services are using their maximum capacity simultaneously, packets might be lost, so transmission can also be guaranteed most of the time. Contrary to the best-effort characteristic of most computer networks, ATM networks are designed so as to allow QoS parameters to be met, at least statistically. ATM networks aim to accommodate very heterogeneous services, thus allowing a customized set of QoS parameters to be selected by each application. The ATM design should be able to support both the best-effort network protocols and circuit-switched connections. However, although the QoS parameters are meaningful for non-real-time data, they are not the only factors to take into account for real-time video transmission. Video differs from other types of data in that acceptable transmission quality can be achieved even if some of the data is lost [6.16]. The effect of packet losses, which in other applications results in retransmission of data, can be reduced in the video case with appropriate encoding strategies combined with error-concealment techniques.

We now examine some of the network design issues, emphasizing those aspects that directly affect video transmission. Admission control is the task of deciding whether a new connection with a given set of requested QoS parameters can be allowed into the network. The connection should be admitted if it can be guaranteed to have the required QoS without degrading the QoS of other ongoing connections. Because video encoders produce variable rates, a key factor in the admission control problem is to find statistical models for the expected bit rates of video sources. A model characterizes the bit rate of a video connection at various time scales and attempts to capture the short- and long-term dependencies in the bit rate as well. Typical models are correlated with the number of bits for the previous frame [6.27] or Markov chains [6.28]. For a given model, the performance of the network model based on different routing and queuing strategies can be examined. The decision on whether to admit a call can be made based on the expected performance of the network. Admission control is much simpler for circuit-switched networks, because the transmission resources are constant throughout the duration of the call. The only issue is to find out whether currently unused resources are sufficient to carry the additional call. If they are sufficient, the call can be completed; otherwise, it will be rejected. In the ATM environment, the resources needed by each of the services change over time. Thus, the main problem is to estimate the likelihood that resources will be insufficient to guarantee QoS. Best-effort networks do not perform explicit admission control, but insufficient QoS during high-load conditions will drive users out or make them delay their connection [6.29]. Admission control is part of the negotiation process between user and network to set up a connection. The result of the negotiation is a contract that will specify the traffic parameters of the connection. Typical traffic parameters are peak cell rate and sustainable cell rate. These are operational measures of the offered bit rate and are implemented with counters.

The function called Usage Parameter Control (UPC), or policing mechanisms, has the goal of preventing sources from maliciously or unwillingly exceeding the traffic parameters negotiated at call setup. Typically, the network will look at policing methods that are directly linked to the negotiated traffic parameters. For instance, if a certain peak cell rate has been agreed upon, the policing mechanism may consist of a counter that tracks the peak rate and verifies that it does not exceed the negotiated value. One of the most popular policing mechanisms, due to its simplicity, is the so-called leaky bucket [6.30]. A leaky bucket is simply a counter incremented with each cell arrival and decremented at fixed intervals such that the decrement is equivalent to an average cell rate of R. The other parameter of the leaky bucket is the size of the bucket, that is, the maximum allowable value for the counter. The network can detect violations by monitoring whether the maximum value of the counter is reached. If the same cells are found to be violating the policing functions, the network can decide to delete them or to just mark them for possible deletion in case of congestion. The choice of policing function is important because it may determine the type of rate that the sources will transmit through the network.

Video-encoding algorithms for ATM transmission need to be robust to packet losses. This can be achieved in part by using a multiresolution encoding scheme along with different levels of priorities for the cells corresponding to each resolution. Additionally, error-concealment techniques can be used to mark to some extent the perceptual effects in the decoded sequence of the loss information.

Multiresolution encoding schemes separate the information into two or more layers or resolutions. The coarse resolution contains a rough approximation of the full resolution image or sequence. The enhancement or details resolution provides the additional information needed to reconstruct at the decoder the full resolution sequence at the targeted quality. The coarse resolution sequence is obtained by reducing the spatial or temporal resolution of the sequence or by having images of lower quality. A survey of multiresolution encoding techniques can be found in Vetterli and Uz [6.31]. To take advantage of the multiresolution encoding, the information is packetized into two classes of packets according to whether the priority bit provided by the ATM format is set or not [6.16]. The coarse resolution will be transmitted using high priority packets while the detail resolution will be sent with the low priority ones. Using the properties so that the packets with lower priority are discarded first in case of congestion is beneficial in terms of the end-to-end quality [6.32].

A further advantage of using multiresolution coding schemes is that they enable efficient error concealment techniques [6.33, 6.34, 6.35, 6.36]. The idea is to use the available information, that is, packets that were not lost, to interpolate the missing information. When multiresolution coding is used, the information decoded from only the lower resolution layer may be sufficiently good. Other approaches that have been proposed involve interleaving the information so that a cell loss causes minor perceptual degradation in several image blocks (10 to 100) rather than severe degradation in just a few blocks (1 to 10).

ATM transmission provides the possibility of transporting a variable number of bits per frame and thus could seem to make the use of rate control unnecessary. This view is not realistic because each connection will be specified by a series of traffic parameters that will be monitored by the network. Transmission over the limits set by the traffic characteristics may result in lost packets, so rate control is still necessary [6.17]. We can make the distinction between rate control and rate shaping. Rate control entails changing the rate produced by the encoder, and rate shaping only affects the times at which cells are re-sent to the network, but not the total amount of information transmitted.

6.2.1 Layered Video Coding

An often-cited approach for coping with receiver heterogeneity in real-time multimedia transmission is the use of layered media streams. In this model, the source distributes multiple levels of quality simultaneously across multiple network channels. In turn, each receiver individually tunes its reception rate by adjusting the number of layers that it receives. The net effect is that the signal is delivered to a heterogeneous set of receivers at different levels of quality using a heterogeneous set of rates. To fully realize this architecture, we must solve two subproblems: the layered compression problem and the layered transmission problem. In other words, we must develop a compression scheme that allows us to generate multiple levels of quality using multiple layers simultaneously with a network delivery mode that allows us to selectively deliver a subset of layers to individual receivers. We first define the layered compression problem.

Layered Compression

Given a sequence of video frames $\{F_1, F_2, ...\}$, for example, $F_k \in [0, 255]^{640 \times 480}$ for gray-scale NTSC video, we want to find an encoding E that maps a given frame into L discrete codes, (that is, into L layers): $E: F_k \rightarrow \{C_k^1, ..., C_k^L\}$ and further a decoding D that maps a subset of $M \leq L$ codes into a reconstructed frame, $\hat{F}_k^M$, $D: \{C_k^1, ..., C_k^M\} \rightarrow \hat{F}_k^M$ with the property that $d\left(F_k, \hat{F}_k^M\right) \geq d\left(F_k, \hat{F}_k^n\right)$ for $0 \leq m \leq n \leq L$ and a suitably chosen metric d (for example, mean squared error or a perceptual distortion measure). With this decomposition, an encoder can produce a set of codes that are striped across multiple network channels $\{N_1, ..., N_L\}$ by sending codes $\{C_1^k, C_2^k, ...\}$ across N_k. A receiver can then receive a subset of the flows $\{N_1, ..., N_M\}$ and reconstruct the sequence $\{\hat{F}_1^M, \hat{F}_2^M, ...\}$. One approach for delivering multiple levels of quality across multiple network connections is to encode the video signal with a set of independent encoders each producing a different output rate. Hence, we can choose a $D = (D_1, ..., D_L)$ where $D_m: C_k^m \rightarrow \hat{F}_k$. This approach, often called simulcast, has the advantage that we can use existing codecs and/or compression algorithms as system components. Figure 6.10 illustrates the simplicity of a simulcast coder. It produces a multirate set of signals that are independent of one another. Each layer provides improved quality, but does not depend on subordinate layers. Here, we show an image at multiple resolutions, but the refinement can occur across dimensions of line frame rate or SNR. A video signal is duplicated across the inputs to a bulk of independent encoders. These encoders compress the signal to a different rate and different quality. Finally, the decoder receives the signal independent of the other layers. In simulcast coding, each layer of video representing a resolution or quality is coded independently. Thus, a single layer (nonscalable)

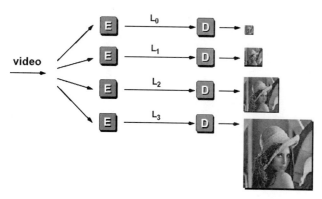

Figure 6.10 Simulcast coder.

decoder can decode any layer. In simulcast coding, total available bandwidth is simply portioned depending on the quality desired for each independent layer that needs to be coded. It is assumed that independent decoders would be used to decode each layer [6.37].

In contrast, a layered coder exploits correlation across subflows to achieve better overall compression. The input signal is compressed into a number of discrete layers, arranged in a hierarchy that provides progressive refinement. For example, if only the first layer is received, the decoder will produce the lowest quality version of the signal. On the other hand, if the decoder receives two layers, it will combine the second-layer information with the first layer to produce improved quality. Overall, the quality progressively improves with the number of layers that are received and decoded.

Figure 6.11 gives a rough sketch of the trade-off between the simulcast and layered approaches from the perspective of rate-distortion theory. Each curve traces the distortion incurred for imperfectly coding an information source at the given rate. Distortion rate functions for an ideal coder $D_I(R)$, a real coder $D_R(R)$, a layered coder $D_L(R)$ and a simulcast coder $D_S(R)$ are presented. The distortion measures the quality degradation between the reconstructed and original signals. The ideal curve $D_I(R)$ represents the theoretical lower bound on distortion

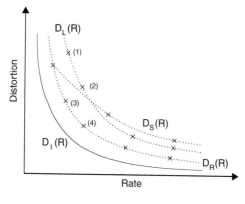

Figure 6.11
Rate distortion characteristics.

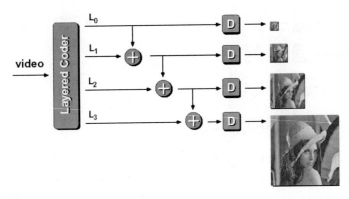

Figure 6.12 Conceptual structure of a layered video.

achievable as a function of rate. A real coder $D_R(R)$ can perform close to the ideal curve, but never better. The advantage of layer representation is that both the encoder and decoder can travel along the distortion rate curve. That is, to move from point (1) to (2) on $D_L(R)$, the encoder carries out incremental computation and produces new output that can be appended to the previous output. Conversely, to move from point (3) to (4) on $D_R(R)$, the encoder must start from scratch and compute a completely new output string. Finally, a simulcast coder $D_S(R)$ incurs the most overhead because each operating point redundantly contains all of the operating points of lesser rate.

The structure of a layered video coder is given in Figure 6.12. The input video is compressed by a layered coder that produces a set of logically distinct output strings. The decoder module D is capable of decoding any cumulative set of bitstrings. Each additional string produces an improvement in reconstruction quality.

Layered Transmission

By combining the approach of layered compression with a layered transmission system, we can solve the multicast heterogeneity problem. In this architecture, the simulcast source produces a layered stream where each layer is transmitted on a different network channel. The network forwards only the number of layers that each physical link can support. Each user receives the best quality signal that the network can deliver. The network must be able to drop layers selectively at each bottleneck link. The concept of layered video coding was first introduced in the context of ATM networks [6.38, 6.39]. The video information is divided into several layers, with lower layers containing low-resolution information and higher layers containing the fine information. Such a model enables integration of video telephony and broadcast video services. In the former case, where a bandwidth is at a premium, lower layers can provide the desired quality. In broadcast applications, a variable number of higher layers can be integrated with the lower ones to provide the quality and the bit rate that is compatible with the receiver.

6.2.2 Error-Resilient Video Coding Techniques

Error-resilience techniques for real-time video transport across unreliable networks include protocol and network environments and their characteristics, encoder error-resilience tools; decoder

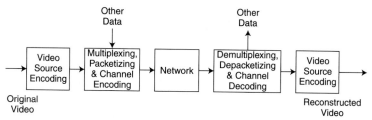

Figure 6.13 Video communication system.

error-concealment techniques and techniques that require cooperation among encoder, decoder and the network. A typical video transmission system involves five steps as shown in Figure 6.13. The video is first compressed by a video encoder to reduce the data rate. The compressed bit stream is then segmented into fixed or variable length packets and multiplexed with other data types such as audio. The packets might be sent directly across the network if the network guarantees bit-error-free transmission. Otherwise, they usually undergo a channel-encoding stage, typically using Forward Error Correction (FEC), to protect them from transmission errors. At the receiver end, the received packets are FEC decoded and unpacked. The resulting bit stream is then input to the video decoder to reconstruct the original video. In practice, many applications embed packetization and channel encoding in the source coder as an adaptation layer to the network.

To make the compressed bit stream resilient to transmission errors, one must add redundancy into the stream so that it is possible to detect and correct errors. Such redundancy can be added in either the source or channel coder. The classical Shannon information theory states that one can separately design the source and channel coders to achieve error-free delivery of a compressed bit stream as long as the source is represented by a rate below the channel capacity. Therefore, the source coder should compress a source as much as possible for a specified distortion. The channel coder can then add redundancy through FEC to the compressed stream to enable the correlation of transmission errors. All the error-resilient encoding methods make the source coder less efficient than it can be so that erroneous or missing bits in a compressed stream will not have a disastrous effect on the reconstructed video quality. This is usually accomplished by carefully designing both the predictive coding loop and variable length coder to limit the extent of error propagation.

Mechanisms devised for combating transmission errors can be categorized into three groups:

- Those introduced at the source and channel encoder to make the bit stream more resilient to potential errors.
- Those invoked at the decoder upon detection of errors to conceal the effect of errors.
- Those that require interactions between the source encoder and decoder so that the encoder can adapt its operations based on the loss conditions detected at the decoder.

We will refer to all of them as error-resilience techniques.

Error-Resilient Encoding

In this approach, the encoder operates so that transmission errors on the coded bit stream will not adversely affect the decoder operation and lead to unacceptable distortions in the reconstructed video quality. Compared to coders that are optimized for coding efficiency, error-resilient coders typically are less efficient in that they use more bits to obtain the same video quality in the absence of any transmission errors. The extra bits are called redundancy bits, and they are introduced to enhance the video quality when the bit stream is corrupted by transmission errors. The design goal in error-resilient coders is to achieve a maximum gain in error resilience with the smallest amount of redundancy.

There are many ways to introduce redundancy in the bit stream. Some of the techniques help to prevent error propagation, and others enable the decoder to perform better error concealment upon detection of errors. Yet another group of techniques is aimed at guaranteeing a basic level of quality and providing a graceful degradation upon the occurrence of transmission errors [6.40, 6.41].

One main cause for the sensitivity of a compressed video stream to transmission errors is that a video coder uses VLC to represent various symbols. Any bit errors or lost bits in the middle of a code word can not only make this code word undecodable, but they can also make the following code words undecodable, even if they are received correctly.

One simple and effective approach for enhancing encoder error resilience is by inserting resynchronization markers periodically. Usually, some header information is attached immediately after the resynchronization information. Obviously, insertion of resynchronization markers will reduce the coding efficiency [6.41]. In practical video-coding systems, relatively long synchronization code words are used.

With RVLC, the decoder can not only decode bits after a resynchronization code word, but also decode the bits before the next resynchronization code word, from the backward direction. Thus, with RVLC, fewer correctly received bits will be discarded, and the affected area by a transmission error will be reduced. RVLC can help the decoder to detect errors that are not detectable when non-RVLC is used, or it can provide more information on the position of the errors and increase the amount of data unnecessarily discarded. RVLC has been adopted in both MPEG-4 and H.263 in conjunction with the insertion of synchronization markers.

Because of the syntax constraint present in compressed video bit streams, it is possible to recover data from a corrupted bit stream by making the corrected stream conform to the right syntax. Such techniques are very much dependent on the particular coding scheme. The use of synchronization codes, RVLC and other sophisticated entropy coding means, such as error-resilient entropy coding, can all make such recovery more feasible and more effective.

Another major cause for the sensitivity of a compressed video to transmission errors is the use of temporal prediction. After an error occurs so that a reconstructed frame at the decoder differs from that assumed at the encoder, the reference frames used in the decoder from there on will differ from those used at the encoder. Consequently, all subsequent reconstructed frames will be in error. The use of spatial prediction for the DC coefficients and motion vectors will also cause error propagation, although it is confined within the same frame. In most video-coding standards, such spatial prediction, and therefore error propagation, is further limited to a subregion in a frame.

One way to stop temporal error propagation is by periodically inserting intracoded pictures. For real-time applications, the use of intraframes is typically not possible due to delay constraints. However, the use of a sufficiently high number of intracoded pictures has turned out to be an efficient and highly scalable tool for error resilience. When applying intracoded pictures for error-resilience purposes, both the number of such intracoded pictures and their spatial placement have to be determined. The number of necessary intraframes is obviously dependent on the quality of the connection. The currently best known way for determining both the correct number and placement of intraframes for error-resilience purposes is the use of a loss-aware rate distortion optimization scheme [6.42].

Another approach to limit the extent of error propagation is to split the data domain into several segments and to perform temporal/spatial prediction only within the same segment. In this way, the error in one segment will not affect another segment. One such approach is to include even-indexed frames in one segment and odd-indexed frames into another segment. Even frames are only predicted from even frames. This approach is called video redundancy coding [6.43]. It can also be considered as an approach for accomplishing multiple description coding. Another approach is to divide a frame into regions, and a region can only be predicted from the same region of the previous frame. This is known as Independent Segment Decoding (ISD) in H.263.

By itself, layered coding is a way to enable users with different bandwidth capacities or decoding powers to access the same video at different quality levels. To serve as an error-resilience tool, layered coding must be paired with UEP in the transport system so that the base layer is protected more strongly, for example, by assigning a more reliable subchannel, using stronger FEC codes or allowing more retransmissions [6.44]. There are many ways to divide a video signal into two or more layers in the standard block-based hybrid video coder. For example, a video can be temporally down-sampled, and the base layer can include the bit stream for the low-frame-rate video, whereas the enhancement layers can include the error between the original video and the up-sampled one from the low frame-rate coded video. The same approach can be applied to the spatial resolution so that the base layer contains a small frame-size video. The base layer can also encode the DCT coefficients of each block with a coarser quantizer, leaving the fine details to be specified in the enhancement layers. Finally, the base layer may include the header and motion information, leaving the remaining information for the enhancement layer. In MPEG and H.263 terminologies, the first three options are known as temporal, spatial and SNR scalabilities, respectively, and the last one is data partitioning.

As with layered coding, Multiple Description Coding (MDC) also codes a service into several substreams, known as descriptions, but the decomposition is such that the resulting descriptions are correlated and have similar importance. For each description to provide a certain degree of quality, all the descriptions must share some fundamental information about the source and must be correlated. On the other hand, this correlation is also the rate of redundancy in MDC. An advantage of MDC over layered coding is that it does not require special provisions in the network to provide a reliable channel. To accomplish their respective goals, layered coding uses a hierarchical, decorrelating decomposition, whereas MDC uses a nonhierarchical, correlating decomposition [6.45].

The objective of error-resilient encoding is to enhance robustness of compressed video to packet loss. The standardized error-resilient encoding schemes include resynchronization marking, data partitioning, and data recovery. For video transmission across the Internet, the boundary of a packet already provides a synchronization point in the variable-length coded bit stream at the receiver side. With MDC, we have robustness to loss of enhanced quality. If a receiver gets only one description (other descriptions being lost), it can still reconstruct video with acceptable quality. If a receiver gets multiple descriptions, it can combine them to produce a better reconstruction than that produced from any one of them. To make each description provide acceptable usual quality, each description must carry sufficient information about the original video. This will reduce the compression efficiency compared to conventional Single Description Coding (SDC). In addition, although more combined descriptions provide a better visual quality, a certain degree of correlation between the multiple descriptions has to be embedded in each description, resulting in further reduction of the compressed efficiency.

Decoder Error Concealment

Decoder error concealment refers to the recovery or estimation of lost information due to transmission errors. Given the block-based hybrid coding paradigm, three types of information may need to be estimated in a damaged MB: the texture information, including the pixel and DCT coefficients values for either an original image block or a predictive error block; the motion estimation, consisting of Motion Vectors (MV) for an MB in either P- or B-mode and, finally, the coding mode of MB. A simple and yet very effective approach to recover a damaged MB in the decoder is by copying the corresponding MB in the previously decoded frame, based on the MV for this MB. The recovery performance by this approach is critically dependent on the availability of the MV. To reduce the impact of the error in the estimated MVs, temporal prediction may be combined with spatial interpolation. Another simple approach is to interpolate pixels in a damaged block from pixels in adjacent correctly received blocks as all blocks or MBs in the same row are put into the same packet. The only available neighboring blocks are those in the current row and the row above. Because most pixels in these blocks are too far away from the missing samples, usually only the boundary pixels in neighboring blocks are used for interpolation [6.46]. Instead of interpolating individual pixels, a simple approach is to estimate the DC coefficient (that is, the mean value) of a damaged block and replace the damaged block by a constant equal to the estimated DC value. The DC value can be estimated by averaging the DC values of surrounding blocks [6.47]. One way to facilitate such spatial interpolation is by an interleaved packetization mechanism so that the loss of one packet will damage only every other block.

A problem with the spatial interpolation approach is how to determine an appropriate interpolation filter. Another shortcoming is that it ignores received DCT coefficients, if any. These problems are resolved in [6.34] by requiring the recovered pixels in a damaged block to be smoothly connected with its neighboring pixels (Zhu, Wang, and Shaw), both spatially in the same frame and temporally in the previous/following frames.

Another way of accomplishing spatial interpolation is by using spatial interpolation using the Projection Onto Convex Set (POCS) method [6.48, 6.49]. The general idea behind POCS-based estimation methods is to formulate each constraint about the unknowns as a convex set. The optimal solution is the intersection of all the convex sets, which can be obtained by recursively projecting a previous solution onto individual convex sets. When applying POCS for recovering an image block, the spatial smoothness criterion is formulated in the frequency domain by requiring the DFT of the recovered block to have energy only in several low frequency coefficients. If the damaged block is believed to contain an edge in a particular direction, one can require the DFT coefficients to be distributed along a narrow strip orthogonal to edge direction, that is, low pass along the edge direction and all pass in the orthogonal direction. The requirement on the range of each DFT coefficient magnitude can also be converted into a convex set. Because the solution can only be obtained through an iterative procedure, this approach may not be suitable for real-time applications.

Error-Resilient Entropy Code

Video coders encode the video data using VLCs. Thus, in an error-prone environment, any error would propagate throughout the bit stream unless we provide a means of resynchronization. The traditional way of providing resynchronization is to insert special synchronization code words into the bit stream. These code words should have a length that exceeds the maximum VLC code length and also be robust to errors. Thus, a synchronization code should be recognized even in the presence of errors. The Error-Resilient Entropy Code (EREC) is an alternative way of providing synchronization. It works by rearranging variable-length blocks into fixed-length slots of data prior to transmission. The EREC is applicable to variable-length codes. For example, these blocks can be macroblocks in H.263. Thus, the output of the coding scheme is variable-length blocks of data. Each variable-length block must be a prefix code. This means that, in the presence of errors, the block can be decoded without reference to previous or future blocks. The decoder should also be able to know when it has finished decoding a block. The EREC frame structure consists of N slots of length s_i bits. This, the total length of the frame, is $T = \sum_{i=1}^{N} s_i$ $[bits]$. It is assumed that the values of T, N and s_i are known to both the encoder and the decoder. Thus, the N slots of data can be transmitted sequentially without the risk of loss of synchronization. EREC reorganizes the bits of each block into the EREC slots. The decoding can be performed by relying on the ability to determine the end of each variable-length block. Figure 6.14 shows an example of the operation of the EREC algorithm. There are six blocks of lengths 11, 9, 4, 3, 9, 6 and six equal length slots with s_i=7 bits. In the first stage of the algorithm, each block of data is allocated to a corresponding

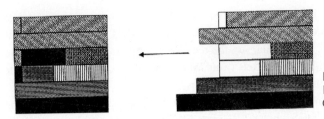

Figure 6.14 An example of the EREC algorithm [6.48]. ©1996 IEEE.

EREC slot [6.48]. Starting from the beginning of each variable-length block, as many bits as possible are placed into the corresponding slot. In the following stages of the algorithm, each block with data yet to be coded searches for slots with space remaining. If there is space available in the slot searched, all or as many bits as possible are placed into that slot. If there is enough space in the slots, the reallocation of the bits will be completed within N stages of the algorithm. The final result of the EREC algorithm is shown in Figure 6.14. In the absence of errors, the decoder starts decoding each slot. If it finds the block end before the slot end, it knows that the rest of the bits in that slot belong to other blocks. If the slot ends before the end of the block is found, the decoder has to look for the rest of the bits in another slot. In case one slot is corrupted, the location of the beginning of the rest of the slots is still known, and the decoding of them can be attempted.

6.2.3 Scalable Rate Control

The main challenge in designing a multimedia application across a communication network is how to deliver multimedia streams to users with minimal replay jitters. In general, a network-based multimedia system can be conceptually viewed as a layer-structure system, which consists of application layer on the top, compression layer, transport layer and transmission layer, as shown in Figure 6.15 [6.49]. To diminish the impact on the video quality due to the delay jitter and available network resources (for example, bandwidth and buffers), traffic shaping and SRC are qualified candidates at two different system levels. Traffic shaping is a transport-layer approach, and SRC is a compression-layer approach.

The basic concept behind the traffic-shaping approach is that, before the encoded video bit stream is injected into the network for transmission, the traffic pattern is already shaped with the desired characteristics, such as maximal delay bounds and peak instantaneous rate [6.50]. Therefore, all the system components along the network path from the sender to the receiver can be configured to meet the QoS as desired by allocating the appropriate resources a priori. On the other hand, the SRC approach is a compression-layer technique where the source video sequence is compressed according to the application's requirement and available network resource.

In the development of an SRC scheme, we need to consider a common feature of employing an Internet-frame coding between two consecutive video frames in several widely used video-compression schemes such as MPEG-1, MPEG-2 and H.263. Although the interframe coding scheme exploits the similarity usually found in encoding two consecutive video frames and achieves significant coding efficiency, the output with a variable-length video bit stream is

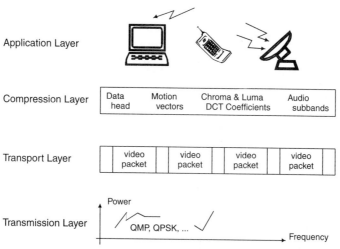

Figure 6.15 Layer structure of a network-based multimedia system [6.49]. ©2000 IEEE.

not well suited for a fixed-rate communication channel. To better use network resources and to transmit coded video bit stream as accurately as possible, the network parameters and encoding parameters should be jointly considered, and their relationship should be modeled accurately. Technically speaking, rate control is a decision-making process where the desired encoding rate for a source video can be met accurately by properly setting a sequence of Quantization Parameters (QP). To cope with various requirements of different coding environments and applications, a rate-control scheme needs to provide sufficient flexibility and scalability. For example, multimedia applications are categorized into two groups, which are VBR application and CBR application. For VBR applications, rate control attempts to achieve the optimum quality for a given target rate. In CBR and real-time applications, a rate-control scheme must satisfy low-latency and buffer constraints. In addition, the rate-control scheme has to be applicable to a variety of sequences and bit rates. Thus a rate-control scheme must be scalable for various bit rates, various spatial resolutions, various temporal resolutions, various coders (DCT and wavelet) and various granularities of VO.

The purpose of rate control is consequently to enforce the specification of the bit stream. The general system is shown in Figure 6.16. The bit stream from the coder is fed into a buffer at a rate $R'(t)$, and it is served at some rate $\mu(t)$ so that the output rate $R(t)$ meets the specified behavior. The bit stream is smoothed by the buffer whenever the service rate is below the input rate. The size of the buffer is determined by delay and implementation constraints. In the encoder, the compression rate is increased when buffer overflow is at risk. The issue is to reduce the variability of the rate function $R(t)$ while minimizing the effects of the consistency of the perceptual quality [6.51, 6.52]. The joint problem of traffic characterization and rate control is to find a suitable description of the bit stream that is sufficiently useful to the network and that can be enforced without overly throttling the compression rate. Provided that a model has been cho-

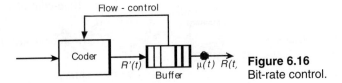

Figure 6.16
Bit-rate control.

sen, the user should not estimate the parameters for it and find a way to regulate the service rate μ(t) so that R(t) strictly obeys the specification. The network would then have the possibility to verify that the traffic is in accordance with its specifications.

Rate Control Techniques

In developing a rate-control technique, there are two widely used approaches:

- Analytical model-based approach
- Operational rate distortion R(D)-based approach

In the model-based approach, various distribution characteristics of signal source along with associated guarantees are considered. Based on the selected model, a closed-form solution is derived using optimization theory. Such a theoretical optimization solution cannot be implemented easily because there is only a finite discrete set of quanitzers and the source signal model varies spatially. Alternatively, an operational R(D)-based approach is used in a practical coding environment. For example, to minimize the overall coding distortion subject to a total bit budget constraint, lots of techniques based on dynamic programming or Lagrangian multiplier for optimization solutions have been developed [6.53, 6.54, 6.55, 6.56]. These methods share the similar concepts of data preanalysis. By analyzing the R(D) characteristics of future frames, the bit-allocation strategy is determined afterward. The Lagrangian multiplier is a well-known technique for optimal bit allocation in image and video coding, but with an assumption that the source consists of statistically independent components. Thus, an interframe-based coding may not find the Lagrangian multiplier approach applicable because of the temporal dependency.

Frame dependencies are taken into account in bit-rate control [6.54]. However, potentially high complexity with increasing operating R(D) points make this method unsuitable for the applications requiring interactivity or low encoding delay. In [6.57], Ding investigated a joint encoder and channel rate-control scheme for VBR video across ATM networks and claimed that the rate control scheme has to balance both issues of consistent video quality in the encoder side and bit-stream smoothness for statistical multiplexing gain in the network side. A parametric R(D) model for MPEG encoders, especially for the picture-level rate control, has been proposed [6.58]. Based on the bit rate "m quant" model, the desired "m quant" is calculated and used for encoding every MB by combining with appropriate quantization matrix entry in a picture. A normalized R(D) model-based approach has been also developed for H.263-compatible video codecs. By providing good approximation of all 32 rate-distortions relations, the authors claim that the proposed model offers an efficient and less-memory-requirements approach to approximate the rate and distortion characteristics for all QPs [6.59]. Rate-control techniques for MPEG-4 object-level and MB level video coding were proposed in [6.60, 6.61]. However, most of the aforementioned techniques only focus on a single coding environment, either frame level, object level or macro level. None of these techniques demonstrates its applicability to MPEG-4 video coding, including the previous three coding granularities simultaneously.

In Lee, Chiang and Zhang [6.49], based on a revised quadratic R(D) model, SRC proposes a single framework that is designed to meet both VBR without delay constraints and CBR with low latency and buffer constraints. With this scalable framework based on a new R(D) model and several new concepts, not only more accurate bit rate control with buffer regulation is achieved, but scalability is also preserved for all test video sequences in various applications [6.62]. By considering video contents and coding complexity in the quadratic R(D) model, the rate-control scheme with joint buffer control can dynamically and appropriately allocate the bits among VOs to meet the overall bit-rate requirement with uniform video quality.

Because of the precision of the R(D) model and ease of implementation, the rate-control scheme with the following new concepts and techniques has been adopted as part of the rate-control scheme in MPEG-4 standard:

- A more accurate second order R(D) model for target bit-rate estimation
- A sliding-window method for smoothing the impact of scene change
- An adaptive selection criterion of data points for a better model updating process
- An adaptive threshold shape control for better use of bit budget
- A dynamic bit-rate allocation among VOs with different coding complexities

This rate-control scheme provides a scalable solution, meaning that the rate-control technique offers a general framework for multiple layers of control for objects, frames and MBs in various coding contexts.

Theoretical Foundation of the SRC

In the R(D) model, the distortion is measured in terms of quantization parameter [6.49]. The block diagram is presented in Figure 6.17.

The rate control consists of four stages: initialization stage, pre-encoding stage, encoding stage and post-encoding stage. Assuming that the source statistics are Laplacian distributed [6.63]:

$$P(x) = \frac{\alpha}{2} e^{-\alpha |x|}, \quad \text{where} -\infty < x < \infty \tag{6.5}$$

The distortion measure is defined as

$$D\left(x, \tilde{x}\right) = \left| x - \tilde{x} \right| \tag{6.6}$$

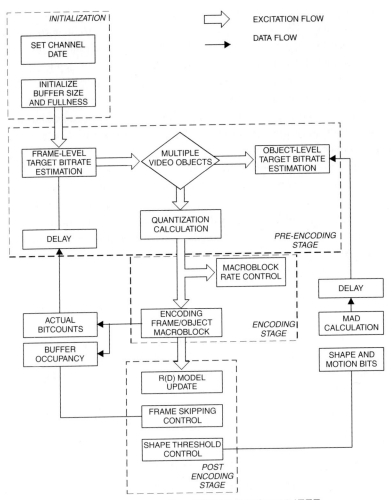

Figure 6.17 Block diagram of the SRC [6.49]. ©2000 IEEE.

There is a closed-form solution for the R(D) functions derived in Viterbi and Omura [6.64]:

$$R(D) = \ln\left(\frac{1}{\alpha D}\right) \qquad (6.7)$$

where

$$D_{\min} = 0, \quad D_{\max} = \frac{1}{\alpha}, \quad 0 < D < \frac{1}{\alpha} \qquad (6.8)$$

The R(D) function is expanded into a Taylor series

$$R(D) = \left(\frac{1}{\alpha D} - 1\right) - \frac{1}{2}\left(\frac{1}{\alpha D} - 1\right)^2 + R_3(D) = -\frac{3}{2} + \frac{2}{\alpha}D^{-1} - \frac{1}{2\alpha^2}D^{-2} + R_3(D) \qquad (6.9)$$

Based on this observation, a new model to evaluate the target bit rate before performing the actual encoding process is presented in Lee, Chiang and Zhang [6.49]. The new model is formulated as follows

$$R_i = \alpha_1 Q_i^{-1} + \alpha_2 Q_i^{-2} \qquad (6.10)$$

where R_i is the total number of bits used for encoding the current frame i, Q_i is quantization level used for the current frame i, and α_1 and α_2 represent first and second order coefficients. Although this model provides the theoretical foundation for the rate-control scheme, the major drawback is its lack of considering two factors. At first, the R(D) model is not scalable for video contents. Second, the R(D) model does not exclude the bit counts used for coding the overhead including video/frame syntax, motion vectors and shape information.

To enhance the R(D) model with more accuracy, a simple prediction is used to predict those bits using the last coded frame as a reference. These bits used for nontexture information are considered as constant numbers irrespective of this distortion and are excluded from the target bit-rate estimation. To accurately estimate the target bit rate with scalability, the original quadratic R(D) formula is modified by introducing two new parameters: Mean Absolute Difference (MAD) and nontexture overhead (H).

$$\frac{R_i - H_i}{M_i} = \alpha_1 Q_i^{-1} + \alpha_2 Q_i^{-2} \qquad (6.11)$$

where R_i, Q_i and α_1, α_2 are previously defined, H denotes the bits used for header, motion vectors and shape information and M_i represents MAD, computed using motion-compensated residual for the luminance component (that is, Y component).

To solve the target bit rate, it is assumed the video is encoded first as an I-frame and subsequently as P-frames. The scheme has been extended to variable GOP structure and B-frames. Let T_i be the bit budget used for the first I-frame, N_p the number of P-frames, H_p the bit budget used for nontexture information and T_p the bit budget used for all P-frames. Then, the total bit budget is

$$R = T_i + N_p T_p$$

$$\frac{T_p - H_p}{M_p} = \alpha_1 Q_p^{-1} + \alpha_2 Q_p^{-2} \qquad (6.12)$$

Then, the T_p and Q_p can be obtained based on the technique described in Chang and Zhang [6.63].

Let

$$X_{n\times 2} = [1, 1/Q_p(i)], \text{ and } Y_{n\times 1} = [Q_p(i) T_p(i)], \qquad (6.13)$$

where i = 1,2,…,n, and n is the number of selected data samples. Then

$$\begin{bmatrix} \alpha_1 \\ \alpha_2 \end{bmatrix} = (X^T X)^{-1} X^T Y \qquad (6.14)$$

Based on these two model parameters α_1 and α_2, the quantization level Q_p and target bit rate T_p can be computed before encoding the next frame.

In the initialization stage, the major tasks that the encoder has to complete with respect to the rate control include the following:

- Initializing the buffer size based on the latency requirement
- Subtracting the bit counts of the first frame from the total bit counts
- Initializing the buffer fullness in the middle level

Without loss of generality, we assume that the video sequence is encoded first as an I-frame and subsequently as P-frames. In this stage, the encoder encodes the first I-frame using an initial Q_p value specified as an input parameter.

In the pre-encoded stage, the tasks of the rate control scheme include target bit estimation, further adjustment of the target bit based on the buffer status for each VO and QP calculation. The target bit count is estimated in the following phases: frame-level bit rate; object level, if desired and MB level bit-rate estimation, if desired.

In the encoding stage, the major tasks that the encoder has to complete include the following: encoding the video frame (object), recording all actual bit rates and activating the MB layer rate control if desired. In the encoding stage, if either the frame or object-level rate control is activated, the encoder compresses each video frame or VO using Q_p as computed in the pre-encoding stage. However, some low-delay applications may require strict buffer regulations, less accumulated delay and better spatial perceptual quality. An MB level rate control is necessary. However, an MB level rate control is costly at low rates because there is additional overhead if the QP is changed frequently within a frame.

In the postencoding stages, the encoder needs to complete the following tasks: updating the corresponding quadratic R(D) model for the entire frame or an individual VO, performing the shape-threshold control to balance the bit usage between the shape information and texture information and performing the frame-skipping control to prevent the potential buffer overflow and/or underflow.

6.2.4 Streaming Video across the Internet

Real-time transport of live video or stored video is the predominant part of real-time multimedia. On the other hand, video streaming refers to real-time transmission of stored video. There are two modes for transmission of stored video across the Internet: the download mode and the streaming mode. In the download mode, a user downloads the entire video file and then plays back the video file. However, full file transfer in the download mode usually suffers long and perhaps unacceptable transfer time. In the streaming mode, the video content need not be downloaded in full, but is being played out while parts of the content are being received and decoded. Due to its real-time nature, video streaming has bandwidth, delay and loss requirements. Designing mechanisms and protocols for streaming video pose many challenges. Streaming video has six key areas: video compression, application-layer QoS control, continous media distribution services, streaming servers, media synchronization mechanisms and protocols for streaming media. Each of the six areas is a basic building block with which an architecture for streaming video can be built. The relations among these basic building blocks are illustrated in Figure 6.18 [6.64, 6.65]. Raw video and audio data are precompressed by video compression and audio compression algorithms and saved in storage devices. Upon the client's request, a streaming server retrieves compressed audio/video data from storage devices. Then, the application layer QoS control module adapts the audio-video bit streams according to the network states and QoS requirements. After the adaptation, the transport protocols packetize the compressed bit streams and send the audio-video packets to the Internet. Packets may be dropped as they experience excessive delays inside the Internet due to congestion. To improve the quality of audio-video transmission, continuous media distribution services are developed for the Internet. For packets that are successfully delivered to the receiver, they first pass through the transport layers and then are processed by the application layer before being decoded at the audio-video decoder. To achieve synchronization between video and audio presentations, media synchronization mechanisms are required. As it can be seen, these areas are closely related, and they are coherent constituents of the video streaming architecture.

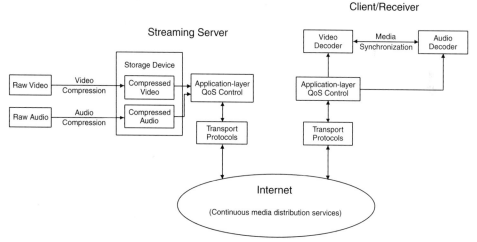

Figure 6.18 Video-streaming architecture [6.65]. ©2001 IEEE.

Video Compression

Video streaming is an important component of many Internet multimedia applications, such as distance learning, digital libraries, home shopping and video on demand. The best-effort nature of the current Internet poses many challenges to the design of streaming-video systems. Our objective is to give the reader a perspective on the range of options available and the associated trade-off among performance, functionality and complexity of existing approaches. To provide insights on design of streaming-video systems, we begin with video compression. Raw video must be compressed before transmission to achieve efficiency. Video-compression schemes can be classified into two categories: scalable and nonscalable video coding. Scalable video is capable of gracefully coping with the bandwidth fluctuations on the Internet [6.66]. Because raw video consumes a large amount of bandwidth, compression is usually employed to achieve a transmission efficiency. The primary objectives of ongoing research on scalable video coding are to achieve high compression efficiency, high flexibility (bandwidth scalability) and/or low complexity. Due to the conflicting nature of efficiency, flexibility and complexity, each scalable video-coding scheme seeks a trade-off among these three factors. Designers of video-streaming service need to choose an appropriate scalable video-coding scheme that meets the target efficiency and flexibility at an affordable cost/complexity. For simplicity, we only show the encoder and decoder in intramode and only use DCT. Intramode coding refers to coding a video unit (for example, an MB) without reference to previously coded data. For wavelet-based scalable video coding, we recommend references [6.67, 6.68, 6.69]. A nonscalable video encoder and decoder are presented in Figure 6.19.

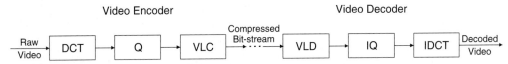

Figure 6.19 Nonscalable video encoder and video decoder.

In contrast, a scalable video encoder (Figure 6.20) compresses a raw video sequence into multiple substreams. One of the compressed substreams is the base substream, which can be independently decoded and can provide coarse visual quality. Other compressed substreams are enhanced substreams, which can only be decoded together with the base substream and which can provide better visual quality. The complete bit stream (that is, combination of all the substreams) provides the highest quality.

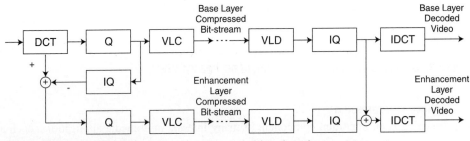

Figure 6.20 SNR scalable video encoder and video decoder.

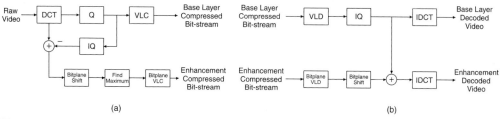

Figure 6.21 FGS: a) encoder and b) decoder [6.65]. ©2001 IEEE.

To provide more flexibility in meeting different demands of streaming (for example, different access link bandwidths and different latency requirements), a scalable coding mechanism called Fine Granularity Scalability (FGS) was proposed to MPEG-4 [6.70, 6.71, 6.72]. As shown in Figure 6.21 an FGS encoder compresses a raw video sequence into two substreams: a base layer bit stream and an enhancement layer bit stream. The FGS encoder uses bit-plane coding to represent the enhancement stream. Bit-plane coding uses embedded representations [6.73]. With bit-plane coding, an FGS encoder is capable of achieving combination rate control for the enhancement stream. This is because the enhancement bit stream can be truncated anywhere to achieve the target bit rate. For example, a DCT coefficient can be represented by 7 bits (that is, its value ranges from 0 to 127). There are 64 DCT coefficients in an 8x8 block. Each DCT coefficient has a Most Significant Bit (MSB), and all the MSBs from the 64 DCT coefficients form bit plane 0. Similarly, all the second-most significant bits form bit plane 1. Bit planes of enhancement DCT coefficients are shown in Figure 6.22.

A version of FGS is Progressive FGS (PFGS). It shares the good features of FGS, such as fine granularity bit-rate scalability and error resilience. Unlike FGS, which only has two layers, PFGS can have more than two layers. The essential difference between FGS and PFGS is that FGS only uses the base layer as a reference for motion prediction while PFGS uses multiple layers as references to reduce the prediction error, resulting in higher coding efficiency [6.74].

Requirements Imposed by Streaming Applications

These requirements are bandwidth, delay, loss, VCR (video-cassette-recorder) functions and decoding complexity.

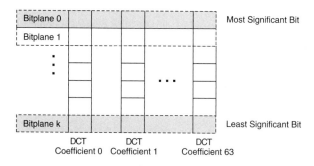

Figure 6.22 Bit planes of enhancement DCT coefficients [6.65]. ©2001 IEEE.

To achieve acceptable perceptual quality, a streaming application typically has a minimum bandwidth requirement. However, the current Internet does not provide bandwidth reservation to support this requirement. In addition, it is desirable for video-streaming applications to employ congestion control to avoid congestion, which happens when the network is heavily loaded. For video streaming, congestion control takes the form of rate control, that is, adapting the sending rate to the available bandwidth in the network. Compared with nonscalable video, scalable video is more adaptable to the varying available bandwidth in the network.

Streaming video requires bounded end-to-end delay so that packets can arrive at the receiver in time to be decoded and displayed. If a packet does not arrive on time, the playout process will pause, which is annoying to human eyes. A video packet that arrives beyond its delay bound (playout time) is useless and can be regarded as lost. Because the Internet introduces time-varying delay, a buffer at the receiver is usually introduced before decoding [6.75].

Packet loss is inevitable on the Internet. It can damage pictures, which is displeasing to human eyes. Thus, it is desirable that a video stream be robust to packet loss. Multiple description coding is such a compression technique to deal with packet loss.

Some streaming applications require VCR-like functions, such as stop, pause/resume, fast forward, fast backward and random access. A dual-bit-stream least-cost scheme to provide VCR-like functionality efficiently for MPEG video streaming is proposed in Lin et al. [6.76].

Some devices, such as cellular phones and PDAs, require low power consumption. Therefore, streaming-video applications running on these devices must be simple. In particular, low decoding complexity is desirable.

So far, we have discussed various compression mechanisms and requirements imposed by streaming applications on the video encoder and decoder. Next, we present the applications-layer QoS control mechanisms, which adapt the video bit streams according to the network status and QoS requirements.

Application Layer QoS Control

The objective of application layer QoS control is to avoid congestion and to maximize video quality in the presence of packet loss. The application layer QoS control techniques include congestion control and error control. These techniques are employed by the end systems and do not require any QoS support from the network.

Burst loss and excessive delay have devastating effects on video presentation quality, and they are usually used by network congestion control. Thus, congestion control mechanisms are necessary to help reduce packet loss and delay. For streaming video, congestion control takes the form of rate control [6.77]. There are three kinds of rate control: source-based, receiver-based and hybrid rate control. The source-based rate control is suitable for unicast. The receiver-based and hybrid rate control are suitable for multicast because both can achieve good trade-off between bandwidth efficiency and service flexibility for multicast video.

Under the source-based rate control, the sender is responsible for adapting the video transmission rate. Feedback is employed by source-based control mechanisms. Based on the feedback information about the network, the sender could regulate the rate of the video stream. The

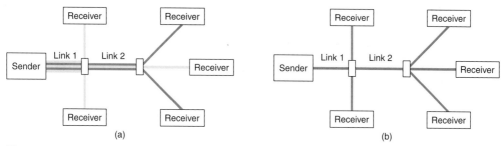

Figure 6.23 (a) Unicast video distribution using multiple point-to-point connections and (b) multicast video distribution using point-to-multipoint transmission [6.65]. ©2001 IEEE.

source-based rate control can be applied to both unicast [6.78] and multicast [6.79]. Unicast video distribution using multiple point-to-point connections as well as multicast video distribution using point-to-multipoint transmission are shown in Figure 6.23. For unicast video, source-based rate control mechanisms follow two approaches: a probe-based and a model-based approach [6.77]. The probe-based approach is based on probing experiments. Specifically, the services probe for the available network bandwidth by adjusting the sending rate in a way that could maintain the packet loss ratio p below a certain threshold p_{th} [6.78]. There are two ways to adjust the sending rate:

- Additive increase and multiplicative decrease [6.78]
- Multiplicative increase and multiplicative decrease [6.80]

The model-based approach is based on a throughput model of a TCP connection. Specifically, the throughput of a TCP connection can be characterized by the following formula:

$$\lambda = \frac{1.22 \, MTU}{RTT \, \sqrt{p}} \tag{6.15}$$

where λ is a throughput of a TCP connection, MTU is the packet size used by the connection, RTT is the round-trip time for the connection and p is the packet-loss ratio experienced by the connection. Under the model-based rate control, this expression is used to determine the sending rate of the video stream. Thus, the video connection could avoid congestion in a way similar to that of TCP and it can compete fairly with TCP flows. For this reason, the model-based rate control is called TCP-friendly rate control [6.81]. For multicast under the service-based rate control, the sender uses a single channel to transport video to the receivers. Such multicast is called single-channel multicast. For single-channel multicast, only the probe-based rate control can be employed [6.79]. Single-channel multicast is efficient because all the receivers share one channel. However, single-channel multicast is unable to provide flexible services to meet the different demands from receivers with various access link bandwidths. In contrast, if multicast video was to be delivered through individual unicast streams, the bandwidth efficiency is low but the services could be differentiated because each receiver can negotiate the parameters of the services with the source. Under the receiver-based rate control, the receivers regulate the receiving rate of video streams by adding/dropping channels while the sender does not participate in rate control [6.77]. Receiver-based rate control is used in multicasting scalable video where there are several layers in the scalable video and each layer corresponds to one channel in the multicast tree.

Similar to the source-based rate control, the existing receiver-based rate-control mechanisms follow two approaches: a probe-based and a model-based approach. The basic probe-based rate control consists of two parts:

- When no congestion is detected, a receiver probes for the available bandwidth by joining a layer/channel, resulting in an increase of its receiving rate. If no congestion is detected after the joining, the joint experiment is successful. Otherwise, the receiver drops the newly added layer.
- When congestion is detected, a receiver drops a layer (that is, leaves a channel), resulting in a reduction of its receiving rate [6.66].

Unlike the probe-based approach that implicitly estimates the available network bandwidth through probing experiments, the model-based approach uses explicit estimation for the available network bandwidth.

Under the hybrid rate control, the receivers regulate the receiving rate of video streams by adding/dropping channels while the sender also adjusts the transmission rate of each channel based on feedback from the receivers [6.82].

An error-control mechanism includes FEC, retransmission, error-resilient encoding and error concealment. There are three kinds of FEC: channel coding, source coding-based FEC and joint source/channel coding. The advantage of all FEC schemes over retransmission-based schemes is reduction in video transmission latency. Source coding-based FEC can achieve lower delay than channel coding, and joint source/channel coding could achieve optimal performance in a rate-distortion sense. The disadvantages of all FEC schemes are increase in transmission rate and inflexibility to varying loss characteristics. Unlike FEC, which add redundancy regardless of correct receipt or loss, a retransmission-based scheme only resends the packets that are lost. Thus, a retransmission-based scheme is adaptive to varying loss characteristics, resulting in efficient use of network resources. The limitation of delay-constrained retransmission-based schemes is that their effectiveness diminishes when the RTT is too large. Currently, an important direction is to combine FEC with retransmission [6.65]. In addition, FEC can be used in layered video multicast so that each client can individually trade off latency for quality based on specific requirements. MDC is a recently proposed mechanism for error-resilient coding. The advantage of MDC is its robustness to loss. The cost of MDC is reduction in compression efficiency. The current research effort gears toward finding a good trade-off between the compression efficiency and the reconstruction quality from one description. Error concealment is performed by the receiver when packet loss occurs and can be used in conjunction with other techniques, for example, congestion control and other error-control mechanisms.

Continuous Media Distribution Services

In order to provide quality multimedia presentations, adequate support from the network is critical. This is because network support can reduce transport delay and packet loss ratio. Streaming video and audio are classified as continuous media because they consist of a sequence of media quanta (such as audio samples or video frames), which convey meaningful information only when presented in time. Built on top of the Internet (IP), continuous media distribution devices are designed with the aim of providing QoS and of achieving efficiency for streaming video/audio across the best-effort Internet. Continuous media distribution services include the following:

- Network filtering
- Application-level multicast
- Content replication.

As a congestion-control technique, network filtering is aimed at maximizing video quality during network congestion. Figure 6.24 illustrates an example of placing filters in the network. The nodes labeled R denote routers that have no knowledge of the format of the media streams and that may randomly discard packets. The filter nodes receive the client's requests and adapt the stream sent by the server accordingly. This solution allows the service provider to place filters on the nodes that connect to network bottlenecks. Furthermore, multiple filters can be placed along the path from a server to a client.

To illustrate the operation of filters, a system model is depicted in Figure 6.25. The model consists of the server, the client, at least one filter and two virtual channels between them. One channel is for control, and the other is for data. The same channels exist between any pair of filters. The control channel is bidirectional, which can be realized by TCP connections. The model allows the client to communicate with only one host (the lost filter), which will either forward the requests or act upon them. The operations of a filter on the data plane include receiving video stream from a server or previous filter and sending video to the client or the next filter at the target rate. The operations of a filter on the control plane include receiving requests from the client or the next filter, acting upon requests and forwarding the requests to its previous filter. Typi-

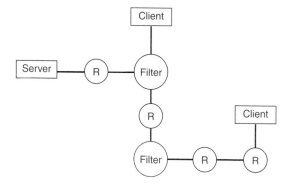

Figure 6.24 Filters placed inside the network [6.65]. ©2001 IEEE.

Figure 6.25 A system model of network filtering [6.65]. ©2001 IEEE.

cally, frame-dropping filters are used as network filters. The receiver can change the bandwidth of the media stream by sending requests to the filter to increase or decrease the frame-dropping rate. The advantages of using frame-dropping filters inside the network include improved video quality and bandwidth efficiency. This is because the filtering can help to save network resources by discarding those frames that are late.

As an extension to the IP layer, IP multicast is capable of providing efficient multipoint packet delivery. The efficiency is achieved by having only one copy of the original IP packet sent by the multicast source and transmitted along any physical path in the IP multicast tree. However, there are many barriers in deploying IP multicast. These problems include scalability, network management, deployment and support for higher layer functionality, for example, error flow and congestion control. The application-level multicast is aimed at building a multicast service on top of the Internet. It enables independent Content Service Providers (CSPs), ISPs or enterprises to build their own Internet multicast networks and to interconnect them into larger, world-wide media multicast networks. The advantage of the application level multicast is that it breaks the barriers such as scalability, network management, and support for congestion control, which have prevented ISPs from establishing "IP multicast" peering arrangements.

An important technique for improving scalability of the media delivery system is content media replication. The content replication takes two forms: caching and mirroring [6.83], which are deployed by publishers, CSPs and ISPs. Both caching and mirroring seek to place content closer to the clients and both share the following advantages:

- Reduced bandwidth consumption on network links
- Reduced load on streaming servers
- Reduced latency for clients
- Increased availability

Mirroring is to place copies of the original multimedia files on other machines scattered around the Internet, that is, the original multimedia files are stored on the main server while copies of the original multimedia files are placed on the duplicate servers. In this way, clients can retrieve multimedia data from the nearest duplicate server, which gives the clients the best performance. On the other hand, caching makes local copies of contents that the client retrieves. Clients in a single organization retrieve all contents from a single local machine, called a cache. The cache retrieves a video file from the original server, storing a copy locally and then passing it on to the client who requests it. If a client asks for a video file that the cache has already stored, the cache will return the local copy rather than going all the way to the original server where the video file resides.

Streaming Servers

Streaming servers are essential in providing streaming services. To offer quality streaming services, streaming servers are required to process multimedia data under timing constraints in order to prevent artifacts, for example, jerkiness in video motion and pops in audio during playback on the clients. In addition, streaming servers also need to support VCR-like control operations, such as stop, pause/resume, fast forward and fast backward. Furthermore, streaming servers have to retrieve media components in a synchronous fashion. A streaming server consists of three subsystems: communicator, operating system and storage system. A communicator involves the application layer and transport protocols implemented on the server. Through a communicator, the clients can communicate with a server and can retrieve multimedia contents in a continuous and synchronous manner. Different from traditional operating systems, an operating system for streaming services needs to satisfy real-time requirements for streaming applications. A storage system for streaming services has to support continuous media storage and retrieval. In what follows, we discuss synchronization mechanisms for streaming media.

Media Synchronization

A major feature that distinguishes multimedia applications from other traditional data applications is the integration of various media streams that must be presented in a synchronized fashion. For example, in distance learning, the presentation of slides should be synchronized with the commenting audio stream as shown in Figure 6.26. Otherwise, the current slide being displayed on the screen may not correspond to the lecturer's explanation heard by the listeners. With media synchronization, the application at the receiver side can present the media in the same way as they were originally captured. Media synchronization refers to maintaining the temporal relationships within one data stream and between various media streams. There are three levels of synchronization: intrastream, interstream and interobject synchronization. The three levels of synchronization correspond to three semantic layers of multimedia data [6.84].

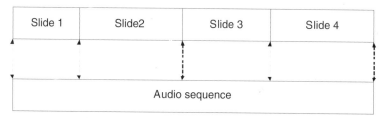

Figure 6.26 Synchronization between the slides and the commenting audio stream [6.65]. ©2001 IEEE.

The lowest layer of continuous media or time-dependent data (such as video or audio) is the media layer. The unit of media layer is a logical data unit such as an audio-video frame, which adheres to strict temporal constraints to ensure acceptable user perception at playback. Synchronization at this layer is referred to as intrastream synchronization, which maintains the continuity of logical data units. Without intrastream synchronization, the presentation of the stream may be interrupted by pauses or gaps.

The second layer of time-dependent data is the stream layer. The unit of the stream layer is a whole stream. Synchronization at this layer is referred to as interstream synchronization, which maintains temporal relationships among different continuous media. Without interstream synchronization, the skew between the streams may become intolerable. For example, users could be annoyed if they notice that the movements of the lips of a speaker do not correspond to the presented audio.

The highest layer of a multimedia document is the object layer, which integrates streams and time-dependent data, such as text and still images. Synchronization of this layer is referred to as interobject synchronization. The objective of interobject synchronization is to start and stop the presentation of the time-independent data within a tolerable time interval, if some previously defined points of the presentation of a time-dependent media object are reached. Without interobject synchronization, the audience of a slide show could be annoyed if the audio is commenting on one slide while another slide is being presented. For more information on media synchronization, we recommend [6.84] and [6.85].

Protocols for Streaming Video

Several protocols have been standardized for communication between clients and streaming servers. According to their functionalities, the protocols directly related to Internet streaming video can be classified into the following three categories: network layer protocol, transport protocol and session control protocol. Network layer protocol provides basic network service support such as network addressing. The IP serves as the network layer protocol for Internet video streaming.

Transport protocol provides end-to-end network transport functions for streaming applications. Transport protocols include UDP, TCP, RTP, and RTCP. UDP and TCP are lower-layer transport protocols, and RTP and RTCP are upper-layer transport protocols that are implemented on top of UDP and TCP.

Session control protocol defines the messages and procedures to control the delivery of the multimedia data during an established session. The RTSP [6.86] and the SIP [6.87] are such session control protocols.

To illustrate the relationship among the three types of protocols, protocol stacks for media streaming are presented in Figure 6.27. For the data plane at the sending side, the compressed audio-video data is retrieved and packetized at the RTP layer [5.21]. The RTP packetized streams provide timing and synchronization information, as well as sequence numbers [6.88]. The RTP packetized streams are then passed to the UDP/TCP layer and the IP layer. The resulting IP packets are transported across the Internet. At the receiver side, the media streams are pro-

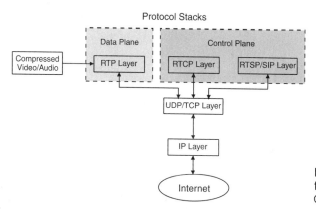

Figure 6.27 Protocol stacks for media streaming [6.65]. ©2001 IEEE.

cessed in the reversed manner before their presentations. For the control plane, RTCP packets and RTSP packets are multiplexed at the UDP/TCP layer and are moved to the IP layer for transmission across the Internet.

Basic building blocks for a streaming video architecture tie together a broad range of technologies from signal processing, networking and server design. A full understanding of the whole architecture is essential for developing the particular signal-processing techniques suitable for streaming video. Furthermore, an in-depth knowledge of both signal-processing and networking technologies helps to make effective design and use of application-layer QoS control, continuous media distribution services and protocols. Finally, a clear understanding of the overall architecture is instrumental in the design of efficient, scalable and/or fault-tolerant streaming servers.

6.3 Multimedia Transport Across ATM Networks

Multimedia itself denotes the integrated manipulation of at least some information represented as continuous media data, as well as some information encoded as discrete media data (text and graphics). Multimedia communication deals with the transfer, protocols, services and mechanisms of discrete media data and continuous media data (audio or video) on and across digital networks. Such communication requires that all involved components be capable of handling a well-defined QoS. The most important QoS parameters are required capacities of the involved resources and compliance to end-to-end delay and jitter as timing restrictions and restriction of the loss characteristics. A protocol designed to reserve capacity for continuous media data, transmitted in conjunction with the discrete media data over, for example, an ATM/LAN, is certainly a multimedia communication issue [6.89]. The success of ATM for multimedia communications depends on the successful standardization of its signaling mechanisms, its ability to attract the development of the native ATM applications and the integration of the ATM with other communications systems. The integration of ATM into the Internet world is under investigation. If there will be ATM applications such as video on demand, there is also the need for a side-by-side integration of ATM and Internet protocols. The success of wireless ATM (WATM) relies on the success of ATM/BISDN in wired networks. When ATM networks become a standard in the wired area, the success of WATM will be realized.

6.3.1 Multiplexing in ATM Networks

In order to transfer the information to the destination, the network performs the generic functions of multiplexing and routing. The routing functions, in order to provide connectivity, are not dependent on the information type in the transfers. On the other hand, multiplexing is highly dependent on the requirements by the information type and application context because multiplexing determines much of the transfer quality on the network. The optimization criteria for the transfer are to minimize the queuing and to maximize the utilization. A joint optimization is possible if the multiplexed streams are shaped to minimize the temporal variability.

Asynchronous time division multiplexing enables statistical multiplexing, but does not mandate it. Statistical multiplexing has been successfully used for data communication for three decades and more recently also in radio networks by means of spread spectrum techniques. The network provides fair access to the transmission capacity and routing. The end equipment is responsible for the quality of the transmission by means of retransmission and forward error correction. The choice of multiplexing mode for asynchronous transfers depends on several issues, as illustrated in Figure 6.28 [6.90]. Here, the link has capacity C_{link}, and the source has peak rate $\hat{R}$, mean rate $\overline{R}$ and maximum burst length $\hat{b}$. The required quality is denoted by Q.

As for general service classes, we can define three classes:

- Deterministic multiplexing with fixed quality guarantees
- Statistical multiplexing with probabilistic quality guarantees
- Statistical multiplexing without quality guarantees

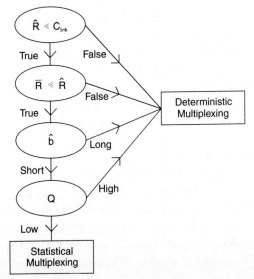

Figure 6.28 Choice of multiplexing mode [6.90]. ©2000 Prentice Hall.

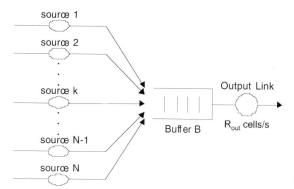

Figure 6.29 Multiplexer model [6.90]. ©2000 Prentice Hall.

The multiplexer can be modeled as a finite capacity queuing system with buffer size B (in ATM cells) and one server with fixed output rate R_{out}. The input of the multiplexer consists of various video sources. The model of the multiplexer is illustrated in Figure 6.29. There are different interleaving schemes for placing data from various sources into the common buffer at two time scales: frame time and cell time [6.91, 6.92, 6.93, 6.94].

In a frame interleaving scheme, the information of each video source is multiplexed in the unit of a frame. It is assumed that each video source has a buffer that can store one frame of information, and all sources are synchronized in the frame boundary. At each frame time, the multiplexer scans each source and puts the information into the common buffer.

In a cell interleaving scheme, the multiplexing process is performed in the unit of a cell. It is assumed that all sources are transmitted at their peak rates from the sources to the multiplexer and that the cells in each frame are uniformly spaced. Each video source is synchronized in frame. An example of a frame-based and cell based interleaving scheme is given in Figure 6.30.

The queuing model of the two-layer coding multiplexer is shown in Figure 6.31. The multiplexing queue is managed by a push-out strategy that allows the buffer to be fully shared by both traffic layers. Cells at the secondary layer are lost if the number of cells from both the primary and secondary layers in the buffer are greater than the buffer size. The cells at the primary layer are lost when the total number of cells from the primary layer is greater than the buffer size [6.95]. We can assume that the cell spacing is uniformly distributed across a frame interval by means of a smoothing scheme for a video source [6.96]. The multiplexer places the incoming cells in a common buffer with capacity B and then transmits them across a 155.5 Mb/s channel. When a large number of video sources are multiplexed, a Poisson arrival process can be assumed.

6.3.2 Video Delay in ATM Networks

Information is delayed in ATM networks, and because we consider asynchronous transfers, these delays will not be constant, not even for deterministic multiplexing. This delay has to be considered end-to-end because delay limits are posed by the application. The video signal is

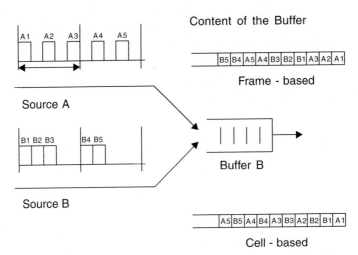

Figure 6.30 Example of multiplexing: frame-based and cell-based interleaving scheme [6.90]. ©2000 Prentice Hall.

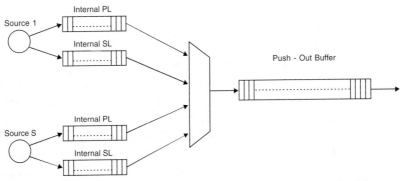

Figure 6.31 ATM video multiplexer for two-layer video streams [6.90]. ©2000 Prentice Hall.

delayed as protocol functions are executed and when the signal is transmitted across the network. The following instances may cause the end-to-end delays:

- Acquisition and display of the video
- Encoding, rate-control and decoding
- Segmentation and reassembly
- Protocol processing
- Wave propagation and transmission
- Queuing

The acquisition is the time that it takes to capture a field or a frame depending on scanning, to digitize it and to perform color and scanning (interlaced to progressive) conversions. The reciprocal functions are performed before display. If there is no scanning conversion, the delay can be of the order of a single pixel instance.

The functions closer to the network are the segmentation of service data units or streams into protocol data units and their reassembly. The time to fill a packet or a cell might be excessive at low rates. For example, it takes 125 µs per octet at 64 Kb/s. If the rate is temporarily or constantly low, it may be necessary to enforce a time limit or to send partially filled cells or packets of restricted length. Reassembly delay depends on message length and transfer rate. For instance, there is no delay for unstructured stream-oriented data. There may be restrictions on the MTU to achieve acceptable delay. The MTU is then dependent on the transfer rate, or the minimum acceptable rate is determined by the MTU size.

Protocol processing is a major cause of delay. It includes framing of information, calculation of check sums and address lookup in hosts and switches and routers. In general, protocols should be implemented to reduce maximum delay and to maximize throughput.

Wave propagation is limited by the speed of light. It takes roughly 100 ms to reach half way around the globe (5 µs per kilometer in fiber). The transmission time is the length of the packet or cell on the transmission line. The wave propagation determines when the first bit of a packet reaches the end of a transmission line, and the transmission time specifies how much later the last bit arrives. The transmission delay is reduced by increasing the line capacity and by reducing the number of links per route.

Because the multiplexing is asynchronous, there will be queuing in the network. Queuing delays in the network vary dynamically from cell to cell and packet to packet for a given route. The delay depends on the instantaneous load in each multiplexer, number of multiplexing hops on the route, amount of buffer space per node and whether deterministic or statistical multiplexing is used. The scheduling discipline affects the distribution of the delays.

Cell Delay Variation (CDV) or jitter can have a significant impact on the quality of a video stream. To keep the encoder and decoder in synchronization with each other, the encoder places PCRs periodically in the TS. These are used to adjust the system clock at the decoder as necessary. If there is jitter in the ATM cells, the PCRs will also experience jitter. Jitter in the PCRs will propagate to the system clock, which is used to synchronize the other timing functions of the decoder. This will result in picture quality degradation.

There are two control issues regarding delay:

- The variations must be equalized to maintain the isochronal sample rate.
- The absolute value must be limited for interactive applications.

Equalization at the network interface of the receiver is not sufficient unless all subsegment protocol processing and data transfers within the end system are fully synchronous. This means that equalization will basically always be needed at the application layer. It is, in fact, the most

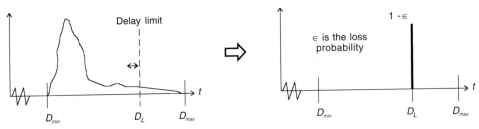

Figure 6.32 Equalization of delay variations.

appropriate location because the bit stream can be synchronized to the display system (the digital-to-analog converter). It should be noted that each stage of equalization introduces more delay.

In Figure 6.32, the delay is equalized by buffering data up to the acceptable limit D_L. Segments that are delayed by more than the limit are treated as if they were lost. Equalization of delay variations is done by buffering data delivered by the network to a predetermined limit before delivery. Late data is discarded (there is no loss if $D_L \geq D_{max}$). The general problem with this approach concerns the choice of D_L to find a proper trade-off between delay and loss and to determine that each pixel has been delayed by D_L when displayed.

A common simplification is to equalize queuing delay at the reassembly point. It is at the adaptation layer in case of ATM and at the transport layer in case of IP. Jitter introduced in the end system is then removed before or after the decoding to obtain signal synchronization.

The delay equalization requires the end system to have a clock that is synchronized in frequency to the sending clock. Usually the clocks at the sender and the receiver will have the same nominal frequency. The jitter is of much larger magnitude than the clock difference. Synchronization could be obtained by locking both clocks to a common reference clock, as carried by the global positioning system and by synchronous digital networks, or by using the network time protocol [6.97]. If a clock reference is not available, then the sender clock has to be estimated from the arriving packet stream. Such a technique uses a phase-locked loop presented in Figure 6.33. The input signal to the loop can be either time stamps carried in the cells or packets or the buffer full level [6.98]. After the clocks are sufficiently synchronized and the data stream is sent completely isochronously, the delay is equalized by simply reading the application frames from the buffer with the same time intervals as sent. Variable rate video complicates the equalization because it is difficult to know how much of the time between arrivals is due to the generating process and how much is due to queuing in the network. Therefore, time stamps in every cell or packet are needed to mark their generating instances.

Signal synchronization is finally obtained after decoding. The frame buffer absorbs much of the delay variations in the decoding, and the residual could be eliminated by repeating or skipping frames to make adjustments to fit the display's clock. If the display allows an external clock, finer adjustments can be made by stretching and shortening the vertical and horizontal tracing times of the cathode ray tube's display. When several video streams emanating from dif-

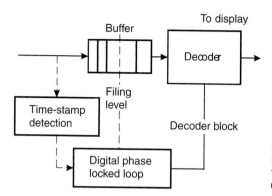

Figure 6.33 Phase-locked loop for the sender-clock frequency estimation [6.90]. ©2000 Prentice Hall.

ferent sources are displayed together, only one of the received signals can be used to synchronize the display system. The other signals must be stretched or contracted to fit that timebase.

6.3.3 Errors and Losses in ATM

The encoding process introduces controlled amounts of distortion in order to compress the signal. The video signal will also be exposed to bit errors induced in the electronics and in the optics. The probability of bit error is low, below 10^{-8}, but not negligible. More troublesome is the information loss in the ATM network when full stretches of the signal are deleted. The causes of loss are transmission burst errors, loss of cells and packets due to multiplexing overload, misrouting due to inaccurate addresses or entries in address tables and delay greater than the acceptable threshold. Undetected loss in a signal can place encoders and decoders out of phase. Burst errors caused by loss of synchronization and by equipment failures have durations of 20 to 40 ms. Their probability of occurrence has been estimated to be below 10^{-7} [6.94]. Loss, especially due to multiplexing overloads, appears to be the most common signal corruption caused by the ATM network.

Error recovery is based on limited error propagation and correction or concealment of the missing portion of the signal. Error propagation is restricted by proper framing of the bit stream so that errors and loss can be detected [6.99].

Generally speaking, in the ATM network, a cell can be lost due to two reasons:

- Channel errors
- Limitations of network capacity and statistical multiplexing

A communication channel is subject to different impairments. If an uncorrectable error occurs in the address field of an ATM cell, the cell will not be delivered to the right destination. This cell is considered to be lost. This is a rare cause of loss in ATM networks.

An ATM network takes advantage of statistical multiplexing, but also takes the risk of simultaneous traffic peaks of multiple users. Although a buffer can be used to absorb the instantaneous traffic peak to some extent, there is still a possibility of buffer overflow in case of congestion. In the case of network congestion or buffer overflow, the network congestion control protocol will drop cells. The malfunction or inefficient network management will also cause the cell loss. For example, loss of synchronization and lack of recovery measures in the physical layer would result in a stream of cell losses in the resynchronization/acquisition phase.

In an ATM network, cell discarding can occur on the transmitting side if the number of cells generated are in excess of allocated capacity, or it can occur on the receiving side if a cell has not been received within the delay time of the buffer memory. Cells can be discarded in the ATM network by the congestion control procedure.

If the incoming traffic exceeds allocated capacity and causes the buffer overflow, the sender could be informed by the network traffic control protocol to reduce the traffic flow or to switch to a lower grade service mode by subsampling and interlacing.

If the network becomes congested and the input buffer overflows, it will drop some cells to reduce the traffic and to assume the normal communication phase.

If the error occurs in the cell header, especially in the address field, the cell may be misdelivered or go astray in the network. In the receiver, if a cell is not received within the maximum time out window, the cell is considered to be lost. The loss of a cell leads to the loss of 384 consecutive bits, which may cause a serious degradation in picture quality for VBR compressed video signals. If the cell loss is caused by network congestion, a few consecutive cells, which contain thousands of bits of information, may be lost. Furthermore, the cell loss may affect the subsequent frames if an interframe coding scheme is employed. Therefore, cell loss is a major problem encountered in VBR coding in the ATM environment. A cell loss may cause the loss of code synchronization. Because a variable number of data is packed into a cell, there is no way of knowing how much information is lost when a cell loss occurs unless some side information is available. Cell loss can occur unpredictably in ATM networks. It is assumed to be random with the probability of cell loss depending only on whether a previous cell of the same priority was lost.

Asynchronously multiplexed networks, such as those based on ATM, have cells and packets as multiplexing units that are shorter than a full cell (session). The multiplexing unit in traditional Time Division Multiplexing (TDM) networks is a call (a session). Network framing means that appropriate control information is added to each multiplexing unit. An example of application framing is the MPEG slice layer that packs bits together for 16 consecutive lines. The purpose of the network framing is to detect and to possibly correct lost and corrupted multiplexing units. Errors and loss handling are shown in Figure 6.34 [6.15]. Errors may be detected by a CRC of sufficient length [6.100]. Loss is detected by means of sequence members, which turn it into erasures (known location and unknown values).

It is important that the sequence number is based on the number of transferred data octets. Knowing that a cell or a packet has been lost does not tell how much data it contained.

Errors and loss can be identified by a CRC on the application frame after reassembly. It is important that frame length is known a priori because the length of a faulty frame cannot be ascertained. The failed CRC could be caused by a bit error, which would not be affected by its length or by a lost packet or cell.

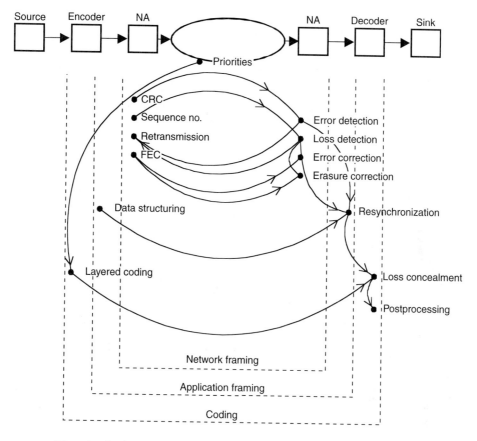

NA - network adaptation

Figure 6.34 Errors and loss handling [6.90]. ©2000 Prentice Hall.

A lost or corrupted network frame would, in case of regular data communication, be retransmitted. There are complications with the use of retransmissions for video. First, the delay requirements might not allow it because it adds at least another round-trip delay that is likely to violate end-to-end delay requirements for conversational services. Second, the jitter introduced is much higher than that induced by queuing. Delay equalization is thus further complicated. Even if this would be acceptable, the continuously arriving datastream must be buffered until the missing frame eventually is received.

There are several reasons to be cautious of FEC of cell and packet loss. First, it adds a fairly complex function to the system, which will be reflected in its cost. Second, the interleaving adds delay. Third, loss caused by multiplexing overload is likely to be correlated because the overload is caused by traffic bursts and more loss may occur than what the code can correct. If an interleaving matrix cannot be corrected, then the full matrix is useless, and the loss situation is in fact made worse.

The interleaving matrix should, of course, be made to cope with burst losses but, again, it increases the delay. Fourth, the coding adds overhead.

6.3.4 MPEG Video Error Concealment

To transmit video traffic effectively across ATM networks, we need to study the issues involved in packetizing encoded video sequences. In particular, it is important to study the effect of ATM cell loss and to develop postprocessing techniques that can be used for error concealment. Error-concealment approaches by Wang [6.101] have assumed that both encoding and decoding occur simultaneously with the decoder communicating to the encoder the location of damaged picture blocks. Many of these techniques are not realistic for real-time applications because they require retransmission of ATM cells. Prioritization approaches to ATM cell-loss concealment have been proposed in several records [6.102 through 6.110]. Figure 6.35 shows a block diagram of the packing/error-concealment scheme using ATM. The cell depacketization operation also provides information as to which macroblocks are missing [6.111].

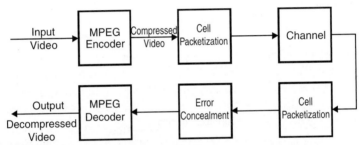

Figure 6.35 Block diagram of the packing/error concealment scheme [6.90]. ©2000 Prentice Hall.

This information is passed to the error concealment algorithm, which attempts to conceal the missing blocks. The goal of video error concealment is to estimate missing MBs in the MPEG data that were caused by dropped ATM cells. The use of spatial, temporal and picture quality concepts are exploited.

Two error-recovery approaches for MPEG encoded video across ATM networks are described in Salama et al. [6.111]. The first approach aims at reconstructing each lost pixel by spatial interpolation from the nearest undamaged pixels. The second approach recovers lost MBs by minimizing intersample variations within each block and across its boundaries.

6.3.5 Loss Concealment

A loss is detected either by means of the network or application-framing information. The corrupted application frame can be considered useless. The application framing should contain sufficient information to allow the next correctly received segment to be decoded. This means that the location within the picture of the information in the segment must be known and that there cannot be any coding dependencies between the information in the segments. The latter condition implies that there cannot be any prediction dependencies across segment boundaries and that variable-length code words are not split by segment boundaries [6.112].

The decoded picture will contain an empty area that corresponds to the lost information. This area can be concealed by using surrounding pixels in time and space [6.102, 6.113]. For instance, the corresponding area in the previous frame can be used [6.112]. It might be best to repeat the full frame if the corruption is severe. When the coded motion vectors are correctly received, they can be used to find the most appropriate replacement in the previous frame. The prediction error is the only remaining error in the area.

The loss concealment can be improved by thoughtful packing of the information, such as separate transfers of motion vectors and prediction errors. A more general framework is often referred to as layered or hierarchical coding.

6.3.6 Video Across WATM Networks

Due to the success of ATM on wired networks, WATM has become the direct result of the ATM anywhere movement. WATM can be viewed as a solution for next-generation personal communication networks or a wireless extension of the BISDN networks. There has been a great deal of interest recently in the area of wireless networking. Issues, such as bit error rates and cell loss rates, are even more important when transmitting video across a wireless network. A very high performance wireless LAN which operates in the 60 GHz millimeter wave band can experience cell loss rates of 10^{-4} to 10^{-2} [6.114]. To provide adequate picture quality to the user, some form of error correction or concealment must be employed. One option is to use the MPEG-2 error-resilience techniques and to modify the MPEG-2 standard slightly when it is used across WATM networks. This technique is known as MB resynchronization [6.114]. In MB resynchronization, the first MB in every ATM cell is coded absolutely rather than differentially. This allows for resynchronization of the video stream much more often than would be possible if resynchronization could only take place at the slice level. It would be relatively simple to incorporate this method with the existing MPEG-2 coding standard by adding an interworking adapter at the boundary between the fixed and wireless networks [6.115]. A second proposal for improving error resilience in wireless networks is to use FEC methods. In addition, improved performance can be achieved by using a two-layer scalable MPEG-2 coding scheme rather than one layer [6.116].

Mobile ATM defines the design functions of control/signaling. In WATM networks, a mobile end-user establishes a VC to communicate with another end-user, either a mobile or ATM end-user. When the mobile end-user moves from one Access Point (AP) to another AP, proper handover is required. To minimize the interruption to cell transport, an efficient switching of the active VCs from the old data path to new data path is needed. Also, the switching should be fast enough to make the new VCs available to the mobile users. During the handover, an old path is released, and a new path is then re-established. In this case, no cell is lost, and cell sequence is preserved. Cell buffering consists of uplink buffering and downlink buffering. If a VC is broken when the mobile user is sending cells to ATM APs (AAPs), uplinking buffering is required. The mobile user will buffer all the outgoing cells. When the connection is up, it sends out all the buffered cells so that no cells are lost unless the buffer overflows. Downlink buffering is performed by APs to preserve the downlink cells for sudden link interruptions, congestion or retransmissions. It may also occur when handover is executed. When the handover occurs, the current QoS may not be supported by the new data path. In this case, a negotiation is required to set up new QoS, because the mobile user may be in the access range of several APs. Therefore, it will select the one that can provide the best QoS.

When a connection is established between a mobile ATM endpoint and another ATM endpoint, the mobile ATM end point needs to be located. There are two basic location management schemes: the mobile scheme and the location register scheme. In the mobile scheme, when a mobile ATM moves, the reachability update information only propagates to the nodes in a limited region. The switches within the region have the correct reachable information for the mobiles. When a call is originated by switching in this region, it can use the location information to establish the connection directly. If a call is originated by a switch outside this region, a connection is established between this switch and the mobile's home agent, which then forwards the cells to the mobile. This scheme decreases the number of signaling messages during a local handover. In the location register scheme, an explicit search is required prior to the establishment of connections. A hierarchy of location registers, which is limited to a certain level, is used.

6.3.7 Heterogeneous Networking

Heterogeneity in networks comes from many sources. Link capacity may vary by several orders of magnitude: from 64 to 128 Kb/s for ISDN lines, several hundred Kb/s for wireless LANs, 10 to 100 Mb/s for LANs such as Ethernet and more for ATM networks. Protocols across these various networks may differ up to the link layer, but also at the network layer. For example, hosts directly connected to ATM networks may run a native ATM stack, but those connected to the Internet may run the TCP/IP stack. End stations can differ in the processing power, available to consume multimedia information. Some may have hardware video-decoding capability, and others may perform the decoding in software. The speed of the processor or the bus architecture may place a bottleneck on the multimedia consumption rate. We will deal with the network heterogeneity in the context of distributing multimedia information through multicast transmission. Multicasting significantly improves the efficiency of network resource use in situations involving one-to-many or many-to-many communications.

The layered video deals with implementing applications that use layered video coding and multicast transmission to handle heterogeneity caused by differences in network link capacity, processing power or display resolution. In such scenarios, receivers express interest in getting higher resolution data by subscribing to the appropriate multicast transport stream (a multicast address in IP or a multicast virtual circuit in ATM) [6.117]. For example, Figure 6.36 shows a near video-on-demand system, transmitting layered video in three layers. All receivers subscribe to the base layer with a bit rate chosen to match the transmission characteristics of the low-bit-rate wireless network. The first enhancement layer has a bit rate suited for receivers on the

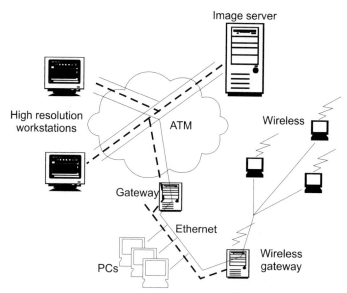

Figure 6.36 Layered networking in heterogeneous networks [6.117]. ©1997 IEEE.

Ethernet. The second enhancement layer is a high-bit-rate layer that the high-powered workstations on the ATM network can receive and decode.

The video source adjusts the bit rates transmitted across each of the layers and dynamically adapts them to the link capacity and receiver processing powers in the current scenario. If the source is transmitting live video, the encoding parameters can be directly manipulated to adjust the bit rates allocated to the different layers. If the information to be transmitted is recorded and pre-encoded video, the source can internally represent the video as a very large number of layers and can map it dynamically to a smaller set of transmission streams based on feedback from the receivers. This mechanism allows the source to match the bit rates of three streams correctly to the characteristics of the wireless, Ethernet and ATM receivers without knowing these characteristics beforehand.

The IP/ATM gateway project addresses the problems of bandwidth and protocol heterogeneity. The gateway is responsible for the following tasks:

- Translating connection set-up messages of the QoS signaling protocols between the two domains. This includes mapping the QoS and traffic parametric between the two domains.
- Forwarding data packets on the IP domain onto appropriate virtual circuits on the ATM side and vice versa. The gateway should perform data forwarding with QoS support, taking QoS parameters into account for scheduling packet transmission.

- Translating session advertisement messages from the two domains to allow a session in one domain to be visible in the other.
- Performing load balancing between multiple gateways connecting the two domains.

6.4 Multimedia Across IP Networks

Multimedia has become a major theme in today's information technology that merges the practices of communications, computing and information processing into an interdisciplinary field. In this Internet era, IP-based data networks have emerged as the most important infrastructure, reaching millions of people anytime and anywhere. They serve as an enabling technology that creates a whole new class of applications to enhance productivity, reduce costs and increase business agility. Anticipating that multimedia across IP will be one of the major driving forces behind the emerging broadband communications of the 21st century, we address the challenges facing the delivery of multimedia applications across IP in a cost-effective, ubiquitous, and quality-guaranteed manner.

6.4.1 Video Transmission Across IP Networks

The problem of sending video across IP has essentially two main components: video data compression and design of communication protocols (Figure 6.37).

Figure 6.37 Structure of a video-streaming system [6.118]. ©2001 IEEE.

One approach consists of designing a low-bit-rate coder, protecting the resulting bit stream with channel codes and using one of the standard Internet transport protocols to transmit the resulting datastream. If the source bit rate is low enough and the channel is not too congested, it is possible to use TCP, in which case no errors occur and therefore there is no need for channel codes. Otherwise, UDP is used with a constant packet injection rate, and low-redundancy channel codes are used to protect against infrequent lost packets. Figure 6.38 illustrates an approach centered around coding problems. In this case, all the intelligence goes into the design of good compression algorithms. It is assumed that the network is a black-box, or a fixed standard pipe, with no differentiation of packets at the socket level or lower.

Figure 6.38 An approach centered around coding problems [6.118]. ©2001 IEEE.

The main drawback of this approach is that it does not deal well with the time-varying nature of the channel. To avoid having to deal with these time variations, the channel is severely underused by using a low-bit-rate coder. This is because, at higher injection rates, fluctuations in the packet-loss rate make it very difficult to guarantee a low probability of decoding error. At higher bit rates, a careful matching of Reed-Solomon (RS) codes to the importance of different portions of an MPEG-2 stream has been proposed [6.118].

Another widely used approach consists of designing new transport protocols, but using either standard video-coding algorithms that generate a fixed syntax for the compressed bit stream or using layered coding techniques. This approach has advantages over the first one discussed previously, the most obvious being that flow control is part of the protocol. In addition, because the bit-stream syntax is known, it is possible to put all the blocks along a given motion trajectory into a single packet so that, if this packet is lost, the entire motion path is lost and therefore error propagation is limited. Also, it is possible to retransmit packets selectively depending on, for example, whether a lost packet contains intracoded blocks or not. These concepts are illustrated in Figure 6.39. In this case, all the intelligence goes into the design of good communication protocols. It is assumed that the coder is typically one of the standards (MPEG-x or H.26x).

Figure 6.39 An approach centered around the design of network protocols [6.118]. ©2001 IEEE.

The main drawback of this approach is it is limited in performance by the nature of the coders used. Video coders based on multiresolution techniques are inherently mismatched to a network that provides no form of packet differentiations. Also, because the modified protocols cannot ensure error-free transmission, when errors do indeed occur, the quality of the decoded signals suffers severely because of lack of robustness in the coders used.

The design of the interface between networks and applications is a problem that has received significant attention. Most of the open literature so far has focused on transmission and traffic regulation across ATM networks, of which perhaps the simplest example is the leaky bucket controller [6.90]. Video across ATM is of interest for two reasons: (a) ATM is one of the candidate transport technologies for future BISDN all the way to the end-user and (b), ATM networks are able to provide QoS guarantees for applications. However, current IP networks are inherently different from ATM networks in that they take a best-effort approach to packet transmission and routing, so no QoS guarantees are provided. As a result, there is no contract to be negotiated between the source and the network. Hence, there are no policing mechanisms applied by the network at its interface with the source.

6.4.2 Traffic Specification for MPEG Video Transmission on the Internet

To promote the evolution of the Internet from a simple data network into a true multiservice network, the IETF Integrated Services WG (ISWG) is defining an Integrated Services Internet, in which traditional best-effort datagram delivery and additional enhanced QoS delivery classes exist [6.119]. Although the IETF has considered various QoS classes, to date, only two of these, Guaranteed Service and Controlled-Load Service, have been formally specified. Guaranteed Service provides an ensured level of bandwidth, a firm end-to-end delay bound and no queuing loss for conforming packets in a data flow. Controlled-Load Service provides a service equivalent to the best-effort delivery on a highly loaded network. Therefore, Guaranteed Service is intended for real-time traffic, and Controlled-Load Service is intended for classes of applications, like adaptive real-time applications, that can tolerate a certain amount of loss and delay, provided that it is kept to a reasonable level.

In order for the new Internet to allow applications to request network packet delivery characteristics according to their needs, sources are expended to declare the offered traffic characteristics. Tspec and admission control rules have to be applied to ensure that requests are accepted only if sufficient network resources are available. Moreover, service-specific policing actions have to be employed within the network to ensure that nonconforming data flows. Policing at the network access point is performed through a token bucket device; packets revealed as nonconforming are marked and forwarded as best-effort traffic.

Traffic specification is a reference point allowing the source and the network to pursue separately two targets:

- To provide the agreed Tspec (source)
- To allocate source requests and to police source traffic (network)

In this context, it is necessary to provide applications with the capability of calculating the Tspec parameters to be declared to the network on the basis of both a limited set of parameters statistically characterizing the data source and the performance of the smoother used in the source to reshape its traffic according to the declared Tspec. The ISWG defined the following Tspec parameters [6.119]:

- Peak rate measured in bytes of IP packets per second, specifying the maximum rate at which the service can inject bursts of traffic into the network.
- Token bucket depth, measured in bytes.
- Bucket rate of the token buckets, measured in bytes of IP packets per second.
- Minimum policed unit (m), measured in bytes, specifying the minimum size of the network packets. All packets of a size less than m will be treated by the policer as being of size m.

- Maximum IP packet size, measured in bytes, specifying the maximum size for a packet that will confirm to the Tspec.

The video source can be modeled as a switched-batch Bernoulli Process (SBBP) taking into account both intra- and inter-GOP correlations [6.120]. Then, a discrete time-queuing system can be used to model video traffic smoothing of the source. After evaluating the loss probability and the average delay suffered in the smoother device, we model the traffic at the output of the smoother, that is, the traffic actually sent across the network. Finally, we model the token bucket at the access point of the network to calculate the marking probability for the packet that does not comply with the specifications.

We can use this paradigm as a tool with the following objectives:

- To design the buffer size of the video server smoother
- To calculate Tspec parameters that are sufficient to guarantee a) both the upper bound for the loss probability and the average delay suffered in the smoother and b) the upper bound for the marking probability in the token bucket at the network access point

6.4.3 Bandwidth Allocation Mechanism

Multimedia applications require the transmission of real-time streams across a network. Pay-per-view movies, distance learning and digital libraries are examples of multimedia applications that require the transmission of real-time streams across a network. Such streams (such as video) can exhibit significant bit rate variability, depending on the encoding system used, and can require high network performance [6.121]. Moreover, these streams require performance guarantees from the network, such as guaranteed bandwidth and loss rate. This poses significant problems when such streams are delivered across the Internet. In fact the real-time network applications that currently run across the Internet achieve a QoS that is far from what is desired. To solve these problems, a small set of differential services has been recently introduced. Among these, Premium Service is suitable for transmitting real-time stored stream (full knowledge of the stream characteristics) [6.122]. It uses a Bandwidth Allocation Mechanisms (BAM) based on the stream peak rate. Due to the variable bandwidth requirement, the peak rate BAM can waste large amounts of bandwidth. One possible approach to reduce bandwidth requirements is to reduce the video VBR. However, even using smoothing techniques, the variability is still present, and, hence, the BAM can still waste a large amount of bandwidth [6.123, 6.124]. Bit-rate variability can also be reduced by modifying the video QoS, but, in this case, the client must settle for a lower QoS [6.122]. Another approach is to allocate the bandwidth in a dynamic way instead of through a fixed bandwidth channel. For instance, one report [6.125] suggests using renegotiation mechanisms to avoid bandwidth waste. Although bandwidth effective, this technique may, at some point in time, require additional bandwidth while transmitting a video. This raises a potential problem because the required additional bandwidth may not be available, leading this technique not to provide the needed bandwidth guarantees. From these considerations, a BAM should not modify while transmitting a video stream. For these reasons, a peak rate BAM is used. A new BAM that uses less bandwidth than the peak rate BAM, while providing the same service, was proposed [6.122]. This BAM does not affect the real-time stream QoS and does not require any modification to the Premium Service Architecture.

To avoid bandwidth waste, the proposed BAM uses dynamic bandwidth allocation and never asks for additional bandwidth. This substantially differs from other dynamic BAMs. For instance, the regeneration mechanisms described in Feng and Rexford [6.126] may make requests for additional bandwidth. They cannot guarantee that these requests will be satisfied by the network. Conversely, the BAM described in Furini and Towsley [6.122] provides bandwidth guarantees as well as the peak rate BAM while using less bandwidth. This is achieved by allocating the peak bandwidth to the premium channel, but progressively reducing this allocation results in the decrease of the peak rate of the remaining stream. This is possible because the streams with known characteristics are considered. Consequently, there is no need to ask for additional bandwidth while transmitting a stream. Hence, this BAM provides the same guarantees as the peak rate BAM while using less bandwidth.

In order to describe BAM, a sender that provides the service and a receiver that desires the service were considered [6.122]. The receiver requests a video, composed of N frames from the sender. Without loss of generality, a discrete time model, where one time unit corresponds to the time between successive frames, was assumed. For a 24-fps full-motion video, the duration of a frame is 1/24 of a second. Denoted by a(i), the amount of data is sent at time i, i=1,2,...,N. We can introduce the bandwidth function, which will be used by the BAM.

$$band(i) = \max\{a(j), j \geq i\}, \quad i = 1,...,N \tag{6.16}$$

If the bandwidth is allocated using this nonincreasing function, there is no need to ask for additional bandwidth. Conversely, it is possible to reduce the allocated bandwidth when it is no longer needed. At time j, just before sending the quantity a(j), a request to deallocate the bandwidth is sent if $band(j)<band(j-1)$, and the new allocated bandwidth will be band(j) instead of band (j-1). The overhead introduced by the deallocation messages is very small compared to the video transmission. Experiments show that at most 20 deallocation messages are sufficient for a 28-minute video. The bandwidth use, U, achieved using our bandwidth allocation mechanisms is

$$U = \sum_{i=1}^{N} a(i) / \sum_{i=1}^{N} band(i) \tag{6.17}$$

It is greater than what is obtained using classic BAM because $band(i) \leq Peak\ rate, i = 1,...,N$.

6.4.4 Fine-Grained Scalable Video Coding for Multimedia Across IP

Multimedia streaming and the set of applications that rely on streaming are expected to continue growing. A primary objective of most researchers in this field is to mature Internet video solutions to the level when viewing of good-quality video of major broadcast television events across the Web becomes a reality. One generic framework that addresses both the video-coding and networking challenges associated with Internet video is scalability. Any scalable Internet video-coding solution has to enable a very simple and flexible streaming framework. Hence, it must meet the following requirements [6.127]:

- The solution must enable a streaming server to perform minimal real-time processing and rate control when outputting a very large number of simultaneous unicast streams.
- The scalable Internet video-coding approach has to be highly adaptable to unpredictable bandwidth variations due to heterogeneous access technologies of the receivers or due to dynamic changes in network conditions.
- The video-coding solution must enable low complexity decoding and low memory requirements to provide common receivers, in addition to powerful computers, the opportunity to stream and decode desired Internet video content.
- The streaming framework and related scalable video-coding approach should be able to support both unicast and multicast applications. This eliminates the need for coding content in different formats to serve different types of applications.
- The scalable bit stream must be resilient to packet loss events, which are quite common across the Internet.

The previous requirements were the primary drivers behind the design of the FGS video coding scheme [6.128]. For example, the 3D wavelet/subband-based coding schemes require large memory at the receiver, and, consequently they are undesirable for low complexity devices [6.73, 6.129]. In addition, some of the methods rely on motion compensation to improve the coding efficiency at the expense of sacrificing scalability and resilience to packet losses [6.129]. Other video-coding techniques totally avoid any motion compensation and consequently sacrifice a great deal of coding efficiency [6.73, 6.130].

The FGS framework strikes a good balance between coding efficiency and scalability while maintaining a very flexible and simple video-coding structure. When compared with other packet-loss-resilient streaming solutions, FGS has also demonstrated good resilience attributes under packet losses [6.131]. After new extensions and improvements to its original framework, FGS has been adopted in the MPEG-4 Video standard as the core video-coding method for MPEG-4 streaming applications [6.132]. Since the first version of the MPEG-4 FGS draft standard [6.133], there have been several improvements introduced to the FGS framework. In particular, there are three aspects of the improved FGS method. First, a very simple residual-computation approach was proposed. Despite its simplicity, this approach provides the same or better performance than the performance of a more elaborate residual computation method. Second, an adaptive quantization approach was proposed, and it resulted in two FGS-based video-coding tools. Third, a hybrid all-FGS scalability structure was also proposed. This novel FGS scalability structure enables quality (that is, SNR), temporal or both temporal-SNR scalable video coding and streaming [6.128].

In order to meet the requirements outlined, FGS encoding is designed to cover any desired bandwidth range while maintaining a very simple scalability structure. Examples of the FGS

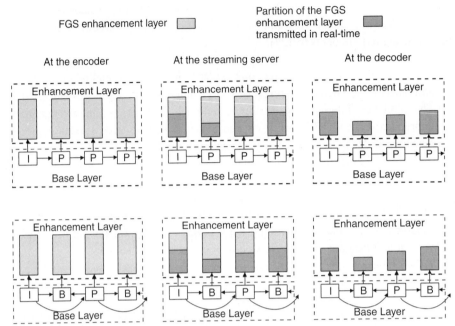

Figure 6.40 Examples of the FGS scalability structure [6.128]. ©2001 IEEE.

scalability structure at the encoder (left), streaming server (center) and decoder (right) for a typical unicast Internet-streaming application are shown in Figure 6.40. The top and bottom rows of the figure represent base layers without and with bidirectional (B) frames, respectively.

The FGS structure consists of only two layers: a base layer coded at a bit rate Rb and a single enhancement layer coded using a fine-granular scheme to a maximum bit rate of Re. This structure provides a very efficient, yet simple, level of abstraction between the encoding and the streaming processes. The encoder only needs to know the range of bandwidth [Rmin = Rb, Rmax = Re] over which it has to code the content. On the other hand, the streaming server has a total flexibility in sending any desired portion of any enhancement layer frame, without the need for performing complicated real-time rate-control algorithms. This enables the server to handle a very large number for unicast streaming sessions and to adapt to their bandwidth variations in real-time. On the receiver side, the FGS framework adds a small amount of complexity and memory requirements to any standard motion-compensation-based video decoder.

For multicast applications, FGS also provides a flexible framework for the encoding, streaming and decoding processes. Identical to the unicast case, the encoder compresses the content using any desired range of bandwidth [Rmin = Rb, Rmax =Re]. Therefore, the same compressed streams can be used for both unicast and multicast applications. At time of transmission, the multicast server positions the FGS enhancement layer into any of the preferred number of multicast channels, each of which can occupy a desired portion of the total bandwidth. Example of an FGS-based multicast scenario is given in Figure 6.41. The distribution of the base layer is

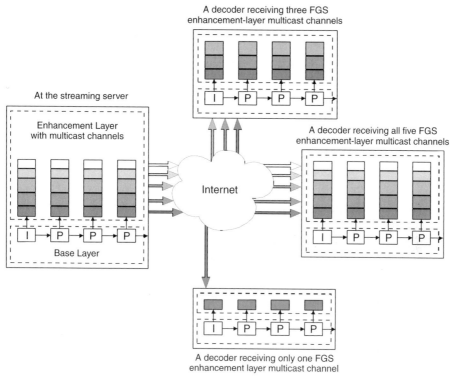

Figure 6.41 Examples of the FGS-based multicast scenario [6.128]. ©2001 IEEE.

implicit and therefore is not shown in the figure. At the decoder side, the receiver can subscribe to the base-layer channel and to any number of FGS enhancement layer channels that the receiver is capable of accessing. It is important to note that, regardless of the number of FGS enhancement layer channels that the receiver subscribes to, the decoder has to decode only a single enhancement layer.

The FGS framework requires two encoders, one for the base layer and the other for the enhancement layer. The base layer can be compressed using any motion-compensation video-encoding method. The DCT-based MPEG-4 Video standard is a good candidate for the base layer encoder due to its coding efficiency especially at low bit rates. Prior to introducing FGS, MPEG-4 included a very rich set of video-coding tools, most of which are applicable for the FGS base layer [6.134].

The FGS enhancement layer encoder can be based on any fine-granular coding method. When FGS was first introduced to MPEG-4, three approaches were proposed for coding the FGS enhancement layer: wavelet, DCT and matching pursuit-based methods. This led to several proposals and extensive evaluation of these and related approaches. In particular, the performances of different variations of bit-plane DCT-based coding and wavelet compression methods were studied, compared and presented [6.135]. Based on an analysis of the FGS enhancement layer SNR, the study concluded that both bit-plane DCT coding and EZW based compression provide very similar results.

6.5 Multimedia Across DSLs

The Internet with all its applications is changing the way we work, live and spend time. However, today the Internet is facing a major problem. Growing demand for access has produced bottlenecks and traffic jams, which are slowing down the Internet. In an attempt to overcome these restrictions, access has pushed the technology of traditional telephony to new and innovative heights with the emergence of Asymmetric DSL (ADSL) technology. High-speed ADSL eliminates bottlenecks, giving all subscribers quick and reliable access to Internet content. Telecom service providers have yet to realize the full potential of ADSL. Traditional telephone and Internet services are only the beginning, but the ability to offer broadcast video services is a reality. Cable TV operators are beginning to offer voice and data services. There is increasing competition from Competitive Local Exchange Carriers (CLEC) and other carriers, making it imperative that traditional telecom service provides video services. By offering a range of services, established service providers can generate additional revenue and can protect their installed base. Direct Broadcast Satellite (DBS) providers, particularly in Europe and Asia, are offering a compelling Multichannel Video Program Distribution (MVPD) service [6.142].

A key factor contributing to the successful deployment of ADSL access systems has been the facility for overlying data services on top of existing voice service without interfering with the voice service. For the users, this offers the following:

- Always-on service capability. There is no need to dial up because the IP connection is always available and so is the office networking model in which network resources are available all the time.
- Virtual second voice line. Unlike when the user is connected through a modem, the voice line remains available for incoming and outgoing calls.

For the operator, the service overlay allows ADSL to be installed throughout the network, irrespective of what types of narrow band switches are installed. After the initial success of ADSL, it became apparent that it could be used to offer multiple phone lines together with a greater range of services (for example, VPNs) targeted at specific markets. This has been made possible by the high bandwidth of ADSL, backed up by progress in voice compression, echo cancelling and digital signal-processing technologies. ADSL offers a high data bandwidth, of which a portion can be used to offer additional voice services integrated with the data services. Symmetric DSL techniques, such as Single Pair High-Speed DSL (SHDSL) cannot be deployed as an overlay to existing analog telephone services, so the delivery of voice and data services using a single facility requires voice to be carried directly on the DSL link. The techniques used to transport voice and data in an integrated way across DSL, whether ADSL or SHDSL, are referred to as Voice over DSL (VoDSL).

With VoDSL, two main market segments are of interest to service providers. The first is small- to medium-sized businesses, a significant percentage of which need to be able to send and receive data of around 500 Kb/s. The voice needs of these customers are typically met by 4 to 12 outgoing lines. Using, for example, ADPCM voice coding, at peak times these phone lines consume only 128 to 256 Kb/s of the ADSL bandwidth, which is typically in excess of 2 Mb/s downstream and more than 500 Kb/s upstream. The second market interested in VoDSL services is residential users who will appreciate the extra two to four voice lines that VoDSL offers [6.143].

From the service provider's perspective, ADSL offers considerable opportunities in terms of providing source of incremental revenue and a way of reducing costs. Regardless of the type of operator, there are compelling reasons for the success of VoDSL services. Advantages for the user include ISDN voice quality; automated provisioning, which greatly reduces the time taken to add or remove services, handle data and voice service with one-stop-shopping; a single bill and a common helpdesk. The keys to success are the bundling of both data and voice lines and pricing flexibility [6.136 through 6.141].

ADSL will be delivering multimedia services to millions of users. The transmission of digital multimedia data requires the existing systems to be augmented with functions that can handle more than just ordinary data. In addition, the high volume of multimedia data can be handled efficiently only if all available system services are carefully optimized.

6.5.1 VODSL Architecture

The architecture deployed by a telecom service provider to deliver video services will vary. A typical example is shown in Figure 6.42. In the access network, the ATM provides layer 2 connectivity across ADSL.

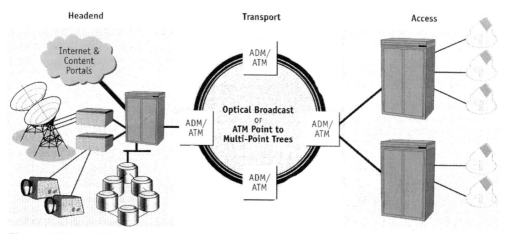

Figure 6.42 Architecture deployed to deliver video services.

Thus, each DSL Access Multiplexer (DSLAM) is either an ATM multiplexer or switch. As a result, video programs must be delivered in either MPEG-over ATM format or MPEG-over-IP-over-ATM format. Both technologies are currently available, but the market appears to be favoring the IP as the network layer delivery vehicle. Although IP adds some overhead to the video stream, it greatly simplifies in-home distribution across Ethernet-compatible media. Also, more applications are available for IP, broadening its audience. In both cases, the headend and transport networks are similar. The term headend originated in the cable industry and is used here to denote a location where content is aggregated for TV channels, VoD, e-commerce portals, Internet access and so on. The location of headend, and even whether it is centralized or distributed, is an architectural choice. As the video content is delivered to the user across the ATM access network, content can be injected into the network at almost any location. In the case of a broadcast TV service, video arrives from various services across diverse media, including DBS, local off-air broadcast and studios. Content from all these services has to be fed to an encoding platform and converted into MPEG format, if not already in this format. Given that the end-delivery network is ADSL, it is highly recommended that the output video signals should be shaped to optimize link use and to ensure that the ADSL line is not overloaded. The output channels are delivered to an ATM network using either MPEG-over-IP-over-ATM or MPEG-over-ATM encapsulation. Interactive services, such as VoD and network-based time-shifted TV, are delivered from servers, which store content in MPEG format and deliver a copy at the subscriber's request. The server must be dimensioned for both the amount of content it must store and the number of active subscribers retrieving data. Single large servers or multiple distributed servers can be used to meet this requirement. The trade-off is between transport costs, replicated server costs and management complexity. Other servers for a variety of video services can also be collected at the headend. As for the headend in a VoDSL architecture, it can be centralized or distributed. Because the content is distributed using IP and/or ATM, connectivity is very flexible.

The role of the transport network is to deliver the content from the headend locations to the appropriate DSLAMs, or their attached switches/routers, in the access network. The network must transport two specific types of traffic: multicast and unicast, corresponding to the broadcast and interactive services.

Broadcast traffic is transported as IP multicast, ATM point-to-multipoint or a combination of the two. IP multicast overlay using ATM is shown in Figure 6.43. Traffic must be delivered to all DSLAM locations in the network, essentially emulating a cable service that delivers all channels at all locations. Given that the traffic is either IP or ATM, the choices for constructing the distribution network are ATM point-to-multipoint or IP multicast. A good solution for an overlay network is to use ATM point-to-multipoint connections in an ATM-switching environment. ATM is stable technology with the proven ability to replicate high bandwidth data. This approach will work across almost any transport network, such as SONET/Synchronous Digital Hierarchy (SONET/SDH) or Dense Division Multiplexing (DDM), and supports native MPEG-over-ATM and MPEG-over-IP encapsulation. The links that carry the broadcast channels can also be used to transport other data, such as interactive content. The downside to this multicast overlay

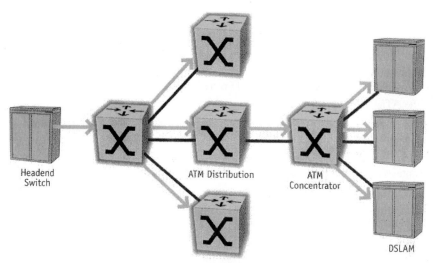

Figure 6.43 IP multicast overlay using ATM.

approach is the increased cost of supporting multiple optical transport links and any intermediate ATM switches required to complete the point-to-multipoint. ATM switches attached to the DSLAMs are not included as extra costs, because these switches will typically exist within the network.

IP multicast-capable routers can also be used to distribute the broadcast TV channels if IP is the network layer chosen for the service (Figure 6.44). If an existing IP network provides the required capacity and performance for multicast replication, then it may be feasible to add broadcast television streams. The end-delivery encapsulation is ATM, so the IP multicast streams must be encapsulated into ATM virtual circuits for the final leg of the journey. To ensure high-quality video, IP networks must also be properly engineered to deliver QoS.

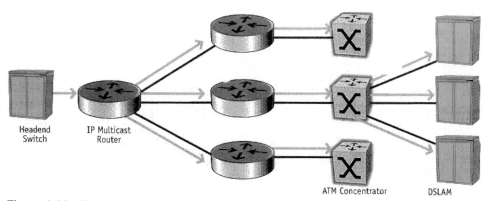

Figure 6.44 IP multicast overlay using routers.

The ADSL access network is well suited to point-to-point IP architectures. Many different architectures can be chosen for building a unicast network, including existing Broadband Remote Access Servers (BRAS), ATM switch/routers and IP cards within the DSLAM connected to routers.

Unicast and broadcast services can be delivered across the same network infrastructure. For example, the ATM concentrator nodes that aggregate the DSLAMs support both point-to-multipoint and point-to-point virtual circuits. Bidirectional, interactive traffic across ATM point-to-point virtual circuits can be aggregated at either BRAS or router, depending on the service requirements. These routing devices, located within certain central offices, can then be connected to a data center across the same optical transport medium that delivers the broadcast traffic.

The DSLAM is the last element in the access network before the subscriber's home. It is responsible for switching the video channels delivered to the subscriber. In the interest of service response (rapid channel changing) and bandwidth savings, the nearer the multicasting device is to the subscriber, means the better the offered service. To meet the performance requirements, the DSLAM must always support multicasting in hardware. DSLAM multicast replication is shown in Figure 6.45.

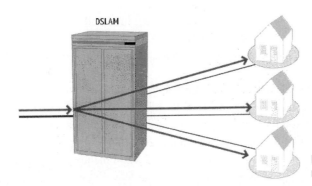

Figure 6.45 DSLAM multicast replication.

However, the integrated DSLAM approach is not ideal in cases where the service provider has an installed base of DSLAMs that do not support such features. Also, broadband Digital Loop Carriers (DLC) with ADSL links are unlikely to provide any multicast switching capabilities. Thus, it is also necessary to offer multicast switching using an external device. Typically, this will be either an IP multicast router or ATM switch supporting logical multicast (multiple point-to-multipoint leaves from the same connection on the same port), or a combination of the two. Multicast for existing DSLAMs is shown in Figure 6.46. Note that the uplink from the DSLAM/DLC to the switching device will constrain the number of video subscribers supported by the DSLAM, because all content channels are treated as unicast from the switching device onward. Unicast interactive traffic must also travel through the DSLAM, so both multiple virtual circuits and QoS guarantees must be available within the DSLAM to support both broadcast and interactive services concurrently. The strength of the ATM access networks lies in its use of virtual circuits (Figure 6.47.)

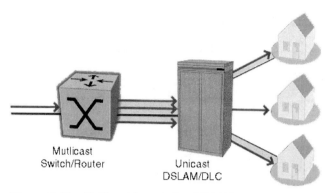

Figure 6.46 Multicast for existing DSLAMs.

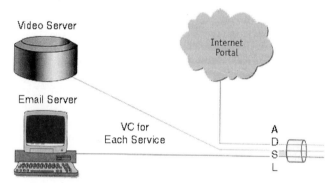

Figure 6.47 Multiple services to a subscriber.

A single subscriber might require multiple services, each of which is best served by a unique device. For example, high-speed Internet access traffic might best be delivered through a BRAS, which provides a rich feature set for accounting. Multiple services to a subscriber are presented in Figure 6.47.

After the VoDSL channel is terminated at the subscriber's premises by a DSL modem, it is necessary to distribute the content to the set-top box so that it can be viewed on the television. This is typically done via Ethernet, which can also be connected to the PC.

When the video is encapsulated as MPEG-over-IP-over-ATM, there are more options for in-home distribution. A variety of Ethernet-compatible media are available or under development, including wireless Ethernet, wired Ethernet, Home Phoneline Networking Alliance (HPNA) and Powerline technologies. Obviously, media that do not require new in-home wiring are very attractive because they considerably reduce the cost of home installations and the need to send an engineer to the subscriber's premises.

Wireless Ethernet is one of the most promising emerging technologies for rapid home installation. Use of wireless Ethernet in the home is represented in Figure 6.48.

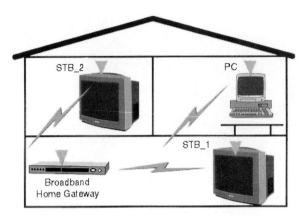

Figure 6.48 Use of wireless Ethernet in the home.

The DSL modem becomes an integral part of a broadband home gateway, which supports home links, such as wireless Ethernet, to communicate with IP devices in the home. A television set-top box could be one of these devices, so the IP video stream is directed from the home gateway to the set-top box. This is possible using the IEEE 802.11b standard, which can support up to 11 Mb/s, sufficient to supply two remote set-top boxes. Future technologies promise to increase the bandwidth of the wireless connection to 20 or 30 Mb/s.

6.5.2 Delivering Voice Services Across DSL

Delivering VoDSL offers a lucrative opportunity for both established and emerging providers. Today ADSL users can receive one or more digital voice lines on top of their existing analog telephone line. Depending on the target ADSL penetration and the network application, it will become economically attractive to implement data-only access networks, based on pure ATM DSLAMs.

Several benefits of VoDSL are reinforced by a network solution that brings about the convergence of voice and data in both the access and core networks. Several architectures have been proposed in the context of VoIP. The term "Voice over Packets (VoP)" is used to refer to these architectures because in many cases the top-level features and advantages remain valid whether voice and data are carried in ATM, IP frames on IP on top of ATM or the access or core network. At the transport level, much of the final architecture will depend on the deployment scheme followed by the service provider. Evolution from VoDSL to next generation network is shown in Figure 6.49. In a next generation network, voice calls are no longer handled by exchanges, but by a central high capacity call server. This server controls the gateways that perform the conversion to VoP. Already next generation network architectures are widely deployed in the core network. Trunk gateways are typically positioned between the local exchange and a packet-based network (ATM or IP). The transit exchanges are then replaced by more cost-effective router equipment. Control servers centralize control, thereby reducing operating costs, and voice compression saves on bandwidth, particularly on long distance and intercontinental links. The initial VoDSL scenario introduces gateway functionality at or in the CPE and access nodes.

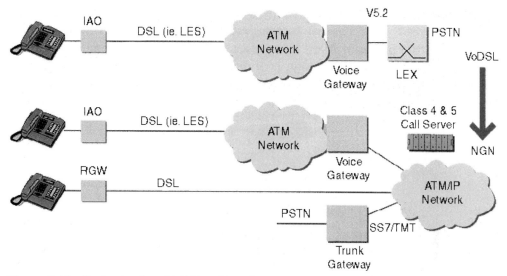

Figure 6.49 Evaluation from VoDSL to Next Generation Network (NGN).

6.5.3 Multimedia Across ADSL

The primary motivation behind ADSL is the delivery of multimedia services. Given the fast growth of Internet and multimedia applications, the widespread acceptance of ADSL systems will depend on the ability to provide efficient delivery and refined quality of multimedia data to the subscribers. Multimedia data has quite different characteristics compared to general data. One major characteristic is the layer coded structure where multimedia data is constructed into separate data streams, each representing a layer. The layers have different QoS requirements, that is, data rate and error performance (Bit Error Rate [BER] and Symbol Error Rate [SER]).

With limited communication resources, for example, bandwidth and transmitted power, a key design issue in multimedia communications is to handle the layers differently and, therefore, efficiently. A larger amount of channel resources should be assigned to the layers with higher importance. For example, it is well known that the use of scalability can enhance the error robustness of a video service. There are two transmission schemes for multimedia data across ADSL: serial transmission and parallel transmission.

Serial Transmission: TDM

ADSL transmission is divided into time slots. In each time slot, only data from a single layer is transmitted. The layers are time-division multiplexed. The design task is finding a time slot for layer assignment to achieve high-efficiency transmission and to provide an acceptable QoS to the users. Such a system is named serial transmission [6.144]. The time slot as well as subchannel structure for the serial transmission is shown in Figure 6.50. For these source layers, time slots 1 and 2 are assigned to transmitting layer 1, and slots 3 and 4 are assigned to transmitting layers 2 and 3, respectively. The subchannel power and bit rate distributions are different in slots

1 and 2 compared to 3 and 4. It is important to note that, within a single time slot, the power and bit rate are allocated so that all the usable subchannels perform at the same error rate. For two time slots transmitting different layers, the subchannels' error performances are completely different. An error performance distribution across the time slots and subchannels is illustrated in Figure 6.51. The system configuration is with four layers from single or multiple sources, 256 subchannels and 10 time slots. The BER is constant within the same time slot and different across the time slots.

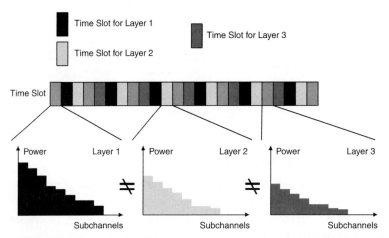

Figure 6.50 Serial transmission for multimedia data across ADSL [6.145]. ©2000 IEEE.

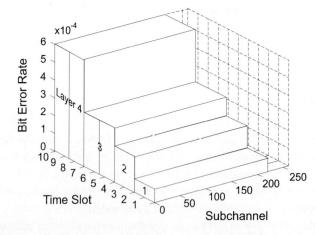

Figure 6.51 Error performance across the time slots and subchannels for serial multimedia data transmission across ADSL [6.145]. ©2000 IEEE.

Parallel Transmission Frequency Division Multiplexing

The essence of the parallel transmission is to use the special ADSL channel characteristics by transmitting multimedia layers simultaneously, each occupying a set of subchannels. As shown in Figure 6.52, frequency division multiplexing the data streams corresponds to different multimedia layers. The three service layers are transmitted in the same slot, each occupying a set of subchannels. Although the time slots are still grouped into frames, for every time slot in a frame, the power and bit-rate allocation remain the same. In general, data from the important layers are transmitted through the subchannels with better channel performance, that is, larger channel gain, lower noise variance or simply the good subchannels. Such assignment provides reliable transmission to the most important layers without large power consumption. In contrast to the serial transmission, the parallel transmission can integrate various traffic flows with different QoS requirements without any frequent channel parameter changes.

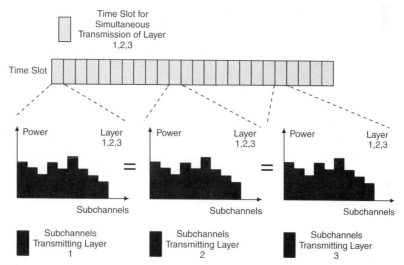

Figure 6.52 Parallel transmission for multimedia data over ADSL [6.145]. ©2000 IEEE.

The error-performance distribution of the parallel transmission is quite different from that of the serial transmission, as can be seen from Figure 6.53. It achieves constant error performance across the time slots, but different error performances at the subchannels transmitting different layers.

6.6 Internet Access Networks

In building an infrastructure for delivering all the services to the consumer, the most critical part is the links to the subscriber's home. This so-called last mile has some unique characteristics. It generally consists of copper wire pairs over which analog voice signals travel to a telco central office and is therefore limited in its information-carrying capacity. Replacing copper pairs with

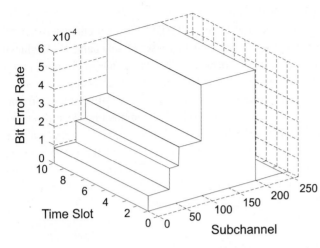

Figure 6.53 Error performance across the time slots and subchannels for parallel multimedia data transmission across ADSL [6.145]. ©2000 IEEE.

optical fibers would be a good way to speed Internet access, but, in fact, the amount of revenue foreseen from each residential and small business customer cannot economically justify such a move. From the sociopolitical point of view, faster Internet access for everyone is considered important. Out of about 120 million United States households, 40% have at least one computer. Of the computerized households, some 60% access the Internet regularly. If it is assumed that 10% of those regular Internet accesses become early adopters of fast Internet access services, 2.88 million households will be involved, each of which may be expected to spend about $40 a month on the second phone line and Internet access. Accordingly, the market for combined services offering a second telephone line only for Internet access over a short term of one year or so may be conservatively estimated at $1.3 billion a year. This is an impressive amount, and it is just the beginning.

ADSL offers asymmetric rates of transfer of data to and from the Internet. The uplink rates can go up to 768 Kb/s and downlink rates are 6 to 8 Mb/s, depending on the length and condition of the local loop, or the wiring between the customer's premises and the telco central office.

Cable companies bring analog TV signals across optical filters to their neighborhood distribution points (headends), where the signals are distributed to residences by coaxial cables. The combination of fiber and coaxial cable, which can carry high-speed data as well as TV signals, is known as hybrid fiber coax (HFC). Each distribution point typically serves 200 to 500 residences. The extent of the network of a cable TV operator is measured in terms of homes passed, that is, the number of homes adjacent to which the operator's cable passes, regardless of whether those homes have been signed up as customers. Realistically, cable modems are capable of passing data upstream at speeds of 200 Kb/s to 2 Mb/s and downstream at speeds up to about 10 Mb/s.

Cable modems, capable of operating at higher speeds than ADSL, have some serious drawbacks. The cable link to a residence is shared among many users, so if many of them decide to log on to the Internet at the same time, achievable communication speeds may plunge.

Because the lines are shared, a hacker may be able to drop on a neighbor's connection to the Internet or on an intranet, which is a security problem that may be serious to some users. Consequently, a customer who happens to be a road warrior will be unable get access into the Internet at airports or hotels through his laptop computer at his usual data rate. If he is able to connect at all, it will be through a dial-up modem at a much lower speed.

The most valuable benefit of wireless services is that they make access possible for people who are on the move. They are also attractive in certain cases where the user is stationary. Digital cellular telephones are quickly becoming the main communication tools for people on the move. Although they are good for retrieving email and checking stock quotes, the R&D is targeting multimedia communications. Multimedia communications are available, but at present only to stationary users. Satellite broadcasts, for example, allow fast download of Internet contents with a return path (that is, the uplink from the user computer to an Internet service provider) on a dial-up modem. For a rural user, it is possible to use Local Multipoint Distribution Services (LMDS). This option uses millimeter-wave radio at frequencies of about 30 GHz. A typical installation has a central base station with an unidirectional antenna serving many residences, each of which has a directional dish aimed at the base station. The service is theoretically capable of sustaining a data transfer rate of about 30 Mb/s. The systems work well provided that the users are within about 3.5 Km of the local transceivers. Yet another viable alternative is to access the Internet across the unlicensed 2.4 GHz band. With this approach, the service provider broadcasts Internet data, using digital spread-spectrum wireless technology from an antenna on a tower. The range is some 60 Km. Computer manufacturers and software developers also have legitimate interests in both methods of fast Internet access: ADSL and cable modems. Faster means of Internet access generate demand for faster computers and newer software.

6.6.1 DSL Networks

Whether by phone or by mail, local phone companies will soon be urging their customers to sign up for an Internet service speedier than anything achievable with today's dial-up modem. A technology known generally as DSL is responsible. It needs only a single twisted wire pair to provide both Internet access and conventional analog telephony. In the future, it may also be used to deliver pay-per-view video. Most of these services will require minor changes to the phone wiring in subscribers' homes [6.146]. The most suitable version of the technology for residential broadband access to the Internet is generally held to be ADSL. Its downstream bandwidth, from the Internet to the home, may reach 6.144 Mb/s and its upstream bandwidth, from the home PC to the Internet, may reach 640 Kb/s [6.147]. The asymmetry in ADSL transmission speeds matches the flow of data to and from the Internet. In a typical Internet session, after all, a Web surfer sends short messages upstream to request data and is bombarded with information in return. It is this downstream transmission rate that limits the usefulness of most connections [6.148].

From a performance point of view, the big difference between DSL technologies and cable modems is that DSL gives each customer a dedicated link to the central office of the local phone company. On the other hand, with cable modems, several users share a single coaxial cable [6.149].

Cable TV operators, ever since they saw the promise of their coaxial cables, have been busily equipping homes with high-speed access to the Internet. Their coax links penetrate more than 90% of residences in the United States, with European and Japanese percentages not far behind and with rapid growth in the nonindustrial world as well. Telephone companies were slower off the mark, having made impressive earlier attempts to capture the residential Internet access market with their ISDN. The sudden threat posed by cable TV companies is now forcing them to deploy ADSL.

ADSL does have its problems, however. Loading coils are one. Because of the distributed capacitance and resistance along their lengths, phone wires attenuate and distort voice signals, with effects that increase with frequency and distance. To equalize the lines across the frequency range of interest for voice communication (up to 4 KHz), phone companies install inductors in their longer lines. These loading coils improve the frequency response within the voice band, but at a price: They increase attenuation for signals above that band. Consequently, they must be removed for high-speed data transmission.

Signal dispersion is another problem with high-frequency signals. The physical characteristics of transmission lines are such that signals of different frequencies propagate at different velocities. Pulses, which represent data and are made up of several frequencies, tend to spread out as they propagate down a line, eventually overlapping with each other. This effect is known as intersymbol interference and limits the data rate that can be supported. Like attenuation, the effects of dispersion get worse with frequency and line length. Both near-end and far-end crosstalk are also problems. The first arises when a receiver is located at the same end of a cable as a transmitter operating in the same frequency band.

The frequency band used for full-rate ADSL is broken into three parts as shown in Figure 6.54. It can be seen that in the FDM used for ADSLs, the frequency band from DC to 1.1 MHz is divided into three subbands (top). The first is used for analog voice. Namely, the 0 to 4 KHz range is reserved for voice telephony; the portion between 25 KHz and 138 KHz for upstream data to the Internet and the rest of the band, up to 1.1 MHz, for downstream data from the Internet to the PCs. If some form of echo cancellation technology is used, the downstream bandwidth may be expanded. Echoes are signals generated by the local transmitter that get fed to the local receiver due to coupling between wires. An echo canceller takes care of echoes because it knows what was transmitted and can subtract it from what was received. Basically, two types of ADSL modems now coexist: Carrierless Amplitude Phase (CAP) and Discrete Multitone (DMT). They differ in how they perform line coding, that is, in how they modulate digital data onto an analog carrier. CAP uses quadrature amplitude where a pair of equal-frequency signals are varied to create between 4 and 1,024 discrete line conditions or symbols. Each symbol represents several bits, and the actual number is dependent on the total number of possibilities.

Example 6.2 In a CAP-4 scheme, with a total of four possible symbols, each symbol represents two bits. A phase of 0 degree could represent 00, 90 degrees could be 01, 180 degrees

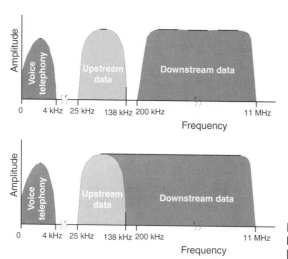

Figure 6.54 The frequency band used for full-rate ADSL [6.150]. ©1999 IEEE.

would be 10 and 270 degrees would be 11. In a more practical example, CAP-16 uses 12 different phases and then adds a further four symbols by repeating four of the phases, but at half the amplitude. In this scheme, then, a total of 16 distinct combinations of bit patterns, starting from 0000 and ending with 1111, can be represented. Because each symbol represents 4 bits, CAP-16 can operate at a line speed of, say 10000 symbols per second, and yet can transmit data at 40 Kb/s.

The business of providing the physical medium for the transport of data to the Internet is to be different from the business of providing Internet services. The first is called a network service provider. Such a provider would deploy the ADSL technology offering users a fast access path to the Internet. After the path is established, the ISP's home page, also known as a portal, is the first to come up on the screen of the user's computer.

6.6.2 Cable Access Networks

Today's cable plants suit one-way communication, from the headend outward, and so are perfectly adapted to broadcast television. For Internet access or telephony, however, traffic must flow in both directions, so a range of frequencies between 5 and 42 MHz is allocated for upstream signals, both analog and digital. Any discussion of access to the Internet must emphasize that, for most residential uses, other than video or desktop conferencing, the exchange of data is asymmetrical. The request an Internet user sends to look up a Web site or search for some information consists of very few packets but may trigger a deluge of data in the other direction. This deluge is one of the main problems that cable modems aim to solve.

Because the connection between a cable modem and the Internet is packet switched, the link is always active. As soon as the computer is turned on, it begins a new session with the Internet, and the ISP dynamically assigns it an IP number, which remains valid until the session is terminated. A large provider with thousands of clients may assign local IP addresses to clients using its proxy server to interface with the outside world, that is, outsiders will see the proxy server's address, instead of the local address, and all communication between a user and the outside world will take place through the proxy server. The telephone service provided by the cable operators is also always on in that it is not interrupted by power outages. Should utility power fail, backup batteries at the cable system's nodes would keep the phones working even when other services go down. This feature is akin to the lifeline service provided by phone companies.

For some years now, multiple-system operators have tempted their clients with Internet and cable TV service. This they did with a splitter box and a cable modem of proprietary design. The devices cost a lot and have only a single source. Hence, there arose a strong desire for a set of specifications that would be widely acceptable to the industry.

Of concern in sending digitized voice across a packetized system is that delay and delay variations (jitter) be kept within fairly tight limits. Small packets are desirable because they keep the delay down, but are undesirable because they increase overhead. Long packets have lower overhead, but take longer to fill. In a multiuse hybrid fiber-coax system, a packet size of an IEEE 802.3/Ethernet frame (1,518 bytes, maximum) would represent a compromise among many requirements.

The question often arises as to what mechanism should be adopted for the MAC layer. Downstream data present no problem as it is broadcast to all subscribers alike. However, upstream data travels from all subscribers to just one headend, risking collisions between packets. In a conventional carrier-coax system, because cable modems do not transmit and receive in the same frequency band, a station cannot sense collisions between its upstream packet and packets sent from other stations. A Time Division Multiple Access (TDMA) method has been chosen to coordinate the upstream hybrid fiber-coax transmissions. In implementing this, the headend regularly broadcasts a message to see if any station that so wishes undergoes an initialization procedure that synchronizes its clock with that of the headend and determines how long it takes for a signal to traverse the path connecting them.

Electrical noise in the upstream direction can give both end-users and the multiple-systems operator a big headache. Noise can seep into the system from many sources, including home appliances. The use of cheap cabling or a poorly designed splitter can make the problem worse. In a hybrid fiber-coax configuration, each drop cable and neighborhood feeder cable acts as spokes of a sprawling antenna, and the headend acts as a giant receiver and accumulates all their noise. The collected noise degrades all spokes, thus worsening the situation. There are various ways to minimize the ingress of spurious noise. The first is to protect the service area thoroughly for potential sources of their effects. The second is to use, as far as possible, the cleanest part of the 5 to 42 MHz band, which is the portion that is between 21 MHz and 27 MHz. The third is to reduce the number of households served by each headend or regional hub. Admittedly, reducing the number of hubs in a given area will raise system costs. However, in the long run, the investment is worthwhile for an operator committed to good customer service.

The cable modem and terminal equipment manufacturer, Com21, has taken an interesting approach to tackling the noise in the upstream, or return, paths. The key is the company's eight-port card, known as the Return Path Multiplexer (RPM). Located at the fiber node, the eight-port card multiplexes eight return paths so that the ComComtroller, at the headend, can schedule such that, at any given instant, it receives data from only one of those eight upstream links. The seven other return paths are blocked so long as one is open. Thus, at no time can the noise from eight return paths accumulates at the headend or hub, obviating the expense of deploying the extra hubs or headends otherwise necessary to reduce the number of subscribers and the noise per hub.

The rapidity of the progress being made in cable modems may seem to imply that every important problem has been solved, but many remain. Some customers complain about low speed, which may be caused by noise, and costs are always an issue.

Of course, there is one additional aspect to accessing the Internet, the selection of the ISP. By now, many surfers have developed preferences for or aversions to particular providers. However, cable companies do not appear to be willing to open their networks to outside service providers, and, unlike telephone companies, they are not required to do so.

6.6.3 Fixed Wireless Routed for Internet Access

Wireless technology is useful even in congested urban areas. For wireless links, construction and equipment costs have a ratio of roughly 20:80, whereas, for a terrestrial optical-fiber link, the ratio would be more than reversed, about 90:10. Thanks to that enormous cost advantage, wireless systems can be a boon in nonindustrial countries with little telephone infrastructure. Further, the systems may be expanded and scaled up incrementally as groups of new customers sign up, a strategy that demands less initial investment [6.151].

In industrial countries, wireless local loops make sense in many situations. Wireless links are increasingly being used for broadband services. A fixed wireless provider often operates in what is referred to as the point-to-multipoint mode. Its antenna communicates with several different clients' antennas installed within a well-defined region. An ISP might use a point-to-point link to connect its hubs to a distant point that, in turn, is connected to the Internet backbone across a high-speed wired link. A radio link that supports analog voice telephony acts as a simple local loop between a user and a telephone company's central office. However, there is another way to use fixed wireless technology to make voice calls, the Internet. In such case, the user's computer would digitize the voice and set up an IP address with the provider, for example, communication between the user and the Internet probably with Ethernet frames. At the provider's facility, the packetized signals would be converted into conventional phone signals and then to the PSTN. For large business customers or for multiple dwelling units, a competitive local exchange carrier might provide a broadband wireless connection by way of a private branch exchange at the customer premises. The competitive local carrier would then provide the gateway to the public phone network [6.152].

In the United States, the Federal Communications Commission (FCC) has set aside 15 frequency bands for use in commercial fixed wireless service at frequencies of 2 to 40 GHz. In other countries, frequencies are also allocated by the national telecommunications regulating authorities. Several frequency bands are available for fixed wireless services. The 2.4000 to 2.4835 GHz band is popular with many operators because it is unlicensed and no fee is required for its use. It is also popular with equipment vendors because it is available internationally and equipment made for it has a relatively large market.

A technology for wireless services does exist in the form of spread-spectrum techniques, and some licenses were assigned a number of years ago for wireless services [6.153]. The licensees are referred to as the band's primary communications user. With time, the FCC was persuaded to allow other operators—any who so desired, in fact—to use this band as well. The band was then declared to be unlicensed, but not totally regulated. Lower than power levels specified by the FCC, newer operators must use spread-spectrum technology so as not to interfere with the primary users. The primary users did not have to switch the spread spectrum and are allowed higher power. It is possible that more than one operator in this band will use the same scheme for spread spectrum in the dense geographical area. Hence, the likelihood of one interfering with the other is small. However, the possibility of interference exists, which is the price one pays for using an unlicensed frequency.

Two more unlicensed Industrial, Scientific, and Medical (ISM) bands in the United States are over the bands spanning 5.725 to 5.875 GHz and 24.0 to 24.25 GHz. The former is also known as Unlicensed National Information Infrastructure (UNII) band. In the United States, the rest of the fixed wireless spectrum is taken up by licensed bands. Probably the least used bands so far for Internet access are those that cover 2.1500 to 2.1620, 2.5960 to 2.6440 and 2.6500 to 2.6800 GHz, providing 13 channels of 6 MHz each. The first two were licensed back in 1970 when they were called Multipoint Distribution Services (MDSs), to broadcast 6 MHz television channels. In 1996, the FCC expanded the band to cover its present range and allowed for multichannel services called Multichannel Multipoint Distribution Services (MMDS).

At present, the market for local multipoint distribution systems is expanding quite rapidly and is expected to exceed $2 billion by 2003. After all, the technology has the bandwidth needed for broadband services, and its recurring cost is modest. It is therefore expected to appeal to a broad range of customers.

From a service provider's point of view, large enterprises would be the most profitable customers in terms of profit per invested dollar. Which system best meets the needs of a user organization has many determinants, including the number of customers expected in the near and long term, the estimated number of users logged on to the Internet at any one time, the purchase price, the cost of running the system and the nature of the terrain across which the service is to be deployed. Perhaps the most important factor to a service provider is the cost of license for using a piece of the spectrum, unless it is in the unlicensed ISM band. The wireless communications market hums with news about purchases of licenses and acquisition of companies handling licenses. The design of the system is critically dependent on the frequency used and the transmitted power. In general, higher the frequency means the more complex that the system is.

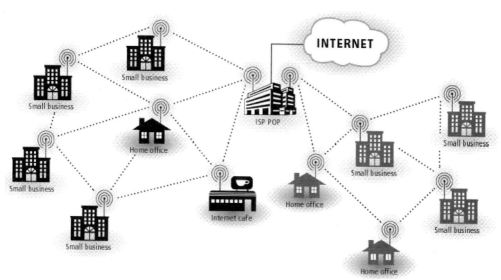

Figure 6.55 Multipoint-to-multipoint wireless scheme.

In a multipoint-to-multipoint wireless environment, multiple logical links exist between one receive/transmit point and its neighboring points in the system (Figure 6.55). Each link between two points may also have different characteristics, such as transmit power, data rate and reliability. All these factors called for a new approach toward the physical media access and network protocols. The next generation architecture attempts to formalize these new protocol modules, the function of each one of them and how information could be exchanged between the modules. The experience gained in the GloMo's Wings project enabled Rooftop to batch its own commercial Internet radio powered by Internet Radio Operating System (IROS) operated in the unlicensed 2.4 GHz IMS band. In a Rooftop Spirit environment, all nodes have same transmit/receive capabilities. However, at least one node is wired to the Internet backbone by a high-speed access line. All of the nodes in the system are aware of all of the other nodes and relay the information forward and backward from node to node. Rooftop calls this node an Airhead. A single Airhead can support 10 to 15 clients and can maintain a reasonable rate of data flow. IROS ensures that all the nodes in the system are aware of all the other nodes and relay information from one to another, as needed, to get packets to their intended destination. One end of every path will invariably be the ISP's point of presence (POP).

6.7 Multimedia Across Wireless

The explosion of technological advancements and the success of the second-generation digital cellular systems (for example, Global System for Mobile (GSM) and Personal Digital Cellular (PDC) have established wireless communications as indispensable in modern life. Because of the low-cost and low-consumption characteristics of emerging wireless products targeted at low- to medium-bit-rate services, these products are expected to play an important role in wireless communications in the next few years. Wireless in multimedia communications (for example, audiovisual telephony and videoconferencing) require medium- to high-bit-rate channels (64 Kb/s to 2Mb/s per user). Therefore, for these applications, it will be necessary to have broadband wireless networks that support bit rates in excess of 2 Mb/s per radio channel, where each radio channel could be shared by multiple users or sessions. In addition, these services have to be provided with some QoS guarantees across their respective, error-prone wireless connections. In order to achieve these goals, one has to address the following key issues:

- How to increase the capacity of wireless channels
- How to provide QoS in a cost-effective way
- How to combat the wireless channel impairments

These questions and related technical issues have been addressed in numerous overview and research papers in the field. For example, an overview of Personal Communication Systems (PCSs) that can provide timely exchange of multimedia information with anyone, anywhere, at any time and at low cost through portable handsets is given in Klee and Oui [6.154]. An overview of various multiple-access schemes is provided. In Hanzo [6.155], various wireless multimedia concepts are described, together with sampling and coding theory, cellular concepts, multiple access, modulation and channel-coding techniques. An overview of various emerging wireless broadband networks in Europe as well as a discussion about the frequency spectrum issue are presented in detail in Mikkonen et al. [6.156]. An overview of the status of wideband wireless local access technologies can be found in Palavan et al. [6.157]. Reference 6.158 gives an overview of wireless broadband communications by addressing some of the applications and services that are foreseen as well as some of the technical challenges that need to be solved.

In addition to the continuing interest in wireless audiovisual communication applications mentioned, a great deal of interest has emerged in higher-end wireless multimedia services. However, current wireless networks, which are primarily low-bit-rate narrow-band systems targeted for voice or data, are inadequate for supporting audiovisual communication applications or high-end multimedia services.

Most multimedia services tend to be real-time in nature, that is, the data being transported need to get to the destination by a certain time in order to be useful. This implies the need to develop techniques for call admission, bandwidth allocation and the handling of real-time variable rate streams. These are problems that apply to wired networks as well and are not, therefore, unique to wireless multimedia communication systems.

The two major protocol-related problems in wireless multimedia concern medium access and QoS. Wireless systems are inherently multiple medium access in nature and therefore need to have a reliable MAC layer that also supports QoS [6.159].

Audio, video and graphics need to be compressed before transport across a bandwidth-constrained wireless channel. Given the emphasis on mobile wireless systems in the past, the media element that has received the most attention in the context of wireless multimedia is speech [6.155]. This is natural because the most widely deployed wireless multimedia system today is cellular telephony, which is a fairly limited bandwidth system. There has also been a great deal of interest in wireless video, given the increased bandwidth capabilities of Universal Mobile Telecommunications Systems (UMTS). The two video compression standards that are most relevant to these systems are MPEG-4 and H.263, both of which have been evaluated for uses in GSM systems. Because of the unreliable nature of wireless networks, it has become important to build source coding schemes that are robust to channel errors. Scalable compression schemes that offer graceful degradation with loss of data have became popular.

Audio and graphics are two source elements that have not received extensive research in the context of wireless systems. There has, however, been some work on handwriting coding [6.159].

Even with scalable and multiple description-based source-coding schemes, there will still be lost data on wireless systems. Error recovery and concealment at the receiver is therefore an important topic and has received some attention again, primarily for video. These error-concealment techniques rely to a large extent on knowing the underlying source compression technique and exploiting some of the tools that are used therein [6.160].

Most of the wireless systems today also support mobility. The velocity associated with mobility has been one of the key parameters that affect system design. For this reason, many of the approaches to solving channel-related problems associated with mobility have been developed for specific classes of mobile systems: pedestrian (velocity of a few meters/sec), vehicular (velocities of about 100 meters/sec) and high speed (velocities of hundreds of kilometers/sec). Mobility also affects routing and addressing, which have received a significant amount of attention.

6.7.1 Wireless Broadband Communication System (WBCS) for Multimedia

Depending on its applications, there are two distinct approaches to the development of WBCS: Wireless LAN (WLAN) and Mobile Broadband System (MBS). Although the core network dilemma is still going strong between IP and ATM for broadband multimedia services, almost all of the WBCS technology demonstrations are based on ATM technology. ATM as a broadband infrastructure has been designed for multimedia communications to accommodate a variety of data rates, QoS requirements and connection and connectionless paradigms. It is quite natural to assume a combination of wireless and ATM-based services at the consumer end of a wired network. In order to deliver multimedia traffic across broadband wireless networks, we need to have sufficient bandwidth and be able to support service-specific QoS requirements concerning delay, delay variation and packet loss on a per-connection basis.

The radio physical layer is essentially the soul of any wireless network. Ideally, one wants to find a radio physical layer technology that is spectrum efficient, minimizes the radio overhead and is robust in both indoor and outdoor environments. Because of various channel impairments, it is very hard to get an optimal radio physical layer.

The wireless broadband air interface will demand a relatively large frequency band to support bit rates in excess of 2 Mb/s. This type of allocation is hard to find lower than 3 GHz, and the availability of bandwidth becomes easier on higher frequencies, but at the cost of complex and expensive techniques [6.156]. Because at higher frequencies the path loss is greater, line of sight operation becomes important and wall penetration becomes a challenge for WLANs.

In the wireless environment, the transmitted radio signal is subject to various time-varying impairments that arise from inherent user mobility and unavoidable changes related to the movement of the surrounding environment. This results in fading and shadowing effects. Another problem is the presence of multipath propagation, leading to fading and time delay spread, which give rise to Intersymbol Interference (ISI) that can strongly increase the BER. One way to overcome fading induced impairments is to use antenna diversity techniques. This is a useful concept for capacity enhancement. The combination of antenna diversity and equalization has the potential to offer significant performance and capacity gains. Another way to improve spectrum efficiency is through power control [6.161].

The wireless transmission medium is a shared radio environment. Therefore, coordinated scheduling of transmission by a control access point can be used to maximize throughput. The major issues are to define a flexible air interface and efficient error control and traffic scheduling algorithms. Data Link Control (DLC) is the core for multiplexing services with varying QoS demands. A generic wireless DLC consists of a flexible packet access interface, delay- oriented and delay-variation-oriented scheduling for multimedia traffic and error control per service requirement.

Multiple Access (MA) protocol is required to minimize or eliminate the chance of collision of different information bursts transmitted from different users. Ideally, the desired MA scheme should be insensitive to the channel impairments. The services provided to the user must satisfy certain quality requirements no matter how bad the channel is. In addition, a good MA protocol can improve system capacity and can lower the system cost. Hence, the MA scheme is a very important design issue for efficient and fair use of the available system resources. In order to provide multimedia services under the limited bandwidth constraint, a sophisticated MA scheme is crucial to cope with various traffic characteristics. Voice is delay sensitive, but relatively loss insensitive; data are loss sensitive, but delay insensitive and voice data rate is generally much higher than either voice or ordinary data and is also delay sensitive. In wireless communication, the multiple access channel can be shared by a large number of users using an MA scheme. There are three types of MA schemes: Frequency Division Multiple Access (FDMA), TDMA and Code Division Multiple Access (CDMA). FDMA assigns a unique frequency band to each user. TDMA assigns access in time slots, and CDMA assigns a unique code using a spread spectrum technique [6.162].

Whereas the DLC layer is used to enhance the transport capability of the physical layer, the LLC layer is used to improve error performance. Error control is typically achieved by coding and/or retransmission. A trade-off between coding and retransmission has to be optimized for the efficient transmission of data across the air interface. For error control, FEC and Automatic Repeat Request (ARQ) are very effective in improving QoS parameters.

Channel allocation is an important issue in radio resource management. It involves allocating the radio resources systemwide to achieve the highest spectrum efficiency. There are two basic forms of channel allocation, namely Fixed Channel Allocation (FCA) and Dynamic Channel Allocation (DCA) [6.154].

6.7.2 Audiovisual Solutions for Wireless Communications

There has been a great deal of standardization and research effort in the area of audiovisual coding for wireless communications. Many papers and reviews have been published covering different aspects of audiovisual coding for wireless communications [6.155, 6.163, 6.164]. In general, previous reviews have focused on wireless speech-coding schemes and error-resilient video standards. Meanwhile, a great deal of research has been conducted in the area of joint source-channel coding for error-prone networks.

In the area of wireless audio, the focus has been on the development of speech-coding solutions for cordless systems, cellular telephony services and emerging personal communication services [6.155]. Because of the time-varying impairments that characterize wireless communication channels, special attention has to be paid when coding any multimedia signal to be delivered across such networks. In particular, compressed image or video signals can experience severe degradation if transmitted across error-prone channels. This is due mainly to the following:

- The use of Variable Length Coding (VLC) in compressed bit stream
- The development of prediction-based coding needed for eliminating both spatial and temporal redundancies in the original signal

Channel errors affecting a VLC could result in a loss of synchronization at the decoder.

Example 6.3 The picture is divided into equal regions of pixels where each region is referred to as a GOB. The example assumes that a synchronization code is used in the bit stream at the beginning of every GOB. Therefore, an error inside the compressed bit stream could damage the remainder of the GOB being decoded until synchronization is achieved at the beginning of the next GOB. An example illustrating the impact of an error in the compressed data on the region of a picture in the pixel domain is shown in Figure 6.56.

Example 6.4 When interpicture prediction-based coding is employed, major degradation in a video sequence can be observed if a reference frame (for example, an intracoded picture) experiences any errors. In this case, when a corrupted picture is used to predict another picture, the error will propagate in the video sequence until the affected area is refreshed by transmitting intracoded blocks or a whole new intracoded picture. An example of an error-propagation scenario in a video sequence is shown in Figure 6.57.

RVLCs enable the receiver to decode the bit stream in a backward manner starting from the next resynchronization marker after an error. Therefore, for any bit stream segment located between two consecutive synchronization markers, RVLCs could assist (Figure 6.58) the decoder in isolating and consequently discarding the region of the bit-stream segment experiencing one or more errors.

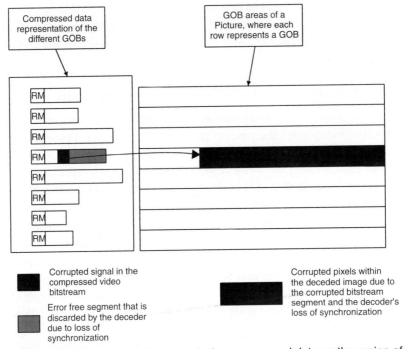

Figure 6.56 The impact of an error in the compressed data on the region of a picture in the pixel domain.

The data segment for the case with RVLC is shown wider (that is, more bits) than the case without RVLCs because some coding efficiency is lost when using RVLCs.

In general, adding any robustness to the compressed signal in the form of resynchronization bits or using RVLCs reduces the coding efficiency. In order to maintain a good balance between coding efficiency and error resilience, other mechanisms have been proposed in conjunction with standard-compliant bit streams. For example in Reyes et al. [6.164], a transcoding scheme is used to increase the robustness of an H.263 signal transmitted across a combined wired-wireless network. The transcoding mechanism, which is employed at the boundary between the wired and wireless segments of the combined network, is designed to improve the resilience of the video signal to errors while minimizing loss of coding efficiency.

In Joint Source Channel (JSC) coding, the time-varying characteristics of an error-prone channel are taken into consideration when designing the source and channel coders of a wireless system [6.165 through 6.172].

Figure 6.59 shows a generic model of a visual source-channel coder. The source signal could be either a still image or a video sequence. For video sequences, the source signal could be either an original picture (for an intracoded frame) or a residual signal representing the difference between an original picture and a prediction of that picture (for example, for motion-compensated prediction). The source signal usually undergoes an orthogonal transform that provides clustering of high-energy coefficients in a compact manner. The DCT is an example of such a

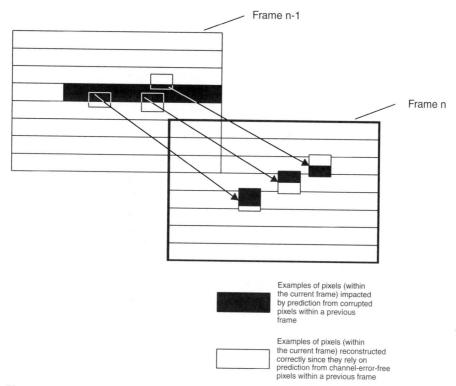

Figure 6.57 Error propagation due to interframe prediction coding.

transform. DCT has been proposed in the context of optimized JSC coding [6.167, 6.173]. Meanwhile, wavelet transform and subband-based video coding have also been proposed for many JSC solutions [6.165, 6.168, 6.169, 6.170]. Moreover, wavelet transform-based image compression has been adopted in MPEG-4 as the basis for a still-image texture-coding tool and also in JPEG 2000. Wireless and mobile applications are among key target application areas for both of these standards.

The second stage of the generic JSC model is the classification and grouping of the transform coefficients. This stage is needed for more efficient and robust quantization and coding (source and/or channel) of the coefficients. In Li and Chen [6.167], a different type of classification and grouping is employed. Although a block-based DCT is used, the DCT coefficients are grouped on the basis of their frequencies into B subband images (subsources). Therefore, for NxN DCT blocks, there are N^2 subsources. Each subsource contains all coefficients with the same frequency (for example, the DC coefficients). This enables the sender to allocate different amounts of bits for the different subsources. Most of wavelet transform-based video-coding solutions proposed for error-prone channels are based on grouping the wavelet coefficients using EZW [6.174, 6.175]. An improved variation of the EZW algorithm is known as Set Partitioning in Hierarchical Trees (SPIHT) [6.175]. In EZW coding, efficiency is achieved on the basis of the

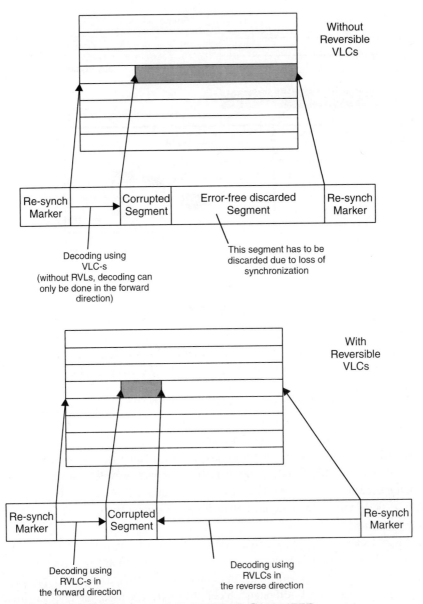

Figure 6.58 Benefits of using RVLCs [6.163]. ©1997 IEEE.

hypothesis of decaying spectrum. The energies of the wavelet coefficients are expected to decay in the direction from the root of a spatial orientation tree toward its descendents. If the wavelet coefficient c_n of a node n is found insignificant (relative to some threshold $T_i = 2^k$), it is highly probable that all descendants D(n) of the node n are also insignificant (relative to the same

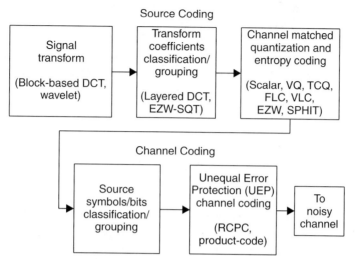

Figure 6.59 Generic model of JSC coding system for a noisy wireless network [6.165]. ©1996 IEEE.

threshold T_k). If the root of a tree and all of its descendents are insignificant, this tree is referred to as a Zero Tree (ZTR). If a node n is insignificant (i.e., $|c_n| < T_k$) but one (or more) of its descendants is (are) significant, then this scenario represents a violation of the decaying spectrum hypothesis. Such a node is referred to as an Isolated Zero Tree (IZT). In the original EZW algorithm, a significant coefficient c_n (i.e., $|c_n| > T_k$) is coded either positive (POS) or negative (NEG) depending on the sign of the coefficient. If $S(n,T_k)$ represents the significance symbol used for coding a node n relative to a threshold $T_k = 2^k$, then

$$S(n,T_k) = \begin{cases} ZTR & if \ |c_n| < T_k \ and \ \max_{m \in D(n)} (|c_n| < T_i) \\ IZT & if \ |c_n| < T_k \ and \ \max_{m \in D(n)} (|c_n| \geq T_k) \\ POS & |c_n| \geq T_k \ and \ c_n > 0 \\ NEG & |c_n| \geq T_k \ and \ c_n < 0 \end{cases} \quad (6.18)$$

The coding procedure used for coding the significance of the wavelet coefficients is referred to as significance map coding.

The third stage of the generic JCS model is the quantization and entropy coding of the classified transform coefficients. Fixed-length entropy coding (that is, using Fixed Length Codes [FLCs]), VLCs, hybrid fixed-variable length coding or arithmetic entropy-based coding mechanisms are normally used in conjunction with some type of quantization. In addition to scalar quantization, VQ [6.176] and TCQ [6.177] are among the popular techniques proposed for wireless video.

After the source encoder generates its symbols, the channel coder provides the necessary protection to these symbols prior to their transmission across error-prone networks. One popular approach that has been used extensively is the UEP paradigm. The UEP paradigm enables the channel coder to use different levels of protection depending on the channel condition. In addition, under UEP, the channel coder can provide different levels of protection for the different source symbols depending on their importance. The importance of a source symbol can be measured on the basis of the amount of distortion when that symbol is corrupted by a channel error. For any source-driven UEP approach, there is a need for a classification process for the source symbols. In Li and Chen [6.167], a simple layered approach is used to classify the different bits of the Uniform Threshold TCQ (UTTCQ) coder. The higher the layer in which the bits are located (within the trellis) means the more important they are. All bits in the same layer are treated in the same way (in terms of importance) and therefore are grouped into one data block. Then different levels of protection are used for the different data blocks. Rate-Compatible Punctured Convolutional (RCPC) codes were invented to provide a viable and practical channel-coding solution for the UEP paradigm [6.178, 6.179]. RCPC codes are generated by a single (channel) encoder decoder puncturing the code at the output of a convolutional coder. Because of its flexibility and low complexity, the RCPC channel coder has been extremely popular for wireless video transmission. Consequently, it has been employed for source driven, channel driven or both source and channel UEP.

6.7.3 Mobile Networks

Broadband wireless communications have gained increased interest during the last few years. This has been fueled by a large demand on high-frequency use as well as a large number of users requiring simultaneous high-data-rate access for the applications of wireless mobile Internet and e-commerce. The convergence of wireless mobile and access will be the next storm in wireless communications, which will use a new network architecture to deliver broadband services in a more generic configuration to wireless customers, and it will support value-added services and emerging interactive multimedia communications. Large bandwidth, guaranteed QoS, and ease of deployment coupled with recent advancements in semiconductor technologies make this converged wireless system a very attractive solution for broadband service delivery.

One of the most interesting activities in mobile communications today is the development and standardization of so-called Third Generation (3G) mobile systems known as UMTS in European Telecommunication Standards Institute (ETSI) and International Mobile Telecommunication (IMT) IMT-2000 in ITU. The first system was deployed in 2001. The 3G systems will provide the user with higher data rates than the current second generation systems, such as GSM. Largely because of the higher data rates, the 3G systems are also expected to enable the use of multimedia applications including, for example, video content. IMT-2000 is expected to play the key role of the mobile telecommunication infrastructure for providing multimedia services supported by user bit rates up to 2 Mb/s.

At the same time, the second generation systems are developing fast, and it is now obvious that many of the 3G applications will be realized already in second generation systems, and the transition to 3G systems, although bringing a performance enhancement, will be smooth from the application point of view.

Multimedia Across Wireless

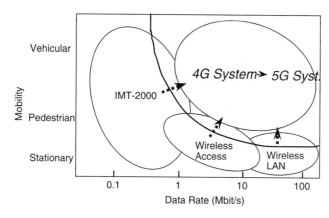

Figure 6.60 Targets of mobile communication systems beyond IMT-2000.

In the history of mobile systems, the 1980s are the analog systems era. Because it was the beginning of mobile communication services, the analog systems are called the first generation systems. Roughly 10 years ago, digital cellular and cordless services were started. The 1990s are called the second generation era for mobile systems. Like in the first generation systems, telephone is the major service in the second generation systems. Multimedia mobile services and worldwide roaming capability are requested for mobile systems. To cope with these demands, IMT-2000 has been standardized as the 3G system, and the services were started in 2001. Figure 6.60 shows the targets of mobile communication systems beyond IMT-2000. IMT-2000 will support the data rates of up to 2 Mb/s. In the high mobility environment, it will be 144 Kb/s. Namely, IMT-2000 offers full mobility capability, but its data rates are limited. As shown, there are different approaches to the fourth generation (4G) system, research on wireless access and WLAN systems. Under the limited mobility condition, the targets of these systems are on even higher data rates corresponding to the broadband services in the fixed networks. The 4G system has to support next generation applications and to be transparent with fixed network. The network of 4G systems may be new one with IP capability. Seamless service area and high-data-rate services should be supported. A flexible mobile terminal with multimode functions will be needed.

Using GSM as an example, at least two clear trends can be seen in the continuous enhancement of the system:

- The basic speech service is developing with enhancements in quality/capacity performance.
- The available data rates are increasing, making live video transmission and multimedia realistic.

Speech Transmission in GSM

The dominant service in mobile communication systems is the basic telephony service. GSM speech service has been improved in two different standardization items since 1995: the Enhanced Full Rate (EFR) and the Adaptive Multirate (AMR) codecs [6.180].

The GSM EFR codec provides speech quality equivalent to that of a wireline telephony reference (ADPCM 32 Kb/s). The EFR codec uses a 12.2 Kb/s bit rate for speech coding and 10.6 Kb/s for error protection adding up to the GSM full rate channel total of 22.8 Kb/s. Speech coding is based on the Algebraic Code Excited Linear Prediction (ACELP) algorithm. The codec provides substantial quality improvement compared to the existing GSM Full Rate (FR) and Half Rate (HR) codecs. The EFR codec provides wireline quality for the most typical error conditions as well as for background noise and mobile-to-mobile calls. Depending on the implementation platform for a mobile handset, the computational complexity of the EFR codec is within 15 to 20 MIPS. The algorithmic delay of the codec is 20 ms [6.181].

The performance of EFR could be improved by using a different capacity allocation between source and channel coding in severe channel error conditions. The GSM AMR codec would operate in FR (22.8 Kb/s) and HR (11.4 Kb/s) channels using multiple bit rates and adapting the source coding and channel-coding bit-rates according to the estimated quality of the radio channel. By switching the codec to operate in the FR channel during good channel conditions, the AMR codec can also provide channel capacity gain over the EFR codec. Compared to the earlier GSM speech codecs, the ARM codec also requires the specification of link adaptation with inband codec mode control and transmission of channel quality measurement data [6.182]. The channel error performance was studied by using a subjective listening test and error-free 16 Kb/s (G.728) Low Delay CELP (LD-CELP) codec as a high-quality reference codec. The GSM FR codec was also included as an additional reference for a subset of test conditions. The mode adaptation algorithm has also been tested with dynamic channel models where the error conditions change during the test sample. The improvement is substantial in all conditions validating the AMR concept.

Video Across GSM

The GSM system presently offers circuit-switched data rates of 9.6 Kb/s which is usually not adequate for real-time transmission of video. However, the specifications for High-Speed Circuit-Switched Data (HSCSD) [6.183] have already been completed, and deployment of these higher data rates for GSM started in 1999. The data-rate enhancements include an increase of the single slot data rate from 9.6 Kb/s to 14.4 Kb/s, as well as the use of multiple data transmission slots for one connection. For example, one practical configuration will use two data slots providing an aggregate full duplex data rate of 28.8 Kb/s. This rate will already be sufficient for simple video telephony as well as for some asymmetric video applications. Even with increased data rates, the acceptance of low-bit-rate video quality will remain an important issue. It is essential to continue improvement in the video compression efficiency. Providing high data rate services for wide area coverage will always carry a cost, and the capacity pressure will favor efficient compression.

In general, the mobile video coder has the same operating principle as most other compressed video coders as shown in Figure 6.61. The prediction error $e_n(x,y)$ is compressed and sent to the decoder together with motion vectors. To indicate that the compression of the prediction error is typically lossy, the compressed prediction error is denoted as $\tilde{E}_n(x,y)$. In the decoder the nth frame of the sequence is reconstructed by predicting each segment and then by adding the received prediction error, that is,

$$\tilde{I}_n(x,y) = P_n(x,y) + \tilde{E}_n(x,y) \qquad (6.19)$$

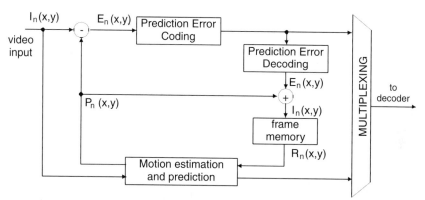

Figure 6.61 Block diagram of the mobile video encoder [6.184]. ©1997 Elsevier.

The block diagram of the motion estimation module is shown in Figure 6.62.

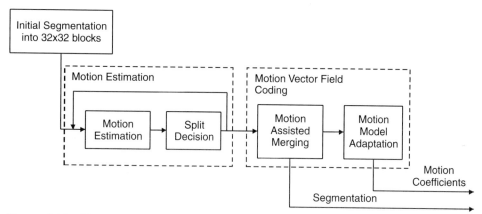

Figure 6.62 Block diagram of the mobile video motion vector field estimation and encoding [6.184]. ©1997 Elsevier.

The segmentation starts by splitting 32x32 blocks in quad-tree fashion as long as sufficient motion estimation performance improvement is obtained. The resulting segments are again merged together based on the motion estimation error. The affine motion model of each segment is then reduced to contain the smallest possible number of motion coefficients without compromising motion estimation performance. The improved compression performance of mobile video is a significant benefit for mobile applications. However, real-time video across cellular phones typically also requires error-resilience enhancement both to the system level as well as to the video codec itself.

Mobile ATM

The item, which defines the design functions of control/signaling, is called mobile ATM. In WATM networks, a mobile end-user establishes a VC to communicate with another user, either a mobile or an ATM end-user. When the mobile end-user moves from one AP to another AP, proper handover is required. To minimize the interruption to cell transport, an efficient switching of the active VC from the old data path to new data path is needed. Also, the switching should be fast enough to make the new VCs available to the mobile users. During the handover, an old path is released, and a new path is then reestablished. In this case, no cell is lost, and cell sequence is preserved. Cell buffering consists of uplink buffering and downlink buffering. If VC is broken when the mobile user is sending cells to APs, unlinking buffering is required. The mobile user will buffer all the outgoing cells. When the connection is up, it sends out all the buffered cells so that no cells are lost unless the buffer overflows. Downlink buffering is performed by APs to preserve the downlink cells for sudden link interruption congestion or retransmissions. It may also occur when the handover is executed [6.185].

When a connection is established between one mobile ATM endpoint and another ATM end point, the mobile ATM endpoint needs to be located. There are two basic location management schemes: the mobile scheme and the location register scheme. In the mobile scheme, when a user moves, the reachability update information only propagates to the nodes in a limited region. When a call is originated by switching in this region, it can use the location information to establish the connection directly. If a switch outside this region originates a call, a connection is established between this switch and the mobile's home agent, which then forwards the calls to the mobile. This scheme decreases the number of signaling messages during a local handover. In a location register scheme, an explicit search is required prior to the establishment of connections. A hierarchy of location registers, which is limited to a certain level, is used [6.186].

Mobile IP

The evolution of mobile networking will differ from that of telephony in some important respects. The end points of a telephone connection are typically human. Computer applications are likely to involve interconnections between machines without human interruption. Obvious examples of this are mobile computing devices on airplanes, ships and automobiles. Mobile networking may well also come to depend on position-finding devices, such as a satellite global positioning system, to work in tandem with wireless access to the Internet. There are still some technical obstacles that must be overcome before mobile networking can become widespread. The most fundamental is the IP, the protocol that connects the networks of today's Internet, and routes packets to their destinations according to IP addresses. These addresses are associated with a fixed network location much as a nonmobile phone number is associated with a physical jack in a wall. When the packet's destination is a mobile node, this means that each new point of attachment made by the node is associated with a new network number and, hence, a new IP address. Mobile IP is a proposed standard protocol that builds on IP by making mobility transparent to applications and higher-level protocols like TCP.

Mobile IP (RFC2002), a standard proposed by a WG within the IETF, was designed to solve this problem by allowing the mobile node to use two IP addresses: a fixed home address and a core address that changes at each new point of attachment [6.187, 6.188]. There is a great deal of interest in mobile computing and apparently in mobile IP as a way to provide for it. Mobile IP is the basis either directly or indirectly of many current research efforts and products. For example, the Cellular Digital Packet Data (CDPD) has created a widely deployed communications infrastructure based on a previous draft specification of the protocol [6.189]. In addition, most major router vendors have developed implementations for mobile IP.

IP routes packets from a source end point to a destination by allowing routers to forward packets from incoming network interfaces to outbound interfaces according to routing tables. The routing tables maintain the next-hop (outbound interface) information for each destination IP address, according to the number of networks to which that IP address is connected. The network number that is derived from the IP address typically carries information with it that specifies the IP node's point of attachment. To maintain existing transport layer connections as the mobile node moves from place to place, it must keep its IP address the same. In TCP, connections are indexed by a quadruplet that contains the IP addresses and port numbers of both connection endpoints. Changing any of these numbers will cause the connection to be disrupted and lost. On the other hand, correct delivery of packets to the mobile node's current point of attachment depends on the network number contained within the mobile node's IP addresses, which change at new points of attachment. To change the routing requires a new IP address associated with the new point of attachment. Mobile IP has been designed to solve this problem by allowing the mobile node to use two IP addresses. In mobile IP, the home address is static and is used, for instance, to identify TCP connections. This takes care of address changes at each new point of attachment and can be thought of as the mobile node's topologically significant address. It indicates the network number and thus identifies the mobile node's point of attachment with respect to the network topology. The home address makes it appear that the mobile node is continually able to receive data on its home network when mobile IP requires the existence of a network node known as the home agent. By home network, we mean the network at which the mobile node seems reachable to the rest of the Internet, by virtue of its assigned IP address. Home agent is a node on the home network that effectively causes the mobile node to be reachable at its home address even when the mobile node is not attached to its home network. Whenever the mobile node is not attached to its home network, the home agent gets all the packets to arrive for the mobile node and arranges to deliver them to the mobile node's current point of attachment.

Whenever the mobile node moves, it registers its new care-of address with its home agent. To get a packet to a mobile node from its home network, the home agent delivers the packet from the home network to the care-of address. Further delivery requires that the packet be modulated so that the care-of address appears at the destination IP address. When the packet arrives at the care-of address, the reserve transformation is applied so that the packet once again appears to have the mobile node's home address as the destination IP address. When the packet arrives at the mobile node, addressed to the home address, it will be processed properly by TCP.

Mobile IP is best understood as the cooperation of three separable mechanisms:

- Discovering the care-of address
- Registering the care-of address
- Timing to the care-of address

Mobile IP discovery does not modify the original fields of existing router advertisements, but simply extends them to associated mobility functions. When the router advertisements also contain the needed care-of address, they are known as agent advertisements, which are the procedures by which a mobile agent becomes known to the mobile node. Home agents and foreign agents typically broadcast agent advertisements at regular intervals, for example, once a second or once every few seconds. An agent advertisement performs the following functions:

- Allows for the data detection of mobile agents
- Lists one or more available care-of addresses
- Informs the mobile node about special features provided by foreign agents, for example, alternative encapsulation techniques
- Lets mobile nodes determine the network number and status of their link to the Internet
- Lets the mobile node know whether the agent is a home agent, a foreign agent or both, and therefore whether it is on its home network or a foreign network

After a mobile node has a care-of address, its home agent must find out about it. Figure 6.63 shows the registration process defined by mobile IP for this purpose. The process begins when the mobile node, possibly with the assistance of a foreign agent, sends a registration request with the care-of address information. When the home agent receives this request, it adds the necessary information to its routing table, approves the request and sends a registration reply back to the mobile node.

In mobile IP, foreign agents are mostly passive, relaying registration requests and replies back and forth between the home agent and the mobile node. The foreign agent also decapsulates traffic from the home agent and forwards it to the mobile node [6.191].

Figure 6.64 shows the tunneling operations in mobile IP. The default encapsulation mechanism that must be supported by all mobile agents using Mobile IP is IP-within-IP [6.192]. By encapsulation, we mean the process of incorporating an original IP packet inside another IP packet, making the fields within the original IP header temporarily lose their effects. Using IP-within-IP, the home agent, or the tunnel's service, inserts a new IP header, or tunnel header, in front of the IP header of any datagram addressed to the mobile node's home address. The new tunnel header uses the mobile node's care-of address as the destination IP address, or tunnel destination. The tunnel source IP address is the home agent, and the tunnel header uses 4 as the

Multimedia Across Wireless

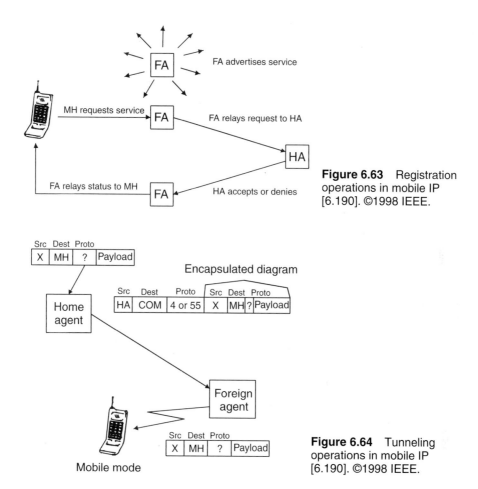

Figure 6.63 Registration operations in mobile IP [6.190]. ©1998 IEEE.

Figure 6.64 Tunneling operations in mobile IP [6.190]. ©1998 IEEE.

higher-level protocol number, indicating that the next protocol header is again an IP header. In IP-within-IP the entire original IP header is preserved as the first part of the payload of the tunnel header. Sometimes the tunnel header uses protocol number 55 as the inner header. This happens when the home agent uses minimal encapsulation instead of IP-within-IP [6.192]. Processing for the minimal encapsulation header is slightly more complicated than that for IP-within-IP because some of the information from the tunnel header is combined with the information in the inner minimal encapsulation header to reconstitute the original IP header. On the other hand, header overhead is reduced. A complete description of the Mobile IP architecture can be found in RFC2002 [6.188]. Related specifications are available in RFCs 2003 through 2006 [6.193]. According to this, mobile node should register with its home agent each time that it changes its care-of address. If the distance between the visited network and the home network of the mobile node is large, the signaling for these registrations may be long. Mobile IP regional registration is shown in Figure 6.65.

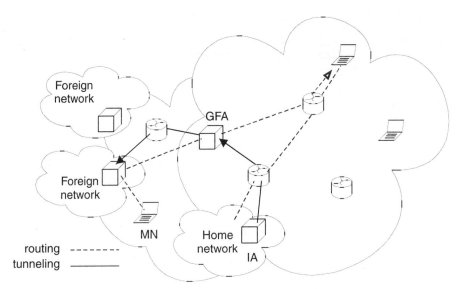

Figure 6.65 Mobile IP regional registration [6.193]. ©2001 IEEE.

With regional registration, when a mobile network arrives for the first time in a new domain, it registers with its Home Agent by means of a local Foreign Agent. However, the Foreign Agent does not register directly with the Home Agent but registers the Mobile Network on a Gateway Foreign Agent (GFA). It is the GFA that registers the Mobile Network with the Home Agent. When the Mobile Network moves from a Foreign Agent to another located within the same domain (behind the same GFA), the new FA simply registers the new care-of address on the GFA. When the regional registration is used, the Corresponding Node (CN) packets are first tunneled from the Home Agent to the GFA and from the GFA to the Foreign Agent. If the reverse tunneling is used, the Foreign Agent forwards Mobile Network datagrams by tunneling to the GFA, and the GFA forwards them by tunneling to the Home Agent that finally forwards them to the CN.

Although the mobile IP architecture has been developed for both IPv4 and IPv6 protocols, some problems arise when mobile IP is used in conjunction with IPv4 private access networks. In the current IPv4-based Internet, a lot of access networks use private address schemes to interconnect the network terminals locally and to connect them to the public Internet. In order to route the packets coming from the private hosts to the public Internet, some mapping/translating mechanisms must be used in order to identify the packets within the public Internet. These schemes are generally called a Network Address Translation (NAT) or in some cases, Network Address and Port Translation (NAPT) [6.194]. There are different reasons for using a private address scheme and NAT mechanisms within a private access network, including the following:

- As a solution to the problem of address depletion in IPv4.

- For security. NAT helps ensure security because each outgoing or incoming request must go through a translation process that also offers the opportunity to authenticate the request or to match it to a previous request.
- For flexibility. NAT conserves the IP addresses within a private network when the network changes its point of attachment to the global Internet. No host reconfiguration is needed.

NAT is included as part of routers and is often part of a corporate firewall. Although NAT has many advantages, it has some problems when it is used in conjunction with nonfriendly protocols and/or applications [6.195]. Particularly, the MIP architecture does not work correctly in a NAT-based access scenario.

Wireless Multimedia Delivery

In the year 2001, the Wireless Multimedia Forum Technical Working Group (WMF TWG) [6.196] published its first document, which recommends technologies, formats and protocols that can be used by the various supply-chain members in the streaming multimedia wireless space. Equipment makers, content developers and service providers that build products conforming to the specifications in Recommended Technical Framework Document (RTFD) Version 1.0 will enable equipment from many vendors to interoperate and will allow software interfaces to be interchangeable across networks [6.197].

Achieving such consensus will accelerate the market for multimedia content and services. For example, having a common technology framework for wireless multimedia delivery will reduce the number of multimedia platforms that content providers will have to support. This should hasten their time to market with new content and services and step up the pace at which their content can reach a broad, far-flung user audience.

The RTFD Version 1.0 standards specification document defines the compression, session initiation/call setup, file format and streaming mechanisms to be used between content-creation subsystems, multimedia distribution servers and wireless multimedia terminals in a streaming multimedia network system. RTFD implementations have resulted in the delivery of interoperable streaming multimedia services to any mobile device in 2001.

New mobile services could include the delivery of news, weather, stock and sports updates to mobile users. In addition, traveling parents could receive clips of a child's soccer game or performance in the school play. Geographic location services could be combined with dating services whereby handheld users could receive a multimedia profile of a dating service candidate who lives in the geographic ballpark of the user's location. Children in Japan are already using cell phones to send animated multimedia greetings to one another, and interactive games that could be streamed among participating users across wireless networks are in development in companies across the globe.

In recommending specifications for common use in wireless multimedia networks, the WMF cooperates with related worldwide standards bodies such as the Third Generation Partner-

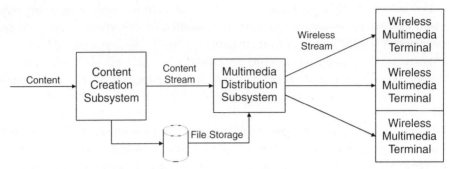

Figure 6.66 Wireless multimedia network system.

ship Project (3GPP) and the ITU. RTFD Version 2.0, already in development, will recommend standards for additional mobile streaming media capabilities, such as QoS, scene description (where graphics and text should appear, relative to multimedia content) and billing capabilities. It will also recommend standards for delivering downloadable multimedia content, such as video email.

RTFD Version 1.0 explicitly addresses the Streaming Multimedia (SMM) application, which includes both on-demand and live streaming using voice and video as the primary media types. The components of an SMM system include the following:

- Content-creation subsystem
- Multimedia distribution
- Wireless multimedia terminals

The content-creation subsystem is responsible for converting raw or compressed media content stored in a file or captured in real time to a content stream suitable for delivery. It then forwards it on to the multimedia distribution server. To do its job, the content-creation subsystem makes use of certain compression technologies and must format the files. A wireless multimedia network system is shown in Figure 6.66.

Content is generated by the content-creation subsystem, distributed to users by the multimedia distribution servers and displayed by wireless multimedia terminals.

The job of the multimedia distribution server, after it has received the multimedia content from the content creation system, is to stream live or stored content to wireless multimedia terminals. RTFD Version 1.0 defines streaming communication between the content-creation subsystem and the multimedia distribution server only for stored content. This multimedia distribution server also can manipulate or repurpose content.

The role of the wireless multimedia terminal is to receive streamed multimedia content from the multimedia distribution server and display it to the user. For streaming media applications, the content may be either live or on demand.

SIP in Mobile Environment

The most important SIP operation is that of inviting new participants to a call. A user first obtains an address where the user is called to translate this address into an IP address where a server may be found. After the server's address is found, the client can send an initiation message to the server. However, as the server that receives the message is not likely to be the host where the user to be invited is actually located, we need to distinguish between different server types that a complete SIP implementation should fulfill. A proxy server receives a request and then forwards it toward the current location of the caller, either directly to the caller or to another server that might be better informed about the actual location of the caller. A redirect server receives requests and informs the caller of the next-hop server. The caller then contacts the next-hop server directly. A user agent server resides on the host where the caller is actually located. It is capable of querying the user about what to do with the incoming call, that is, accept, reject or forward and send the response back to the caller. To assist the end systems in locating their requested communication partner, SIP supports a further server type called register server. It is mainly thought to be a database containing locations as well as user preferences as indicated by the user agents [6.197].

Mobile IP is a widely propagated protocol for supporting mobile communication. With MIP, an end system can be reached in different networks under the same address. However, MIP was primarily designed for TCP communication because it allows for communicating TCP end systems to maintain a connection. This connection is maintained even if there is only one directing the traffic from the calling node to a home agent that maintains location information about the mobile node.

The home agent accepts the TCP connection on behalf of the mobile node and tunnels the traffic either directly toward the mobile node or to a foreign agent that forwards the traffic to the mobile node. The mobile node can communicate directly with the calling node. This kind of communication causes a so-called *triangular routing,* which increases the total end-to-end delay. Further, tunneling the traffic between the home agent and the mobile node increases the amount of consumed bandwidth. This increased complexity is necessary in order to avoid the need for changing the TCP stack at the end systems so as to accommodate for the case of a change in the address of the communicating end systems. A further drawback of MIP is that it uses network terminal addresses to identify the end systems. Although this is appropriate for achieving terminal mobility, it does not easily allow for personal mobility.

SIP provides for personal mobility by using addresses similar to the email addresses that identify a person and not a device. To map this address to a network address, SIP uses the REGISTER method. The same method can be used for providing mobility as well. This means that, whenever the user changes location or device, he can register with the new address. Additionally, the mobile end system informs the other side about the change in the network address. Multimedia communication initiated through SIP is in general based on UDP. Thus, a change in the network address to which to send data does not cause problems with the state of management at the UDP sender as is the case for TCP senders.

Both SIP and Mobile IP share the same problems of authentication and authorization of mobile users. A user roaming into a foreign network needs to authenticate itself to receive access to the network. A harmonization between SIP and Mobile IP might become a promising solution for providing secure and authenticated mobility for both UDP and TCP.

Multicast Routing in Cellular Networks

The most significant trends in today's telecommunications industry are the growth of the cellular network and the rapid rise of the Internet. One of the fastest growing sectors in the telecommunications industry is the cellular service, introducing new demands for user and handset mobility in the future network. A similar explosive increase is observed in the number of Internet subscribers. The current Internet architecture offers a simple point-to-point best-effort service. On the other hand, recently several new classes of distributed applications have been developed, such as remote video, multimedia conferencing, data fusion, visualization, and virtual reality. These applications are not only point-to-point with a single sender and a single receiver of data, but also can often be multipoint-to-multipoint with several senders and several receivers of data. There is a widespread agreement that any new network architecture must be capable of accommodating multicast and a variety of QoS. Multicast enables sources to send a single copy of a message to multiple receivers who explicitly want to receive the information. IP multicasting is a receiver-based concept. Receivers join a particular multicast session group [6.198]. The sender does not need to maintain a list of receivers. The effect of mobility specifically on multicast routing can be summarized as the following:

- When the source of a multicast datagram is a mobile host, a copy of the datagram may not reach all multicast group members, making source-oriented protocol efficient.
- Multicast group members move, requiring an easily reconfigurable multicast tree topology.
- Transient loops may form during tree reconfiguration.
- Channel overhead caused by tree reconfiguration updates increases with mobility, network size and membership size.
- Multicasting necessitates sending copies of a message to multiple locations within the static network. To ensure an acceptable reliable delivery, messages may be buffered at multiple Base Transceiver Stations (BTSs).
- A mobile host may experience a delay in receiving a multicast datagram when it enters a cell that has no other group member located in the same cell.

Core-Based Tree (CBT) is designed to construct and maintain a shared-tree architecture that offers improvements in scalability over source tree architectures by a factor of the number of active sources. Shared trees save bandwidth and state compared with source tree [6.199].

CBT is a backbone within connected group nodes called cores. The backbone is formed by selecting one router, called the primary core, to serve as a connection point for the other cores, called secondary cores. A router willing to participate in the multicast session sends a join request toward the closest core. When the join request reaches a core or an on-tree node, a join acknowledgement is sent back along the reverse path, forming a new branch from the tree to the requesting router. If the core that is reached is a secondary core and is off-tree, it connects to a primary core using the same process. In Brown and Singh [6.200], a new protocol for providing reliable multicast message delivery in mobile networks where the mobile multicast groups experience fragment adds and drops is introduced. Also, a hierarchical network structure, where at the lowest level are the mobile hosts roaming between cells, is proposed. At the next level are the Mobile Support Stations (MSSs), one to a cell, which provide Mobile Hosts with connectivity to the underlying network and with one another. At the top level, groups of MSSs are controlled by a supervisor called the supervisor host (SH). The SH is part of the wired network, maintains connections for MHs and is responsible for maintaining the negotiated QoS. SHs connect mobile networks to the fixed networks and communicate among themselves across the fixed networks.

Broadband Wireless Mobile

Convergence of broadband wireless mobile and access will be the next storm in wireless communications. Fueled by many emerging technologies, including digital signal processing, software-definable radio, intelligent antennas, semiconductor devices and digital transceivers, the future wireless system will be much more compact, with limited hardware and more flexible and intelligent software elements. The compact hardware and the very small portion of software will go the way that the computer industry did in the past. A compact multidimensional broadband wireless model will be adopted for system design and implementation.

Wireless mobile Internet will be the key application of the converged broadband wireless system. The terminal will be compatible with mobile and access services, including wireless multicasting as well as wireless trunking. This new wireless terminal will have the following features:

- At least 90% of traffic will be data.
- The security function will be enhanced.
- A voice recognition function will be enhanced.
- The terminal will support single and multiple users with various service options.
- The terminal will be fully adaptive and software reconfigurable.

As wireless communications evolve to this convergence, 4G mobile wireless communications (4G mobile) will be an ideal mode to support high-data-rate connections from 2 to 20 Mb/s based on the new spectrum requirement for IMT-2000 as well as the coexistence of the current spectrum for broadband wireless access. This 4G mobile system's vision aims at the following:

- Providing a technological response to accelerated growth in demand for broadband wireless connectivity
- Ensuring seamless services provisioning across a multitude of wireless systems and networks, from private to public and from indoor to wide area

- Providing optimum delivery of the user's wanted service through the most appropriate network available
- Coping with the expected growth in Internet-based communications
- Opening new spectrum frontiers

The future wireless network should be an open platform supporting multicarrier, multibandwidth and multitrend air interfaces, with content-oriented Bandwidth-on-Demand (BoD) services dominant throughout the whole network. In this way, packetized transmission will go all the way from one wireless end terminal directly to another. Figure 6.67 shows a network reference model architecture. The major benefits of this architecture are that the network design is simplified and that the system cost is greatly reduced. The BTS is now a smart, open platform with a basic broadband hardware pipe embedded with a Common Air Interface Basic Input-output System (CAI BIOS). Most functional modules of the system are software definable and reconfigurable. The packet switching is distributed in the broadband packet backbone, or core network, called packet-division multiples (PDM). The wireless call processing, as well as other console processing, is handled in this network. The gateway acts as proxy for the core network and deals with any issues for the BTS, and the BTS is an open platform supporting various standards, optimized for full harmonization and convergence. The terminal mobile station can be single or multiuser oriented, supporting converged wireless applications [6.201].

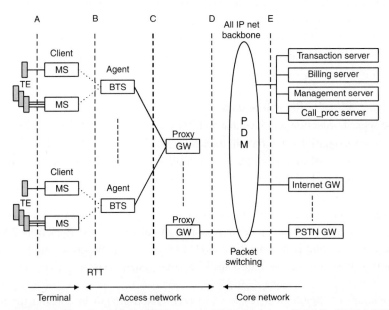

Figure 6.67 A network reference model architecture [6.201]. ©2000 IEEE.

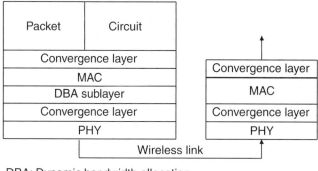

Figure 6.68 General protocol stack [6.201]. ©2000 IEEE.

DBA: Dynamic bandwidth allocation

Considering the signaling protocol, the client/server model is established between a wireless terminal and the core network. The BTS becomes the agent in both directions. Figure 6.68 shows the system protocol stack.

Different services—ATM, IP Synchronous Transfer Mode (STM), and MPEG—can be supported through a service convergence layer. To guarantee wireless QoS and high spectrum use, Dynamic Bandwidth Allocation (DBA) is required through the MAC. The DBA sublayer improves the conventional layer architecture.

The DBA scheduler is the core of the MAC. To realize dynamic resource allocation, this scheduler is essential for the broadband wireless link, which in general helps the following:

- Support class of service offerings
- Provide diagnostic support for all network protocols
- Eliminate the need for traffic shaping
- Increase spectrum use

The transmission convergence layer handles various transmission modulations, error corrections, segmentations and interface mapping of wireless mobile and access in the physical layer.

6.7.4 Broadcasting Networks

A broadcasting system, combined with an interactive channel from a telecommunication system, proves the cheapest and most efficient solution for services that many users share. This holds true as long as the individual information sent to a user stays small compared to the total amount of distributed information. It is more efficient to let the broadcasting system provide a group of users with high capacity rather than divide the capacity into several smaller, individual channels for the same group. Today's market demands global access by multiple users and portable or mobile reception. The cellular phone industry exemplifies this demand. Portable and mobile reception can be offered through a digital terrestrial radio system, or sometimes through satellite. Introducing multimedia in a broadcast system is not straightforward. Existing multimedia systems build on two-way communication links and error-free transmission. However, broadcasters cannot guarantee an error-free or noninterrupted channel. Systems used solely for broadcasting audio or TV were designed with these difficulties in mind. Adding services with multimedia in broadcasting networks demands new concepts that take radio channel characteristics into account.

Figure 6.69 shows an outline of a multimedia system model using broadcast. This model contains three main parts: the content provider, service provider and network provider. Each provider type contains different servers and protocols that include functions, such as an interactive channel, conditional access and system management.

The content provider supplies the information in a multimedia system. The information can consist of audio, video, text, graphics, still pictures or a combination of these. Content providers require appropriate handling of the supplied information in the system and payment for the actual use of the information. Thus, the information can be divided into three different classes:

- Information that the terminal cannot store
- Information that the terminal can store, but not copy
- Information that the terminal can store and copy

A service can consist of a mixture of these information classes. The content provider can offer the same information to a service provider at varying fees, depending on how the service provider uses the information. Content and service providers should agree on the permitted use

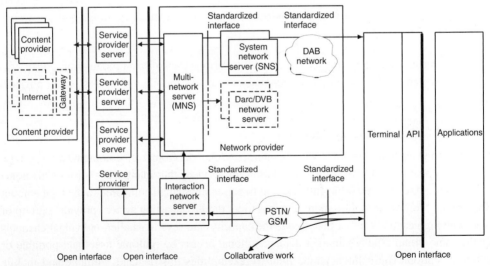

Figure 6.69 Multimedia system model using digital broadcast [6.202]. ©1997 IEEE.

of the information, and service providers can inform end-users of these restrictions. Content providers want to ensure that no one violates the agreed use of the information. Therefore, choosing a system strongly depends on the possibility to protect the content provider's interests.

Service providers create the actual service. They may store the information from the content provider in a server or database and package it into a service (non-real-time services), or they may link the flow of information as a service in itself and distribute it directly without storing it (real-time services). Several layers of service providers could exist, because a service at the terminal end can combine several actual services from different service providers. Service providers must also support subscriber management, service management, information protection mechanisms and billing functions. Service providers receive information that may act as simple channels that transmit whatever content providers supply. In other scenarios, service providers order specific information from content providers, who store the information. If the service is interactive, the service provider may want to access information in distributed databases from various content providers. In other cases, service providers may need to act as buffers to transmit the information, allowing for retransmission. Service providers must also securely handle the content they receive and preserve the content provider's copyright throughout the system.

Selecting which system to use depends on the desired service's capacity requirements. Because network providers can supply several broadcast networks of different types or at different locations, the transport system requires a multinetwork server or router to provide flexible solutions. The multinetwork server can supply system management, select different networks and support other network-independent functions.

Network providers will also have to provide a system network server, which is a unit that translates the information to distribute into the correct format for the transport system. The system network server optimizes the setting of the transport system's available system parameters (such as error protection, repetition and segmentation) in order to adapt to the available transmission channel. It should also support other network-dependent functions.

Network providers must adapt the information transported to several protocol layers according to the system specification. Figure 6.70 illustrates the components of a DAB system's protocol stack. A stream multiplex and fast information channel build the DAB stream. The fast information channel handles multiplex configuration information, which contains information on building the complete multiplex. This includes information such as the number of available audio or data channels, the labels identifying the channels, and descriptions of whether certain channels should link together in the receiver to create a full service. The fast information channel also carries service information describing each service. Streamed audio with program-associated data, a packet mode and proprietary services sit on top of the stream multiplex, possibly encrypted. DAB was developed within the EUREKA147 project and is standardized within the ITU, ETSI and the European Broadcasting Union (EBU). The Coded Orthogonal Frequency Division Multiplex (CODFM) provides the robust transmission channel for mobile reception. The radio frequency signal is wideband (a 1.5 MHz frequency block) with a maximum net bit rate of approximately 1.8 Mb/s, depending on the level of protection.

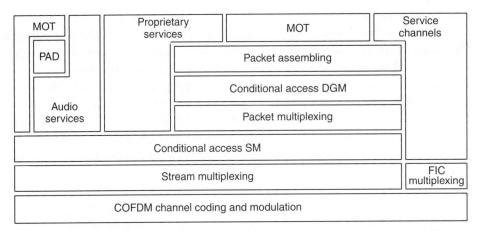

Figure 6.70 Digital audio broadcasting system protocol stack [6.202]. ©1997 IEEE.

DAB has a flexible general-purpose system that reconfigures at any time and supports a wide range of sources, channel-coding options and data services [6.203]. It also incorporates conditional access, such as encryption and addressing, enabling transmission to secluded groups. In conjunction with the DAB specification, a Multimedia Object Transfer (MOT) protocol was developed to support transmitting multimedia objects in the DAB datastream. MOT supports transporting objects and files, segmenting the objects, interleaving on different levels and linking objects in different datastreams. The protocol lets the terminal identify different types of objects, such as JPEG, MPEG or ASCII text so that it can determine whether it has the system resources to handle the object. MOT also includes optional parameters to support applications, such as triggering applications, giving objects time stamps, creating a file name or providing an alternative display mode if a certain decoder is not available.

Network providers require, and sometimes provide, several network-independent functions. These functions should be gathered in an overall logical unit, called a multinetwork server, which supports all the networks. Network providers that manage several networks need to route the information for transmission to the different networks from one or a few central access points. The multinetwork server should support a unified approach so that network providers can bill the service providers and supply the information needed for the service providers to bill the end-users. Scheduling via data carousels, supported by the multinetwork server, can avoid overloading the transmission channels or creating cyclic services. If this functionality accompanies a certain network, it should be distributed to the service network server or a similar unit. Other functions that network providers may support include synchronizing the different data, audio, or video systems before transmission in a single stream in one channel; transmitting the items separately with time stamps (maybe in different channels) and informing the receiver how to assemble all of it. If the latter method is used, the functionality of the transmission system may be chosen to fit each data type work efficiently, but the requirements on the receiver will increase.

In a broadcasting system, service providers have two types of customers: the end-users of the service and the content providers. Service providers should make life as simple as possible for content providers. For example, they should support the content provider's copyright. However, service providers must also try to simplify tasks for end-users. Therefore, determining where to assemble the information becomes an important issue.

A terminal can have many shapes. It can be a receiver dedicated to a particular service, a piece of extraction equipment applied to an ordinary DAB receiver that extracts the service or a DAB receiver (terminal) containing the necessary tools for providing a built-in service. The terminal's basic services include interoperating and presenting data transmitted across air. A large part of the data is presented as soon as it reaches the terminal, like ordinary radio and TV programs. Other parts of the bit stream are immediately disregarded, like the part that represents radio programs other than the ones listened to or data services that the terminal cannot process. Part of the bit stream can be saved onto the hard disk for latter use, either by a broadcast program or interactively by a user.

Digital Video Broadcasting (DVB)

DVB is a transmission scheme based on MPEG-2 video compression utilizing the standard MPEG-2 transmission scheme. DVB provides superior picture quality with the opportunity to view pictures in standard format or wide screen (16:9) format, along with mono, stereo or surround sound. It also allows a range of new features and services including subtitling, multiple audio tracks, interactive content, and multimedia content where, for instance, programs may be *linked* to Web material.

Satellite transmission has led the way in delivering digital TV to viewers. A typical satellite channel has 36 MHz bandwidth, which may support transmission at up to 35 to 40 Mbps using Quadrature Phase-Shift Keying (QPSK) modulation. The audio-video, control data and user data are all formed into fixed-size MPEG-2 transport packets. The complete coding process may be summarized by the following:

- Inverting every eighth synchronization byte
- Scrambling the contents of each packet
- RS coding at 8% overhead
- Interleaved convolutional coding (the level of coding ranges from 1/2 to 7/8, depending on the intended application)
- Modulation using QPSK of the resulting bit stream

The question often arises as to why DVB chose to use MPEG-2. The MPEG-2 coding and compression system was chosen after an analysis and comparison of potential alternatives. Unlike other compression tools that claim to provide greater degrees of compression for given quality, but which are as yet unproven for a wide range of program material or across different broadcasting systems, MPEG-2 is tried and tested. It has been repeatedly shown to be capable of providing excellent quality pictures at bit rates that are practical for the services that will be required. From a commercial point of view, the adaptation of MPEG-2, an existing proven standard, was advantageous because it allowed DVB to concentrate its effort on finding ways of carrying the already well-specified MPEG-2 data packets through a range of different transmission media, including satellite, cable, terrestrial, and so forth. DVB can effectively be regarded as the bridge between broadcasters and the networks across which MPEG-2 data packets can be carried. Another important reason for choosing MPEG-2 was that it includes techniques for the inclusion of Program Specification Information (PSI) for the configuration of decoders. DVB extended these techniques to provide a complete Service Information (SI) capability, enabling receivers to tune automatically to particular services and to decode a mixture of services and service components, including television, sound and data. SI also allows services to be grouped into categories with relevant schedule information, making it possible to provide user-friendly program guides. Another important consideration was that the design of the complete MPEG-2 system makes it possible to freeze the design of a decoder while still retaining the flexibility to make quality improvements at the encoding end of the chain.

DVB has cooperated closely with the Digital Audio Visual Council (DAVIC), whose brief tenure includes the whole range of multimedia transmissions, and many DVB systems have been accepted within the DAVIC standard. MPEG-2 is a video, audio and data coding scheme that can be applied to a range of applications beyond broadcasting, and many multimedia features may, in time, be available from DVB services. DVB systems can naturally carry any or all the items used for multimedia presentations, including text, still pictures, graphics and different types of moving images, and can allow for multimedia extensions to be added. Therefore, DVB members have been focusing on the broadcast market for the immediate commercial future.

The first reason for a broadcaster to select DVB is that DVB makes much better use of available bandwidth. For satellite DTH broadcasters, this advantage is clear. Where a satellite transponder used to carry one analog channel, DVB can offer up to 18 digital channels. DVB-Terrestrial (DVB-T) offers a clearer picture to the end-user, in addition to the capacity for more channels. DVB Cable (DVB-C) offers broadband two-way interactivity. Today, the production, contribution and distribution of television are almost entirely digital. The last step, transmission to the end-user, is still analog. DVB brings this last step into the digital age.

The DVB system has the capability to use a return path between the set-top decoder and the broadcaster. This can be used by a subscriber management system. It requires a modem and the telephone network or a cable TV return path or even a small satellite uplink. This return path can be used for audience participation, such as voting, game playing, teleshopping, telebanking and delivering messages to the decoder. DVB already offers a kind of interactivity without the need for a return path, simply by the breadth of program choices available, for example, multiple sports events and near video on demand.

Data Transmission Using MPEG-2 and DVB

The growing use of multimedia-capable PCs to access the Internet and in particular, the use of the WWW, has resulted in a growing demand for Internet bandwidth. The emphasis has moved from basic Internet access to the exception that connectivity may be provided regardless of the

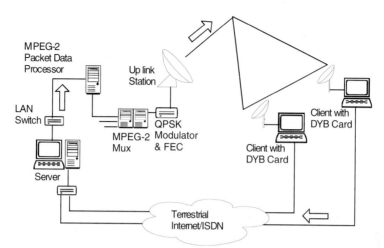

Figure 6.71 Typical configuration for providing DTH Internet delivery using DVB [6.204].

location. This presents challenges to the networking community, particularly as users become familiar with the benefits of high-speed connectivity. Along with an increased use of the Internet, there has been a revolution in TV transmission with the emergence of DVB. The same system may support a high-speed Internet and is being supported on a number of DVB satellite systems. A high-speed (6 to 34 Mb/s) simplex data transmission system may be built using a digital Low Noise Block (LNB) and standard TV antenna connected through an L-band coaxial cable to a satellite data receiver card in a PC (or LAN adaptor box). In many cases, a return link may be established using the available (standard dial-up modem) terrestrial infrastructure, providing the full-duplex communication required for the Internet service. Low-cost satellite return channels are also available. The overall system may provide low cost, high bandwidth Internet access to any location within link coverage of the DVB satellite service.

Data is already being sent across DVB networks using the MPEG-2 transport stream. A variety of proprietary encoding schemes are being used. Data transmission may be simplex or full duplex (using an interaction channel for the return) and may be unicast (point to point), multicast (one to many) or broadcast. Typical configuration for providing DTH Internet delivery using DVB is shown in Figure 6.71. In an effort to standardize services, the DVB specification suggests data may be sent using one of five profiles [6.204].

- *Data piping*—Where discrete pieces of data are delivered using containers to the destination.
- *Data streaming*—Where the data takes the form of a continuous stream that may be asynchronous (that is, without timing, as for Internet packet data), synchronous (that is, tied to a fixed-rate transmission clock, as for emulation of a synchronous

communication link) or synchronized (that is, tied through time stamps to the decoder clock).
- *MPE*—Based on DSM-CC and intended for providing LAN emulation to exchange packet data.
- *Data carousels*—Scheme for assembling datasets into a buffer that is played out in a cyclic manner (periodic transmission). The data sets may be of any format or type. The data is sent using fixed-sized DSM-CC sections.
- *Object carousels*—Resemble data carousels, but primarily intended for data broadcast services. The data sets are defined by the DVB network-independent protocol specification.

At the time DVB was being developed in Europe, a parallel program of standards and equipment development was also going on in the United States by the Advanced Television System Committee (ATSC). Among other things, ATSC adopted a different audio-coding standard and Vestigial Side Band (VSB) modulation. The United States has adopted a system based on ATSC DTV.

The MPEG-2 standards define how to format the various component parts of a multimedia program. They also define how these components are combined into a single synchronous transmission bit stream. The process of combining the streams is known as multiplexing. The multiplexed stream may be transmitted across a variety of links, such as the following

- Radio frequency links (UHF/VHF)
- Digital broadcast satellite links
- Cable TV networks
- Standard terrestrial communication links
- Microwave line of sight links (wireless)
- DSLs
- Packet/cell links (ATM, IP, IPv6, Ethernet)

Many of these formats are being standardized by the DVB project.

Each ES is input to an MPEG-2 processor that accumulates the data into a stream of PES packets. A PES packet may be a fixed or variable-sized block, with up to 65,536 bytes per block, and it includes a 6-byte protocol header. A PES is usually organized to contain an integral number of ES access units. The PES header starts with a 3-byte start code, followed by a 1-byte stream and 2-byte length field.

The MPEG-2 standard allows two forms of multiplexing.

MPEG Program Stream

This is a group of tightly coupled PES packets referenced to the same time base. Such streams are suitable for transmission in a relatively error-free environment and enable easy software pro-

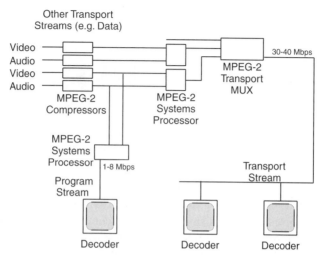

Figure 6.72 Combining ESs from encoders into a transport stream or a program stream [6.204].

cessing of the received data. This form of multiplexing is used for video playback and for some network applications.

MPEG Transport Stream

Each PES packet is broken into fixed-sized transport packets forming a general-purpose way of combining one or more streams, possibly with an independent time base. This is suitable for transmission in which there may be potential packet loss or corruption by noise, and/or where there is a need to send more than one program at a time. Combining ESs from encoders into a transport stream or a program stream is shown in Figure 6.72. The service information component on the transport stream is not shown. The program stream is widely used in digital video storage devices, and also where the video is reliably transmitted across a network. DVB uses the MPEG-2 transport stream over a wide variety of underlying networks. Because both the program stream and transport stream multiplex a set of PES inputs, interoperability between the two formats may be achieved at the PES level [6.205].

A transport stream consists of a sequence of fixed-size transport packets of 188 bytes. Each packet comprises 184 bytes of payload and a 4-byte header. One of the items in this 4-byte header is the 13-bit PID, which plays a key role in the operation of the transport stream. The format of the transport stream is described using Figure 6.73. This figure shows two ESs sent in the same MPEG-2 transport multiplex. Each packet is associated with a PES through the setting of the PID value in the packet header (the values of 64 and 51). The audio packets have been assigned PID64 and the video packets PID51. These are arbitrary, but different, values. As usual, there are more video than audio packets. Note that the two types of packets are not evenly spaced in time. The MPEG transport stream is not a time division multiplex because packets

Figure 6.73 Single program transport stream (audio and video PES) [6.204].

with any PID may be inserted into the transport stream at any time by the transport stream multiplexer. If no packets are available at the multiplexer, it inserts null packets to retain the specified transport stream bit rate.

Although the MPEG-transport stream may be directly used across a wide variety of media, it may also be used across a communication network. It is designed to be robust with short frames, each one being protected by a strong error correction mechanism. It is constructed to match the characteristics of the generic radio or cable channel and expects an uncorrected BER of better than 10^{-10}. The MPEG-2 transport stream is so called, to signify that it is the input to the transport layer in the ISO OSI seven-layer network reference model. MPEG-2 transport requires the underlying layer to identify the transport packets and to indicate in the transport packet header when a transport packet has been erroneously transmitted. The MPEG-transport stream packet size also corresponds to eight ATM cells.

Each MPEG-2 transport stream packet carries 184 bytes of payload data prefixed by a 4-byte (32-bit) header. The format of a transport stream packet is shown in Figure 6.74. The header has the following fields:

- The header starts with a well-known synchronization byte (8 bits).
- A set of three flag bits are used to indicate how the payload should be processed. The first flag indicates a transport error. The second flag indicates the start of payload. The third flag indicates a transport priority bit.
- The flags are followed by a 13-bit PID. This is used to identify the stream to which the packet belongs that was generated by the multiplexer. The PID allows the receiver to differentiate the stream to which each received packet belongs. Some PID values are predefined and are used to indicate various streams of control information. A packet with an unknown PID, or one with a PID that is not required by the receiver, is silently discarded.

The particular PID value is reserved to indicate that the packet is a null packet and is to be ignored by the receiver.

- The two scrambling control bits are used by conditional access procedures to encrypt the payload of some TS packets.
- Two adaptation field control bits may take four values:

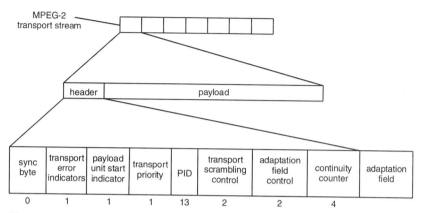

Figure 6.74 Format of a transport stream packet [6.204].

- 01 - No adaptation field, payload only
- 10 - Adaptation field only, no payload
- 11 - Adaptation field, followed by payload
- 00 - Reserved for future use.

- There is a half-byte continuity counter (4 bits).

Two options are possible for inserting PES data into the transport stream packet payload:

- The simple option, from both the encoder and receiver viewpoints, is to send only one PES (or a part of single PES) in a transport stream packet. This allows the transport stream packet header to indicate the start of the PES, but, because a PES packet may have an arbitrary length, it also requires the remainder of the transport stream packet to be padded, ensuring correct alignment of the PES with the start of transport stream packet.
- In general, a given PES packet spans several transport stream packets so that the majority of transport stream packets contain confirmation data in their payloads. When a PES packet is starting, however, the payload unit start indicator bit is set to "1," which means the first byte of the transport stream payload contains the first byte of the PES packet header. Only one PES packet can start in any single transport stream packet. The transport stream header also contains the PID so that the receiver can accept or reject PES packets at a high level without burdening the receiver with too much processing. This has an impact on start PES packets. MEPG PES mapping onto the MPEG-2 TS is shown in Figure 6.75.

DVB transmission may be received through a variety of equipment, as follows:

- A set-top-box
- An in-built decoder forming a part of a DTV set or a digital VCR

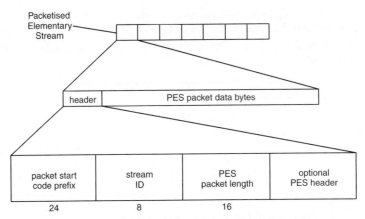

Figure 6.75 MPEG PES mapping onto the MPEG-2 transport stream [6.204].

- A receiver located centrally in a house that feeds DTVs, camcorders and so forth with signals through a digital bus
- A DVB-compliant PC receiver card that displays the various styles of content (TV, audio and data) on the PC screen
- A DVB Multimedia Home Platform (MHP)

The next generation of digital recorders will provide much greater functionality than existing VCRs. By transmitting information as MPEG metadata prior to each program, the digital recorder may itself determine whether the broadcast content should be recorded.

The MHP is defined to enable DVB receivers to be constructed in a common format and to provide common features and interfaces. DVB MHP uses the Java programming language to access an API, which gives access to DVB services. Three application profiles have been defined:

- Enhanced broadcast
- Interactive television
- Internet access

Broadband Multimedia Satellite Systems

DVB transmission by satellite (often known as DVB-S), defines a series of options for sending MPEG-transport stream packets across satellite links. The DVB-S standard requires the 188-byte (scrambled) transport packets to be protected by 16 bytes of RS coding. The resultant bit stream is then interleaved, and convolutional coding is applied. The level of coding may be selected by the service provider. The digital bit stream is then modulated using QPSK. A typical satellite channel has a 36-MHz bandwidth, which may support transmission at up to 35 to 40 Mb/s.

Multimedia is a term that may be applied to digital media-rich content, digital platforms and networks. The full range of multimedia services has been demonstrated over next-generation IP networks, leading to questioning as to the merits of introducing a unified ATM network. Types of multimedia services are the following:

- TV well suited to satellite
- VoD
- Electronic Program Guide (EPG)
- E-commerce services (shopping, gaming and so forth)
- Internet proxy services (selected Web and email)
- Internet access offered only by some current systems
- Games with surprising take up by some users

Most modern broadband systems also permit some level of interactivity. This is true of most current content (TV, radio and so forth), but is especially true of the new digital services, where users often want to sort, manipulate or participate based on the received content. Although it is possible to conceive a one-way Internet service (one-way routed Internet), most services will require some form of return channel to allow two way packet flow.

Multiplatform delivery is also a key concept for many people thinking of broadband multimedia. This is the ability to deliver content (digital video, Web pages and so forth) to a range of network devices, TV sets, PCs multimedia devices, Web-enabled telephones and wireless devices. There are five key players in a typical broadband multimedia system: content owners, middleware developers, service providers, network operators and customers.

Convergence is the term given to the perception that many platforms now have (or soon will have) common capabilities. One advantage of convergence is the ability to access the same information from a TV set as from a PC. In contrast to what is seen as a complexity of a PC, most potential broadband customers perceive the TV as the less difficult-to-use device. Nevertheless, to evolve the set-top-box will need to acquire much more sophistication and may never prove the ideal device for personal access.

The DVB Return Channel System by Satellite (DVB-RCS) was specified by an ad hoc ETSI technical group founded in 1999. This has tracked developments by key satellite operators and followed a number of pilot projects organized by the European Space Agency (ESA). The DVB-RCS system is specified in ETSI EN 301 790. This specifies a satellite terminal, sometimes known as a Satellite Interactive Terminal (SIT) or Return Channel Satellite Terminal (RCST), supporting a two-way DVB satellite system. The use of standard system components provides a simple approach and should reduce the time to market. The transmit capability uses a Multifrequency Time Division Multiple Access (MF-TDMA) scheme to share the capacity available for transmission by the user terminal. The data to be transported may be encapsulated in ATM cells, using ATM adaptation layer 5 (AAL-5) or may use a native IP encapsulation across MPEG-2 transport.

DVB-RCS terminals require a two-way microwave feed arrangement/antenna system to transmit and receive in the appropriate satellite frequency bands. These are typically connected by a cable (or group of cables) to an indoor unit. This unit could be a set-top-box within a network interface integrated in a PC peripheral or may be integrated in a PC expansion card. A key goal of DVB equipment suppliers is to reduce equipment costs.

Multimedia Home Platform

The MHP encompasses the peripherals and the interconnection of multimedia equipment through the in-home digital network. The MHP solution covers the whole set of technologies that are necessary to implement digital interactive multimedia in the home, including protocols, common API languages, interfaces and recommendations. At the beginning of 1996, the UNITEL-universal set-top box project was launched by the Program of the European Commission. The main aim of this project was to raise awareness of the benefits of developing a common platform for user-transparent access to the widest range of multimedia services. Promising progress has since been achieved toward the harmonization of what is now widely called the MHP. The MHP Launching Group was born from the UNITEL initiative in order to open the project to external parties through joint meetings. Key representatives of the High Level Strategy Group took part in this group, and this collaboration eventually led to the transfer of these activities to the DVB Project. Two DVB working groups were subsequently set up [6.206]:

- A commercially oriented group, DVB-MHP, to define the user and market requirements for enhanced and interactive broadcasting in the local cluster (including Internet access)
- A technical group, DVB-TAM (Technical Issues Associated with MHP), to work on the specification of the DVB API.

Different reference models have been defined for each MHP system currently in use. UNITEL used object-modeling tools to define the application classes and functionalities that would ultimately identify the hardware and software resources required by an MHP system. With this system, users would be able to access:

- Enhanced broadcasting services
- Interactive broadcasting services
- Internet services

This model offers system modularity through the use of key interfaces. These surfaces will be able to maintain the stability of MHP systems as they evolve, both in terms of hardware and software. Backward compatibility will be supported to the largest possible extent, for example, by using scalable applications.

The reference model consists of five layers [6.207].

- Application (content and script) and media (audio, video and subtitle) components
- Pipes and streams

- The API and native navigation/selection functions
- Platform/system software or middleware, including the interactive engine, the Runtime Engine (RTE) or virtual machine, the application manager and so forth
- Hardware and software resources and associated software.

Multimedia Car Platform

The Multimedia Car Platform (MCP) project is based on the results and achievements of the two predecessor projects, Multimedia Environments for Mobiles (MEMO) and Mobile Television and Innovative Receivers (MOTIVATE).

The major streams of the MCP projects are the following:

- Service definition and implementation based on user cases and service requirements for the car driver and passenger
- Specification for an open multimedia platform in the car integrating communication, entertainment and navigation
- Implementation of the first multimedia car terminal in the world
- Specification of the architecture of a hybrid network, including service handover and interoperability between different networks

The MEMO system architecture and protocol are used as the starting point for the work in MCP. However, although MEMO only provided specifications for DAB as a broadcast radio interface, in MCP this will be provided also for DVB-T [6.208]. An MCP network reference model is shown in Figure 6.76. MCP started work in January 2000. In a very short time, it attracted major interests of the car industry and has become one of the most important European manufacturers to get involved in MCP. MCP will encourage convergence among telecommunication, broadcasting and media, which actually have partly prohibitive cross regulations.

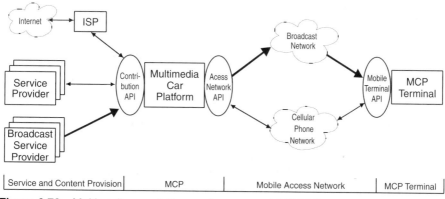

Figure 6.76 Multimedia car platform reference model [6.208].

MCP will actively promote changes in regulation in Europe to allow dynamic usage of time and frequency for data in broadcasting networks.

6.8 Digital Television Infrastructure for Interactive Multimedia Services

DTV technology appears commercially today in hybrid digital-analog systems, such as digital satellite and cable systems, and it serves as the default delivery mechanism for HDTV. All digital sets, such as HDTV, can display higher resolution of digital format and do not require additional external conversion equipment. The digital format for TV broadcast transport has several key advantages over analog transport. For service operators, the key benefit is the high-transport-efficient digital compression that packs five or more times as many channels in a given distribution network bandwidth. This, in turn, increases the operator's revenue potential by delivering more content and pay-per-view events, including Near Video-on-Demand (NVoD) movies with multiple, closely spaced start times. End-users have a larger program selection with spaced start times. End-users have a larger program selection with CD-quality audio and better picture quality potential even when viewed in a hybrid setup with analog TV sets [6.209].

Figure 6.77 shows an example of a satellite-based DTV broadcast system. However, most of the discussion that follows is independent of the physical distribution network and applies to cable and over-the-air digital transmission as well.

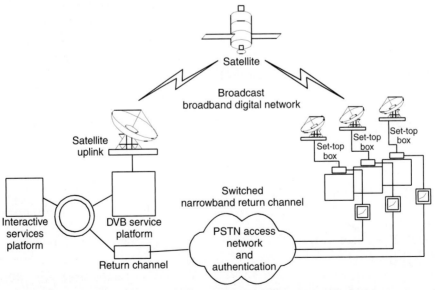

Figure 6.77 Components of a satellite-based digital broadcast system [6.209]. ©1998 IEEE.

A digital video broadcast network distributes audio and video streams to subscribers using a transport protocol. In standard-based implementations, the MPEG-2 transport stream carries digital data across the broadcast network. The MPEG-2 transport structure may contain multiple video and audio channels, as well as private data. MPEG PIDs uniquely identify all program components. A typical digital set-top box contains a control microprocessor and memory, a network interface and tuner, demodulator for cable and satellite, a transport stream demultiplexer and MPEG audio and video decoders. A set-top box also implements user interface components such as the remote-control input and on-screen graphical display capability for output, which are used for controls, and the electronic programming guide. Where appropriate, the conditional access keys distributed to authorized subscribers are used to decrypt the encrypted content in the set-top boxes. Similar functionality can also be packetized into PC-compatible interface cards to let PCs receive digital video, audio and data from the broadcast network. Within the constraints of their service subscriptions, individual users can watch any channel by tuning into the appropriate program within the broadcast multiplex. The individual set-top box that resides on the user's premises handles the channel tuning.

From the data delivery point of view, the DTV infrastructure provides a broadband digital distribution network, data transport protocols and digital terminals (set-top decoders) on the user's premises. As such it provides a powerful platform for delivering information and data services that not only enrich, but fundamentally transform, the television-viewing experience. DTV systems always provide a one-way broadcast path for distributing digital video. Optionally, a return communication link can be provided to allow the upstream flow of data from users to the service center. The return channel often supports the impulse buying of pay-per-view and NVoD events. The return channel is usually implemented through a narrowband communication link such as a PSTN, or an ISDN. Cable systems with two-way-enabled plants can implement a return channel across the cable infrastructure. Because both cable and satellite DTV systems use the same data transport mechanism and protocols, the MPEG-2 transport stream, the physical nature of the underlying distribution network is transparent to data services.

Data broadband technology enables DTV service providers to enhance their customers' television-viewing experience by providing a wide range of interactive services as an incremental add-on to the DTV broadcast infrastructure. Depending on the underlying DVB system infrastructure, the following classes of interactive services are possible:

- Broadcast-only interactive services
- Broadcast with a batch return channel
- Broadcast with an online return channel

Table 6.1 summarizes types of broadcast-based services that a DTV infrastructure with different return channel options can deploy. Note that user-level capability and interactivity is a function of the network and connectivity infrastructure. An important challenge facing DTV service designers lies in dividing data services that operate in the most common, broadcast-only

Table 6.1 Interactive data services as a function of underlying infrastructure capability [6.209].

Functionality	Digital broadcast (one to many) downstream			Point-to-point	
	One-way plant (satellite or cable)			Two-way plant (cable only)	
	No return	Polled return (PSTN)	Real-time return (PSTN)	Cable return (real-time)	
Network	Broadcast One-way	Broadcast Polled return	Broadcast Phone return, dial-up	Broadcast Two-way HFC	Switched Two-way FTTC, ATM
Interactivity	Local	One-way (user response)	Two-way	Two-way	Two-way
User-level function	Browse (view interactively)	Browse plus batch transaction	Browse plus real-time transaction	Browse plus real-time transaction	Full service

FTTC: Fiber To The Curb

©1998 IEEE.

environment and that scale up in user-level functions with the increased capability of the underlying DTV infrastructure, when available.

Interactive Broadcast Data (IDB) Services

Interactive broadcast data services can be broadly categorized as follows:

- *IDB services*—Provide primarily data-only channels, with optional background audio. When a user selects such a service, a data-only screen displays. The user may use hotspot or hyperline-style mechanisms.
- *Interactive Video Broadcast (IVB) services*—Combine video and data channels to provide an enhanced TV-viewing experience. Service delivery and user experience of combined video-data services can be further categorized in terms of their temporal relationship:
  - *Unsynchronized video and data*—Video and data may be typically related or unrelated. Common examples include a simple interactive icon overlay, a partial data overlay such as a ticket display or a data screen with a video (broadcast) insert.
  - *Synchronized video and data*—In this mode, video and data are both typically related and authorized to be synchronized at playback.

Table 6.2 shows a sampling of interactive services.

Table 6.2 Classification of sample services [6.209].

Service	Broadcast (no return channel)	Polled return channel	Real-time return
Primary user-level service capability	Browse (local interactivity)	N/A	N/A
Broadcast Video Services			
Broadcast video	Tune	Tune	Tune
Electronic program guide	View, tune to selection	View, tune to selection	View, tune to selection
Impulse PPV, NVoD	View (order by phone)	View, order (smart card log)	View, order
IDB Services			
Information services	Browse (data carousel)	Browse and acknowledge	Browse and request
Games	Download and play	Download and play (delayed comparison of scores)	Download and play, real-time comparison of scores, multiplayer
Home shopping	Browse (order by phone)	Browse, order (delayed confirmation)	Browse, order
IVB Services			
Enhanced program information	Additional information broadcast, synchronized on current video program	Additional information broadcast, with delayed requests	Fully interactive, ability to request additional information
Interactive advertising	Browse (service information)	Browse, order coupons, brochures, goods (delayed confirmation)	Browse, order online
Play-along programming	Play along, keeps local score	Play along, keeps local score, delayed comparison of scores	Play along, local and networked scoring in real time

Table 6.2 Classification of sample services [6.209]. (Continued)

Service	Broadcast (no return channel)	Polled return channel	Real-time return
Online Services			
Email, forums	Receipt only	Receipt with delayed reply, post	Full interactive receive and replay
Internet access	Intranet broadcast	N/A	Full service
VoD	Not supported	Not supported	Fully interactive

©1998 IEEE.

Many DTV systems use an intermittent batch return channel for billing of Impulse Pay-Per-View (IPPV) events. In this approach, the return channel is usually implemented using a telephone line with a low-speed, dial-up modem. At specified intervals, the control billing system polls individual set-top boxes to retrieve the accumulated information. Optionally, the set-top box might dial up the central system when its local event storage almost fills up or the IPPV viewing credit nearly runs out.

Data Carousel Concept

To support low-end set-top boxes with small amounts of application memory and no additional storage capacity, the datastreams are broadcast cyclically on the DTV network. In effect, the DTV network acts as a large serial disk for storage. This approach gives rise to what is known as the data carousel. The approach allows clients with a local caching capability to find the necessary data and code on the network at any time, with the worst-case access latency equal to the carousel cycle duration. Figure 6.78 shows the basic layout of a data carousel. The carousel datastream consists of an indexing and naming mechanism to locate objects within the data carousel, application code to download to the receiver when the user tunes into the carousel data channel and application data objects that the user terminal retrieves at run time in response of the user's interactive requests.

A DTV network provides an ideal platform for distributing data broadcasts in a carousel fashion. The interactive data services (carousel) are a multiplexed MPEG-2 transport stream. From the management and distribution points of view, data service files can be handled in exactly the same manner as any other DTV stored content (such as NVoD). This lowers system acquisition and operation costs for interactive services because the familiar service center equipment and procedures for TV signal delivery, such as the scheduler and NVoD server, also distribute data services.

For more detailed information on these topics, the references [6.210, 6.211, 6.212] are recommended as research papers dealing with some theoretical concepts of data broadcast. A

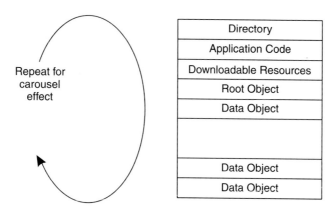

Figure 6.78 Structural layout of a data carousel [6.209]. ©1998 IEEE.

detailed treatment of DTV components and infrastructure can be found in [6.213]. The reference [6.214] deals with user-interaction issues.

6.9 Concluding Remarks

Research on telecommunication networks is focused on post-ISDN architectures and capabilities, such as an integrated packet network and BISDN. The economics and flexibility of integrated networks make them very attractive, and packet network architectures have the potential for realizing these advantages. However, the effective integration of speech and other signals, such as graphics, image and video, into an IPN can rearrange network design priorities. Although processing speeds will continue to increase, it will also be necessary to minimize the nodal per-packet processing requirements imposed by the network design. This is a motivation for new switching concepts like fast packet switching and ATM. Data signals must generally be received error free in order to be useful, but the inherent structure of speech and image signals and the way in which they are perceived allows for some loss of information without significant quality impairment. This presents the possibility of purposely discarding limited information to achieve some other goal, such as the control of temporary congestion.

Layered coding refers to coding a video into a base layer and one or several enhancement layers. The base layer provides a low but acceptable level of quality, and each additional enhancement layer will incrementally improve the quality. By itself, layered coding is a way to enable users with different bandwidth capacities or decoding powers to access the same video at different quality levels.

When transmitting a video signal, residual errors are inevitable regardless of the error-resilience and channel-coding methods used. Thus, ways of mitigating these errors have to be devised. A number of approaches have been proposed in the literature toward the error concealment. Such approaches can be classified into spatial and temporal domain approaches. Compared to coders that are optimized for coding efficiency, error-resilient coders typically are less efficient in that they use more bits to obtain the same video quality in the absence of any transmission errors. These extra bits are called redundancy bits, and they are introduced to enhance the video quality when the bit stream is corrupted by transmission errors. The design goal in error-resilient coders is to achieve a maximum gain in error resilience with the smallest amount of redundancy.

The main challenge in designing a multimedia application across communication networks is how to deliver multimedia streams to users with minimal replay jitters. To diminish the impact of the video quality due to the delay jitter and available network resources (for example, bandwidth and buffers), traffic shaping and SRC are qualified candidates at two different system levels. Traffic shaping is a transport layer approach, and SRC is a compression layer approach. The SRC approach is compressed according to the application's requirement and available network resources.

Streaming video across the Internet faces many technological as well as business challenges, and new codecs, protocols, players and subsystems have been developed to address them. Since its introduction in the early 1990s, the concept of streaming media has experienced a dramatic growth and transformation from a novel technology into one of the mainstream manners in which people experience the Internet today. The concept of streaming media comes at a time when basic multimedia technologies have already established themselves on desktop PCs. Streaming media is a technology that enabled the user to experience a multimedia presentation on-the-fly while it was being downloaded from the Internet.

The provision that bandwidth on demand with strict QoS guarantees is a fundamental property of ATM networks that makes them especially suitable for carrying real-time multimedia traffic. Statistical multiplexing of VBR connections within the backbone network allows effective aggregation and capacity engineering. Several statistical aggregation mechanisms (such as those based on equivalent bandwidth computation) can be used to dimension appropriately the capacity for backbone transport. Scalability is thus possible because per-connection explicit bandwidth renegotiation is not needed within the backbone. Bandwidth renegotiation frequency is high only with the access nodes, where the expected number of high bit-rate multimedia connections is not too large. In addition, the flexibility of soft-QoS control allows applications to specify a wide range of QoS expectations, from best effort to strict guarantees. Because only transport-level parameters are renegotiated, the framework can be applied independently of the network layer. The soft-QoS specification can be used independently by various network technologies to ensure that users' end–to-end service expectations are met.

Anticipating that multimedia across IP will be one of the major driving forces behind the emerging broadband communications, addressing the challenges facing the delivery of multimedia applications across IP is of great importance. In order for the Internet to allow applications to request network packet delivery characteristics according to their needs, sources are expected to declare the offered traffic characteristics. Admission control rules have to be applied to ensure that requests are accepted only if sufficient network resources are available. Moreover, service-specific policing actions have to be employed within the network to ensure that nonconforming data flows do not affect the QoS commitments for already-active data flows. One generic framework that addresses both the video-coding and networking challenges associated with Internet video is scalability. Any scalable Internet video-coding solution has to enable a very simple and flexible streaming framework. The fine-grained scalable framework strikes a good balance between coding efficiency and scalability while maintaining a very flexible and simple video-coding structure.

With the advent of common uses of the Internet, the demands for real-time and low-rate VoIP applications are growing rapidly. Because the delivery of packets is not guaranteed in the IP networks, it is necessary to deal with the audible artifacts, which are caused by burst packet losses. Packet loss seriously degrades the speech quality of the analysis-by-synthesis coders because the loss parameters not only affect the current speech frame, but also produce the so-called error-propagation problem resulting from corrupted filter memory. This packet-loss problem can be solved by using different model parameters.

DSL technology offers unprecedented scalability for interactive video services. It is the basis for the point-to-point architecture that is the key to providing a combination of interactive video and broadcast services. The implementation of video services is a high priority for telecom providers. Delivering voice services across DSLs offers a lucrative opportunity for both established and emerging services.

Data broadcasting in support of multimedia applications requires efficient use of bandwidth resources in order to maximize the availability of playout content. From the data delivery point of view, the DTV infrastructure provides a broadband digital distribution network, data transport protocols and digital terminals on the user premises.

References

Chapter 1
Multimedia Communications

[1.1] H.T.Mouftah, "Multimedia communications: an overview," *IEEE Comm. Magazine*, vol.30, pp.18-19, May 1992.

[1.2] R.V.Cox et al, "On the applications of multimedia processing to communications," *Proc. of the IEEE*, vol.86, pp.755-824, May 1998.

[1.3] J.Rosenberg et al, "Multimedia communications for users," *IEEE Comm. Magazine*, vol.30, pp.20-36, May 1992.

[1.4] Z.S.Bojkovic, "Multimedia communication system: modeling, standardization, requirements," in *Proc. Int. Conference on Multimedia Technology and Digital Telecommunication Services*, ICOMPT'96, pp.5-13, Budapest, Hungary, Oct. 1996.

[1.5] L.C.Wolf, C.Griwadz and R.Steinmetz, "Multimedia communication," *Proc. of the IEEE*, vol.85, pp.1915-1933, Dec. 1997.

[1.6] Z.S.Bojkovic, "Image decomposition and compression in digital multimedia systems," in *Proc. IX Int. Conference on Signal Processing Applications and Technology, ICSPAT'95*, pp.940-944, Boston, MA, USA, Oct. 1995.

[1.7] R.Steinmetz and K.Nahrstedt, *Multimedia Computing, Communications and Applications*, Englewood Cliffs, NJ: Prentice-Hall, July 1995.

[1.8] A.Hoshi et al, "An integrated multimedia desktop communication and collaboration platform for broadband ISDN," *Proc. IEEE Multimedia'92*, pp.28-37, April 1992.

[1.9] A.Watabe et al, "Distributed desktop conferencing system with multimedia interface," *IEEE J. Selected Areas in Comm.*, vol.9, pp.531-539, May 1991.

[1.10] K.R.Rao and Z.S.Bojkovic, *Packet video communications over ATM networks*, Upper Saddle River, NJ: Prentice Hall PTR, 2000.

[1.11] E.Fouques, "Switching and routing: providing value with multimedia services," *Alcatel Telecomm. Review*, pp.198-204, 3rd quarter 1999.

[1.12] A.A.Lazar et al, "MAGNET: Columbia's integrated network testbed," *IEEE J. Selected Areas in Comm.*, vol.11, pp.859-871, Nov. 1993.

[1.13] P.Gonet, "Fast packet approach to integrated broadband networks," *Networks*, vol.9, pp.292-298, Dec. 1986.

[1.14] J.Sidron and J.S.Gopal, "PARIS: An approach to integrated high speed private networks," *Int. J. Digital Analog Cable Syst.*, vol.1, pp.77-85, 1988.

[1.15] L.Chiariglione and L.Corguier, "System consideration for picture communication," in *Proc. ICC'84*, pp.245-249, Amsterdam, May 1984.

[1.16] Z.S.Bojkovic, "Some issues in packet video: modeling, coding and compression," 2nd *Int. Workshop on Image and Signal Processing: Theory, Methodology, Systems and Applications,* pp.2-23, Budapest, Hungary, Nov. 1995.

[1.17] Z.S.Bojkovic, "Recent trends in packet video transmission on ATM networks," in *Proc. Basis of Electronics Workshop '94,* pp.17-105, Cluj, Romania, Oct. 1994.

[1.18] G.Karlsson, *Asynchronous transfer of video,* Swedish Institute of Computer Science (SICS), Research report R95:14, Kist, Sweden, 1997.

[1.19] G.Karlsson, "ATM adaptation for video," in *Proc. of 6th Int. Workshop on Packet Video,* pp.E.3.1-5, Portland, OR, 1994.

[1.20] P.White and J.Groweroft, "The integrated services in the Internet: State of the art," *Proc. of the IEEE,* vol.85, pp.1934-1946, Dec. 1997.

[1.21] Z.S.Bojkovic, "Digital HDTV system compression," *J. on Comm.,* vol.45, pp.2-10, May-June 1994.

[1.22] U.Black, *ATM: Foundation for broadband networks,* Englewood Cliffs, NJ: Prentice Hall, 1995.

[1.23] J.Hecke, "Statistical multiplexing gain for variable bit rate video codecs in ATM networks," *Int. J. Digital Analog Comm. Systems,* vol.4, pp.261-268, 1991.

[1.24] E.Binder and B.Schaffer, "Vision O.N.E.-optimized network evolution: an architecture for the evolution of public communication networks into the universal broadband ISDN," *Telecom. Rep. Int,* vol.14, pp.12-19, 1991.

Chapter 2
Audio-Visual Integration

[2.1] T.Chen and R.R.Rao, "Audio-visual integration in multimodal communication," *Proc. IEEE,* vol.86, pp.837-852, May 1998.

[2.2] L.R.Rabiner and B.H.Juang, *Fundamentals of speech recognition,* Englewood Cliffs, New Jersey: Prentice Hall, 1993.

[2.3] L.R.Rabiner, B.H.Juang and C.H.Lee, "An overview of automatic speech recognition," *Automatic speech and speaker recognition, advanced topics,* C.H.Lee, F.K.Soong and K.K.Paliwal, Eds., Norwell, Massachusetts: Kluwer, pp.1-30, 1996.

[2.4] E.Levin and R.Pieraccini, "CHRONUS, the next generation," in *Proc. ARPA Spoken Language Systems Technology Workshop,* pp.269-271, Austin, TX, 1995.

[2.5] S.Morishima, K.Aizawa and H.Harashima, "An intelligent facial image coding driven by speech and phoneme," in *Proc. IEEE ICASSP'89,* pp.1795-1798, Glasgow, United Kingdom, 1989.

[2.6] F.Lavagetto, "Converting speech into lip movements: A multimedia telephone for hard of hearing people," *IEEE Trans. Rehab. Eng.,* vol.3, pp.1-14, March 1995.

[2.7] P.Griffin and H.Noot, "The FERSA project for lip-sync animation," *Lecture Notes Comput. Sci.,* vol.1024, pp.528-529, 1995.

[2.8] S.Voran and S.Wolf, "Proposed framework for subjective audiovisual testing," *ANSI Working Group* T1A1.5, vol.T1A1.5, pp.93-151, Nov. 1993.

[2.9] Bellcore, *Experimental combined audio/video subjective test method,* ITU-T Study Group 12, SGC/12-01, Feb.1994.

[2.10] H.McGurk and J.MacDonald, "Hearing lips and seeing voices," *Nature,* pp.746-748, Dec. 1976.

[2.11] R.D.Easton and M.Basala, "Perceptual dominance during lip reading," *Perception Psychophys.*, vol.32, pp.746-748, Dec. 1976.

[2.12] D.Burnham and B.Dodd, " Auditory-visual speech perception as a direct process: The McGurk effect in infants and across languages," *Speechreading by humans and machines,* D.Stork and M.Hennecke, Eds., Berlin, Germany: Springer Verlag, pp.103-114, 1996.

[2.13] K.Green, "The use of auditory and visual information in phonetic perception," *Speechreading by humans and machines,* D.Stork and M.Hennecke, Eds., Berlin, Germany: Springer Verlag, pp.55-77, 1996.

[2.14] C.Fisher, "Confusions among visually perceived consonants," *J. Speech Hearing Res.*, vol.11, pp.796-804, 1968.

[2.15] B.Dodd and R.Campbell, Eds., *Hearing by Eye: The psychology of lipreading*, London, England: Lawrence Erlbaum, 1987.

[2.16] K.Neely, "Effect of visual factors on the intelligibility of speech," *J. Acoust. Soc. Amer*, vol.28, pp.1275-1277, Nov. 1956.

[2.17] B.Walden et al, "Effects of training on the visual recognition of consonants," *J. Speech Hearing Res.*, vol.20, pp.130-145, 1977.

[2.18] K.Berger, *Speechreading: Principles and methods*, National Educational Press, Baltimore, Maryland, 1972.

[2.19] A.P.Benguerel and M.Pichora-Fuller, "Coarticulation effects in lip reading," *J. Speech Hearing Res.*, vol.25, pp.600-607, 1982.

[2.20] K.E.Finn and A.A.Montgomery, "Automatic optically-based recognition of speech," *Pattern Recognition Lett.*, vol.8, pp.159-164, March 1988.

[2.21] B.P.Yuhas, M.H.Goldstein and T.J.Sejnowski, "Integration of acoustic and visual speech signals using neural networks," *IEEE Comm. Magazine,* vol.27, pp.65-71, Nov. 1989.

[2.22] B.Carlson and M.Clements, "A projection-based likelihood measure for speech recognition in noise," *IEEE Trans. Speech and Audio Processing,* vol.2, pp.97-102, Jan. 1994.

[2.23] K.Otani and T.Hasegawa, "The image input microphone – A new nonacoustic speech communication system by media conversion from oral motion images to speech," *IEEE J. Selected Areas in Comm.*, vol.13, pp.42-48, Jan. 1995.

[2.24] P.Duchnowski et al., "Toward movement-invariant automatic lip reading and speech recognition," *Proc. IEEE ICASSP'95*, pp.109-122, 1995.

[2.25] K.Waters and T.M.Levergood, DECface: *An automatic lip synchronization algorithm for synthetic faces*, DEC Cambridge Research Lab. Tech. Rep., Cambridge, MA, Sept. 1993.

[2.26] G.Wolberg, *Digital image warping*, New York, New York: IEEE Press, 1990.

[2.27] F.I.Parke, "Parameterized models for facial animation," *IEEE Comput. Graph. Application Magazine*, vol.12, pp.61-68, Nov. 1982.

[2.28] M.Rudfalk, CANDIDE: *A parameterized face*, Linkoping Univ., Sweden, Rep. LITH-ISY-I-0866, Oct. 1987.

[2.29] T.Chen, H.P.Graf and K.Wang, "Lip-synchronization using speech-assisted video processing," *IEEE Signal Processing Lett.*, vol.2, pp.57-59, Apr. 1995.

[2.30] R.Rao and R.Mersereau, "On merging hidden Markov models with deformable templates," *Proc. IEEE ICIP'95*, pp.556-559, Washington, DC, Oct. 1995.

[2.31] ITU-T Draft Recommendation H.263 Version 2, H.263+ *Video coding for low bit rate communication*, Sept. 26, 1997.

[2.32] E.D.Petajan, "Automatic lipreading to enhance speech recognition," in *Proc. IEEE GLOBECOM*, pp.265-272, Atlanta, GA, Nov. 1984.

[2.33] K.Prasad, D.Stork and G.Wolff, *Preprocessing video images for neural learning of lipreading*, Ricoh California Research Center, Tech. Rep. CRC-TR-9326, Menlo Park, CA, Sept. 1993.

[2.34] C.Bregles and Y.K.Cao, "Eigenlips for robust speech recognition," *IEEE ICASSP'94*, pp.669-672, Adelaide, Australia, 1994.

[2.35] G.I.Chion and J.N.Hwang, "Lipreading from color video," *IEEE Trans. Image Processing*, vol.6, pp.1192-1195, Aug. 1997.

[2.36] E.Dubois and T.S.Huang, "Motion estimation," *IEEE Signal Processing Magazine*, vol.15, pp.35-40, March 1998.

[2.37] R.Chellappa and T.Chen, "Audio-visual integration in multimodal communication," *IEEE Signal Processing Magazine*, vol.14, pp.37-38, July 1997.

[2.38] T.Chen and R.R.Rao, "Audio-visual interaction in multimedia communication," *IEEE ICASSP'97*, vol.1, pp.179-182, Munich, Germany, April 1997.

[2.39] A.J.Viterbi and J.K.Omura, *Principles of digital communication and coding*, New York, New York: McGraw-Hill, 1979.

[2.40] P.L.Silsbee and A.C.Bovik, "Computer lipreading for improved accuracy in automatic speech recognition," *IEEE Trans. Speech and Audio Processing*, vol.4, pp.337-351, Sept. 1996.

[2.41] K.Mase and A.Pentland, "Automatic lipreading by optical-flow analysis," *Syst. Comput. Jpn.*, vol.22, pp.67-76, Jan. 1991.

[2.42] M.Turk and A.Pentland, "Face recognition," *J. Cognitive Neurosci.*, vol.3, pp.71-86, Jan. 1991.

[2.43] L.D.Cohen and I.Cohen, "Finite-element methods for active contour models and balloons for 2D and 3D images," *IEEE Trans. PAMI*, vol.15, pp.1131-1141, Nov. 1993.

[2.44] P.L.Silsbee, "Sensory integration in audiovisual automatic speech recognition," *Proc. 28th Ann. Asilomar Conf.*, vol.1, pp.561-565, Pacific Grove, California, Nov. 1994.

[2.45] G.I.Chion and J.N.Hwang, "Image sequence classification using a neural network based active contour model and a hidden Markov model," *Proc. IEEE ICIP'94*, vol.3, pp.926-930, Austin, Texas, Nov. 1994.

[2.46] S.Y.Kung, J.N.Hwang, "Neural networks for intelligent multimedia processing," *Proc. IEEE*, vol.86, pp.1244-1272, June 1998.

[2.47] B.P.Yuhas et al, "Neural networks models for sensory integration for improved vowel integration," *Proc. IEEE*, vol.78, pp.1658-1668, Oct. 1990.

[2.48] P.Duclinowski, V.Meir and A.Waibel, "See me, hear me: Integrating automatic speech recognition and lipreading," *Proc. ICSLP'95*, pp.547-550, Yokohama, Japan, 1995.

[2.49] C.Bregler, S.M.Omohundro and Y.Konig, "A hybrid approach to bimodal speech recognition," *Proc. 28th Asilomar Conf. Signals, Systems and Computers*, pp.572-577, Pacific Grove, California, Nov. 1994.

[2.50] H.Hermansky et al, "RASTA-RLP speech analysis technique," *Proc. IEEE ICASSP'92*, vol.1, pp.121-124, San Francisco, California, 1992.

[2.51]   M.R.Civanlar and T.Chen, "Password-free network security through joint use of audio and video," *Proc. SPIE Photonic East*, pp.120-125, Nov. 1996.

[2.52]   J.Luettin, N.A.Thacker and S.W.Beet, "Speaker identification by lip reading," *Proc. Int. Conf. Spoken Language Processing*, pp.62-65, Philadelphia, Pennsylvania, Oct. 1996.

[2.53]   P.L.Nikias, "Riding the new integrated media systems wave," *IEEE Signal Processing Magazine*, vol.14, pp.32-33, July 1997.

[2.54]   J.Nam and A.H.Tewfik, "Combined audio and visual streams analysis for videosequence segmentation," *Proc. IEEE ICASSP'97*, vol.4, pp.2665-2668, Munich, Germany, April 1997.

[2.55]   Y.Wang et al, "Multimedia content classification using motion and audio information," *Proc. IEEE Int. Symp. Circuits and Systems*, pp.1488-1491, Hong Kong, June 1997.

[2.56]   ISO/IEC Standard DIS 13818-1, *Generic coding of moving pictures and associated audio: Systems*, April 1995.

[2.57]   C.Saraceno and R.Leonardi, "Audio as a support to scene change detection and characterization of video sequences," *Proc. IEEE ICASSP'97*, pp.2597-2600, Munich, Germany, April 1997.

[2.58]   D.J.Dekle, C.A.Fowler and M.G.Funnel, "Audio-visual integration in perception of real words," *Perception Psychophys.*, vol.51, pp.355-362, 1992.

[2.59]   S.Health, *Multimedia and communication technology*, Stoneham, Massachusetts: Focal, 1996.

[2.60]   B.Shen et al, *Microsystems technology for multimedia applications: An introduction*, Piscataway, New Jersey: IEEE Press, 1995.

[2.61]   L.Chiariglione, "The development of an integrated audiovisual coding standard: MPEG," *Proc. IEEE*, vol.83, pp.151-157, Feb. 1995.

[2.62]   K.R.Rao and P.Yip, *Discrete cosine transform: Algorithms, advantages, applications*, New York, New York: Academic Press, 1990.

[2.63]   K.R.Rao and J.J.Hwang, *Techniques and standards for image, video and audio coding*, Upper Saddle River, New Jersey: Prentice Hall PTR, 1996.

[2.64]   B.Furth, "Multimedia systems: An overview," *IEEE Multimedia Magazine*, vol.1, pp.47-59, Spring 1994.

[2.65]   M.Blather and R.Deneburg, Eds., *Multimedia interface design*, Reading, Massachusetts: Addison-Wesley, 1991.

[2.66]   L.C.Wolf, *Resource management for distributed multimedia systems*, Boston, Massachusetts: Kluwer Academic, 1996.

[2.67]   L.C.Wolf and R.Steinmetz, "Concepts for resource reservation in advance," *J. Multimedia Tools/Applications*, vol.4, pp.255-278, May 1997.

[2.68]   P.N.Belhumeur, J.P.Hespanha and D.J.Kriegman, " Eigenfaces vs. Fisherfaces: Recognition using class specific linear projection," *IEEE Trans.PAMI*, vol.19, pp.711-720, July 1997.

[2.69]   S.J.Moon and J.N.Hwang, "Robust speech recognition based on joint model and feature space optimization of hidden Markov models," *IEEE Trans. Neural Networks*, vol.8, pp.194-204, March 1997.

[2.70]   E.Y.H.Tseng, J.N.Hwang and F.Seehan, "Three-dimensional object representation and invariant recognition using continuous distance transform neural network," *IEEE Trans. Neural Networks*, vol.8, pp.141-147, Jan. 1997.

[2.71] E.Owens and B.Blazek, "Visemes observed by hearing-impaired and normal-hearing adult viewers," *J. Speech Hearing Res.*, vol.28, pp.381-393, Sept. 1985.

Chapter 3
Multimedia Processing in Communications

[3.1] D.McLeod et al, "Integrated media systems," *IEEE Signal Processing Magazine*, vol.16, pp.33-43, Jan. 1999.

[3.2] A.N.Venetsanopoulos and A.Dumitras, "Multimedia signal processing applications and systems," Invited keynote paper at *World Multiconference on Circuits, Systems, Communications and Computers,* Athens, Greece, July 2000.

[3.3] A.Krikelis, "Multimedia signal processing architectures," *IEEE Concurrency*, vol.53, pp.5-7, July-Sept. 1997.

[3.4] K.C.Pohlmann, *Principles of digital audio*, New York, NY: McGraw-Hill, 1995.

[3.5] A.N.Netravali and B.G.Haskell, *Digital pictures*, New York: Plenum Press, 1998.

[3.6] Z.S.Bojkovic et al, *Advanced topics in digital image compression*, Timisoara, Romania: Editing Politechnica, 1997.

[3.7] N.Jayant, "High quality networking of audio-visual information," *IEEE Comm. Magazine*, vol.31, pp.84-95, Sept. 1993.

[3.8] A.K.Jain, *Fundamentals of digital image processing*, Englewood Cliffs, NJ: Prentice Hall, 1989.

[3.9] A.J.Viterbi and J.V.Omura, *Principles of digital communications and coding*, New York, NY: McGraw-Hill, 1997.

[3.10] J.L.Mitchell et al, *MPEG video compression standard*, New York, NY: Chapman and Hall, 1996.

[3.11] W.B.Pennebaker and J.L.Mitchell, *JPEG still image compression standard*, NewYork, NY: Van Nostrand Rheinhold, 1993.

[3.12] N.Jayant, "Signal compression: Technology targets and research directions," *IEEE J. Selected Areas in Communications*, vol.10, pp.796-818, May 1992.

[3.13] Z.Bojkovic, "Some issues in packet video: modeling, coding, compression," *2nd Int. Workshop on Image and Signal Processing: Theory, Methodology, Systems and Applications*, pp.2-23, Budapest, Hungary, Nov. 1995.

[3.14] A.G.Tescher, "Multimedia is the message," *IEEE Signal Processing Magazine*, vol.16, pp.44-54, January 1999.

[3.15] N.K.Bose, *Applied multidimensional systems theory*, New York, NY: Van Nostrand Rheinhold, 1982.

[3.16] D.E.Dudgeon and R.M.Mersereau, *Multidimensional digital signal processing*, Englewood Cliffs, NJ: Prentice-Hall, 1984.

[3.17] Y.S.Lim, *Two-dimensional signal and image processing*, Englewood Cliffs, NJ: Prentice-Hall 1990.

[3.18] R.M.Mersereau, "The processing of hexagonally sampled two-dimensional signals," *Proc. of the IEEE*, vol.67, pp.930-979, June 1979.

[3.19] T.S.Huang, " Two dimensional windows," *IEEE Trans. Audio Electroacust.*, vol.20, pp.88-90, Jan. 1972.

[3.20] G.K.Rivard, "Direct fast Fourier transform of bivariate functions," *IEEE Trans. ASSP*, vol.25, pp.250-252, March 1977.

[3.21] P.P.Vaidyanathan, *Multirate systems and filter banks*, Englewood Cliffs, NJ: Prentice Hall, 1993.

[3.22] G.Sharma and R.Chellappa, "A model-based approach for the estimation of 2D maximum entropy power spectra," *IEEE Trans. Info. Theory*, vol.31, pp.90-99, Jan. 1985.

[3.23] Y.Cafon, "High-resolution frequency wave number spectrum analysis," *Proc. of the IEEE*, vol.57, pp.1,408-1,418, Sept. 1969.

[3.24] A.Zakhor and A.V.Oppenheim, "Reconstruction of two-dimensional signals from threshold crossings," *Proc. of the IEEE*, vol.78, pp.31-55, Jan. 1990.

[3.25] M.I.Sezan, M.K.Ozkan and S.V.Forgeh, "Temporally adaptive filtering of noisy image sequences using robust motion estimation algorithms," *Proc. IEEE ICASSP*, vol.IV, pp.2,429-2,432, May 1991.

[3.26] Y.T.Tse and R.L.Beker, "Global zoom/pan estimation and compensation for video compression," *Proc. IEEE ICASSP*, vol.IV, pp.2725-2728, May 1991.

[3.27] C.Horne, "Improving block based motion estimation by the use of global motion," *SPIE* vol.2,094, pp.576-587, 1993.

[3.28] H.C.Reeve and Y.S.Lim, "Reduction of blocking effects in image coding," *Optical Eng.*, vol.23, pp.34-37, Jan. 1984.

[3.29] B.Ramamurthi and A.Gersho, "Nonlinear space-variant post-processing of block coded images," *IEEE Trans. ASSP*, vol.34, pp.1,258-1,268, Oct. 1986.

[3.30] R.L.Stevenson, "Reduction of coding artifacts in transform image coding," *Proc. IEEE ICASSP*, vol.V, pp.401-404, April 1993.

[3.31] L.Yan, "A nonlinear algorithm for enhancing low bit-rate coded motion video sequence," *Proc. IEEE ICASSP*, vol.II, pp.923-927, Nov. 1994.

[3.32] Y.Yang, N.P.Galatsanos and A.K.Katsaggelos, "Iterative projection algorithms for removing the blocking artifacts of block-DCT compressed images," *Proc. IEEE ICASSP*, vol.V, pp.401-408, April 1993.

[3.33] B.Macq et al, "Image visual quality restoration by cancellation of the unmasked noise," *Proc. IEEE ICASSP*, vol.V, pp.53-56, Nov. 1994.

[3.34] T.Chen, "Elimination of subband-coding artifacts using the dithering technique," *Proc. IEEE ICASSP*, vol.II, pp.874-877, Nov. 1994.

[3.35] W.Li, O.Egger and M.Kunt, "Efficient quantization of noise reduction device for subband image coding schemes," *Proc. IEEE ICASSP*, vol. IV, pp.2,209-2,212, Detroit, MI, May 1995.

[3.36] T.Ebrahimi and M.Kunt, "Visual data compression for multimedia applications," *Proc. IEEE*, vol.86, pp.1,109-1,125, June 1998.

[3.37] I.D.Markel and A.H.Gray, *Linear prediction of speech*, Berlin, Germany: Springer-Verlag, 1976.

[3.38] B.S.Atal and S.L.Hanaker, "Speech analysis and synthesis by linear prediction of the speech wave," *J. Acoust. Soc. Amer.*, vol.50, pp.637-655, April 1971.

[3.39] R.Bracewell, *The Fourier transform and its applications*, New York, NY: McGraw-Hill, 1986.

[3.40] A.Buzo et al, "Speech coding based upon vector quantization," *IEEE Trans. ASSP*, vol.28, pp.562-574, Oct. 1980.

[3.41] D.Y.Wang, B.H.Juang and A.H.Gray, "An 800 Bits/s vector quantization LPC vocoder," *IEEE Trans. ASSP*, vol.30, pp.770-780, Oct. 1982.

[3.42] H.Duddley, "Phonetic pattern recognition vocoder for narrowband speech transmission," *J. Acoust. Soc. Amer.*, vol.30, pp.733-739, Aug. 1958.

[3.43] B.Atal and M.Schroeder, "Predictive coding of speech and subjective error criteria," *IEEE Trans. ASSP*, vol.27, pp.247-254, June 1979.

[3.44] J.Chen et al, "A low-delay CELP coder for the CCITT 16 Kbit/s speech coding standard," *IEEE J. Selected Areas in Comm.*, vol.10, pp.830-849, May 1992.

[3.45] ITU-T Recomm. G.729, *Coding of speech at 8 kbit/s using conjugate-structure algebraic-code-excited linear-prediction (CS-ACELP)*, March 1996.

[3.46] I.A.Gerson and M.A.Jasink, "Vector sum excited linear prediction (VSELP)," in *Advances in speech coding*, Cuperman and Gersho, (Eds.), pp.67-79, Hingham, Massachusetts: Kluwer Academic Publishers, 1991.

[3.47] T.Honkanen et al, "Enhanced full rate coder for IS-136 digital cellular system," *Proc. IEEE ICASSP*, pp.731-734, Munich, Germany, 1997.

[3.48] P.Kron, E.F.Deprettere and R.J.Sluyter, "Regular pulse excitation - A novel approach to effective and efficient multipulse coding of speech," *IEEE Trans. ASSP*, vol.34, pp.1,054-1,063, 1986.

[3.49] J.P.Campbell, V.C.Welch and T.E.Termin, " The new 4800 bps voice coding standard," *Proc. Military Speech Tech.*, pp.64-70, 1989.

[3.50] M.M.Sondhi, "A hybrid time-frequency domain articulatory speech synthesizer," *IEEE Trans. ASSP*, vol.35, pp.955-967, July 1987.

[3.51] R.Sproat and J.Olive, "An approach to text-to-speech synthesis," in *Speech coding and synthesis*, Kleijn and Paliwal, (Eds.), pp.613-633, Amsterdam, The Netherlands: Elsevier, 1995.

[3.52] E.Moulines and W.Verhelst, "Time-domain and frequency domain techniques for prosodic modification of speech," in *Speech Coding and Synthesis*, Kleijn and Paliwal, (Eds.), pp.519-555, Amsterdam, The Netherlands: Elsevier, 1995.

[3.53] M.M.Sondhi, D.R.Morgan and J.L.Hall, "Stereophonic acoustic echo cancellation - An overview of the fundamental problem," *IEEE Signal Processing Letters*, vol.2, pp.148-151, Aug. 1995.

[3.54] G.W.Elko et al, "Adjustable filter for differential microphones," U.S. Patent 5,303,307, April 1994.

[3.55] J.L.Flanagan et al, "Autodirective microphone systems," *Acoustica*, vol.73, pp.58-71, 1991.

[3.56] M.M.Sondhi, "An adaptive echo canceller," *The Bell Systems Technical J.,* vol.6, pp.497-510, March 1967.

[3.57] B.G.Haskell, A.Puri and A.Netravali, *Digital video: an introduction to MPEG-2*, Norwell, Massachusetts: Chapman and Hall, 1997.

[3.58] A.M.Tekalp, *Digital video processing*, Englewood Cliffs, NJ: Prentice Hall, 1995.

[3.59] A.J.Patti, M.J.Sezan and M.Tekalp, " Robust methods for high-quality stills from video in the presence of dominant motion," *IEEE Trans. CSVT*, vol.7, pp.328-342, May 1997.

[3.60] R.R.Schultz and R.L.Stevenson, "Extraction of high-resolution frames from video sequences," *IEEE Trans. Image Processing*, vol.5, pp.996-1,011, July 1996.

[3.61] A.J.Patti, M.J.Sezan and M.Tekalp, "Super-resolution video reconstruction with arbitrary sampling lattices and non-zero aperture time," *IEEE Trans. Image Processing*, vol.6, pp.1,064-1076, Aug. 1997.

[3.62] H.J.Zhang, A.Kankanhalli and S.W. Smoliar, "Automatic partitioning of full-motion video," *Multimedia Systems*, vol.1, pp.10-28, Jan. 1993.

[3.63] I.Y.A.Wang and E.H.Adelson, "Representing moving images by layers," *IEEE Trans. Image Processing*, vol.3, pp.625-638, April 1994.

[3.64] G.Ahanger and T.D.C.Little, "A survey of technologies for parsing and indexing digital video," *J. Visual Comm. and Image Representation*, vol.7, pp.28-43, Jan. 1996.

[3.65] B.L.Yeo and B.Liu, "Rapid scene analysis of compressed video," *IEEE Trans. CSVT*, vol.5, pp.533-544, Nov. 1995.

[3.66] A.Akutsu and V.Tonomura, "A video tomography: an efficient method for camera work extraction and motion analysis," *ACM Multimedia 94*, pp.349-356, Oct. 1994.

[3.67] Y.D.Courtney, "Automatic video indexing via object motion analysis," *Pattern Recognition*, vol.30, pp.607-626, Sept. 1997.

[3.68] A.N.Venetsanopoulos and A.Dumitras, "Multimedia signal processing applications and systems," Invited keynote paper at World Multiconference on Circuits, Systems, Communications and Computers, Athens, Greece, July 2000.

[3.69] M.Sonka, V.Hlavec and R.Boyle, *Image processing, analysis and machine vision*, Norwell, Massachusetts: Chapman and Hall, 1993.

[3.70] P.Raghu and Y.Yegnanarayana, "Segmentation of Gabor-filtered textures using deterministic relaxataion," *IEEE Trans. Image Processing*, vol.5, pp.1,625-1,636, Dec. 1996.

[3.71] R.M.Haralick, K.Shanmugam and I.Dinstein, "Textural features for image classification," *IEEE Trans. SMC*, vol.3, pp.610-621, Nov. 1973.

[3.72] A.K.Jain and K.Karu, "Learning texture discrimination masks," *IEEE Trans. PAMI*, vol.18, pp.195-205, Feb. 1996.

[3.73] J.P.Eakins, J.M.Boardman and M.E.Graham, "Similarity retrieval of trademark images," *IEEE Multimedia*, vol. 5, pp.53-63, April-June 1998.

[3.74] F.D.Acqua and P.Gamba, "Simplified model analysis and search for reliable shape retrieval," *IEEE Trans. CSVT*, vol.5, pp.654-666, Sept. 1998.

[3.75] P.E.Trahanias, D.Karakas and A.N.Venetsanopoulos, " Directional filtering of color images: theory and experimental results," *IEEE Trans. Image Processing*, vol.5, pp.868-880, July 1996.

[3.76] K.Plataniotis et al, "Color image processing using adaptive multichannel filters," *IEEE Trans. Image Processing*, vol.6, pp.933-949, July 1997.

[3.77] K.Brandenburg, "OCF - A new coding algorithm for high quality sound signals," *Proc. IEEE ICASSP*, pp.5.1.1-5.1.4, May 1987.

[3.78] J.Johnston, "Transform coding of audio signals using perceptual noise criteria," *IEEE J. Selected Areas in Comm.*, vol.6, pp.314-323, Feb. 1988.

[3.79] W.Y.Chan and A.Gersho, "High fidelity audio transform coding with vector quantization," *Proc. IEEE ICASSP*, pp.1,109-1,112, May 1990.

[3.80] Y.F.Deheri et al, "A MUSICAM source codec for digital audio broadcasting and storage," *Proc. IEEE ICASSP*, pp.3,605-3,608, May 1991.

[3.81] M.Iwadare et al, "A 128 Kb/s Hi-Fi audio codec based on adaptive transform coding with adaptive block size MDCT," *IEEE J. Selected Areas in Comm.*, vol.10, pp.138-144, Jan. 1992.

[3.82] T.Painter and A.Spanias, "A review of algorithms for perceptual coding of digital audio signals," *Int. Conference on Digital Signal Processing*, vol.1, pp.175-208, Santorini, Greece, July 1997.

[3.83] J.Johnston, "Perceptual transform coding of wideband stereo signals," *Proc. IEEE ICASSP*, pp.1,993-1,996, May 1989.

[3.84] J.Zwislocki, "Analysis of some auditory characteristics," in *Handbook of Mathematical Psychology*, R.Luce et al, (Eds.), New York, NY: John Wiley, 1965.

[3.85] B.Scharf, "Critical bands," in *Foundations of modern auditory theory*, New York, NY: Academic Press, 1970.

[3.86] E.Zwicker and H.Fastl, *Psychoacoustic facts and models*, Berlin: Springer-Verlag, 1990.

[3.87] J.Johston, "Estimation of perceptual entropy using noise masking criteria," *Proc. IEEE ICASSP*, pp.2,524-2,527, May 1988.

[3.88] H.Fletcher, "Auditory patterns," *Rev. Mod. Phys.*, pp.47-65, Jan. 1940.

[3.89] N.Jayant et al, " Signal compression based on model of human perception," *Proc. IEEE*, vol.81, pp.1,385-1,422, Oct. 1993.

[3.90] P.Noll, " Wideband speech and audio coding," *IEEE Comm. Magazine*, vol.31, pp.34-44, Nov. 1993.

[3.91] N.Jayant and P.Noll, *Digital coding of waveforms*, Englewood Cliffs, NJ: Prentice-Hall, 1984.

[3.92] Y.Mahieux et al, "Transform coding of audio signals using correlation between successive transform blocks," *Proc. IEEE ICASSP*, pp.2021-2024, May 1989.

[3.93] Y.Malhieux and J.Petit, "Transfom coding of audio signals at 64 Kbits/s," *Proc. IEEE GLOBECOM*, pp.405.2.1-405.2.5, Nov. 1990.

[3.94] ISO/IEC JTC1/SC29/WG11 MPEG IS13813-3, *Information technology-generic coding of moving pictures and associated audio, Part3: Audio, MPEG-2*, 1994.

[3.95] K.Brandenburg and J.D.Johnston, "Second generation perceptual audio coding: the hybrid vocoder," *Proc. 88th Conv. Aud. Eng. Soc.*, reprint #2937, March 1990.

[3.96] M.Paraskevas and J.Mourjopoulos, "A differential perceptual audio coding method with reduced bitrate requirements," *IEEE Trans. ASSP*, vol.43, pp.490-503, Nov. 1995.

[3.97] D.Schulz, "Improving audio codecs by noise substitution," *J. Audio. Eng. Soc.*, pp.593-598, July/Aug. 1996.

[3.98] W.Y.Chan and A.Gersho, "Constrained-storage vector quantization in high fidelity audio transform coding," *Proc. IEEE ICASSP*, pp.3,597-3,600, May 1991.

[3.99] J.Princen and A.Bradley, "Analysis/synthesis filter bank design based on time domain aliasing cancellation," *IEEE Trans. ASSP*, vol.34, pp.1,153-1,161, October 1986.

[3.100] N.Iwakani et al, "High-quality audio coding at less than 64 Kb/s by using transform-domain weighted interleave vector quantization (TWINVQ)," *Proc. IEEE ICASSP*, pp.3,095-3,098, May 1995.

[3.101] T.Moriya et al, "Extension and complexity reduction of TWIN VQ audio coder," *Proc. IEEE ICASSP*, pp.1,029-1,032, May 1996.

[3.102] P.Duhamel et al, "A fast algorithm for the implementation of filter banks based on time-domain aliasing cancellation," *Proc. IEEE ICASSP*, pp.2,209-2,212, May 1991.

[3.103] J.W.Woods and S.D.O'Neil, "Subband coding of images," *IEEE Trans. ASSP*, vol.34, pp.1,278-1,288, Oct. 1986.

[3.104] P.P.Vaidyanathan, *Multirate systems and filter banks*, Englewood Cliffs, NJ: Prentice-Hall, 1993.

[3.105] P.H.Westerink, Y.Biemond and D.E.Boekee, "Evaluation of image subband coding scheme," *EUSIPCO'88*, pp.1,149-1,152, Grenoble, France, Sep. 1988.

[3.106] R.M.Gray and D.L.Neuhoff, "Quantization," *IEEE Trans. Info. Theory.*, vol.44, pp.1-63, Oct. 1998.

[3.107] P.Voros, "High-quality sound coding within 2x64 Kb/s using instantaneous dynamic bit allocation," *Proc. IEEE ICASSP*, pp.2,536-2,539, May 1988.

[3.108] A.Charbonnier and J.P.Petit, "Subband ADPCM coding for high quality audio signals," *Proc. IEEE ICASSP*, pp. 2,540-2,543, May 1988.

[3.109] R.N.J.Veldhuis, "Subband coding of digital audio signals without loss of quality," *Proc. IEEE ICASSP*, pp.2,009-2,012, May 1989.

[3.110] D.Teh et al, "Subband coding of high-fidelity quality audio signals at 128 Kbps," *Proc. IEEE ICASSP*, pp.II.197-200, May 1990.

[3.111] I.Daubechies, *Ten lectures on wavelets*, Philadelphia, Pennsylvannia: Society for Industrial and Applied Mathematics, 1992.

[3.112] D.Sinha and A.H.Tewfik, "Low bit rate transparent audio compression using a dynamic dictionary and optimized wavelets," *Proc. IEEE ICASSP*, pp.I-197-200, May 1993.

[3.113] D.Sinha and A.H.Tewfik, "Low bit rate transparent audio compression using adapted wavelets," *IEEE Trans. Signal Proc.*, vol.41, pp.3,463-3,479, Dec. 1993.

[3.114] A.H.Tewfik and M.Ali, "Enhanced wavelet based audio coder," *Conf. Rec. of the 27th Asilomar Conf. on Signals, Systems and Computers*, pp.896-900, Pacific Grove, California, Nov. 1993.

[3.115] K.Hamdy et al, "Low bit rate high quality audio coding with combined harmonic and wavelet representations," *Proc. IEEE ICASSP*, pp.1,045-1,048, May 1996.

[3.116] M.Black and M.Zeytinoglu, "Computationally efficient wavelet packet coding of wideband stereo audio signals," *Proc. IEEE ICASSP*, pp.3,075-3,078, May 1995.

[3.117] P.Kudimakis and M.Sandler, "On the performance of wavelets for low bit rate coding of audio signals," *Proc. IEEE ICASSP*, pp.3,087-3,090, May 1995.

[3.118] P.Kudumakis and M.Sandler, "On the compression obtainable with four-tap wavelets," *IEEE Signal Processing Let.*, vol.10, pp.231-233, Aug. 1996.

[3.119] S.Boland and M.Deriche, "High quality audio coding using multipulse LPC and wavelet decomposition," Proc. IEEE ICASSP, pp.3067-3069, May 1995.

[3.120] S.Boland and M.Derichle, "Audio coding using the wavelet packet transform and a combined scalar vector quantization," *Proc. IEEE ICASSP*, pp.1,041-1,044, May 1996.

[3.121] R.V.Cox et al, "On the applications of multimedia processing to communications," *Proc. IEEE*, vol.86, pp.755-824, May 1998.

[3.122] M.Markovic and Z.S.Bojkovic, "On speech compression standards in multimedia videoconferencing: implementation aspects," *Int. Workshop on Image and Signal Processing IWISP'96*, pp.541-544, Manchester, UK, Nov. 1996.

[3.123] A.N.Netravali and B.G.Haskell, *Digital pictures-representation, compression and standards*, New York, NY: Plenum Press, 1995.

[3.124] H.M.Hang and J.W.Woods, *Handbook of visual communications*, New York, NY: Academic Press, 1995.

[3.125] ITU-T Recomm. H.261, *Video codec for audiovisual services at px64 Kb/s*, March 1993.

[3.126] ITU-T Recomm. H.263, *Video coding for low bit rate communications*, March 1996.

[3.127] K.Rijkse, "H.263: Video coding for low-bit-rate communications," *IEEE Trans. Comm.*, vol.34, pp.42-45, Dec. 1996.

[3.128] ISO/IEC IS11172 (MPEG-1), *Information technology-coding of moving pictures and associated audio for digital storage media up to about 1.5 Mb/s*, 1993.

[3.129] A.Puri, "Video coding using the MPEG-2 compression standard," *SPIE/VCIP*, Boston, MA, pp.1,701-1,713, Nov. 1993.

[3.130] T.Homma, *MPEG contribution: Report of the adhoc group on MPEG-2 applications for multi-view point pictures*, ISO/IEC SC29/WG11 Doc. 861, March 1995.

[3.131] J.Ni, T.Yang and D.H.K.Tsang, "CBR transportation on VBR MPEG-2 video traffic for video-on-demand in ATM networks," *Proc. IEEE ICC*, pp.1,391-1,395, Dallas, TX, June 1996.

[3.132] B.L.Tseng and D.Anastassiou, "Multiview point video coding with MPEG-2 compatibility," *IEEE Trans. CSVT*, vol.6, pp.414-419, Aug. 1996.

[3.133] K.R.Rao and P.Yip, *Discrete cosine transform: algorithms, advantages, applications*, New York, NY: Academic Press, 1990.

[3.134] B.Girod et al, "Image and video coding," *IEEE Signal Processing Magazine*, vol.15, pp.40-48, March 1998.

[3.135] M.Shapiro, "Embedded image coding using zerotrees of wavelet coefficients," *IEEE Trans. Signal Processing*, vol.41, pp.3,445-3,462, Dec. 1993.

[3.136] A.Said and W.Pearlman, "Image compression using the spatial orientation tree," *Proc. IEEE ISCAS*, pp.279-282, Chicago, IL, 1993.

[3.137] K.Horn, T.Wiegand and B.Girod, "Bit allocation methods for closed-loop coding of oversampled pyramid decomposition," *Proc. IEEE ICIP*, pp.17-20, Santa-Barbara, CA, 1997.

[3.138] ISO/IEC 13818, *Generic coding of moving pictures and associated audio information, Part 2: Video*, Geneva 1996.

[3.139] M.T.Orchard and G.J.Sullivan, "Overlapped block motion compensation: an estimation-theoretic approach," *IEEE Trans. Image Processing*, vol.3, pp.693-699, Sept. 1994.

[3.140] B.Girod, "Motion-compensation prediction with fractional-pel accuracy," *IEEE Trans. Comm.*, vol.41, pp.604-612, May 1993.

[3.141] B.Girod, "Rate-constrained motion estimation," in *Visual Comm. and Image Proc. VCIP'94*, A.V.Katsaggelos, (Ed.), Proc. SPIE 2308, pp.1,026-1,034, 1994.

[3.142] D.R.Stinson, *Cryptography: theory and practice*, Boca Raton, Florida: CRC Press, 1995.

[3.143] L.T.Cox et al, *Secure spread spectrum watermarking for multimedia*, Tech. Report, NEC Research Institute, no.95-10, p.1, 1995.

[3.144] W.Bender et al, "Techniques for data hiding," *IBM Systems Journal*, vol.35, pp.313-336, 1996.

[3.145] M.Swanson, B.Zhu and A.H.Tewfik, "Data hiding for video in video," *Proc. IEEE ICIP*, vol.II, pp.676-679, Santa Barbara, California, Oct. 1997.

[3.146] M.Swanson, M.Kobayashi and A.H.Tewfik, "Multimedia data-embedding and watermarking technologies," *Proc. IEEE*, vol.86, pp.1,064-1,087, June 1998.

[3.147] N.Jayant, J.Johnston and R.Safranek, "Signal compression based on models of human perception," *Proc. IEEE*, vol.81, pp.1,385-1,422, Oct. 1993.

[3.148] A.Pive et al, "DCT-based watermark recovering without resorting to the uncorrupted original image, *Proc. IEEE ICIP*, vol.1, pp.520-523, Santa Barbara, CA, Oct. 1997.

[3.149] J.O'Ruanaidh and T.Pun, "Rotation, scale and translation invariant digital image watermarking," *Proc. IEEE ICIP*, vol.1, pp.536-539, Santa Barbara, CA, Oct. 1997.

[3.150] D.Kundor and D.Hatzinakos, "A robust digital image watermarking method using wavelet-based fusion," *Proc. IEEE ICIP*, vol.1, pp.544-547, Santa Barbara, CA, Oct. 1997.

[3.151] X.G.Xia, C.G.Boncelet and G.R.Arce, "A multiresolution watermark for digital images," *Proc. IEEE ICIP*, vol.1, pp.548-551, Santa Barbara, CA, Oct. 1997.

[3.152] J.O'Ruanaidh, W.Dowling and F.Boland, "Watermarking digital images for copyright protection," *IEE Proc. Vision, Image and Signal Processing*, vol.143, pp.250-256, Aug. 1996.

[3.153] L.T.Cox et al, "Secure spread spectrum watermarking for multimedia," *IEEE Trans. Image Processing*, vol.6, pp.1,673-1,687, Dec. 1997.

[3.154] J.O'Ruanaidh, F.Douling and F.Boland, "Phase watermarking of digital images," *Proc. IEEE ICIP*, vol.III, pp.239-242, Sept. 1996.

[3.155] C.I.Podilchuk and E.J.Delp, "Digital watermarking: algorithms and applications," *IEEE Signal Processing Magazine*, vol.18, pp.33-46, July 2001.

[3.156] C.T.Hsu and J.L.Wu, "Hidden signature in images," *Proc. IEEE ICIP*, vol.III, pp.223-226, Sept. 1996.

[3.157] ISO/IEC JTC1/SC29/WG11 Doc. M5804, *Application domains for watermarking standards*, March 2000.

[3.158] M.Sid-Ahmed, *Image processing*, Tokyo, Japan: McGraw-Hill, 1995.

[3.159] A.Dumitras, "State-of-the-art and trends in content-based access and manipulation of visual data," Technical Report, National Research Council of Canada, Dec. 1998.

[3.160] B.Erol, A.Dumitras and F.Kossentini, "Emerging MPEG standards: MPEG-4 and MPEG-7," *Handbook of Image and Video Processing*, (A. Bovik, Ed.), Orlando, Florida: Academic Press 2000.

[3.161] X.Lee, Y.Q.Zhang and A.Leon-Garcia, "Information loss recovery for block-based image coding techniques - a fuzzy logic approach," *IEEE Trans. Image Processing*, vol.4, pp.259-273, March 1995.

[3.162] M.Ghanbari and C.J.Hughes, "Packing coded video signals into ATM cells," *Trans. Networking*, vol.1, pp.505-509, October 1993.

[3.163] S.Y.Kung, *Digital neural networks*, Englewood Cliffs, NJ: Prentice Hall, 1993.

[3.164] M.Arbib, *The handbook of brain theory and neural networks*, Cambridge, Massachusetts: MIT Press, 1995.

[3.165] L.Rothkrantz, V.R.Van and E.Kerckholfs, "Analysis of facial expressions with artificial neural networks," European Simulation Multiconference, Prague, Czech Republic, pp.790-794, June 1995.

[3.166] V.Shostri, L.C.Rabelo and E.Onjeyenve, "Device-independent color correction for multimedia applications using neural networks and abductive modeling approaches," *IEEE Int. Conference on Neural Networks*, pp.2,176-2,181, Washington, DC, June 1996.

[3.167] J.Rish et al, "Interactive information visualization for exploratory intelligence data analysis," *Proc. of IEEE 1996 Virtual Reality Annual Int. Symposium*, pp.230-238, Santa Clara, CA, April 1996.

[3.168] E.Andre, G.Herzog and T.Rist, "From visual data to multimedia presentations," *IEEE Colloquium Grounding Representations: Integration of Sensory Information in Natural Language Processing, Artificial Intelligence and Neural Networks*, pp.1:1-3, May 1995.

[3.169] K.Langer and F.Bodendorf, "Flexible user-guidance in multimedia CBT-applications using artificial neural networks and fuzzy logic," *Proc. Int. ICSC Symposia on Intelligent Industrial Automation and Soft Computing*, pp.B9-13, March 1996.

[3.170] T.Mandl and H.C.Womser, "Soft computing-vague query handling in object oriented information systems," *Proc. HIM'95 (Hypertext-Information Retrieval-Multimedia)*, pp.277-291, Konstanz, Germany, April 1995.

[3.171] Y.H.Tseng, J.N.Hwang and F.Sheehau, "Three-dimensional object representation and invariant recognition using continuous distance transform neural networks," *IEEE Trans. Neural Networks, Special Issue on Pattern Recognition*, vol.8, pp.141-147, Jan. 1997.

[3.172] G.I.Chiou and J.N.Hwang, "Image sequence classification using a neural network based active contour model and a hidden Markov model," *Proc. IEEE ICIP*, vol.II, pp.926-930, Austin, Texas, Nov. 1994.

[3.173] Y.Matsuyama and M.Tan, "Multiple descent competitive learning as an end for multimedia image processing," Proc. of Joint Int. Conference on Neural Networks, pp.2,061-2,064, Oct. 1993.

[3.174] T.Chen and R.R.Rao, "Audio-visual interaction in multimedia communication," *Proc. IEEE ICASSP*, vol.1, pp.179-182, Munich, Germany, April 1997.

[3.175] A.Pedolti, G.Ferigno and M.Redolfi, "Neural networks in multimedia speech recognition," *Proc. of the Int. Conference on Neural Networks and Expert Systems in Medicine and Healthcare*, pp.167-173, Plymouth, UK, Aug. 1994.

[3.176] Y.Nakagawa, E.Hirote and W.Pedrycz, "The concept of fuzzy multimedia intelligent communication system," *Proc. of the Fifth IEEE Int. Conference on Fuzzy Systems*, pp.1,476-1,480, New Orleans, LA, Sep. 1996.

[3.177] C.Fan, N.Namazi and P.Penafiel, "New image motion estimation algorithm based on the EM technique," *IEEE Trans. PAMI*, vol.18, pp.348-352, March 1996.

[3.178] Y.K.Chen, Y.Lin and S.Y.Kung, "A feature tracking algorithm using neighborhood relaxation with multi-candidate pre-screening," *Proc. IEEE ICIP*, vol.II, pp.513-516, Lausanne, Switzerland, Sept. 1996.

[3.179] S.Y.Kung and J.N.Hwang, "Neural networks for intelligent multimedia processing," *Proc.of the IEEE*, vol.86, pp.1,244-1,272, June 1998.

[3.180] I.J.Cox, J.Ghosh and P.Yianilos, *Feature-based face recognition using distance*, Tech. Report 95-09, NEC Research Institute, Princeton, NJ, 1995.

[3.181] S.H.Lin, Y.Chan and S.Y.Kung, "A probabilistic decision-based neural network for location of deformable objects and its applications to surveillance system and video browsing," *Proc. IEEE ICASSP*, pp.3,554-3,557, Atlanta, GA, May 1996.

[3.182] T.S.Huang et al, "A neuronet approach to information fusion," *Proc. IEEE First Workshop on Multimedia Signal Processing*, pp.45-50, Princeton, NJ, June 1997.

[3.183] H.H.Yu and W.Wolf, "A hierachical, multi-resolution method for dictionary-driven content-based image retrieval," *Proc. IEEE ICIP*, pp.823-826, Santa Barbara, CA, Oct. 1997.

[3.184] S.H.Lin, S.Y.Kung and L.J.Lin, "Face recognition/detection by probabilistic decision-based neural networks," *IEEE Trans. Neural Networks*, vol.8, pp.114-132, Jan. 1997.

[3.185] Z.S.Bojkovic, D.A.Milovanovic and N.Mastorakis, "Neural networks applications for multimedia processing," *Proc. of 5th Seminar on Neural Network Applications in Electrical Engineering*, pp.87-91, Belgrade, Yugoslavia, Sep. 2000.

[3.186] *The chromatic media processor*, http://www.chromatic.com/.

[3.187] A.Peleg, S.Wilkic and V.Weiser, "Intel MMX for multimedia PCs," *Comm. of the ACM*, vol.40, pp.25-38, Jan. 1997.

[3.188] P.Lapsley et al, *DSP processor fundamentals*, Piscataway, New Jersey: IEEE Press, 1997.

[3.189] T.Chen, "The past, present and future of image and multi-dimensional signal processing," *IEEE Signal Processing Magazine*, vol.15, pp.21-58, March 1998.

[3.190] A.Peleg and U.Weiser, "MMX technology extension to the Intel architecture," *IEEE Micro*, vol.16, pp.42-50, Jan. 1996.

[3.191] S.Wolfram, *The mathematica book*, III edition, Port Chester, New York: Wolfram Media, Inc. and Cambridge University Press, 1996.

[3.192] *MATLAB Image Processing Toolbox*, http://www.mathworks.com/products/image/.

[3.193] *National Instruments LabView*, http://www.natinst.com/.

[3.194] *NIH Image*, http://rsb.info.nih.gov/nih-image/.

[3.195] I.Kuroda and T.Nishitani, "Multimedia processors," *Proc. of the IEEE*, vol.86, pp.1,203-1,221, June 1998.

[3.196] ISO/IEC JTC1/SC29/WG11, Recomm. H.262, ISO/IEC 13818-2, *Information technology - Generic coding of moving pictures and acoustic audio information: Video*, Geneva, Switzerland, 1994.

[3.197] M.Johnson, *Superscalar multiprocessor design*, Englewood Cliffs, NJ: Prentice Hall, 1991.

[3.198] R.Bannon and Y.Saito, "The alpha 21164 PC microprocessor," *Proc. of Compcon*, pp.20-27, New York, NY: IEEE Computer Science Press, 1997.

[3.199] R.B.Lee, "Subword parallelism with MAX-2," *IEEE Micro*, vol.16, pp.51-59, Aug. 1996.

[3.200] M.Tremblay, D.Greenley and K.Normoyle, "The design of the microarchitecture of Ultra Sparc," *Proc. of the IEEE*, vol.83, pp.1,653-1,663, Dec. 1995.

[3.201] C.Hansen, "Architecture of a broadband media processor," *Proc. of Compcon*, pp.334-340, New York, NY: IEEE Computer Science Press, 1996.

[3.202] Y.Oshime, B.J.Shen and S.H.Jen, "High-speed memory architectures for multimedia applications," *IEEE Circuit Design. Magazine.*, vol.13, pp.8-13, Jan. 1997.

[3.203] T.Robinson et al, "Multigigabyte/sec DRAMs with the MicroUnity mediachannel interface," *Proc. of Compcon*, pp.378-381, New York, NY: IEEE Computer Science Press, 1996.

Chapter 4
Distributed Multimedia Systems

[4.1] K.R.Rao and Z.S.Bojkovic, *Packet video communications over ATM networks*, Upper Saddle River, NJ: Prentice-Hall PTR, 2000.

[4.2] V.O.K.Li and W.Liao, "Distributed multimedia systems," *Proc. of the IEEE*, vol.85, pp.1,063-1,108, July 1997.

[4.3] B.Furth, "Multimedia systems: An overview," *IEEE Multimedia Magazine*, vol.1, pp.47-59, Spring 1994.

[4.4] A.Campell, G.Carlson and H.Hutchinson, "A quality of service architecture," *Comput. Comm. Review*, vol.24, pp.6-27, April 1994.

[4.5] A.Vogel et al, "Distributed multimedia and QoS: A survey," *IEEE Multimedia Magazine*, vol. 2, pp.10-19, Summer 1995.

[4.6] K.Nahrsted and R.Steinmetz, "Resource management in networked multimedia systems," *IEEE Comp. Mag.*, vol.28, pp.52-63, May 1995.

[4.7] K.Nahrsted and J.M.Smith, "The QoS broker," *IEEE Multimedia Mag.*, vol. 2, pp.53-67, Spring 1995.

[4.8] L.C.Wolf, C.Griwodz and R.Steinmetz, "Multimedia communication," *Proc. of the IEEE*, vol.85, pp.1,915-1,933, Dec. 1997.

[4.9] H.Schulzrinne et al, "RTR: A transport protocol for real-time applications," Internet Engineering Task Force, RFC 1889, Jan. 1996.

[4.10] A.Rayes and K.Sage, "Integrated management architecture for IP-based networks," *IEEE Communications Mag.*, vol.38, pp.48-53, April 2000.

[4.11] H.Hegering, S.Abeck and B.Neumair, *Integrated management of networked systems*, San Francisco, CA: Morgan Kaufmann, 1998.

[4.12] M.Guizani and A.Rayes, *Designing ATM switching networks*, New York, New York: McGraw-Hill, 1998.

[4.13] S.Aidarous and T.Plevyak, *Telecommunications network management*, Piscataway, New Jersey: IEEE Press, 1998.

[4.14] Ch.H.Wu and I.D.Irwin, "Multimedia and multimedia communication: A tutorial," *IEEE Trans. Industrial Electronics*, vol.45, pp.4-14, Feb. 1998.

[4.15] X.Forum (1996). XTP: The Xpress Transport Protocol. http://www.ca.saudia.gov/xtp/xtp.html.

[4.16] ATM Forum, *ATM Traffic Management Specification, Version 4.0*, Upper Saddle River, NJ: Prentice-Hall, 1996.

[4.17] L.Zhang et al, "RSVP: a new resource reservation protocol," *IEEE Network Mag.*, vol.7, pp.8-18, Sept. 1993.

[4.18] M.Decina and V.Trecordi, "Convergence of telecommunications and computing to networking models for integrated services and applications," *Proc. IEEE*, vol.85, pp.1,887-1,914, Dec. 1997.

[4.19] P.Newman et al, "IP switching and gigabit routers," *IEEE Comm. Magazine*, vol.35, pp.64-69, Jan. 1997.

[4.20] Y.Reuhter et al, RFC2105, Internet Engineering Task Force, Feb. 1997.

[4.21] C.M.Pazos, M.R.Kotelba and A.G.Malis, "Real-time multimedia over ATM: RMOA," *IEEE Comm. Mag.*, vol.38, pp.82-87, April 2000.

[4.22] G.Carlson, G.S.Blair and P.Robin, "Microkernel support for continuous media in distributed systems," *Comput. Networks ISDN Syst.*, vol.26, pp.1,323-1,341, 1994.

[4.23] C.L.Lin and J.W.Layland, "Scheduling algorithms for multiprogramming in a hard real-time environment," *J. ACM*, vol.20, pp.46-61, Jan. 1973.

[4.24] R.Steinmetz, "Analyzing the multimedia operating system," *IEEE Multimedia Mag.*, vol. 2, pp.68-84, Spring 1995.

[4.25] H.Tokuda, "Operating system support for continuous media applications," in *Multimedia Systems*, Reading, MA: Addison-Wesley, pp.201-220, 1994.

References

[4.26] A.Bouloutas and D.N.Serpanos, "A comparison of multimedia severs," IBM Research Division, T.J.Watson Research Center, Tech. Rep. RC19162, Sept. 1993.

[4.27] D.J.Gemmell et al, "Multimedia storage servers: A tutorial," *Computer*, pp.40-49, May 1995.

[4.28] A.Dan et al, "Buffering and catching in large scale video servers," IBM Research, Tech. Rep. RC19903, Jan.1995.

[4.29] K.D.Huynth and T.M.Khoshgoftaar, "Performance analysis of advanced I/O architectures for PC-based video servers," *ACM/Springer Multimedia Syst.*, vol.2, pp.36-50, June 1994.

[4.30] T.D.C.Little and D.Venkatesh, "Popularity-based assignment of movies to storage devices and video-on-demand system," *ACM/Springer Multimedia Syst.*, vol.2, pp.280-287, Jan 1995.

[4.31] D.Meliksetian, F.F.Kuo Yu and C.Y.Ryer Chen, "Methodologies for designing video servers," *IEEE Trans. Multimedia*, vol.2, pp.62-69, March 2000.

[4.32] D.N.Serpanos, L.Georgiadis and T.Bouloutas, "Multimedia packing: a load and storage balancing algorithm for distributed multimedia servers," *IEEE Trans. CSVT*, vol.8 pp.13-17, Feb. 1998.

[4.33] B.G.Haskall, A.Puri and A.N.Netravali, *Digital video: An introduction to MPEG-2*, New York, NY: Chapman and Hall, 1997.

[4.34] V.O.K.Li et al, "Performance model of interactive video-on-demand systems," *IEEE Journal Selected Areas Comm.*, vol.14, pp.1,099-1,109, Aug. 1996.

[4.35] Y.H.Chang et al, "An open-systems approach to video on demand," *IEEE Commun. Mag.*, vol.32, pp.68-80, May 1994.

[4.36] D.Deloddere, W.Verbiest and H.Verhille, "Interactive video on demand," *IEEE Commun. Mag.*, vol.32, pp.82-88, May 1994.

[4.37] W.J.Liao and V.O.K.Li, "The split-and-merge (SAM) protocol for interactive video-on-demand systems," Proc. *IEEE INFOCOM'97*, pp.1,351-1,358, Kobe, Japan, 1997.

[4.38] L.Crutcher and J.Grinham, "The networked video jukebox," *IEEE Trans. CSVT*, vol.4, pp.105-120, April 1994.

[4.39] C.L.Lin and S.V.Guan, "The design and architecture of a video library system," *IEEE Comm. Magazine*, vol.34, pp.84-91, Jan. 1996.

[4.40] V.Leon and E.Miller, "Subscriber terminal units for video dial tone systems," *IEEE Network Mag.*, vol.9, pp.48-57, Sept/Oct. 1995.

[4.41] G.H.Petit, D.Delodere and W.Verbiest, "Bandwidth resource optimization in video on demand network architectures," *Proc. Int. Workshop Community Networking Integrated Multimedia Services to the Home*, pp.91-97, New York, NY, 1994.

[4.42] F.Schaffa and J.P.Nussbaumer, "On bandwidth and storage tradeoffs in multimedia distribution networks," *Proc. Int. Conf. Multimedia Computing Systems*, pp.1020-1026, Washington, D.C., 1995.

[4.43] H.Ishii and N.Miyake, "Toward an open shared workspace: Computer and video fusion approach to team workstation," *Commun. ACM*, vol.34, pp.37-50, Dec. 1991.

[4.44] R.Kling, "Cooperation, coordination and control in computer-supported work," *Comm. ACM*, vol.34, pp.83-88, Dec.1991.

[4.45] C.A.Elias, S.J.Gibbs and G.L.Rein, "Groupware: Some issues and experiences," *Comm. ACM*, vol.34, pp.38-58, Jan. 1991.

[4.46] K.Watabe et al, "Distributed desktop conferencing system with multiuser multimedia interface," *IEEE J. Selected Areas in Comm.*, vol.9, pp.531-539, May 1991.

[4.47] W.Reinhard et al, "CSCW tools: Concepts and architectures," *IEEE Comput. Mag.*, vol.27, pp.28-36, May 1994.

[4.48] R.Bently et al, "Architectural support for cooperative multiuser interface," *IEEE Comput. Mag.*, vol.27, pp.37-46, May 1994.

[4.49] M.Stefik et al, "WYSIWIS revised: Early experiences with multiuser interfaces," *ACM Trans. Office Inform. Sys.*, vol.5, pp.147-167, Apr. 1987.

[4.50] S.Sarin and I.Grief, "Computer-based real-time conferencing systems," *IEEE Comput. Mag.*, vol.18, pp.33-45, Oct. 1985.

[4.51] R.Fish et al, "Quilt: A collaborative tool for cooperative writing," in *Proc. COIS'88*, pp.23-25, Palo Alto, CA, 1988.

[4.52] I.Grief, R.Seliger and W.Weihl, "Atomic data abstractions in a distributed collaborative editing system," in *Proc. ACM CSCW'86*, pp.160-172, Seattle, WA, 1986.

[4.53] K.L.Kreamer and I.L.King, "Computer-based systems for cooperative work and group decision making," *ACM Computing Surveys*, vol.20, pp.115-146, June 1988.

[4.54] L.M.Applegate, B.R.Konsynski and J.F.Numamcker, "A group decision support system for idea generation and issue analysis in organization planning," in *Proc. ACM CSCW'86*, pp.16-34, Seattle, WA, 1986.

[4.55] K.R.Rao and P.Yip, *Discrete cosine transform: algorithms, advantages, applications*, New York, NY: Academic Press, 1990.

[4.56] I.C.Lauwers and K.A.Lantz, "Collaborative awareness in support of collaboration transparency: requirements for the next generation of shared window systems," *Proc. CHI'90*, pp.303-311, Seattle, WA, 1990.

[4.57] B.Reljin and I.Reljin, "Multimedia: The impact on the teletraffic," *Advances in Physics, Electronics and Signal Processing Applications*, N.E.Mastorakis (Ed.), pp.366-373, World Scientific, 2000.

[4.58] M.Strintzis and S.Mallassiotis, "Object-based coding of stereoscopic and 3D image sequence," *IEEE Signal Proc. Magazine*, vol.16, pp.14-28, May 1999.

[4.59] J.B.Smith and S.F.Weiss, "Hypertext," *Commun. ACM*, vol.31, pp.816-819, July 1988.

[4.60] J.Nielsen, "The art of navigation," *Commun. ACM*, vol.33, pp.296-310, March 1990.

[4.61] R.M.Akscyn, D.L.McCracken and E.A.Yoder, "KMS: A distributed hypermedia system for managing knowledge in organizations," *Commun. ACM*, vol.31, pp.820-835, July 1988.

[4.62] B.Campbell and J.M.Goodman, "HAM: A general purpose hypertext abstract machine," *Commun. ACM*, vol.32, pp.856-861, July 1988.

[4.63] F.Garzotto, P.Paolini and D.Schwabe, "HDM - A model for the design of hypertext applications," *Proc. ACM Hypertext'91*, pp.313-328, San Antonio, TX, 1991.

[4.64] F.Halasz and M.Schwartz, "The Dexter hypertext reference model," *NIST Hypertext Standardization Workshop*, pp.95-133, Gaithersburg, MD, 1990.

Chapter 5
Multimedia Communications Standards

[5.1] R.Chellappa, T.Chen and A.Katsaggelos, "Audio-visual interaction in multimodal communication," *IEEE Signal Processing Magazine*, vol.14, pp.37-38, July 1997.

[5.2] T.Sikora, "MPEG digital video-coding standards," *IEEE Signal Processing Mag.*, vol.14, pp.82-100, Sept. 1997.

[5.3] ISO/IEC SC29/WG11, *MPEG Requirements Group, MPEG-7: Context and objectives*, Doc.N1678, MPEG Bristol Meeting, April 1997.

[5.4] F.Pereira, *First proposal for MPEG-7 visual requirements*, ISO/IEC SC29/WG11 Doc.M1941, MPEG Bristol Meeting, April 1997.

[5.5] S.F.Chang, "Content-based indexing and retrieval of visual information," *IEEE Signal Processing Magazine*, vol.14, pp.45-48, July 1997.

[5.6] K.R.Rao and Z.S.Bojkovic, *Packet video communications over ATM networks*, Upper Saddle River, NJ: Prentice Hall PTR, 2000.

[5.7] R.Koenen, "Profiles and levels in MPEG-4: Approach and overview," *Signal Processing: Image Comm.*, vol.15, pp.463-478, Jan. 2000.

[5.8] F.Pereira, "MPEG-4: Why, what, how and when?," *Signal Processing: Image Comm.*, vol.15, pp.271-279, Jan. 2000.

[5.9] L.Chiarglione, "MPEG and multimedia communications," *IEEE Trans. CSVT*, vol.7, pp.5-17, Feb. 1997.

[5.10] L.Chiarglione, "Impact of MPEG standards on multimedia industry," *Proc. of the IEEE*, vol.86, pp.1,222-1,227, June 1998.

[5.11] ISO/IEC IS11172 (MPEG-1), *Information Technology - Coding Of Moving Pictures and Associated Audio for Digital Storage Media up to About 1.5 Mb/s*, 1993.

[5.12] ISO/IEC IS11172-2 (MPEG-1), *Information Technology - Coding of Moving Pictures and Associated Audio for Digital Storage Media up to About 1.5 Mb/s - Part 2: Video*, 1993.

[5.13] ISO/IEC IS11172-3 (MPEG-1), *Information Technology - Coding of Moving Pictures and Associated Audio for Digital Storage Media up to About 1.5 Mb/s - Part 3: Audio*, 1993.

[5.14] ISO/IEC IS11172-4 (MPEG-1), *Information Technology - Coding of Moving Pictures and Associated Audio for Digital Storage Media up to About 1.5 Mb/s - Part 4: Conformance*, 1993.

[5.15] ISO/IEC IS11172-5 (MPEG-1), *Information Technology - Coding of Moving Pictures and Associated Audio for Digital Storage Media up to About 1.5 Mb/s, Part 5: Software Simulation*, 1993.

[5.16] ISO/IEC IS13818-3 (MPEG-1), *Information Technology - Generic Coding of Moving Pictures and Assoicated Audio - Part 3: Audio*, 1995.

[5.17] K.H.Brandenburg et al, "ISO MPEG-1 Audio: A generic standard for coding of light quality digital audio," *J. of Acoustic Engineering Society*, vol.42, pp.780-792, Oct. 1994.

[5.18] B.L.Tseng and D.Anastassiou, "Multiview point video coding with MPEG-2 compatibility," *IEEE Trans. CSVT*, vol.6, pp.414-419, Aug. 1996.

[5.19] T.Homma, "MPEG contribution: Report of the Ad Hoc Group on MPEG-2 applications for multi-view point pictures," ISO/IEC SC29/WG11 Doc.861, March 1995.

[5.20] A.Puri, "A video coding using the MPEG-2 compression standard," *SPIE/VCIP*, vol.1,199, pp.1,701-1,713, Boston, MA, Nov. 1993.

[5.21] J.Ni, T.Yang and D.H.K.Tsang, "CBR transmission on VBR MPEG-2 video traffic for video-on-demand in ATM networks," International Conference on Communications, Dallas, TX, pp.1,391-1,395, June 1996.

[5.22] M.Orzessek and P.Sommer, *ATM and MPEG-2 Integration of Digital Video into Broadband Networks*, Upper Saddle River, NJ: Prentice Hall PTR, 1998.

[5.23] ISO/IEC IS 13818-1, *Generic Coding of Moving Pictures and Associated Audio, Part 1: Systems*, 1995.

[5.24] ISO/IEC IS 13818-2, *Generic Coding of Moving Pictures and Associated Audio, Part 2: Video*, 1995.

[5.25] ISO/IEC IS 13818-6, *Generic Coding of Moving Pictures and Associated Audio, Part 6: DSM/CC*, 1995.

[5.26] J.De Lameilieure and G.Schamel, "Hierarchical coding of TV/HDTV within the German HDTV project," *Proc. Int. Workshop on HDTV*, pp.8A.1.1-1.8, Ottawa, Canada, Oct. 1993.

[5.27] ISO/IEC JTC1/SC29/WG11 N0702Rev, *Information Technology - Generic Coding of Moving Pictures and Associated Audio, Recommendation H.262, Draft International Standard*, Paris, France March 1994.

[5.28] P.J.Burt and E.H.Adelson, "The Laplacian pyramid as a compact image code," *IEEE Trans. Comm.*, vol.COM-31, pp.532-540, April 1983.

[5.29] C.Gonzales and E.Viscito, "Flexibly scalable digital video coding," *Signal Processing: Image Comm.*, vol.5, pp.1-2, Feb. 1993.

[5.30] T.Sikora, T.K.Tan and K.N.Ngan, "A performance comparison of frequency domain pyramid scalable coding schemes," Proc. Picture Coding Symposium, Lausanne, Switzerland, pp.16.1-16.2, March 1993.

[5.31] A.W.Johnson et al, "Filters for drift reduction in frequency scalable video coding schemes," *Electronics Letters*, vol.30, pp.471-472, March 1994.

[5.32] ITU-T Rec I.610, *OAM Principle of the BISDN Access*, 1995.

[5.33] R.Orfali, D.Harkey and J.Edwards, *The essential client/server survival guide*, New York, NY: Wiley Computer Publishing, John Wiley, 1996.

[5.34] F.Pereira, "Tutorial issue on the MPEG-4 standard," *Signal Processing: Image Comm.*, vol.15, pp.269-270, Jan. 2000.

[5.35] Official MPEG Website, http://www.cselt.it/mpeg.

[5.36] ISO/IEC JTC1/SC29/WG11, Doc.N4030, *MPEG-4 Overview v18.0*, Singapore, March 2001.

[5.37] ISO/IEC Doc.N2724, *MPEG-4 Applications*, Seoul, Korea, March 1999.

[5.38] ISO/IEC Doc.N2501, *Text of FDIS 14496-1: Systems*, Atlantic City, NJ, Oct. 1998.

[5.39] ISO/IEC Doc.N2502, *Text of FDIS 14496-2: Visual*, Atlantic City, NJ, Oct. 1998.

[5.40] ISO/IEC Doc.N2503, *Text of FDIS 14496-3: Audio*, Atlantic City, NJ, Oct. 1998.

[5.41] ISO/IEC Doc.N2550, *Text of FDIS 14496-4: Conformance testing*, Rome, Italy, Dec. 1998.

[5.42] ISO/IEC Doc.N2805, *Text of FDIS 14496-5: Reference software*, Vancouver, Canada, July 1999.

[5.43] ISO/IEC Doc.N2506, *Text of FDIS 14496-6: DMIF*, Atlantic City, NJ, Oct. 1998.

[5.44] MPEG Website on quality tests, http://www.cselt.it/mpeg/qualitytests.htm.

[5.45] L.Chiarglione, "The development of an integrated audiovisual coding standard: MPEG," *Proc. of the IEEE*, vol.83, pp.151-157, Feb. 1995.

[5.46] ISO/MPEG N1375, *Verification Model (VM) Development and Core Experiments*, Chicago, IL, Sept. 1996.

[5.47] ISO/MPEG N998, *Proposal Package Description (PPD) - Revision 3*, Tokyo, Japan, July 1995.

[5.48] ISO/MPEG N3930, *MPEG-4 Requirements*, Pisa, Italy, Jan. 2001.

[5.49] O.Avaro et al, "MPEG-4 Systems: Overview," *Signal Processing: Image Comm.*, vol.15, pp.281-298, Jan. 2000.

[5.50] ISO/MPEG N2723, *MPEG-4 Requirements*, March 1999.

[5.51] ISO/MPEG N2806 FDIS 14496-6, *Delivery multimedia integration framework, Part 6*, Vancouver, Canada, July 1999.

[5.52] C.Herpel and A.Eleftheriadis, "MPEG-4 Systems: Elementary stream management," *Signal Processing: Image Comm.*, vol.15, pp.299-320, Jan. 2000.

[5.53] ISO/MPEG N1713, *FAQ on MPEG-4 Video*, Bristol, England, April 1997.

[5.54] G.Franceshini, "The delivery layer in MPEG-4," *Signal Processing: Image Comm.*, vol.15, pp.347-363, Jan. 2000.

[5.55] E.D.Scheirer, R.Vaananen and J.Huopaniem, "Audio BIFS: Describing audio scenes with the MPEG-4 multimedia standard," *IEEE Trans. Multimedia*, vol.1, pp.237-250, Sept. 1999.

[5.56] ISO/IEC 14772-1, *The Virtual Reality Modeling Language*, 1997. http://www.vrml.org/Specifications/VRML97.

[5.57] ISO/MPEG N1162, *Report on the Ad Hoc Group on Evaluation of Tools and Algorithms of Video Submissions for MPEG-4*, Munich, Germany, Jan. 1996.

[5.58] Z.S.Bojkovic and D.A.Milovanovic, "Audiovisual integration in multimedia communications based on MPEG-4 facial animation," *Circuits, Systems and Signal Processing*, vol.20, pp.311-339, May-June 2001.

[5.59] W.G.Gardner, "Reverberation algorithms," in Applications of Digital Signal Processing to Audio and Acoustics, M.Khars and K.Brandenburg (Eds.), New York, NY: Kluwer, 1998.

[5.60] J.P.Jullien, "Structured model for the representation and the control of room acoustical quality," *Proc. 15th Int. Conf. Acoustic*, Trondheim, Norway, pp.517-520, 1995.

[5.61] J.M.Jot, "Efficient models for reverberation and distance rendering in computer music and virtual audio reality," *Proc. Int. Computer Music Conf.*, Thessaloniki, Greece, pp.236-243, 1997.

[5.62] ISO/MPEG N2502, *Text of FDIS 14496-2, Part 2: Video*, Nov. 1998.

[5.63] ISO/MPEG N2503, *Text of FDIS 14496-3, Part 3: Audio*, Nov. 1998.

[5.64] ISO/MPEG N2501, *Text of FDIS 14496-1, Part 1: Systems*, Nov. 1998.

[5.65] T.Ebrahimi and C.Horne, "MPEG-4 natural video coding - an overview," *Signal Processing: Image Comm.*, vol.15, pp.365-385, Jan. 2000.

[5.66] ITU-T experts group on very low bitrate visual telephony, *ITU-T Rec. H.263 Video Coding for Low Bitrate Communication*, Dec. 1995.

[5.67] R.V.Cox et al, "On the applications of multimedia processing to communications," *Proc. of the IEEE*, vol.86, pp.755-824, May 1998.

[5.68] T.Sikora, "The MPEG-4 video standard verification model," *IEEE Trans. CSVT*, vol.7, pp.19-31, Feb. 1997.

[5.69] K.R.Rao and P.Yip, *Discrete Cosine Transform: Algorithms, Advantages and Applications*, New York, NY: Academic Press, 1990.

[5.70] C.LeBuhan et al, "Shape representation and coding of visual objects in multimedia applications: an overview," *Ann. Telecomm.,* vol.53, pp.164-178, May 1998.

[5.71] J.L.Mitchell et al, *MPEG Video Compression Standard,* New York, NY: Chapman and Hall, 1996.

[5.72] ISO/MPEG N2552, *MPEG-4 Video Verfication Model VM12.0,* Rome, Italy, Dec. 1998.

[5.73] J.Hartman and J.Wernecke, *The VRML Handbook,* Reading, MA: Addison-Wesley, 1996.

[5.74] A.M.Tekalp et al, "Two-dimensional mesh-based visual-object representation for interactive synthetic/natural digital video," *Proc. of the IEEE,* vol.86, pp.1,029-1,051, June 1998.

[5.75] J.Y.Wang and E.H.Adelson, "Representing moving images with layers," *IEEE Trans. Image Processing,* vol.3, pp.625-638, Sept. 1994.

[5.76] G.Wolberg, *Digital image warping,* Los Alamitos, CA: Computer Society Press, 1990.

[5.77] A.M.Tekalp and J.Osterman, "Face and 2D mesh animation in MPEG-4," *Signal Processing: Image Comm.,* vol.15, pp.387-421, Jan. 2000.

[5.78] P.Klara et al, "Simulation of facial muscle actions based on rational free form deformations," *Proc. of Eurographics,* pp.59-69, 1992.

[5.79] G.A.Abrantes and F.Pereira, "MPEG-4 facial animation technology: survey, implementation and results," *IEEE Trans. CSVT,* vol.9, pp.290-305, March 1999.

[5.80] F.Lavagetto and R.Pockaj, "The facial animation engine: toward a high-level interface for the design of MPEG-4 compliant animated faces," *IEEE Trans. CSVT,* vol.9, pp.277-289, March 1999.

[5.81] L.Chen, J.Ostermann and T.Huang, "Adaptation of a generic 3D human face model to 3D range data," *First Workshop on Multimedia Signal Processing,* pp.274-279, Princeton, NJ, June 1997.

[5.82] J.Ostermann, "Animation of synthetic faces in MPEG-4," *Computer Animation,* pp.49-52, June 1998.

[5.83] H.Tao et al, "Compression of facial animation parameters for transmission of talking heads," *IEEE Trans. CSVT,* vol.9, pp.264-276, April 1999.

[5.84] S.Moushima and H.Harashima, "A media conversion from speech to facial image for intelligent man-machine interface, *IEEE J. Selected Areas in Comm.,* vol.9, pp.594-600, May 1991.

[5.85] K.Waters and T.Levergood, "An automatic lip-synchronization algorithm for synthetic faces," *Proc. of the Multimedia Conference,* pp.149-156, San Francisco, CA, Sept. 1994.

[5.86] R.Koenen, F.Pereira and L.Chiarglione, "MPEG-4: Context and objectives," *Image Comm. J.,* vol.9, pp.295-304, May 1997.

[5.87] A.M.Tekalp, *Digital Video Processing,* Englewood Cliffs, NJ: Prentice-Hall, 1995.

[5.88] M.deBerget et al, *Computational Geometry – Algorithms and Applications,* Berlin, Germany: Springer, 1997.

[5.89] Y.Altunbasak and A.M.Tekalp, "Occlusion-adaptive, content-based mesh design and forward tracking," *IEEE Trans. Image Processing,* vol.6, pp.1,270-1,280, Sept. 1997.

[5.90] Y.Nakaya and H.Harashima, "Motion compensation based on spatial transformations," *IEEE Trans. CSVT,* vol.4, pp.339-356, June 1994.

[5.91] Y.Altunbasak and A.M.Tekalp, "Closed-form connectivity preserving solutions for motion compensation using 2D meshes," *IEEE Trans. Image Processing,* vol.6, pp.1,255-1,269, Sept. 1997.

[5.92] Y.Wang and O.Lee, "Active mesh – a feature seeking and tracking image sequence representation scheme," *IEEE Trans. Image Processing,* vol.3, pp.610-624, Sept. 1994.

[5.93] C.Toklu et al, "Tracking motion and intensity variations using hierarchical 2D mesh modeling," *Graphical Models Image Process*, vol.58, pp.553-573, Nov. 1996.

[5.94] P.J.L.vanBeek et al, "Hierarchical 2D mesh representation, tracking and compression for object-based video," *IEEE Trans. CSVT*, vol.9, pp.353-369, March 1999.

[5.95] P.Doenges et al, "MPEG-4 audio/video synthetic graphics/audio for mixed media," *Signal Processing: Image Comm.*, vol.9, pp.433-464, May 1997.

[5.96] S.R.Quackenbush, "Coding of natural audio in MPEG-4," *Proc. IEEE ICASSP*, pp.3,797-3,800, Seattle, WA, 1997.

[5.97] E.D.Sheirer, "The MPEG-4 structured audio standard," *Proc. IEEE ICASSP*, pp.3,801-3,804, Seattle, WA, 1998.

[5.98] E.D.Sheirer, Y.Lee and J.W.Yang, "Synthetic audio and SNHC audio in MPEG-4," in *Advances in Multimedia: Systems, Standards and Networks*, A.Puri and T.Chen (Eds.), New York, NY: Marcel Dekker, 1999.

[5.99] G.A.Soulodre et al, "Subjective evaluation of state of the art two-channel audio codecs," *J. Audio Eng. Soc.*, vol.46, pp.164-177, 1998.

[5.100] N.Jayant, J.Johnston and R.Safranek, "Signal compression based on models of human perception," *Proc. of the IEEE*, vol.81, pp.1,385-1,422, Oct. 1993.

[5.101] M.Bosi et al, "ISO/IEC MPEG-2 advanced audio coding," *J. Audio Eng. Soc.*, vol.45, pp.789-814, Oct. 1997.

[5.102] B.S.Atal and M.R.Shroeder, "Predictive coding of speech signals and subjective error criteria," *IEEE Trans. ASSP*, vol.27, pp.247-254, March 1979.

[5.103] A.Gersho, "Advances in speech and audio compression," *Proc. of the IEEE*, vol.32, pp.900-918, June 1994.

[5.104] M.Nishiguchi and J.Matsumoto, "Harmonic and noise coding of LPC residuals with classified vector quantization," *Proc. of the IEEE ICASSP*, pp.484-487, Detroit, MI, 1995.

[5.105] E.D.Sheirer and B.L.Vercoe, "SAOL: the MPEG-4 structured audio orchestra language," *Comp. Music J.*, vol.23, pp.35-51, Feb. 1999.

[5.106] B.L.Vercoe, W.G.Gardner and E.D.Sheirer, "Structured audio: the creation, transmission and rendering of parametric sound representations," *Proc. of the IEEE*, vol.85, pp.922-940, May 1998.

[5.107] E.D.Sheirer, "Structured audio and effects processing in the MPEG-4 multimedia standard," *Multimedia Syst.*, vol.7, pp.11-22, Jan. 1999.

[5.108] ISO/IEC JTC1 SC29/WG11 N2725, *Overview of the MPEG-4 Standard*, Seoul, Korea, 1999.

[5.109] K.Brandenburg et al, "MPEG-4 natural audio coding," *Signal Processing: Image Comm.*, vol.15, pp.423-443, Jan. 2000.

[5.110] D.Schulz, "Improving audio codecs by noise substitution," *J. Audio Eng. Soc.*, vol.46, pp.593-598, July/Aug. 1996.

[5.111] N.Iwakami and T.Moriya, "Transform domain weighted interleave vector quantization (Twin VQ)," presented at the 101st Convention of the Audio Engineering Society, preprint 4377.

[5.112] N.Iwakami and T.Moriya, "The integrated filter bank based scalable MPEG-4 audio coder," presented at the 105th Convention of the Audio Engineering Society, preprint 4810.

[5.113] ISO/IEC JTC1/SC29/WG11 Doc.N2424, *Report on the MPEG-4 speech codec verification test*, Oct. 1998.

[5.114] P.Kroon et al, "Regular-pulse excitation - A novel approach to effective and efficient multiple coding of speech," *IEEE Trans. ASSP*, vol.31, pp.1,054-1,063, Oct. 1986.

[5.115] K.Brandenburg, "Perceptual coding of high quality digital audio," *Applications of Digital Signal Processing to Audio and Acoustics*, M.Kahrs and K.Brandenburg (Eds.), pp.39-83, New York, NY: Kluwer Academic, 1998.

[5.116] D.Johnston et al, "Current and experimental applications of speech technology for telecom services in Europe," *Speech Comm.*, vol.23, pp.5-16, Feb. 1997.

[5.117] E.D.Sheirer, Y.Lee and J.W.Yang, "Synthetic and SNHC audio in MPEG-4," *Signal Processing: Image Comm.*, vol.15, pp.445-461, Jan. 2000.

[5.118] M.Kitai et al, "ASR and TTS telecommunications applications in Japan," *Speech Comm.*, vol.23, pp.17-30, Feb. 1997.

[5.119] E.D.Sheirer, "Structured audio and effects processing in the MPEG-4 multimedia standard," *Multimedia Systems*, vol.7, pp.11-22, Jan. 1999.

[5.120] J.Signes, Y.Fisher and A.Eleftheriadis, "MPEG-4's binary format for scene description," *Signal Processing: Image Comm.*, vol.15, pp.321-345, Jan. 2000.

[5.121] ISO/IEC WD 14469-8, Doc.3852, *Carriage of MPEG-4 Contents over IP Networks*, Pisa, Italy, Jan. 1999.

[5.122] A.N.Skodras, C.Christopoulos and T.Ebrahimi, "JPEG2000: The upcoming still image compression standard," *Proc. of the 11th Portugese Conference on Pattern Recognition*, pp.359-366, Porto, Portugal, May 2000.

[5.123] *IEEE Signal Processing Magazine*, vol.1, pp.36-58, Sept. 2001.

[5.124] A.Zandi et al, "CREW: Compression with reversible embedded wavelets," *Proc. of IEEE Data Compression Conference*, pp.212-221, Snowbird, Utah, March 1995.

[5.125] ISO/IEC JTC1/SC29/WG1 Doc.N505, *Call for contributions for JPEG2000: image coding systems*, March 1997.

[5.126] M.W.Marcellin and T.R.Fischer, "Trellis coded quantization of memoryless and Gauss-Markov sources," *IEEE Trans. Commun.*, vol.38, pp.82-93, Jan. 1990.

[5.127] J.H.Kasuer, M.W.Marcellin and B.R.Hunt, "Universal trellis coded quantization," *IEEE Trans. Image Processing*, vol.8, pp.1,677-1,687, Dec. 1999.

[5.128] J.Shapiro, "Embedded image coding using zerotrees of wavelet coefficients," *IEEE Trans. Signal Processing*, vol.41, pp.3,445-3,462, Dec. 1993.

[5.129] D.Taubman and A.Zakhor, "Multirate 3D subband coding of video," *IEEE Trans. Image Processing*, vol.3, pp.572-588, Sept. 1994.

[5.130] A.Said and W.Pearlman, "A new fast and efficient image codec based on set partitioning in hierarchical trees," *IEEE Trans. CSVT*, vol.6, pp.243-250, June 1996.

[5.131] J.Li, P.Cleng and C.C.J.Kuo, "On the improvements of embedded zerotree wavelet (EZW) coding," *Proc. SPIE, Visual Comm. and Image Processing*, vol.2,601, pp.1,490-1,501, Taipei, Taiwan, May 1995.

[5.132] P.J.Sementilli et al, "Wavelet TCQ: submission to JPEG-2000," *Proc. of SPIE, Appl. of Digital Image Proc.*, pp.2-12, July 1998.

[5.133] A.Said and W.A.Pearlman, "An image multiresolution representation for lossless and lossy compression," *IEEE Trans. Image Processing*, vol.5, pp.1,303-1,310, Sept. 1996.

References

[5.134] ISO/IEC JTC1/SC29/WG1 N988, C.Christopoulos, *JPEG-2000 Verification Model 2.0 (Technical Description)*, Oct. 1998.

[5.135] W.B.Pennebaker and J.L.Mitchell, "JPEG still image data compression standard," New York, NY: Van Nostrand Reinhold, 1993.

[5.136] ISO/IEC JTC1/SC29/WG1 N482r, D.Speck, *New Options in Radix-255 Arithmetic Coder*, March 1997.

[5.137] ISO/IEC JTC1/SC29/WG11 N1020r, D.Taubman, *Report on Coding Experiment EBCOT (Embedded Block Coding with Optimized Truncation)*, Oct. 1998.

[5.138] ISO/IEC JTC1/SC29/WG11 N1577, *JPEG2000 Part II Working Draft Version 1.0 Pre-Release A*, Jan. 2000.

[5.139] ISO/IEC CD15444-1, *JPEG2000 Committee Draft v1.0*, Dec. 1999.

[5.140] M.W.Marcellin et al, "An overview of JPEG2000," *Proc. of IEEE Data Compression Conference*, pp.523-541, Snowbird, 2000.

[5.141] ISO/IEC 14496-2, Information technology - *Coding of Audio Visual Object - Part 2: Visual*, Dec. 1999.

[5.142] ISO/IEC 14495-1, Information technology - *Lossless and Near-Lossless Compression of Continuous-Tone Still Images: Baseline*, Dec. 1999.

[5.143] W3C, *PNG (Portable Network Graphics) Specification*, Oct. 1996, http://www.w3.org/TR/REC-png.

[5.144] ISO/IEC 11544-1, Information technology, *Coded Representation of Picture and Audio Information - Progressive Bi-Level Image Compression*, March 1993.

[5.145] ISO/IEC FCD14492, Information technology - *Coded Representation of Picture and Audio Information - Lossy/Lossless Coding of Bi-Level Images*, July 1999.

[5.146] D.Santa-Cruz and T.Ebrahimi, "An analytical study of JPEG2000 functionalities," *Proc. IEEE ICIP*, vol.2, pp.49-52, Vancouver, Canada, Sept. 2000.

[5.147] D.Nister and C.Christopoulos, "Lossless region of interest with embedded wavelet image coding," *Signal Processing*. vol.78, pp.1-17, Jan. 1999.

[5.148] D.LeGall and A.Tabatabai, "Subband coding of digital images using symmetric short kernel filters and arithmetic coding techniques," *Proc. IEEE ICASSP*, pp.761-765, New York, NY, April 1988.

[5.149] ISO/IEC JTC1/SC29/WG1, Doc.N1595, *USNB Comments on JPEG2000 CD1.0*, Jan. 2000.

[5.150] M.Antonini et al, "Image coding using the wavelet transform," *IEEE Trans. Image Processing*, vol.1, pp.205-220, April 1992.

[5.151] N.S.Jayant and P.Noll, *Digital Coding of Waveforms: Principles and Applications to Speech and Video*, New Jersey: Prentice-Hall, 1984.

[5.152] M.Rabbani and P.W.Jones, *Digital Image Compression Techniques*, Bellington, Washington, SPIE Optical Engineering Press, 1991.

[5.153] D.Santa-Cruz and T.Ebrahimi, "A study of JPEG2000 still image coding versus other standards," *Proc. of the EUSIPCO*, vol.2, pp.673-676, Tampere, Finland, Sept. 2000.

[5.154] C.Christopoulos and A.N.Skodras, "The JPEG2000 still image coding system: an overview," *IEEE Trans. Consumer Electronics*, vol.46, pp.1,103-1,127, November 2000.

[5.155] R.Koenen and F.Pereira, "MPEG-7: A standardized description of audiovisual content," *Signal Processing: Image Comm.*, vol.16, pp.5-13, Sept. 2000.

[5.156] ISO/IEC JTC1/SC29/WG11, Doc.N3752, *Overview of the MPEG-7 standard*, La Baule, France, Oct. 2000.

[5.157] ISO/IEC JTC1/SC29/WG11, Doc.N3751, *Introduction to MPEG-7*, La Baule, France, Oct. 2000.

[5.158] ISO/IEC JTC1/SC29/WG11, Doc.2727, *MPEG-7 Requirements*, Seoul, Korea, March 1999.

[5.159] ISO/IEC JTC1/SC29/WG11, Doc.2460, *MPEG-7 Context and Objectives V.10*, Atlantic City, Oct. 1998.

[5.160] ISO/IEC JTC1/SC29/WG11, Doc.2691, *Description of Core Experiments for MPEG-7 Color/Texture Descriptors*, Seoul, Korea, March 1999.

[5.161] ISO/IEC JTC1/SC29/WG11, Doc.2690, *Description of Core Experiments for MPEG-7 Shape/Motion Descriptors*, Seoul, Korea, March 1999.

[5.162] P.Salambier and F.Marques, "Image and video segmentation tools for new multimedia services," *IEEE Trans. CSVT*, vol.9, pp.1,147-1,169, Dec. 1999.

[5.163] H.K.Kim and J.D.Kim, "Region-based shape descriptor invariant to rotation, scale and translation," *Signal Processing: Image Comm.*, vol.16, pp.87-93, Sept. 2000.

[5.164] S.Jeannin et al, "Motion descriptors for content-based video representation," *Signal Processing: Image Comm.*, vol.16, pp.59-85, Sept. 2000.

[5.165] W.Zhao et al, "A reliable descriptor for face objects in visual content," *Signal Processing: Image Comm.*, vol.16, pp.123-136, Sept. 2000.

[5.166] ISO/IEC JTC1/SC29/WG11 Doc.2728, *Applications for MPEG-7*, Seoul, Korea, March 1999.

[5.167] *Special Issue on MPEG-7, IEEE Trans. CSVT*, vol.11, June 2001.

[5.168] ISO/IEC N4041, *MPEG-21 Overview*, Singapore, March 2001.

[5.169] ISO/IEC M7490, *Relationship Between the MPEG-21 and Mediacom 2004 Projects*, Sydney, Australia, July 2001.

[5.170] K.Asatani and S.Nogami, "Trends in the standardization of telecommunications on GII, multimedia and other network technologies and services," *IEEE Comm. Magazine*, vol.34, pp.32-46, June 1996.

[5.171] S.Okubo et al, "ITU-T standardization of audiovisual communication systems in ATM and LAN environments," *IEEE J. Selected Areas in Comm.*, vol.15, pp.965-982, Aug. 1997.

[5.172] ITU-T Recommendation H.261, *Video Codec for Audio-Visual Services at px64 Kb/s*, March 1993.

[5.173] ITU-T Recommendation H.320, *Narrow-Band Visual Telephone Systems and Terminal Equipment*, March 1996.

[5.174] ITU-T Recommendation H.263, *Video Coding for Low Bit Rate Communication*, March 1996.

[5.175] K.Rijkse, "H.263: Video coding for low bit-rate communications," *IEEE Trans. Comm.*, vol.34, pp.42-45, Dec. 1996.

[5.176] ITU-T Draft Recommendation H.263 Version 2, *H.263+ Video Coding for Low Bitrate Communication*, Sept. 1997.

[5.177] ITU-LBC-97-094, *Draft 10 of H.263+*, H.263+ Video group, Nice, France, Feb. 1997.

[5.178] B.Erol, M.Gallant and F.Kossentini, "The H.263+ video coding standard: complexity and performances," DCC, Snowbird, UT, March 1998.

[5.179] M.Yamashita, N.D.Kenyon and S.Okubo, "Standardization of multipoint audiovisual systems in CCITT," *Proc. IMAGE COM,* pp.154-159, Bordeaux, France, March 1993.

[5.180] ITU-T Recommendation H.324, *Terminal for low bit rate multimedia communications*, 1995.

[5.181] Draft ITU-T Recommendation H.323, *Visual Telephone Systems and Equipment for Local Area Networks Which Provide a Non-Guaranteed Quality of Service*, May 1996.

[5.182] ITU-T Recommendation H.310, *Broadband Audio-Visual Communications Systems and Terminal Equipment*, 1995.

[5.183] ITU-T Recommendation G.723.1, *Dual Rate Speech Coder for Multimedia Communication Transmitting at 5.3 and 6.3 Kbit/s*, 1996.

[5.184] ITU-T Recommendation H.223, *Multiplexing Protocol for Low Bitrate Multimedia Communication*, 1996.

[5.185] M.H.Willebeek-Le Mair and Z.Y.Shae, "Videoconferencing over packet-based networks," *IEEE J. Selected Areas in Comm.*, vol.15, pp.1,101-1,114, Aug. 1997.

[5.186] H.K.Phyffer, "ISDN with broadband capabilities," *Telecomm. J.*, vol.57, pp.45-50, Jan. 1990.

[5.187] E.Bingioni, E.Cooper and R.Sansom, "Designing a practical ATM LAN," *IEEE Network Mag.*, vol.7, pp.32-39, March 1993.

[5.188] ITU-T Recommendation G.114, *One-Way Transmission Time*, 1993.

[5.189] ITU-T Recommendation H.321, *Adaptation of H.320 Visual Telephone Terminals to BISDN Environments*, 1996.

[5.190] ITU-T Recommendation H.245, *Control Protocol for Multimedia Communication*, 1996.

[5.191] *IEEE Standard for Local and Metropolitan Area Networks - Supplement to Integrated Services (IS) LAN Interface at the Medium Access Control (MAC) and Physical (PHY) Layers: Specification of ISLAN 16-T*, IEEE Standard 802.90-1995.

[5.192] ITU-T Recommendation H.321, *Multipoint Control Units for Audiovisual Systems Using Digital Channels up to 2 Mbit/s*, 1996.

[5.193] ITU-T Recommendation H.331, *Broadcasting Type Audio-Visual Multipoint Systems and Terminal Equipment*, 1993.

[5.194] ISO/IEC 8802-5 (ANSI/IEEE Std.802.5-1992), *Information technology - local and metropolitan area networks - Part 5: Token ring access method and physical layer specifications*, 1992.

[5.195] IEEE Standard 802.1i, *Local area network MAC bridges - fiber distributed data interface (FDDI)*, 1992.

[5.196] ITU-T Recommendation V.70, *Procedures for the simultaneous transmission of data and digitally encoded voice signals over the general switched telephone network or over 2-wire leases point-to-point telephone type circuit*, 1996.

[5.197] ITU-T Recommendation H.225.0, *Media system packetization and synchronization on non-guaranteed quality of service LANs*, 1996.

[5.198] S.Wenger, et al., "Eror resilience support in H.263+," *IEEE Trans. CSVT*, vol.8, pp.867-877, Nov. 1998.

[5.199] ITU-T Recommendation H.261, *Video codec for audiovisual services at px64 Kbit/s, Geneva, 1990, revised in Helsinki*, March 1993.

[5.200] K.R.Rao and P.Yip, *Discrete cosine transform*, New York, NY: Academic Press, 1990.

[5.201] A.N.Netravali and J.D.Robbins, "Motion-compensated television coding: Part I," *Bell Systems Technical J.*, vol.58, pp.631-670, March 1979.

[5.202] ITU-T Recommendation H.242, *System for establishing communication between audiovisual terminals using digital channels up to 2 Mbit/s*, 1993.

[5.203] J.H.Witten, R.M.Neal and J.G.Cleary, "Arithmetic coding for data-compression," *Communications of the ACM*, vol.30, pp.520-540, June 1987.

[5.204] M.T.Orchard and G.J.Sullivan, "Overlapped block motion compensation - an estimation theoretic approach," *IEEE Trans. Image Processing*, vol.3, pp.693-699, Sept. 1994.

[5.205] M.Willebeek-Le Mair, Z.Y.Shae and Y.C.Chang, "Robust H.263 video coding for transmission over the Internet," IBM Res. Rep. RC20532, Poughkeepsie, New York, Aug. 1996.

[5.206] T.Chen, C.T.Swain and B.G.Haskell, "Coding of sub-regions for content-based scalable video," *IEEE Trans. CSVT*, vol.7, pp.256-260, Feb. 1997.

[5.207] ITU-T, Study Group 16, Video Coding Experts Group (Question 15), Doc.Q15F09, *Report of the Ad Hoc Committee H.263++ Development*, Seoul, Korea, Nov. 1998.

[5.208] ITU-T, Study Group 16, Video Coding Experts Group (Question 15), Doc.Q15D62, *Recommended Simulation Conditions for H.263v3*, Tampere, Finland, April 1998.

[5.209] ITU-T, Study Group 16, Video Coding Experts Group (Question 15), Doc.Q15D65, *Video Codec Test Model, Near-Term, Version 10 (TMN10)*, Draft 1, Tampere, Finland, April 1998.

[5.210] ITU-T, Study Group 16, Video Coding Experts Group (Question 15), Doc.Q15F10, *Report of the Ad Hoc Committee H.26L development*, Seoul, Korea, Nov. 1998.

[5.211] ITU-T, Study Group 16, Video Coding Experts Group (Question 15), Doc.Q15D62, *Call for Proposals for H.26L Video Coding*, Geneva, Switzerland, Jan. 1998.

[5.212] ITU-T, Study Group 16, Video Coding Experts Group (Question 6), Doc.VCEG-L24, *Multihypothesis Motion Pictures for H.26L*, Erlangen, Germany, Jan. 2001.

[5.213] R.V.Cox and P.Kroon, "Low bit-rate speech coders for multimedia communication," *IEEE Comm. Magazine*, vol.34, pp.34-41, Dec. 1996.

[5.214] ITU-T Recommendation H.222.1, *Multimedia multiplex and synchronization for audiovisual communication in ATM environments*, 1996.

[5.215] ITU-T Recommendation H.221, *Frame structure for a 64-1920 Kbit/s channel in audiovisual teleservices*, 1997.

[5.216] ITU-T Recommendation H.223, *Multiplexing protocol for low bit rate multimedia communication*, 1997.

[5.217] H.Schulzine et al, *RTP: A transport protocol for real-time applications*, RFC1889, Jan. 1996.

[5.218] ITU-T Recommendation X.680, *Information Technology - Abstract Syntax Notation One (ASN.1) - Specification of Basic Notation*, 1994.

[5.219] M.Ohta, "IETF and Internet standards," *IEEE Comm. Magazine*, vol.36, pp.126-129, Sept. 1998.

[5.220] B.Carpenter, *Architectural principle of the Internet*, RFC1958, June 1996.

[5.221] R.Hevey and S.Bradner, *The organization involved in the IETF standards process*, RFC2028, Oct. 1996.

[5.222] C.Huitema, J.Postel and S.Crocker, "Not all RFCs are standards," RFC1796, April 1995.

[5.223] J.Postel, T.Li and Y.Renhter, *Best current practices*, RFC1818, Aug. 1995.

[5.224] S.Braduer Ed, *The Internet standards process - Revision 3*, RFC2026, Oct. 1996.

References

[5.225] W.Stallings, "IPv6: the new Internet protocol," *IEEE Comm. Magazine*, vol.34, pp.96-108, July 1996.

[5.226] M.Decina and V.Trecordi, "Convergence of telecommunications and computing to networking models for integrated services and applications," *Proc. of the IEEE*, vol.85, pp.1,887-1,914, Dec. 1997.

[5.227] M.Handley and V.Jacobson, *SDP: Session description protocol*, IETF MMUSIC Group, Internet Draft, draft-ietf-mmusic-sdp-04.txt.

[5.228] M.Handly, *SAP: Session enhancement protocol*, IETF MMUSIC Group Draft, draft-ietf-mmusic-sap-00.txt.

[5.229] E.Schooler, H.Schulzinne, and M.Handley, *SIP: Session initiation protocol*, IETF MMUSIC Group, Internet Draft, draft-ietf-mmusic-sip-04.txt.

[5.230] H.Schulzinne, A.Rao and R.Lanthier, *Real time streaming protocol (RTSP)*, IETF MMUSIC Group, Internet Draft, draft-ietf-mmusic-rtsp-05.txt.

[5.231] M.Handley et al, *The Internet multimedia conferencing architecture*, IETF MMUSIC Group, Internet Draft, draft-ietf-mmusic-confarc-00.txt.

[5.232] H.Schulzinne, *RTP: A transport protocol for real-time applications*, RFC1889, Jan. 1996.

[5.233] H.Schulzinne, *RTP profile for audio and video conferences with minimal control*, RFC1890, Jan. 1996.

[5.234] D.Clark and D.Tennenhouse, "Architecture considerations for a new generation of protocols," Proc. of ACM SIGCOM'90, pp.201-208, Sept. 1990.

[5.235] D.Mills, *Network time protocol version 3*, IETF RFC1305, March 1992.

[5.236] D.E.Commer, *Interworking with TCP/IP*, New Jersey: Prentice-Hall, 1991.

[5.237] R.Civanlar, A.Basso and C.Herpel, *RTP payload format for MPEG-4 streams*, IETF Internet Draft, draft-ietf-avt-rtp-mpeg401.txt, Feb. 1999.

[5.238] ISO/IEC JTC1/SC29/WG11 N4081, *Carriage of MPEG-4 contents over IP networks*, March 2001.

[5.239] D.Milovanovic and Z.S.Bojkovic, "MPEG-4 video transmission over Internet," *Proc. TELSIKS'99*, pp.309-312, Nis, Yugoslavia, Oct. 1999.

[5.240] T.D.C.Little and A.Ghafoor, "Multimedia synchronization protocols for broadband integrated services," *IEEE J. Selected Areas in Comm.*, vol.9, pp.1,368-1,382, Dec.1991.

Chapter 6
Multimedia Communications Across Networks

[6.1] N.Jayant, "High quality networking of audio-visual information," *IEEE Commun. Magazine*, vol.31, pp.84-95, Sept. 1993.

[6.2] P.Gonet, "Fast packet approach to integrated broadband networks," *Networks*, vol.9, pp.292-298, Dec. 1986.

[6.3] M.Devanlt et al, "The 'prelude' ATD experiment assessments and future prospects," *IEEE J. Selected Areas in Comm.*, vol.4, pp.1,528-1,532, Dec. 1986.

[6.4] J.Sidron and J.S.Gotal, "PARIS: An approach to integrated highspeed private networks," *Int. J. Digital Analog Cable Syst.*, vol.1, pp.77-85, Jan.-March 1988.

[6.5] P.Brady, "A model for generating on-off patterns in two-way communications," *Bell Syst. Tech. J.*, vol.48, pp.2,445-2,472, Sept. 1969.

[6.6] M.Listanti and F.Villani, "An X.25 compatible protocol for packet voice communications," *Comput. Comm.*, vol.6, pp.23-31, Feb. 1983.

[6.7] B.Gold, "Digital speech networks," *Proc. of the IEEE*, vol.65, pp.1,630-1,658, Dec. 1977.

[6.8] D.Minoli, "Optimal packet length for packet voice communication," *IEEE Trans. Commun.*, vol.COM-27, pp.607-611, 1979.

[6.9] T.Bially et al, "A technique for adaptive voice flow control in integrated packet networks," *IEEE Trans. Comm.*, vol.COM-28, pp.325-333, March 1980.

[6.10] T.S.Chen, Y.Walrand and D.G.Messerchmitt, "Dynamic priority protocols for packet voice," *IEEE J. Selected Areas in Comm.*, vol.7, pp.632-643, June 1989.

[6.11] W.Mounteomery, "Techniques for packet video synchronization," *IEEE J. Selected Areas in Comm.*, vol.1, pp.1,022-1,028, Dec. 1983.

[6.12] G.Barbeis and D.Pazzaglia, "Analysis and optimal design of a packet voice receiver," *IEEE Trans. Comm.*, vol.COM-28, pp.217-227, Feb. 1980.

[6.13] T.Suda et al, "Performance evaluation of a packetized voice system - simulation study," *IEEE Trans. Comm.*, vol. COM-32, pp.97-102, Jan. 1984.

[6.14] D.W.Patr, L.A.DaSilva and V.S.Frost, "Priority discarding of speech in integrated packet networks," *IEEE J. Selected Areas in Commun.*, vol.7, pp.644-656, June 1989.

[6.15] G.Karlsson, *Asynchronous Transfer of Video*, SICS Research Report R95:14, Sweden, 1997.

[6.16] ATM Forum, *ATM User-Network Interface Specification, Version 3.0*, Mountain View, California.

[6.17] A.R.Reibman and B.G.Haskell, "Constraints on variable bit rate video for ATM networks," *IEEE Trans. CSVT*, vol.2, pp.361-372, Dec. 1992.

[6.18] B.G.Haskell, "Buffer and channel sharing by several interframe picturephone coders," *Bell Systems Tech. J.*, vol.51, pp.261-289, Jan. 1972.

[6.19] Y.Zdepsky, D.Raychaudhuri and K.Joseph, "Statistically based buffer control policies for constant rate transmission of compressed digital video," *IEEE Trans. Comm.*, vol.39, pp.947-957, June 1991.

[6.20] J.P.Leduc and S.D'Agostino, "Universal VBR video codecs for ATM networks in the Belgian broadband experiment," *Signal Processing: Image Comm.*, vol.3, pp.157-165, June 1991.

[6.21] C.T.Chen and A.Wong, "A self-governing rate buffer control strategy for pseudoconstant bit rate video coding," *IEEE Trans. Image Proc.*, vol.2, pp.50-59, Jan. 1993.

[6.22] K.H.Tzou, "An intrafield DCT-based HDTV coding for ATM networks," *IEEE Trans. CSVT*, vol.1, pp.184-196, June 1991.

[6.23] A.Ortega, K.Ramchandran and M.Vetterli, "Optimal trellis-based buffered compression and fast approximations," *IEEE Trans. Image Proc.*, vol.3, pp.26-40, Jan. 1994.

[6.24] A.Eleftheriadis, S.Petajan and D.Anastassiou, "Algorithms and performance evaluation of the Xphone multimedia communication system," *Proc. of the ACM Multimedia Conf.*, Anaheim, CA, pp.311-320, Aug. 1993.

[6.25] Y.C.Bolot and T.Turletti, "A rate control mechanism for packet video in the Internet," *Proc. of Infocom'94*, Toronto, Canada, pp.1,216-1,223, June 1994.

[6.26] M.Macedonia and D.Brutzman, "MBONE provides audio and video across the Internet," *Computer*, vol.27, pp.30-36, April 1994.

[6.27] B.Maglaris et al, "Performance models of statistical multiplexing in packet video communications," *IEEE Trans. Commun.*, vol.30, pp.834-843, July 1988.

[6.28] D.P.Heyman, A.Tabatabai and T.Lakshman, "Statistical analysis and simulation study of video teleconferencing traffic in ATM networks," *IEEE Trans. CSVT*, vol.2, pp.49-59, March 1992.

[6.29] A.Ortega, "Video transmission over ATM networks," *Microsystems Technology for Multimedia Applications,* (Shen et al, Eds.), Piscataway, New Jersey: IEEE Press, May 1995.

[6.30] E.P.Rathgeb, "Modeling and performance comparison of policing mechanisms for ATM networks," *IEEE J. Selected Areas in Comm.*, vol.9, pp.325-334, April 1991.

[6.31] M.Vetterli and K.M.Uz, "Multiresolution coding techniques for digital television: A review," Special issues on Multidimensional Processing of Video Signals, Multidimensional Systems and Signal Processing, pp.161-187, March 1992.

[6.32] W.M.Garrett and M.Vetterli, "Joint source/channel coding of statistically multiplexed real time services on packet networks," *IEEE/ACM Trans. Networking*, vol.1, pp.71-80, Feb. 1993.

[6.33] K.Ramchandran et al, "Multiresolution broadcast for digital HDTV using joint source-channel coding," *IEEE J. Selected Areas in Comm.*, vol.11, pp.6-23, Jan. 1993.

[6.34] Q.F.Zhu, Y.Wang and L.Shaw, "Coding and cell-loss recovery in DCT-based packet video," *IEEE Trans. CSVT*, vol.3, pp.248-258, June 1993.

[6.35] M.Ghanbari and V.Seferidis, "Cell-loss concealment in ATM video codecs," *IEEE Trans. CSVT*, vol.3, pp.238-247, June 1993.

[6.36] S.R.McCanne, "Scalable compression and transmission of Internet multicast video," Ph.D. Thesis, University of California, Berkeley, 1996.

[6.37] T.Chiang and D.Anastassiou, "Hieararchical coding of digital television," *IEEE Comm. Magazine*, vol.32, pp.38-45, May 1994.

[6.38] W.Werbiest, "Video coding in an ATM environment," *Third Int. Conference on New Systems and Services in Telecommun.*, Liege, Belgium, Nov. 1986.

[6.39] M.Dutoncheel and W.Verbiest, *Simulation results for a hybrid transform video coding algorithm,* RACE Project 2023, BTM-A 11-05-PR, 1986.

[6.40] Y.Wang and Q.Zhu, "Error control and concealment for video communication: a overview," *Proc. of the IEEE*, vol.86, pp.974-997, May 1998.

[6.41] Y.Wang et al, "Error resilient video coding techniques," *IEEE Signal Proc. Magazine*, vol.17, pp.61-82, July 2000.

[6.42] G.Cote, S.Shirami and F.Kossentini, "Optimal mode selection and synchronization for robust video communications over error phone networks," *IEEE J. Selected Areas in Comm.*, vol.18, pp.952-965, June 2000.

[6.43] S.Wenger et al, "Error resilience support in H.263+," *IEEE Trans. CSVT*, vol.8, pp.867-877, Nov. 1998.

[6.44] I.Kondi, F.Ishtiag and A.K.Katsaggelos, "Joint source-channel coding for scalable video," *Proc. 2000 SPIE Conf. Visual Communications and Image Processing,* pp.324-335, San Jose, CA, Jan. 2000.

[6.45] Q.Zhu and Y.Wang, "Error concealment in visual communication," *Compressed Video Networks,* (A.R.Reibman and M.T.Sun, Eds.), New York, NY: Marcel Dekker, 2000.

[6.46] S.Agni, "Error concealment for MPEG-2 video," *Signal Recovery Techniques for Image and Video Compression and Transmission,* A.K.Katsaggelos and N.P.Galatsanos, (Eds.), pp.235-268, Nowell, MA: Kluwer, 1998.

[6.47] M.C.Hong et al, "Video error concealment techniques," *Signal Processing: Image Comm.*, vol.14, pp.437-492, May 1999.

[6.48] D.W.Redmill and N.G.Kingsbury, "The EREC: an error-resilient technique for coding variable-length block of data," *IEEE Trans. Image Processing*, vol.5, pp.565-574, April 1996.

[6.49] H.J.Lee, T.Chiang and Y.Q.Zhang, "Scalable rate control for MPEG-2 video," *IEEE Trans. CSVT*, vol.10, pp.878-894, Sept. 2000.

[6.50] D.Ferrari and D.Verma, "A scheme for real-time channel establishment in wide-area networks," *IEEE J. Selected Areas in Comm.*, vol.8, pp.368-379, Apr. 1990.

[6.51] M.R.Pickering and J.F.Arnold, "A perceptually efficient VBR rate control algorithm," *IEEE Trans. Image Proc.*, vol.3, pp.527-531, Sept. 1994.

[6.52] A.Ortega et al, "Rate constraints for video transmission over ATM networks based on joint source/network criteria," *Annales des Telecommunications,* vol.50, pp.603-616, July-Aug. 1995.

[6.53] Y.Shoham and A.Gersho, "Efficient bit allocation for an arbitrary set of quantizers," *IEEE Trans. ASSP*, vol.36, pp.1,445-1,453, Nov. 1988.

[6.54] K.Ramchandran, A.Ortega and M.Vetterli, "Bit allocation for dependent quantization with applications to multiresolution and MPEG video coders," *IEEE Trans. Image Process.*, vol.3, pp.533-545, Sept. 1994.

[6.55] J.Choi and D.Park, "A stable feedback control of the buffer state using the controlled Lagrange multiplier method," *IEEE Trans. Image Process.*, vol.3, pp.546-558, Sept. 1994.

[6.56] Y.L.Liu and A.Ortega, "Bit rate control using piecewise approximated rate-distortion characteristics," *IEEE Trans. CSVT*, vol.8, pp.446-459, Aug. 1998.

[6.57] W.Ding, "Rate control of MPEG video coding and recording by rate quantization modeling," *IEEE Trans. CSVT*, vol.6, pp.12-20, Feb. 1996.

[6.58] B.Tao, H.A.Peterson and B.W.Dickinson, "A rate-quantization model for MPEG encoders," *Proc. IEEE ICIP*, vol.1, pp.338-341, Oct. 1997.

[6.59] K.H.Yang, A.Jacquin and N.S.Jayant, "A normalized rate distortion model for H.263-compatible codecs and its application to quantizer selection," *Proc. IEEE ICIP*, vol.1, pp.41-44, Oct. 1997.

[6.60] A.Vetro, H.F.Sun and Y.Wang, "MPEG-4 rate control for multiple video objects," *IEEE Trans. CSVT*, vol.9, pp.186-199, Feb. 1999.

[6.61] J.Ribas-Corbera and S.M.Lei, "Contribution to rate control Q2 experiment: A quantization control tool for achieving target bitrate accurately," *Coding Moving Pictures and Associated Audio*, MPEG96/M1812 ISO/IEC JTC1/SC29/WG11, Sevilla, Spain, Feb. 1997.

[6.62] H.J.Lee, T.Chiang and Y.Q.Zhang, "Scalable rate control for very low bit rate video," *Proc. IEEE ICIP*, vol.2, pp.768-771, Oct. 1997.

[6.63] T.Chang and Y.Q.Zhang, "A new rate control scheme using a new rate-distortion model," *IEEE Trans. CSVT*, vol.7, pp.246-250, Feb. 1997.

[6.64] A.Viterbi and J.Omura, "A new rate control scheme using a new rate distortion model," in *Principles of Digital Communication and Coding*, New York, NY: McGraw-Hill, 1979.

[6.65] D.Wu et al, "Streaming video over the Internet: approaches and directions," *IEEE Trans. CSVT*, vol.11, pp.1-20, Feb. 2001.

[6.66] S.McCanne, V.Jacobson and M.Vetterli, "Receiver-driven layered multicast," *Proc. ACM SIGCOMM'96,* pp.117-130, Aug. 1996.

[6.67] D.Taubman and A.Zakhor, "A common framework for rate and distortion based scaling of highly scalable compressed video," *IEEE Trans. CSVT*, vol.6, pp.329-354, Aug. 1996.

[6.68] B.J.Kim, Z.Xiong and W.A.Pearlman, "Low bit-rate scalable video coding with 3D set partitioning in hierarchical trees (3D SPIHT)," *IEEE Trans. CSVT*, vol.10, pp.1,374-1,387, Dec. 2000.

[6.69] X.Wu, S.Cheng and Z.Xiong, "On packetization of embedded multimedia bitstreams," *IEEE Trans. Multimedia, Special Issue on Multimedia over IP*, vol.3, pp.132-140, March 2001.

[6.70] S.Li, F.Wu and Y.Q.Zhang, "Study of a new approach to improve FGS video coding efficiency," ISO/IEC JTC1/SC29/WG11, MPEG99/M5583, Dec. 1999.

[6.71] W.Li, "Bit-plane coding of DCT coefficients for fine granularity scalability," ISO/IEC JTC1/SC29/WG11 MPEG98/M3989, Oct. 1998.

[6.72] W.Li, "Streaming video profile in MPEG-4," *IEEE Trans. CSVT*, vol.11, Feb. 2000. Special session on Media Streaming, International Conf. on Information Technology: Coding and Computing 2001, Las Vegas, Nevada, April 2001.

[6.73] W.Tan and A.Zakhor, "Real-time Internet video using error resilient scalable comrpession and TCP-friendly transport protocol," *IEEE Trans. Multimedia*, vol.1, pp.172-186, June 1999.

[6.74] F.Wu, S.Li and Y.Q.Zhang, "A framework for efficient progressive fine granularity scalable video coding," *IEEE Trans. CSVT*, vol.11, pp.332-344, March 2001.

[6.75] G.J.Conklin et al, "Video coding for streaming media delivery on the Internet," *IEEE Trans CSVT*, vol.11, pp.269-281, March 2001.

[6.76] C.W.Lin et al, "MPEG video streaming with VCR functionality," *IEEE Trans CSVT*, vol.11, pp.415-425, March 2001.

[6.77] D.Wu, Y.T.Hou and Y.Q.Zhang, "Transporting real-time video over the Internet: challenges and approaches," *Proc. IEEE*, vol.88, pp.1,855-1,877, Dec. 2000.

[6.78] D.Wu et al, "On end-to-end architecture for transporting MPEG-4 video over the Internet," *IEEE Trans. CSVT*, vol.10, pp.923-941, Sept. 2000.

[6.79] J.C.Bolot, T.Turletti and J.Wakeman, "Scalable feedback control for multicast video distribution in the Internet," Proc. ACM SIGCOMM'94, London, UK, pp.58-67, Sept. 1994.

[6.80] T.Turletti and C.Huitene, "Videoconferencing on the Internet," *IEEE/ACM Trans. Networking*, vol.4, pp.340-351, June 1996.

[6.81] S.Floyd and K.Fall, "Promoting the use of end-to-end congestion control in the Internet," *IEEE/ACM Trans. Networking*, vol.7, pp.458-472, Aug. 1999.

[6.82] S.Y.Cheng, M.Ammar and X.Li, "On the use of destination set grouping to improve fairness in multicast video distribution," *Proc. IEEE INFOCOM'96*, pp.553-560, March 1996.

[6.83] L.Fan et al, "Summary cache: a scalable wide-area web cache sharing protocol," *IEEE/ACM Trans. Networking*, vol.8, pp.281-293, June 2000.

[6.84] R.Steinmetz and K.Nahrstedt, *Multimedia: Computing, Communications and Applications*, Upper Saddle River, NJ: Prentice Hall, 1995.

[6.85] G.Blakowski and R.Steinmetz, "A media synchronization survey: reference model, specification, and case studies," *IEEE J. Selected Areas in Comm.*, vol.14, pp.5-35, Jan. 1996.

[6.86] H.Schulzrinne, A.Rao and R.Lanphier, *Real Time Streaming Protocol (RTSP)*, Internet Engineering Task Force, RFC2326, April 1998.

[6.87] M.Handley et al, *SIP: Session Initiation Protocol*, Internet Engineering Task Force, RFC2543, March 1999.

[6.88]   H.Schulzrinne et al, *RTP: A Transport Protocol for Real-Time Applications*, Internet Engineering Task Force, RFC1889, Jan. 1996.

[6.89]   M.Orzessek and P.Sommer, *ATM and MPEG-2: Integrating Digital Video into Broadband Networks*, Upper Saddle River, NJ: Prentice Hall PTR, 1998.

[6.90]   K.R.Rao and Z.S.Bojkovic, *Packet Video Communications over ATM networks*, Upper Saddle River, NJ: Prentice Hall PTR, 2000.

[6.91]   Y.Feng, H.Mehrpour and R.T.Lo, "Statistical multiplexing schemes for MPEG video sources," *Proc. IEEE ICCS/ISPACS'96*, pp.1,501-1,505, Singapore, Nov. 1996.

[6.92]   D.Reininger et al, "Statistical multiplexing of VBR MPEG compressed video on ATM networks," *Proc. IEEE INFOCOM*, vol.3, pp.919-926, San Francisco, CA, March 1993.

[6.93]   M.Krunz, R.Sass and H.Hughes, "Statistical characteristics and multiplexing of MPEG streams," *Proc. IEEE INFOCOM*, pp.455-462, April 1995.

[6.94]   M.R.Ismail et al, "Modeling prioritized MPEG video using TES and a frame scheduling strategy for transmission in ATM networks," *Proc. INFOCOM*, pp.762-770, April 1995.

[6.95]   C.Gao and S.S.Meditch, "Two-layer video coding and priority statistical multiplexing over ATM networks," *Proc. IEEE ICC*, pp.127-136, Dallas, TX, June 1996.

[6.96]   N.Schroff and M.Schwartz, "Video modeling within networks using deterministic smoothing at the source," *Proc. IEEE INFOCOM*, vol.1, pp.342-349, June 1994.

[6.97]   D.L.Mils, "Improved algorithms for synchronizing computer network clocks," *IEEE/ACM Trans. Networking*, vol.3, pp.245-254, June 1995.

[6.98]   P.P.Singh et al, "Jitter and clock recovery for periodic traffic in broadband packet networks," *IEEE Trans. Commun.*, vol.42, pp.2,189-2,196, May 1994.

[6.99]   J.M.Simmons and R.G.Gallagher, "Design of error detection scheme for class C service in ATM," *IEEE/ACM Trans. Networking*, vol.2, pp.80-88, Feb. 1994.

[6.100] ITU-T SG XVIII, *Performance evaluation results on cell loss in ATM networks*, Doc.1047, Nov. 1990.

[6.101] M.Wada, "Selective recovery of video packet loss using error concealment," *IEEE J. Selected Areas in Comm.*, vol.7, pp.207-214, June 1989.

[6.102] M.Ghanbari and V.Sferidis, "Cell-loss concealment in ATM networks," *IEEE Trans. CSVT*, vol.3, pp.238-247, June 1993.

[6.103] M.Ghanbari and C.Hughes, "Packing coded video signals into ATM cells," *IEEE/ACM Trans. Networking*, vol.1, pp.505-508, Oct. 1993.

[6.104] Q.Zhu, Y.Wang and L.Shaw, "Coding and cell loss recovery in DCT based packet video," *IEEE Trans. CSVT*, vol.3, pp.248-258, June 1993.

[6.105] A.S.Tom, C.L.Yeh and F.Chu, "Packet video for cell loss protection using deinterleaving and scrambling," *Proc. IEEE ICASSP*, pp.2857-2860, Toronto, Canada, May 1991.

[6.106] Y.Wang, Q.Zhu and L.Shaw, "Maximally smooth image recovery in transform coding," *IEEE Trans. Commun.*, vol.41, pp.1,544-1,551, Oct. 1993.

[6.107] Y.Wang and Q.Zhu, "Signal loss recovery in DCT-based image and video coders," *Proc. SPIE Conf. on Visual Communications and Image Processing*, pp.667-678, Boston, MA, Nov. 1991.

[6.108] H.Sun and J.Zdepski, "Adaptive error concealment algorithm for MPEG compressed video," *Proc. SPIE Conf. on Visual Communications and Image Processing*, pp.814-824, Boston, MA, Nov. 1992.

[6.109] W.Kwok and H.Sun, "Multidirectional interpolation for spatial error concealment," *IEEE Trans. Consumer Electronics*, vol.3, pp.455-460, Aug. 1993.

[6.110] H.Sun and W.Kwok, "Concealment of damaged block transform coded images using projections onto convex sets," *IEEE Trans. Image Process.*, vol.4, pp.470-477, April 1995.

[6.111] P.Salama et al, "Error concealment techniques for encoded video streams," *Proc. IEEE ICIP*, vol.1, pp.9-12, Washington, DC, Oct. 1995.

[6.112] J.Lee and B.W.Dickinson, "Temporally adaptive motion interpolation exploiting temporal masking in visual perception," *IEEE Trans. Image Process.*, vol.3, pp.513-526, Sept. 1994.

[6.113] L.H.Kien and K.N.Ngan, "Cell-loss concealment techniques for layered video codecs in an ATM network," *IEEE Trans. Image Process.*, vol.3, pp.666-677, Sept. 1994.

[6.114] J.Zhang et al, "MPEG-2 video services for wireless ATM networks," *IEEE J. Selected Areas in Comm.*, vol.15, pp.119-128, Jan. 1997.

[6.115] E.Ayanoglu et al, "Performance improvement in the broadband networks using forward error correction for lost packets recovery," *J. of High-Speed Networks*, vol.1, pp.287-303, June 1993.

[6.116] E.Ayanoglu et al, "Forward error control for MPEG-2 video transport in a wireless ATM LAN," *Proc. IEEE ICIP*, vol.2, pp.833-836, Lausanne, Switzerland, 1996.

[6.117] M.Sudan and N.Shacham, "Gateway-based approach for managing multimedia sessions over heterogeneous signaling domains," *Proc. IEEE INFOCOM'97*, pp.702-711, IEEE Computer Soc. Press, Los Alamitos, CA, 1997.

[6.118] S.D.Servetto and K.Nahrstedt, "Broadcast quality video over IP," *IEEE Trans. Multimedia*, vol.3, pp.162-173, March 2001.

[6.119] A.Lombardo, G.Schembra and G.Morabito, "Traffic specifications for the transmission of stored MPEG video on the Internet," *IEEE Trans. Multimedia, Special Issue on Multimedia over IP*, vol.3, pp.5-17, March 2001.

[6.120] O.Hashida, Y.Takahashi and S.Shimogawa, "Switched batch Bernoulli process (SBBP) and the discrete-time SBBP/G/1 queue with application to statistical multiplexer," *IEEE J. Selected Areas in Comm.*, vol.9, pp.394-401, April 1991.

[6.121] M.Garnet and W.Willinger, "Analysis, modeling and generation of self-similar VBR video traffic," *Proc. ACM/SIGCOM*, pp.269-280, Aug. 1994.

[6.122] M.Furini and D.F.Towsley, "Real-time traffic transmission over the Internet," *IEEE Trans. Multimedia*, vol.3, pp.33-40, March 2001.

[6.123] W.C.Feng and S.Sechrest, "Critical bandwidth allocation for the delivery of compressed video," *Comp. Commun.*, vol.18, pp.709-717, Oct.1995.

[6.124] J.D.Sclahi et al, "Supporting stored video: Reducing rate variability and end-to-end resource requirements through optimal smoothing," *IEEE/ACM Trans. Networking*, vol.6, pp.397-410, Aug. 1998.

[6.125] M.Grossglanser, S.Keshow and D.Tse, "PCBR: A simple and efficient service for multiple time-scale traffic," *IEEE/ACM Trans. Networking*, vol.5, pp.741-755, Nov. 1997.

[6.126] W.C.Feng and J.Rexford, "A comparison of bandwidth smoothing techniques for the transmission of prerecorded compressed video," *Proc. IEEE INFOCOM 1997*, pp.58-66, Kobe, Japan, Apr. 1997.

[6.127] H.M.Radha et al, "Scalable internet video using MPEG-4," *Signal Processing: Image Comm.*, vol.15, pp.95-126, Sept. 1999.

[6.128] H.M.Radha, M.Van der Schaar and Y.Chen, "The MPEG-4 fine-grained scalable video coding method for multimedia streaming over IP," *IEEE Trans. Multimedia*, vol.3, pp.53-68, March 2001.

[6.129] S.J.Choi and J.W.Woods, "Motion-compensated 3D subband coding of video," *IEEE Trans. Image Processing*, vol.8, pp.155-167, Feb. 1999.

[6.130] S.McCanne, M.Vetterli and V.Jacobson, "Low-complexity video coding for receiver-driven layered multicast," *IEEE J. Selected Areas in Comm.*, vol.16, pp.983-1001, Aug. 1997.

[6.131] B.Girod et al, "Packet loss resilient Internet video streaming," VCIP'99, *Proc. SPIE*, vol.3,653, pp.833-844, Jan. 1999.

[6.132] ISO/IEC 14496-2, "MPEG-4 video FGS v.4.0," Noordwijkerhout, Netherlands, Proposed Draft Amendment (PDAM), March 2000.

[6.133] ISO/IEC 14496-2, "FGS Amendment WD v.10.0," Vancouver, BC, Canada, 48th MPEG Meeting, July 1999.

[6.134] ISO/IEC 14496-2, "Information technology coding of audio-visual objects: Visual," March 2000.

[6.135] M.van der Schaar, Y.Chen and H.M.Radha, "Embedded DCT and wavelet methods for fine granular scalable video: analysis and comparison," *IVCP2000, Proc. SPIE*, vol.2,974, pp.643-653, Jan. 2000.

[6.136] M.Yong, "Study of voice packet reconstruction methods applied in CELP speech coding," *Proc. IEEE ICASSP*, vol.2, pp.125-128, San Francisco, CA, March 1992.

[6.137] T.W.Leng, W.P.Blanc and S.A.Mahmoud, "Speech coding over frame relay network," *Proc. IEEE Int. Workshop on Speech Coding*, pp.75-76, Oct. 1993.

[6.138] A.Husain and V.Cuperman, "Reconstruction of missing packets for CELP-based speech coders," *Proc. IEEE ICASSP*, vol.1, pp.245-248, 1995.

[6.139] C.R.Watkins and J.H.Chen, "Improving 16kbps G.728 LD-CELP speech coder for frame erasure channels," *Proc. IEEE ICASSP*, vol.1, pp.241-245, 1995.

[6.140] J.F.Wang et al, "A voice-driven packet loss recovery algorithm for analysis-by-synthesis predictive speech coders over Internet," *IEEE Trans. Multimedia*, vol.3, pp.98-107, March 2001.

[6.141] W.B.Kleijn and K.K.Paliwal, *Speech Coding and Synthesis*, Amsterdam, Netherlands: Elsevier Science, 1995.

[6.142] P.Merriman, "Video over DSL architecture," Alcatel Telecommunications Review, pp.250-257, 4th Quarter 2000.

[6.143] M.Verhoeyen, "Delivering voice services over DSL," Alcatel Telecommunications Review, pp.244-249, 4th Quarter 2000.

[6.144] H.Zhang and K.J.R.Lin, "Robust image and video transmission over spectrally shaped channels using multicast modulation," *IEEE Trans. Multimedia*, vol.1, pp.88-103, March 1999.

[6.145] H.Zhang and K.J.R.Lin, "Multimedia services over digital subscriber lines," *IEEE Signal Processing Magazine*, vol.17, pp.44-60, July 2000.

[6.146] T.Starr, J.Cioffi and P.Silverman, *Understanding Digital Line Technologies*, Upper Saddle River, NJ: Prentice Hall PTR, 2000.

[6.147] K.Maxwell, *Residential broadband, A User's Guide to the Battle for the Last Mile*, New York: John Wiley, 1999.

[6.148] W.Goralski, *ADSL and DSL Technologies*, New York, NY: McGraw-Hill, 1998.

[6.149] A.D.Roy, "Cable: it's not just for TV," *IEEE Spectrum*, vol.36, pp.53-59, May 1999.

[6.150] A.D.Roy, "Ringing from the Internet," *IEEE Spectrum*, vol.36, pp.32-38, May 1999.

[6.151] R.L.Freeman, *Telecommunications Transmission Handbook*, New York, NY: JohnWiley, 1998.

[6.152] R.G.Winch, *Telecommunications Transmission Systems*, New York, NY: McGraw-Hill, 1998.

[6.153] K.Feher, *Wireless Communications – Modulation and Spread Spectrum Applications*, Upper Saddle River, NJ: Prentice Hall, 1995.

[6.154] V.O.Klee and X.X.Oui, "Personal communication systems (PCS)," *Proc. of the IEEE*, vol.83, pp.1,210-1,243, July 1995.

[6.155] L.Hanzo, "Bandwidth efficient wireless multimedia communications," *Proc. of the IEEE*, vol.86, pp.1,342-1,382, July 1998.

[6.156] J.Mikkonen et al, "Emerging wireless broadband networks," *IEEE Comm. Magazine*, vol.36, pp.112-117, Feb. 1998.

[6.157] K.Palavan et al, "Wideband local access: wireless LAN and wireless ATM," *IEEE Commun. Magazine*, vol.35, pp.34-40, Nov.1997.

[6.158] L.Correia and R.Prasad, "An overview of wireless broadband communications," *IEEE Commun. Magazine*, vol.35, pp.28-33, Jan. 1997.

[6.159] D.J.Goodman and D.Raychaudhuri, (Eds.), *Mobile Multimedia Communications*, New York: Plenum Press, 1997.

[6.160] Y.Waugand and Q.F.Hu, "Error control and concealment for video communication: a review," *Proc. of the IEEE*, vol.86, pp.974-977, May 1998.

[6.161] M.Chelouche et al, "Digital wireless broadband corporate and private networks: RNET concepts and applications," *IEEE Comm. Magazine*, vol.35, pp.42-51, Jan. 1997.

[6.162] M.Naghshinehand and M.W.LeMair, "End-to-end QoS provisioning multimedia wireless mobile networks using an adaptive framework," *IEEE Comm. Magazine*, vol.35, pp.72-81, Nov. 1997.

[6.163] G.Wen and J.Villansenor, "A class of reversible variable length codes for robust image and video coding," *Proc. IEEE ICIP'97*, pp.65-68, Oct. 1997.

[6.164] G.Reyes et al, "Video transcoding for resilience in wireless channels," *Proc. IEEE ICIP*, pp.338-342, Oct. 1998.

[6.165] G.Cheung and A.Zakhor, "Joint source/channel coding for scalable video over noisy channels," *Proc. IEEE ICIP'96*, pp.767-770, Sept. 1996.

[6.166] A.A.Alatan and J.W.Woods, "Joint utilization of fixed and variable-length codes for improving synchronization immunity for image transmission," *Proc. IEEE ICIP*, pp.319-323, Oct. 1998.

[6.167] H.Li and C.W.Chen, "Joint source and channel optimized block TCQ with layered transmission and RCPC," *Proc. IEEE ICIP*, pp.644-648, Oct. 1998.

[6.168] P.G.Sherwood and K.Zeger, "Progressive image coding for noisy channels," *IEEE Signal Processing Letters*, vol.4, pp.189-191, July 1997.

[6.169] P.G.Sherwood and K.Zeger, "Error protection for progressive image transmission over memoryless and fading channels," *Proc. IEEE ICIP'98*, pp.324-328, Oct. 1998.

[6.170] H.Man, F.Kossentini and J.T.Smith, "Robust EZW image coding for noisy channels," *IEEE Signal Processing Letters*, vol.4, pp.227-229, Aug. 1997.

[6.171] Z.Xiong, B.J.Kim and W.A.Pearlman, "Progressive video coding for noisy channels," *Proc. IEEE ICIP'98*, pp.334-337, Oct. 1998.

[6.172] D.W.Rednull and N.G.Kingsbury, "The EREC: an error-resilient technique for coding variable length blocks of data," *IEEE Trans. Image Process.*, vol.5, pp.565-574, April 1996.

[6.173] R.Chandramouli, N.Ranganathan and S.J.Ramadoss, "Joint optimization of quantization and on-line channel estimation for low bit-rate video transmission," *Proc. IEEE ICIP'98*, pp.649-653, Oct. 1998.

[6.174] J.M.Shapiro, "Embedded image coding using zerotrees of wavelet coefficients," *IEEE Trans. Signal Processing*, vol.41, pp.3,445-3,462, Dec. 1993.

[6.175] A.Said and W.Pearlman, "A new fast and efficient image codec based on set partitioning in hierarchical trees," *IEEE Trans. CSVT*, vol.6, pp.243-250, June 1996.

[6.176] A.Gersho and R.M.Gray, *Vector Quantization and Signal Compression*, Boston, MA, Kluwer, 1992.

[6.177] M.W.Marcellin and T.R.Fischer, "Trellis coded quantization of memoryless and Gauss-Markov sources," *IEEE Trans. Comm.*, vol.38, pp.82-93, Jan. 1990.

[6.178] J.Hagenamer, "Rate compatible punctured convolutional codes (RCPC) and their applications," *IEEE Trans. Comm.*, vol.36, pp.389-400, April 1988.

[6.179] L.H.C.Lee, "New rate-compatible punctured convolutional codes for Viterbi decoding," *IEEE Trans. Comm.*, vol.42, pp.3,037-3,079, Dec. 1994.

[6.180] D.Milovanovic and Z.Bojkovic, "Audio/video transmission in mobile communications," *Proc. ETAI*, pp.43-46, Ohrid, Republic of Macedonia, Sept. 2000.

[6.181] GSM 06.55, *Digital Cellular Telecommunication System, Performance Characterization of the GSM Enhanced Full Rate EFR Speech Codec*, ETSI Tech. Rep., ETR305, Aug. 1996.

[6.182] ETSI STC SMG11, Version1, *Adaptive Multirate AMR*, Study Phase Rep., Oct. 1997.

[6.183] GSM03.34, *Digital Cellular Telecommunication System (Phase2+): High Speed Circuit Switched Data (HSCSD)-Stage2*, ETSI Tech. Specification, TS101138, April 1997.

[6.184] M.Karcewicz, J.Nieeglowski and P.Haavisto, "Video coding using motion compensation with polynomial motion vector fields," *Signal Processing: Image Comm.*, vol.10, pp.63-91, July 1997.

[6.185] D.Raychaudhury and N.Wilson, "ATM based transport architecture for multiservices wireless personal communication networks," *IEEE J. Selected Areas in Comm.*, vol.12, pp.1,401-1,414, Oct. 1994.

[6.186] K.Duantl, *Location Management for Mobile Networks*, ATM Forum/97-0322, 97-0087, Mountain View, California, Feb. 1997.

[6.187] G.Bautz, *Addressing in Wireless ATM Networks*, ATM Forum/97-0322, Mountain View, California, April 1997.

[6.188] C.Perkins, (Ed.), *Mobility Support*, IETF RFC2002, Reston, Virginia, Oct. 1996.

[6.189] CDPD Consortium, *Cellular Digital Packet Data Specification*, POB809320, Chicago, IL, July 1993.

[6.190] Ch.E.Perkins, "Mobile networking through mobile IP," *IEEE Internet Computing*, vol.2, pp.58-69, Jan.-Feb. 1998.

[6.191] S.E.Deering, Ed., *ICMP Router Discovery Messages*, IETF RFC1256, Reston, Virginia, Sept. 1991.

[6.192] C.Perkins, *IP Encapsulation Within IP*, IETF RFC2003, Reston, Virginia, May 1996.

[6.193] L.Veltri, "IP mobility support in private access networks: an interworking scenario between mobile IP and NAPT," *Proc. IEEE ICT*, vol.1, pp.1,215-1,230, Bucharest, Romania, June 2001.

[6.194] M.Holdrege and P.Srisuresh, *IP Network Address Translator (NAT) Terminology and Considerations*, IETF RFC2663, Reston, Virginia, Aug. 1999.

[6.195] M.Holdrege and P.Srisuresh, *Protocol Complications with the IP Network Address Translator*, IETF RFC3027, Reston, Virginia, Jan. 2001.

[6.196] Wireless Multimedia Forum, *Delivering Streaming Media to the Mobile Masses*, Los Gatos, California, Feb. 2001.

[6.197] D.Sisalem, "SIP and mobile communication," *IEEE ICT, Proc. Special Sessions*, pp.447-450, Bucharest, Romania, June 2001.

[6.198] A.Ballardie, *Core Based Trees (CBT) Multicast Routing Architecture*, IETF RFC2201, Reston, Virginia, 1997.

[6.199] K.L.Calvert, E.Zegura and M.J.Donohoo, *Core Selection Methods for Multicast Routing*, GIT-CC-95/15 *Proc., IC3N'95*, Atlanta, GA, 1995.

[6.200] K.Brown and S.Singh, "A network architecture for mobile computing," *Proc. IEEE INFOCOM'96*, pp.1,388-1,396, March 1996.

[6.201] W.W.Lu, "Compact multidimensional broadband wireless: the convergence of wireless mobile and access," *IEEE Comm. Magazine*, vol.38, pp.119-123, Nov. 2000.

[6.202] R.Rebhan et al, "Multimedia goes mobile in broadcast networks," *IEEE Multimedia*, Vol. 4, pp.14-21, April-June 1997.

[6.203] D.Pan, "A tutorial on MPEG Audio compression," *IEEE Multimedia*, vol.2, pp.60-74, Summer 1995.

[6.204] European Telecommunications Standards Institute, Sophia Antipolis Cedex, France, ETS300421 *DVB-Satellite (1999)*, ETS300429 *DVB-Cable (1998)*, ETS300744 *DVB-Terrestrial (1999)*.

[6.205] S.Pekowsky and A.Andorfer, "Multimedia data broadcasting strategies," *IEEE Comm. Magazine*, vol.39, pp.138-145, April 2001.

[6.206] G.Luettene, "The DVB multimedia home platform," *MUST'98,* May 1998.

[6.207] J.van der Meer and C.M.Huizer, *Interoperability Between Different Interactive Engines for Digital Television, Problems and Solutions*, Eindhoven, The Netherlands, Philips, June 1997.

[6.208] E.Stare et al, *The Multimedia Car Platform*, European Commission Project, Jan., 2000.

[6.209] M.Milenkovic, "Delivering interactive services via a digital TV infrastructure," *IEEE Multimedia*, pp.34-43, Oct.-Dec. 1998.

[6.210] A.Acharya, M.Franklin and S.Zdomi, "Prefetching from a broadcast disk," *Proc. 12th Int. Conf. Data Engineering,* IEEE CS Press, pp.276-285, Los Alamitos, CA, 1996.

[6.211] T.E.Browen et al, "The data cycle architecture," *Comm. ACM,* vol.30, pp.71-81, Dec. 1992.

[6.212] T.Imielinski, S.Viswanathan and B.R.Badrinath, "Data on air: organization and access," *IEEE Trans. Knowledge Data Eng.*, vol.9, pp.353-372, March 1997.

[6.213] D.Minoli, *Video Dialtone Technology*, New York, NY: McGraw-Hill, New York, 1995.

[6.214] J.Nielsen, *Usability Engineering*, Chestnut Hill, MA: Academic Press, 1993.

Index

A

Access point (AP) 421
Access unit (AU) 204
Adaptive DPCM (ADPCM) 53
Advanced audio coding (AAC) 167
    low complexity profile 171
    main profile 171
    scalable sampling rate profile 171
Advanced Synchronization Model (FlexTime) 188
Advanced Television Systems Committee (ATSC) Digital TV (DTV) 480
Algebraic Code Excited Linear Prediction (ACELP) 460
Application Level Framing (ALF) 361
Application programming interface (API) 486
Application-level multicast 408
Application-specific integrated circuits (ASIC) 266
Asymmetric Digital Subscriber Line (ADSL) 432
Asynchronous transfer mode (ATM)
    ATM layer 118
    virtual connection (VC) 116
    virtual path (VP) 116
ATM and IP integration 119
    IP switching 120
    tag switching 120
ATM QoS parameters
    cell loss ratio 119
    maximum burst length 119
    maximum cell delay variation 119
    maximum end-to-end delay 119
    peak rate 118
    sustainable rate 118
ATM types of adaptation
    available bit rate (ABR) 119
    connection-oriented data service 119
    constant bit rate (CBR) 119
    LAN emulation 119
    unspecified bit rate (UBR) 119
    variable bit rate (VBR) 119
Audio and video integration
    automated lip reading 20, 28
    bimodality of human speech 16
    dubbing 16
    facial expressions 22
    joint audio-video coding 31
    lip synchronization 23
    lip tracking 24
    person verification 30
    speech driven talking heads 21
Audio-to-visual mapping 27
    McGurk effect 16
    phoneme 17
    speech parametarization 24
    viseme 17
    visual analysis 25

B

Base transceiver stations (BTS) 470
Bi-directional interpolated pictures (B-pictures) 152
BIFS for Facial Animation 236
Binary Format for Scene (BIFS) 205
    Anim protocol 206
    command protocol 206
Body animation parameter (BAP) set 185
Body definition parameter (BDP) set 185
Broadband Remote Access Servers (BRAS) 436

C

Cable modems 442, 445
Cell delay variation (CDV) 415
Cellular networks 470
Code Division Multiple Access (CDMA) 452
Code Excited Linear Predictive Coding (CELP) 251
Coded orthogonal frequency division multiplex (COFDM) 475
Coexist-carrierless amplitude-phase (CAP) 444
Common intermediate format (CIF) 320
Competitive Local Exchange Carriers (CLEC) 432
Complete timing information (CTI) 376

Compression with reversible embedded wavelet (CREW) 263
Conditional replenishment 152
Comfort noise generation (CNG) 337
Constant bit rate (CBR) 383
Content media replication
  caching 408
  mirroring 408
Content-based video indexing 225
Core based tree (CBT) 470
Custom Pixel Aspect Ratios (PAR) 328
Cyclic-redundancy check (CRC) 418

D

Data carousel 492
Data link control (DLC) 452
Data partitioning 165, 223
Deblocking Filter mode 329
Decoder error concealment
  spatial interpolation 393
Delaunay triangulation 239
Delivery multimedia integration framework (DMIF) 196, 210
  communication architecture 211
  computational model 213
Dense Division Multiplexing (DDM) 434
Differential pulse coding modulation (DPCM) 75
Digital audio coding
  absolute threshold of hearing 53
  ASPEC 62
  compact disc audio 71
  critical band frequency analysis 54
  DCT with VQ 65
  DFT noise substitution 64
  DPAC 63

hybrid coder 61
modified DCT 65
perceptual entropy 57
psychoacoustic model 52
simultaneously masking 55
subband coder 66
temporal masking 56
time-frequency analysis 52
transform coder 58
wavelets decomposition 67
Digital Audio Visual Council (DAVIC) 478
Digital Video Broadcast DVB DVB-T 487
Digital Video Broadcasting (DVB)
  data carousels 480
  data piping 479
  data streaming 479
  DVB Return Channel System via Satellite (DVB-RCS) 485
  DVB-MHP 486
  DVB-S 484
  multi-protocol encapsulated (MPE) 480
  object carousels 480
Discrete multitone (DMT) 444
Distributed multimedia applications
  computer-supported cooperative work 131
  interactive television 127
  telecooperation 131
  telemedicine 135
  video-on-demand 128
Distributed multimedia systems
  networking 108
  resource management 107
DMIF Applications interface (DAI) 196
DSL access multiplexer (DSLM) 434
Dynamic resolution conversion (DRC) 185

E

Electronic program guide (EPG) 485
Embedded block coding with optimized truncation (EBCOT) 264
Embedded zero-tree wavelet (EZW) 75
Environmental spatialization tools 187
Error concealment 333
  MPEG-4 video 420
Error protection (EP) tool 186
Error resilience 223, 388
  data partitioning 392
  data recovery 392
  encoding 390
  independent segment decoding (ISD) 391
  resynchronization marking 390
  reversible variable length coding (RVLC) 390
  video redundancy coding 391
Error resilient entropy code (EREC) 393
European Telecommunication Standards Institute (ETSI)
  Digital Audio Broadcasting (DAB) 475
  Digital Video Broadcasting (DVB) 477
Extensible markup language (XML) 289

F

Face 232
Face animation parameters units 227
Face model 231
Facial animation parameters (FAPs) 227
Facial definition parameters (FAPs) 226

Index

FAP interpolation table (FIT) 235
Fast Fourier transformation (FFT) 169
Field pictures 160
Filtering
    convolution-based 272
    lifting-based 272
Fine granularity scalability (FGS) 403
Forward error correction (FEC) 389, 419
    channel coding 406
    joint source/channel 406
    source coding-based 406
Frame 158
Frame pictures 160
Frequency Division Multiple Access (FDMA) 452

G

Gatekeeper 317
Gateway 316
General audio coding (GA) 244
Global motion compensation (GMC) 184
Global systems for mobile communications (GSM) 458
Graceful degradation 390
Group of blocks (GOB) 321
Group of pictures (GOP) 159
Group of video object planes (GOV) 216

H

Harmonic Vector Excitation Coding (HVXC) 250
Hierarchical coding 421
High speed circuit-switched data (HSCSD) 460

Hyper Text Markup Language (HTML) 351

I

Image coding 71
    human visual system (HVS) 72
    just noticeable distortion (JND) 72
    short-term spectral analysis 73
    spatial redundancy 72
Integrated Layer Processing (ILP) 361
Integrated packet network (IPN) 377
Intellectual property management and protection (IPMP) 305
    descriptors 202
    framework 202
    streams 202
Intellectual property rights (IPR) 88, 269
Interactive broadcast data services 490
Inter-frame prediction (P-picture) 150
International mobile telecommunication (IMT-2000) 458
International Organization for Standardization (ISO) 309
International Telecommunications Union (ITU)
    Recommendations 310
    standardization process 308
Internet Engineering Task Force (IETF) 344
    Internet Architecture Board (IAB) 345
    Internet Draft 346
    Internet Engineering Steering Group (IESG) 345
    Internet Society (ISOC) 344

    Request for comments (RFC) 346
    standardization 344
    Working Groups (WG) 345
Internet networking
    accounting and billing management 116
    architecture 347
    capacity planning 113
    configuration management 115
    fault management 114
    network dimensioning 115
    performance management 114
    round-trip time (RTT) 109
    security management 115
    traffic management 114
Internet protocols 348
    Border Gateway Protocol (BGP) 348
    Distance vector multicast protocol 111
    File Transfer Protocol (FTP) 350
    Hyper Text Transfer Protocol (HTTP) 351
    Internet Control Message Protocol (ICMP) 348
    Internet Protocol (IP) 109
    Internet Protocol V6 351
    multicast routing 110
    Network News Transfer Protocol (NNTP) 350
    Open Shortest Path First (OSPF) 348
    Real-Time Streaming Protocol (RTSP) 410
    Real-Time Transport Protocol (RTP) 109, 112, 361
    Resource Reservation Protocol (RSVP) 111, 381
    Routing Information Protocol (RIP) 347

RTP control protocol
(RTCP) 112
Session Initiation Protocol
(SIP) 410
Simple Mail Transfer
Protocol (SMPT) 350
Simple Network
Management Protocol
(SNMP) 351
Transmission control
protocol (TCP) 109
User Datagram Protocol
(UDP) 109
Voice over Internet protocol
(VoIP) 109
Intraframe coding (I-picture) 150
IP protocol data unit (PDU) 349
ISLAN 16-T 316
ISO/IEC JPEG standard 267
ISO/IEC JPEG-2000 standard 262
   bitstream 275
   file format (JP2) 269
   Irreversible component
     transformation (ICT) 273
   markers 276
   Mixed Raster Content
     (MRC) 266
   MotionJPEG 265
   Reversible component
     transformation (RCT) 273
   Verification model (VM) 263
ISO/IEC JPEG-LS standard 267
ISO/IEC MPEG-1 standard 146
   Constrained parameter set
     (CPS) 148
   rate control 153
   specific storage media
     functionalities 152
ISO/IEC MPEG-2 standard
   Audio 167
   Audio Layer III 168
   coding parameters 155
   Digital storage media-
     command and control
     (DSM-CC) 175
   Levels 155, 166
   Profiles 155, 166
   scalability extensions 163
   semantic description 157
   Systems 171
   video decoding process 157
   video syntax 163
ISO/IEC MPEG-21 standard 301
   content management and
     usage 305
   Digital Item declaration 304
   Digital Item identification
     and description 305
   event reporting 306
   terminals and networks 302
   Users 301
ISO/IEC MPEG-4 standard
   Audio 243
   audiovisual scene 179
   content-based interactivity
     194
   elementary stream
     management 199
   Levels 256
   media objects 180
   MPEG-J 183
   object-based representation
     model 179
   Profiles 189, 259
   requirements 194
   shape coding tools 216
   standardization process 193
   streaming environment 206
   Synthetic Audio (SA) 252
   Systems 195
   texture coding tools 218
   verification testing 192
   Video 214
ISO/IEC MPEG-7 standard 282
   audio descriptors tools 289
   Binary format for MPEG-7
     data (BiM) 287
   color descriptors 291
   Description definition
     language (DDL) 289
   Description scheme (DS) 297
   Descriptor (D) 283
   experimental model (XM)
     299
   motion descriptors 294
   normative interface 287
   shape descriptors 293
   Systems 286
   time series descriptors 291
   visual descriptors tools 290
Isochronous Ethernet (ISO-Enet)
   315
ITU-T audiovisual systems
   H.310, H.321 312
   H.320 312
   H.322 315
   H.323 317
   H.324 319
ITU-T multiplex and
   synchronization standards
   H.221 341
   H.223 341
   H.225 341
   H.245 343
ITU-T speech coding standards
   G.723.1 337
   G.729 336
ITU-T video coding standards
   H.261 319
   H.262 323
   H.263 V2 327
   H.263 V3 332
   H.26L 333

J

Joint Photographic Experts Group
   (JPEG) 262
   development process 263
Joint source channel coding 454

Index

L

Layerd video coding
  compression 386
  transmission 388
Layered video coding 386
Level of detail (LOD) scalability 186
Local multipoint distribution services (LMDS) 443
Logic link control (LLC) 452
Loss detection
  application framing 420
  network framing 418

M

Macroblock (MB) 159
Macroblocks (MB) 150
Markov chains 384
Maximum transfer unit (MTU) 415
Mesh coding 237
Mesh object plane (MOP) 237
Mobile ATM 421, 462
Mobile IP 462
Mobile support station (MSS) 471
Modified Discrete Cosine Transform (MDCT) 168
Motion compensation (MC) 151
Motion vectors (MV) 392
Moving Picture Experts Group (MPEG)
  standardization process 142
MPEG-4 over Internet 365
  client 369
  server 368
  system architecture 366
MPEG-4 textual format (XMT), extensible, 188
Multi program transport streams (MPTS) 174
Multichannel multipoint distribution services (MMDS) 448

multifunctional coding 220
Multihypothesis (MH) 334
Multihypothesis block pattern (MHBP) 334
Multimedia 37, 140
  different media types 13
  material 3
  media interaction 14
  standard 140
Multimedia Car Platform (MCP) 487
Multimedia communications 1
  model 4
  network requirements 7
  services 1
  systems 4
  terminals 11
  user requirements 5
Multimedia object transfer (MOT) protocol 476
Multimedia operating systems
  control processing unit management 123
  file system management 124
  input/output management 124
  memory management 123
Multimedia over ADSL
  parallel transmission 441
  serial transmission 439
Multimedia over ATM 120
  multiplexing 412
  real-time multimedia over ATM (RMOA) WG 121
  video delay 413
Multimedia over broadcasting networks 473
  content providers 474
  network providers 475
  service providers 475
Multimedia over DSL 432
Multimedia over IP 360, 424
  bandwidth allocation mechanisms (BAM) 427

Controlled-load Service 426
Fine-granular scalability (FGS) 429
Guaranteed Service 426
session control 360
traffic specification 426
transport 361
Multimedia over Wireless 449
Multimedia over wireless
  broadband communications 451
  dynamic channel allocation 453
  fixed channel allocation 453
  mobile networks 458
  multiple access (MA) protocol 452
Multimedia processing
  face detection and recognition 91
  face-based video indexing and browsing 94
  neural networks 89
  personal authentication 92
  subject-based retrieval 93
Multimedia processors 94
  arithmetic operations 97
  bit manipulation function 97
  complex instruction set computer (CISC) 97
  digital signal processors (DSP) 94
  DSPs for PC acceleration 97
  embedded microprocessors 101
  general-purpose microprocessor 99
  Intel MMX processors 96
  memory access 97, 99
  multimedia extensions 100
  power dissipation 99
  real-time task switching 98
  reduced instruction set computer (RISC) 97
  stream data input/output 97

Index

superscalar microprocessor 98
Multimedia servers 124
   centralized 125
   distributed 125
   I/O bandwidth requirements 125
   initiate/terminate/switch service 125
   multimedia packing 125
   query/video transactions 125
   session 125
   storage requirements 125
   system resources 124
   traffic load balancing 126
   video streams 124
Multimedia signal processing
   algorithms 43
   color processing 42
   content-based image retrieval 48
   digital media 37
   human vision 42
   network echo cancellation 46
   non-linear video and audio mapping 41
   pre- and post-processing 44
   projection techniques 42
   quantization 41
   speech and audio signal compression 45
   speech synthesis 45
   statistical charcterization 41
   visual and audio models 41
Multiple alpha channels 185
Multiple description coding (MDC) 391
Multipoint control unit (MCU) 312
Multiresolution coding 385

N

Network address translation (NAT) 466
Network filtering 407
Networked multimedia 88
   adaptive signal compression 88
   added pointers 89
   compressed stream segments 89
Null timing information (NTI) 375

O

Object content information (CI) 199
Object descriptor (OD) 199
Overlapped block motion compensation (OBMC) 325

P

Packet identifier (PID) 173, 482
Packet transfer concept 8
   statistical multiplexing 9
   switching 8
   virtual circuit switching 9
Packet video 380
Packet voice 374
Packetized Elementary Stream (PES) 172
Policing mechanisms
   leaky bucket 385
Portable Network Graphics (PNG) 268
Priority breakpoint 166
Program clock reference (PCR) 175
Program specification information (PSI) 478
Progressive fine granularity scalability (PFGS) 403
Projection onto convex sets (POCS) 393
Proposal package description (PPD) 194
Pulse code modulation (PCM) 170

Q

Quadrature phase shift keying (QPSK) modulation 477
Quality of service (QoS) 3
   application layer control 404
   descriptor 203

R

Rate control 385
   analytical model-based 396
   rate distortion approach 396
Rate shaping 385
Rate-compatible punctured convolutional (RCPC) codes 458
Reed-Solomon (RS) codes 425
Region of interest (ROI) 268
Return path multiplexer (RPM) 447

S

Satellite direct-to-home (DTH) 478
Scalable rate control
   compression-layer approach 394
   theoretical foundation 397
Scalable shape coding 185
Scale factor 161
Service information (SI) 478
Set partitioning in hierarchical trees (SPIHT) 455
Shannon information theory 389
Shape-adaptive DCT (SA-DCT) 184
Signal-to-mask ratio (SMR) 169
Signal-to-noise ratio (SNR) scalability 164
Simulation Model 142
Simulcast 387
Single description coding (SDC) 392

Index

Single pair High Speed DSL (SHDSL) 432
Slices 160
Spatial scalability 164
Speech assisted interpolation scheme 24
Sprite coding 221
Statistical multiplexing gain (SMG) 384
Streaming
   audio 85
   buffer management 88
   error-resilient coding 88
   error-resilient transmission 88
   media control 88
   multiple clients 87
   network congestion 87
   packet loss 86
   server 87
   transaction 87
   variable delays 86
   video 87
Streaming media synchronization
   media layer 410
   object layer 410
   stream layer 410
Streaming servers
   communicator 409
   operating system 409
   storage system 409
Supervisor host (SH) 471
Switched batch Bernoulli process (SBBP) 427
Sync layer (SL) 197
Synchronized Multimedia Integration Language (SMIL) 188
Synchronous Optical Network/ Synchronous Digital Hierarchy (SONET/SDH) 434
Synthetic and natural hybrid coding (SNHC) 194
System decoder model (SDM) 204

T

Temporal scalability 164
Terms of Reference (ToR) 338
Test Model 142
Text-to-speech interface (TTSI) 253
Time Division Multiple Access (TDMA) 446, 452
Time stamps
   decoding (DTS) 160
   reference 159
Traffic parameters 385
Traffic-shaping
   transport-layer approach 394
Transport stream packet (TSP) 173
Tspec parameters 426
Twin VQ 248

U

Unequal error protection (UEP) 391, 457
Unequal Protection (UEP) 186
Uniform threshold trellis coded quantization (UTTCQ) 458
Universal mobile telecommunications systems (UMTS) 451
Usage parameters control (UPC) 384

V

Variable length code (VLC) 150
Verification Model 142
Video buffer (VB) 152
Video coding
   hybrid coder 77
   low bit rate telephony 74
   motion compenzation 76
   predictive coder 75
   storing broadcast video on DVD 74
   storing movies on CDROM 74
   subband coder 74
   teleconferencing 74
   transform coder 74
Video object (VO) 216
Video object plane (VOP) 216
Video object tracking 238
Video signal processing
   camera model 47
   cut detection 47
   deblocking filters 47
   format conversion 46
   mosaic 47
   motion analysis 47
   object tracking 47
   storage, archiving and browsing 47
   temporal segmentation 47
   video indexing 48
   visual summarization 48
Video sources
   multiplexing 413
Video streaming 400
   bandwidth 404
   complexity 404
   congestion control 407
   delay 404
   error concealment 406
   FEC 406
   model-based rate control 406
   probe-based rate control 406
   protocols 410
   receiver based rate control 405
   retransmission 406
   scalable coding 402
   source based rate control 404
   VCR like functions 404
Virtual Reality Modeling Language (VRML) 207
Visual object sequence (VS) 216

Visual Texture Coding 267
Voice activity detector (VAD) 337
Voice over DSL (VoDSL) 432, 438

W

Watermarking 78
    associated key 82
    authentication, verification 84
    automated detection/search 82
    broadcast monitoring 84
    copyright protection 79
    different access levels 79
    embedding algorithm 79
    perceptual invisibility 81
    robustness 82
    statistical invisibility 82
    steganography 84
    trustworthily detection 81
Wavelet tiling 185
Wavelets
    Discrete wavelet transform (DWT) 270
    filterbanks 68
    irreversible transform 271
    reversible transform 271
    wavelet packets 68
Wireless ATM networks 421
Wireless Multimedia Forum (WMF) 467

Z

Zig-zag scanning 150

About the Authors

K. R. Rao is Professor of Electrical Engineering at the University of Texas at Arlington. He has authored or co-authored several additional leading texts in the field, including *Techniques and Standards for Image/Video/Audio Coding,* and *Packet Video Communications over ATM Networks* (Prentice Hall). In 1975, with two other researchers, he introduced the Discrete Cosine Transform, one of today's most powerful digital signal processing techniques. He is a Fellow of the IEEE.

Zoran S. Bojkovic received his Ph.D. degree in electrical engineering from the University of Belgrade, Yugoslavia, Faculty of Electrical Engineering. He is currently a professor of electrical engineering at the University of Belgrade. He has taught a wide range of courses in communication networks and signal processing and supervised postgraduate students worldwide. He has published 15 textbooks and more than 300 papers in international books, in peer-reviewed journals and conference proceedings. He is also an active reviewer and a member of the Scientific committee of numerous journals and conferences, and serves as chairman for international conferences, symposiums and workshops. He has conducted workshops/tutorials on multimedia worldwide and participated in many communication, scientific, and industrial projects. He is a member of IEEE Communication Society and EURASIP.

Dragorad A. Milovanovic received the Dipl. Electr. Eng. and Master of Science degree from the University of Belgrade, Yugoslavia Faculty of Engineering. From 1987 to 1991, he was a Research Assistant at the Department of Electrical Engineering, where his research interest includes analysis and design of digital communications systems. He has been working as R&D engineer for DSP software development in digital television industry. Also, he is serving as a consultant for developing standard-based and secure solutions for media coding, streaming and distribution. He has participated in numerous scientific projects and published more than 150 papers in international journals and conference proceedings.

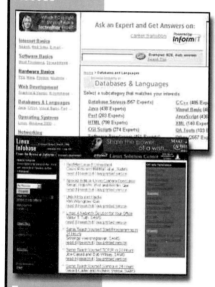